LECTURES ON GAS THEORY

Ludwig Boltzmann

LECTURES ON GAS THEORY

Ludwig Boltzmann

Translated by Stephen G. Brush

DOVER PUBLICATIONS, INC.
NEW YORK

Copyright

Bibliographical Note

This Dover edition, first published in 1995, is an unabridged, unaltered republication of the English translation originally published by the University of California Press, Berkeley, 1964. The original work, in German, was published in two parts by J. A. Barth, Leipzig, Germany, 1896 (Part I) and 1898 (Part II), under the title *Vorlesungen über Gastheorie*.

The Dover edition is published by special arrangement with the University of California Press, 2120 Berkeley Way, Berkeley, California 94720.

This translation was originally "published with the assistance of a grant from the National Science Foundation."

Library of Congress Cataloging-in-Publication Data

Boltzmann, Ludwig, 1844–1906.
[Vorlesungen über Gastheorie. English]
Lectures on gas theory / Ludwig Boltzmann ; translated by Stephen G. Brush. — Dover ed.
p. cm.
Originally published: Berkeley : University of California Press, 1964.
Includes bibliographical references and index.
ISBN-13: 978-0-486-68455-0
ISBN-10: 0-486-68455-5 (pbk.)
1. Kinetic theory of gases. I. Title.
QC175.B7213 1995
533'.7—dc20 94-41221
CIP

Printed in Canada
68455511 2025
www.doverpublications.com

CONTENTS

PART I

THEORY OF GASES WITH MONATOMIC MOLECULES, WHOSE DIMENSIONS ARE NEGLIGIBLE COMPARED TO THE MEAN FREE PATH.

CHAPTER I

THE MOLECULES ARE ELASTIC SPHERES. EXTERNAL FORCES AND VISIBLE MASS MOTION ARE ABSENT.

CHAPTER II

THE MOLECULES ARE CENTERS OF FORCE. CONSIDERATION OF EXTERNAL FORCES AND VISIBLE MOTIONS OF THE GAS.

CHAPTER III

THE MOLECULES REPEL EACH OTHER WITH A FORCE INVERSELY PROPORTIONAL TO THE FIFTH POWER OF THEIR DISTANCE.

PART II

VAN DER WAALS' THEORY; GASES WITH COMPOUND MOLECULES; GAS DISSOCIATION; CONCLUDING REMARKS.

CHAPTER I

FOUNDATIONS OF VAN DER WAALS' THEORY.

CHAPTER II

PHYSICAL DISCUSSION OF THE VAN DER WAALS' THEORY.

CHAPTER III

PRINCIPLES OF GENERAL MECHANICS NEEDED FOR GAS THEORY.

CHAPTER IV

GASES WITH COMPOUND MOLECULES.

CHAPTER V

DERIVATION OF VAN DER WAALS' EQUATION BY MEANS OF THE VIRIAL CONCEPT.

CHAPTER VI

Theory of dissociation.

CHAPTER VII

Supplements to the laws of thermal equilibrium in gases with compound molecules.

TRANSLATOR'S INTRODUCTION

Gas Theory

Boltzmann's *Lectures on Gas Theory* is an acknowledged masterpiece of theoretical physics; aside from its historical importance, it still has considerable scientific value today. It contains a comprehensive exposition of the kinetic theory of gases by a scientist who devoted a large part of his own career to it, and brought it very nearly to completion as a fundamental part of modern physics. The many physicists who are already familiar with *Gastheorie* in the original German edition (1896–1898) will not need this Introduction to remind them of its stature. But perhaps some scientists who scarcely have time to keep up with the latest publications in their subject may want to know why they should bother to read a book published more than 60 years ago; and a brief account of the place of the book in the development of modern physics may be of interest.

Ludwig Boltzmann (1844–1906) played a leading role in the nineteenth-century movement toward reducing the phenomena of heat, light, electricity, and magnetism to "matter and motion" —in other words, to atomic models based on Newtonian mechanics. His own greatest contribution was to show how that mechanics, which had previously been regarded as deterministic and reversible in time, could be used to describe irreversible phenomena in the real world on a statistical basis. His original papers on the statistical interpretation of thermodynamics, the H-theorem, transport theory, thermal equilibrium, the equation of state of gases, and similar subjects, occupy about 2,000 pages in the proceedings of the Vienna Academy and other societies. While some of his discoveries attracted considerable attention and controversy in his own time, not even the handful of experts on kinetic theory could claim to have read everything he wrote. Realizing that few scientists of later generations were likely to study his long original memoirs in detail, Boltzmann decided to publish his lectures, in which the most important parts of the theory, including his own contributions, were carefully explained. In addition, he included his mature reflections and speculations on such questions as the nature of irreversibility and the justification for using statistical methods in physics. His *Vorlesungen über Gastheorie* was in fact the standard reference work for advanced researchers, as well as a popular textbook for students,* for the first quarter of the present century, and is frequently cited even now. A recent example is the following statement by Mark Kac:† "Boltzmann summarized most (but not all) of his work in a two volume treatise *Vorlesungen über Gastheorie.* This is one of the greatest books in the history of exact sciences and the reader is strongly advised to consult it. It is tough going but the rewards are great."

The modern reader will rightly assume that some parts of Boltzmann's theory must have been rendered obsolete by later discoveries. In particular, we know that one cannot expect to develop an adequate theory of atomic phenomena without using quantum mechanics. However, it turns out that almost all the properties of gases at ordinary temperatures and densities can be described by the classical theory developed in the nineteenth century, and for this reason there has been a considerable revival of interest in classical kinetic theory during the last few years, in connection with rarified gas dynamics, plasma physics, and neutron transport theory. The reason why the classical theory works is that, while the internal structure of molecules must be described by quantum mechanics, the interaction between two molecules can be fairly well described by a classical model which ignores this structure and simply uses a postulated force law whose parameters can be chosen to fit experimental data. Aside from phenomena at very high densities or very low temperatures, the only property that the classical theory fails to account for is the ratio of specific heats. However, this failure was already well known at the time Boltzmann wrote this book, and he is therefore quite cautious on this point. He simply concludes that for some unknown reason all the possible internal motions of a molecule do not have an equal share in the total energy, and takes this into account as an empirical fact.

---

* In his endeavors to satisfy the demands of all his readers, Boltzmann has not forgotten even the faculty adviser of graduate students in physics. At the end of Part II, §86, he mentions a class of problems that would be suitable for doctoral dissertations!

† M. Kac, *Probability and Related Topics in Physical Sciences* (New York: Interscience, 1959), p. 261. Reprinted by permission.

As for various questions of a more fundamental nature, such as the relation between entropy and the direction of time, quantum theory has thrown new light on these problems but has not solved them. Boltzmann's remarks are therefore still of interest. Thus Hans Reichenbach proposes to *define* the direction of increasing time as the direction in which entropy increases, so that in a universe in which entropy fluctuates,

> we cannot speak of a direction of time as a whole; only certain sections of time have directions, and these directions are not the same. The first to have the courage to draw this conclusion was Ludwig Boltzmann [*Gas Theory*, Part II, §90]. His conception of alternating time directions represents one of the keenest insights into the problem of time. Philosophers have attempted to derive the properties of time from *reason;* but none of their conceptions compares with this result that a physicist derived from *reasoning* about the implications of mathematical physics. As in so many other respects, the superiority of a philosophy based on the results of science has become manifest . . . .*

The only respects in which the book may be considered out of date are, first, the emphasis on the inverse fifth-power repulsive force model for calculating transport properties, and, second, the molecular model used to explain dissociation phenomena. The reason for choosing the inverse fifth-power force law was that Maxwell (1866)† had discovered that the calculation of transport properties is particularly simple for this case, because the collision integrals do not depend on the relative velocity of the colliding molecules, so that one does not need to know the velocity distribution. Maxwell and Boltzmann therefore believed that it would be more worthwhile to develop an exact theory for this model, rather than attempt to work out a more complicated approximate theory for other force laws. However, it was later shown by Chapman and Enskog (1916–1917) that the important phenomenon of thermal diffusion does not occur for this particular force law, and a quantitative treatment of the general case does require a determination of the velocity distribution in a nonequilibrium state, although it is still not necessary to use the quantum theory. Boltzmann's model for dissociation, employing molecules with small "sensitive regions"

* H. Reichenbach, *The Direction of Time* (Berkeley and Los Angeles: University of California Press, 1956), p. 128.

† References for this and other works cited here will be found in the Bibliography at the end of this volume.

on parts of their surfaces, is an interesting historical curiosity, but the need for such *ad hoc* assumptions to explain the directional character of chemical bonds has been eliminated by quantum mechanics.

History of the kinetic theory

The development of the kinetic theory of gases has been discussed at length elsewhere* so only a brief summary will be given here.

The modern concept of air as a fluid exerting mechanical pressure on surfaces in contact with it goes back to the early seventeenth century, when Torricelli, Pascal, and Boyle first established the physical nature of the air. By a combination of experiments and theoretical reasoning they persuaded other scientists that the earth is surrounded by a "sea" of air that exerts pressure in much the same way that water does, and that air pressure is responsible for many of the phenomena previously attributed to "nature's abhorrence of a vacuum." We may view this development of the concept of air pressure as part of the change in scientific attitudes which led to the mechanico-corpuscular view of nature, associated with the names of Galileo, Boyle, Newton, and others. Instead of postulating "occult forces" or teleological principles to explain natural phenomena, scientists started to look for explanations based simply on matter and motion.

Robert Boyle is generally credited with the discovery that the pressure exerted by a gas is inversely proportional to the volume of the space in which it is confined. From Boyle's point of view that discovery by itself was relatively insignificant, and though he had provided the experimental evidence for it he readily admitted that he had not found any general quantitative relation between pressure and volume before Richard Towneley suggested the hypothesis that pressure is inversely proportional

* See the papers by S. G. Brush and E. Mendoza listed in the Bibliography; also D. ter Haar, *Elements of Statistical Mechanics* (New York: Rinehart, 1954), Appendix I (*H*-theorem and ergodic theorem), and Rev. Mod. Phys. **27**, 289 (1955) (recent developments). A series of selected annotated reprints of the most important original papers on the kinetic theory is to be published by Pergamon Press.

to volume. Robert Hooke also provided further experimental confirmation of the hypothesis. It was long known as "Mariotte's law" on the Continent, but there are good reasons for believing that Mariotte was familiar with Boyle's work even though he does not mention it, so that he does not even deserve the credit for independent (much less simultaneous) discovery: see any good book on history of science for details on this point.

The generalization that pressure is proportional to absolute temperature at constant volume was stated by Amontons and Charles, but it remained for Gay-Lussac to establish it firmly by experiments at the beginning of the nineteenth century.

Boyle's researches were carried out to illustrate not just a quantitative relation between pressure and volume, but rather the qualitative fact that air has elasticity ("spring") and can exert a mechanical pressure strong enough to support 34 feet of water in a pump, or 30 inches of mercury in a barometer. His achievement was to introduce a new dimension—pressure—into physics; he could well afford to be generous about giving others the credit for perceiving the numerical relations between this dimension and others. He also proposed a theoretical explanation for the elasticity of air—he likened it to "a heap of little bodies, lying one upon another" and the elasticity of the whole was simply due to the elasticity of the parts. The atoms were said to behave like springs which resist compression. To a modern scientist this explanation does not seem very satisfactory, for it does no more than attribute to atoms the observable properties of macroscopic objects. It is interesting to note that Boyle also tried the "crucial experiment" which was to help overthrow his own theory in favor of the kinetic theory two centuries later, though he did not realize its significance: he placed a pendulum in an evacuated chamber and discovered, to his surprise, that the presence or absence of air makes hardly any difference to the period of the swings or the time needed for the pendulum to come to rest. In 1859, Maxwell deduced from the kinetic theory that the viscosity of a gas should be independent of its density—a property which would be very hard to explain on the basis of Boyle's theory.

Newton discusses very briefly in his *Philosophiae Naturalis Principia Mathematica* (1687) the consequences of various hypotheses about the forces between atoms for the relation between pressure and volume. One particular hypothesis, a repulsive force inversely proportional to distance, leads to Boyle's law. It seems plausible that Newton was trying to put Boyle's theory in mathematical language and that he thought of the repulsive forces as being due to the action of the atomic springs in contact with each other, but there seems to be no direct evidence for this. Neither Boyle nor Newton asserted that the hypothesis of repulsive forces between atoms is really responsible for gas pressure; both were willing to leave the question open. Boyle mentions Descartes' theory of vortices, for example, which is somewhat closer in spirit to the kinetic theory since it relies more heavily on the rapid motion of the parts of the atom as a cause for repulsion. Incidentally, it is important to realize that there is more to the kinetic theory than just the statement that heat is atomic motion. That statement was frequently made, especially in the seventeenth century, but usually by scientists who did not make the important additional assumption that in gases the atoms move freely most of the time. It was quite possible to accept the "heat is motion" idea and still reject the kinetic theory of gases—as did Humphry Davy early in the nineteenth century.

Despite the tentative way in which it was originally proposed, the Boyle-Newton theory of gases was apparently accepted by most scientists until about the middle of the nineteenth century, when the kinetic theory finally managed to overcome Newton's authority. Thus a theory which was decidedly a step forward from older ideas nevertheless was able to retard further progress for a considerable time.

It is difficult to understand the relative lack of progress in gas physics during the eighteenth century as compared to the seventeenth. Several brilliant mathematicians refined and clarified Newton's principles of mechanics and applied them to the analysis of the motions of celestial bodies and of continuous solids and fluids, but there was little interest in the properties of systems of freely moving atoms. The atoms in a gas were still conceived as being suspended in the ether, although they could vibrate or rotate enough to keep other atoms from coming too close. This model was rather awkward to formulate mathematically, as may be seen from an unsuccessful attempt by Leonhard Euler (1727).

One contribution from this period has been generally recognized as the first kinetic theory of gases. This is Daniel Bernoulli's derivation of the gas laws from a "billiard ball" model—much like the one still used in elementary textbooks today—in 1738. His kinetic theory is only a small part of a treatise on hydrodynamics, a subject to which he made important contributions. From the viewpoint of the historical development of the kinetic theory, his formulation and successful applications of the principle of conservation of mechanical energy were really more important than the fact that he actually proposed a kinetic theory himself. Bernoulli's kinetic theory, while in accord with modern ideas, was a century ahead of its time, for scientists were not yet ready to accept the physical description of a gas that it employed. Heat was still generally regarded as a substance, even though it was also recognized that heat might have some relationship to atomic motions; Bernoulli's assumption that heat is *nothing but* atomic motion was unacceptable, especially to scientists who were interested in the phenomena of radiant heat. The assumption that atoms could move freely through space until they collided like billiard balls was probably regarded as too drastic an approximation, since it neglected the drag of the ether and oversimplified the interaction between atoms.

Late in the eighteenth century, Lavoisier and Laplace developed a systematic theory of chemical and thermal phenomena based on the assumption that heat is a substance, called "caloric." Laplace and Poisson later worked out a quantitative theory of gases, in which the repulsion between atoms was attributed to the action of "atmospheres" of caloric surrounding the atoms; they were able to explain such phenomena as adiabatic compression and the velocity of sound, as well as the ideal gas laws.

The caloric theory was being brought to its final stage of perfection by Laplace, Poisson, Carnot and Clapeyron at about the same time that John Herapath (1790–1868) proposed his kinetic theory (1820, 1821). Herapath attempted to explain not only the properties of gases but also gravity, changes of state, and many other phenomena. His memoir was rejected by the Royal Society, mainly because Humphry Davy (then its president) thought Herapath's theory was too speculative, even though Davy himself had advocated the view that heat is molecular motion. Herapath published several papers on his theory, but did not succeed in converting other scientists to his views. Another British scientist, J. J. Waterston (1811–1883), submitted a memoir on the kinetic theory to the Royal Society in 1845; it was not only rejected, but remained buried in the Society's archives until 1891 when it was discovered by Lord Rayleigh.

During the period 1840–1855, the equivalence of heat, work, and other forms of energy was established experimentally by Joule, and the general principle of conservation of energy was formulated by Mayer and others. Joule himself revived Herapath's kinetic theory (1848) but did not make any further developments or applications of it aside from calculating the velocity of a hydrogen molecule. Krönig independently proposed a simple kinetic theory in 1856, and soon afterwards Rudolf Clausius (1822–1888) and James Clerk Maxwell (1831–1879) started to make substantial progress towards the modern theory. Clausius introduced the idea of the "mean free path" of a molecule between successive collisions (1858), and Maxwell, Clausius, O. E. Meyer (1834–1915), and P. G. Tait (1831–1901) developed the theory of diffusion, viscosity, and heat conduction on this basis. One of the most surprising results—whose confirmation by experiment helped to establish the kinetic theory—was Maxwell's prediction (1859) that the viscosity of a gas should be independent of density and should increase with temperature. The mean free path method is still the easiest way to understand transport phenomena, but it does not yield accurate numerical results and, as mentioned above, it failed to predict thermal diffusion.

The foundations of the modern theory of transport were laid by Maxwell in his great memoir of 1866; it is essentially this method which Boltzmann used to make his discoveries, and which he presents in this book. Maxwell proceeds by writing down general equations for the rate of change of any quantity (such as molecular velocity) at any point in space and time resulting from molecular motions and collisions. These "microscopic" equations are then compared with the corresponding "macroscopic" equations (for example, the Navier-Stokes equations of hydrodynamics) and the coefficients of viscosity, etc., are thus related to certain characteristic functions of molecular collisions. Boltzmann's transport equation, derived in 1872, is a special case of

Maxwell's general equation in which the quantity of interest is the number of molecules having a certain velocity—in other words, the velocity distribution function. As soon as this function has been found, all the transport coefficients can also be computed. Much of modern research in statistical mechanics is based on attempts to solve either the Boltzmann equation or similar equations for other kinds of distribution functions. In a sense there are two alternative ways of developing transport theory: one based on solutions of Maxwell's equations, and the other on the solution of Boltzmann's equation, and these two approaches were followed by Chapman (1916) and Enskog (1917) respectively, the final results being essentially identical. However, Boltzmann was able to deduce immediately a very important result from his equation, which was not at all obvious from Maxwell's original formulation: a quantity called H, which can be identified with the negative of the entropy, must always decrease or remain constant, if one assumes that the velocity distributions of two colliding molecules are uncorrelated. The molecular interpretation of the law of increasing entropy is thus intimately related to the assumption of molecular chaos and the relation between entropy and probability. H remains constant only when the gas attains a special velocity distribution which had previously been deduced by Maxwell (1859) in a less convincing manner.

There is apparently a contradiction between the law of increasing entropy and the principles of Newtonian mechanics since the latter do not recognize any difference between past and future times. If one sequence of molecular motions corresponds to increasing entropy, one should obtain a sequence corresponding to decreasing entropy by simply reversing all the velocities. This is the so-called reversibility paradox (*Umkehreinwand*) which was advanced as an objection to Boltzmann's theory by Loschmidt (1876, 1877) and others. Another difficulty is the fact that a conservative dynamical system in a finite space will always return infinitely often to a state close to any arbitrary initial state; hence the entropy cannot continually increase but must go through cyclic variations. This is the recurrence paradox (*Wiederkehreinwand*) based on a theorem of Poincaré (1890) and used against the kinetic theory by Zermelo (1896). Some scientists held that if there is any contradiction between macroscopic thermodynamics and the kinetic theory, the latter should be rejected since it is based on a hypothetical atomic model, whereas thermodynamics is firmly grounded on empirical observations. Boltzmann's reply to these objections is given in this book.

Another broad field of theoretical research was opened up by J. D. van der Waals (1837–1923), who published his celebrated investigation of the continuity of the liquid and gaseous states in 1873. Although van der Waals' equation of state was supposedly deduced from the virial theorem of Clausius (1870), and was therefore considered an application of the kinetic theory, its derivation relied partly on concepts of the older Laplace-Poisson theory of capillarity, in which molecular motions were ignored. Nevertheless the van der Waals equation of state gives an excellent qualitative description of condensation and critical phenomena, and Boltzmann devotes a substantial portion of Part II of these lectures to an exposition of van der Waals' theory. He also describes his calculation of the next correction term to the equation of state of hard spheres—i.e., what we now call the third virial coefficient.

The failure of the kinetic theory to give a satisfactory explanation of the specific heats of polyatomic gases—and spectroscopic evidence that even monatomic gas molecules must have internal degrees of freedom which apparently do not have their proper share of energy as required by the equipartition theorem—stimulated several theoretical studies of mechanical models later in the nineteenth century. These studies were directed toward attempting to prove rigorously the necessary and sufficient conditions for equipartition. It was generally agreed that a sufficient condition would be that the system passes eventually through every point on the "energy surface" in phase space before returning to its starting point; in other words, no matter what might be the initial velocities and positions of the particles, they would eventually take on all possible values consistent with their fixed total energy. This statement is now known as the "ergodic" hypothesis, and it is frequently asserted that Maxwell and Boltzmann believed that real physical systems have this property. If the hypothesis were true for a system, then it would be legitimate to replace a time-average over the trajectory of a single system (which is what one can actually observe) by a "phase" or "ensemble" average over many replicas of the same system with all possible sets of positions and velocities consistent with a specified energy. From a mathematical point of view, the existence of an ergodic system would imply that an n-dimensional manifold (the energy surface in phase space) can be mapped continuously onto a one-dimensional manifold (the range of the time variable) so that every point in the former corresponds to at least one point in the latter.

While Boltzmann did make certain statements in some of his papers (1871, 1884, 1887) to the effect that a mechanical system might exhibit such behavior—the Lissajous figure with incommensurable periods was one alleged example—there is no foundation for the belief that Boltzmann thought real systems are ergodic. Moreover, Boltzmann did not clearly distinguish between the "ergodic hypothesis" (which we now know to be false) and the so-called "quasi-ergodic hypothesis," that the trajectory of a system may pass *arbitrarily close* to every point on the energy surface. The latter hypothesis does indeed provide a reasonable basis for statistical mechanics, even though it has not yet been proved that it is valid for real physical systems. To anyone acquainted with the modern mathematical theory of sets of points and infinite classes, the distinction between the two hypotheses is obvious, but we must remember that Boltzmann did not accept this theory, although he may have been aware of it (the fundamental papers of Georg Cantor were published during the same period that Boltzmann was publishing his papers on gas theory, some of them in the same journal). On the contrary, Boltzmann often stated that infinitesimal or infinite numbers have no meaning except as limits of finite numbers, and for this reason he gave alternative derivations of some of his theorems using sums of discrete quantities instead of integrals. To say that a trajectory passes through every point in a region would (to him) be meaningless unless one meant by this that the trajectory passes arbitrarily close to every point, and in at least one paper Boltzmann uses the two statements interchangeably.*

Part of the confusion regarding Boltzmann's views is due to the fact that the Ehrenfests, in their classic encyclopedia article

* Boltzmann, Sitzungsber. math.-phys. Classe K. Bay. Akad. Wiss., München ["Mun. Ber."] 22, 329 (1892); the statement referred to is on p. 157 of the English translation in Phil. Mag. [5] 35 (1893).

(1911), and later writers following their terminology, took Boltzmann's word "ergodic" and used it to refer to a single mechanical system whose trajectory passes through every point on the energy surface, whereas Boltzmann himself had used the word in a completely different sense, to mean an aggregate or ensemble of systems whose positions and velocities are distributed over all possible sets of values on the energy surface. The Ehrenfests recognized this distinction, and used the term "ergodic distribution" for the latter concept, but since then Gibbs' terminology has come into general use, and we now use the phrase "microcanonical ensemble" for what Boltzmann called an *Ergode.* (Boltzmann's justification for using *Ergoden* in gas theory is given in Part II, §35, and in other remarks scattered through the rest of this book.)

The statistical mechanics of J. Willard Gibbs (1839–1903) represents the next level of abstraction and refinement of gas theory, following directly on Boltzmann's introduction of *Ergoden,* and relying heavily on the methods of analytical mechanics which Boltzmann applied to gas theory (see Part II, Chapter III). Instead of considering an ensemble of systems with different initial positions and velocities but the same total energy, Gibbs proposed to consider the more general case of a "canonical ensemble" of systems having all possible energies, the number of systems having energy E being proportional to $e^{-\beta E}$ (where $\beta = l/kT$). Gibbs also introduced the "grand canonical ensemble" in which the number of particles in each system may vary, the number of systems with N particles being proportional to $e^{-\mu N}$, where μ (called the "chemical potential") is chosen to give the desired density. These generalized ensembles are easier to deal with mathematically, and enable one to treat systems in contact with a heat reservoir at constant temperature, or systems in which chemical reactions may take place.

There is not space here to describe the modifications of statistical mechanics by quantum mechanics, associated with the names of Planck, Einstein, Bose, Fermi, and Dirac. It may however be worth mentioning that Planck's discovery of the quantum laws of radiation owed much to Boltzmann's statistical theory of entropy. At the same time it must be remembered that during the years preceding the introduction of the quantum theory, the inadequacies of classical mechanics had become increasingly more evident, as theoretical physicists tried in vain to construct satisfactory classical models of atoms and of the ether. There was also a growing reaction against "scientific materialism," and a movement to replace atomic theories by purely descriptive theories based only on macroscopic observables. The most famous leader of this movement was Ernst Mach (1838–1916), whose criticism of mechanics prepared the way for Einstein's theory of relativity. In a series of articles and books, beginning with a critical study of the law of conservation of energy (1872), Mach attacked the use of atomic models in physics, and asserted that the basic purpose of science is to achieve "economy of thought" in describing natural phenomena, not to explain them in terms of hypothetical concepts such as atoms or ether. Gustav Kirchhoff (1824–1887) expressed similar views in his lectures on mechanics (1874) but did not develop them as fully, and indeed in his lectures on heat (published posthumously in 1894) he included a section on kinetic theory. One of the striking features of Boltzmann's *Gas Theory* is the large number of references to Kirchhoff's exposition.

The school of Energetics included many members of the positivist movement, in particular Wilhelm Ostwald (1853–1932), Georg Helm (1851–1923), Pierre Duhem (1861–1916) and others. The Energetists proposed to develop thermodynamics, and the theory of energy transformations in general, on a purely macroscopic basis without any reference to atomic theories. Their attacks on the kinetic theory were particularly strong during the period when Boltzmann was writing this book, and this circumstance helps to explain its polemical flavor. There are many passages justifying various assumptions, defending the theory against possible criticisms, and attacking rival theories. Boltzmann knew he could not get away with vague "heuristic" arguments or appeals to authority, since many of his readers would not be convinced. (It might well be argued that a text written in such a style is really more effective—provided that it contains no serious errors—than a modern text expounding a noncontroversial theory which the student is not encouraged to doubt.)

According to Lenin* the arguments of Mach and other "idealistic" (in the philosophical sense) scientists subverted many Marxist philosophers, who accepted them as the latest results of

* *Materialism and Empirio-Criticism*, Izdanie "Zveno," Moscow, 1909.

science. Lenin quotes with approval the views of Boltzmann, who, while "afraid to call himself a materialist" nevertheless had a theory of knowledge which was "essentially materialistic." Consequently Boltzmann now has the dubious distinction of being a hero of scientific materialism in the eyes of the Marxists.*

As the leader of the atomist school in the 1890's, Boltzmann frequently had to engage in debate with the Energetists. One such debate took place at the meeting of the Naturforschergesellschaft at Lübeck in 1895. As Arnold Sommerfeld recalls it,†

> The champion for Energetics was Helm; behind him stood Ostwald, and behind both of them the philosophy of Ernst Mach (who was not present in person). The opponent was Boltzmann, seconded by Felix Klein. The battle between Boltzmann and Ostwald was much like the duel of a bull and a supple bullfighter. However, this time the bull defeated the toreador in spite of all his agility. The arguments of Boltzmann struck through. We young mathematicians were all on Boltzmann's side; it was at once obvious to us that it was impossible that from a single energy equation could follow the equations of motion of even one mass point, to say nothing of those for a system of an arbitrary number of degrees of freedom. On Ostwald's behalf, however, I must mention his remark on Boltzmann in his book *Grosse Männer* (Leipzig, 1909, p. 405): there he calls Boltzmann "the man who excelled all of us in acumen and clarity in his science."

What did Boltzmann really think about the existence of atoms? One commentator‡ has asserted that "even the most convinced adherents of this [kinetic] theory, such as Boltzmann, ascribe to it merely the value of a mechanical model which imitates certain properties of gases and can afford the experimenter certain useful indications, and are very far from believing bodies to be in reality composed of small particles . . . " Indeed, a curious feature of Boltzmann's attitude toward the molecular model of gases is his attempt to justify the use of asymptotic limits in which the number of molecules in a volume becomes infinite,

* See, e.g., the article by Bogolyubov and Sanochkin, Usp. Fiz. Nauk **61**, 7 (1957), and Davydov's introduction to the Russian translation of *Gas Theory*.

† A. Sommerfeld, Wiener Chem. Zeitung **47**, 25 (1944); see also G. Jaffé, J. Chem. Educ. **29**, 230 (1952).

‡ A. Aliotta, *La Reazione idealistica contro la scienza* (Palermo: Casa Editrice "Optima," 1912); quoted from p. 375 of the English translation by Agnes McCaskill (London, 1914).

while their size becomes infinitesimal. He is perfectly aware of the fact that one can deduce finite numerical values for these molecular parameters from experiment by using gas theory (Part I, §12; Part II, §22); and he refers to Maxwell's theory of rarefied gas phenomena (radiometer effect, etc.) as a triumph of molecular theories over the phenomenological approach (Part I, §1). Yet his desire to obtain a more intuitively clear (*anschaulich*) presentation of his theory impels him to assume that each infinitesimal volume element in the gas contains a very large number of molecules (Part I, §6) and, in attempting to justify this assumption, he goes so far as to say that it is futile to expect to observe fluctuations from the asymptotic laws resulting from the fact that there may be only a finite number of atoms in a small space (Part II, §38). This statement was of course contradicted only a few years later by experiments on Brownian motion and Millikan's oil-drop experiment.

We cannot give here a comprehensive analysis of Boltzmann's views on the philosophy of science, but must refer to his extensive writings and the studies by Broda and Dugas cited in the Bibliography. The following quotation is chosen only because it was originally published in English (in a letter to *Nature*, February 28, 1895) so that there is no danger of distorting Boltzmann's meaning by translation (assuming of course that he is expressing his own views accurately in English):

I propose to answer two questions:—

(1) Is the Theory of Gases a true physical theory as valuable as any other physical theory?

(2) What can we demand from any physical theory?

(The first question I answer in the affirmative, but the second belongs not so much to ordinary physics (let us call it orthophysics) as to what we call in Germany metaphysics. For a long time the celebrated theory of Boscovich was the ideal of physicists. According to his theory, bodies as well as the ether are aggregates of material points, acting together with forces, which are simple functions of their distances. If this theory were to hold good for all phenomena, we should be still a long way off what Faust's *famulus* hoped to attain, viz., to know everything. But the difficulty of enumerating all the material points of the universe, and of determining the law of mutual force for each pair, would be only a quantitative one; nature would be a difficult problem, but not a mystery for the human mind.

When Lord Salisbury says that nature is a mystery [Presidential

Address to the British Association meeting at Oxford, 1894], he means, it seems to me, that this simple conception of Boscovich is refuted almost in every branch of science, the Theory of Gases not excepted. The assumption that the gas-molecules are aggregates of material points, in the sense of Boscovich, does not agree with the facts. But what else are they? And what is the ether through which they move? Let us again hear Lord Salisbury. He says:

"What the atom of each element is, whether it is a movement, or a thing, or a vortex, or a point having inertia, all these questions are surrounded by profound darkness. I dare not use any less pedantic word than entity to designate the ether, for it would be a great exaggeration of our knowledge if I were to speak of it as a body, or even as a substance."

If this be so—and hardly any physicist will contradict this—then neither the Theory of Gases nor any other physical theory can be quite a congruent account of facts, and I cannot hope with Mr. Burbury, that Mr. Bryan will be able to deduce all the phenomena of spectroscopy from the electromagnetic theory of light. Certainly, therefore, Hertz is right when he says: "The rigour of science requires, that we distinguish well the undraped figure of nature itself from the gay-coloured vesture with which we clothe it at our pleasure." [*Untersuchungen über die Ausbreitung der elektrischen Kraft*, p. 31. Leipzig, Barth, 1892]. But I think the predilection for nudity would be carried too far if we were to forego every hypothesis. Only we must not demand too much from hypotheses.

It is curious to see that in Germany, where till lately the theory of action at a distance was much more cultivated than in Newton's native land itself, where Maxwell's theory of electricity was not accepted, because it does not start from quite a precise hypothesis, at present every special theory is old-fashioned, while in England interest in the Theory of Gases is still active; *vide*, among others, the excellent papers of Mr. Tait, of whose ingenious results I cannot speak too highly, though I have been forced to oppose them in certain points.

Every hypothesis must derive indubitable results from mechanically well-defined assumptions by mathematically correct methods. If the results agree with a large series of facts, we must be content, even if the true nature of facts is not revealed in every respect. No one hypothesis has hitherto attained this last end, the Theory of Gases not excepted. But this theory agrees in so many respects with the facts, that we can hardly doubt that in gases certain entities, the number and size of which can roughly be determined, fly about pell-mell. Can it be seriously expected that they will behave exactly as aggregates of Newtonian centres of force, or as the rigid bodies of our Mechanics?

And how awkward is the human mind in divining the nature of things, when forsaken by the analogy of what we see and touch directly?

Nevertheless, Boltzmann's pessimism about the future of the kinetic theory, indicated in his Foreword to Part II, deepened in the following years, and led to fits of severe depression, culminating in his suicide in 1906. This suicide must be ranked as one of the great tragedies in the history of science, made all the more ironic by the fact that the scientific world made a complete turnabout in the next few years and accepted the existence of atoms, following Perrin's experiments on Brownian motion and an accumulation of other evidence. By 1909 even Ostwald himself had been converted, as he admits in the Foreword to the fourth edition of his textbook on chemistry. Atomism had triumphed, though only at the cost of denying to atoms almost all the properties with which they had originally been endowed.

PART I

Theory of gases with monatomic molecules, whose dimensions are negligible compared to the mean free path.

NOTE ON LITERATURE CITATIONS

All the numbered footnotes correspond to Boltzmann's notes in the original text, but the numbers start with 1 in each section rather than each page, and the form of citation of some journals has been changed for the sake of consistency with modern style. Footnotes indicated by asterisks, daggers, etc., have been added by the translator. More complete information including titles of articles cited will be found in the Bibliography at the end of the book.

NOTE ON GOTHIC LETTERS

To forestall possible difficulties in recognition, here are the alphabetical equivalents:

A B C D E F G H I J K L M N O P Q R S T U V W X Y Z
𝔄 𝔅 ℭ 𝔇 𝔈 𝔉 𝔊 ℌ ℑ 𝔍 𝔎 𝔏 𝔐 𝔑 𝔒 𝔓 𝔔 ℜ 𝔖 𝔗 𝔘 𝔙 𝔚 𝔛 𝔜 ℨ

a b c d e f g h i j k l m n o p q r s t u v w x y z
𝔞 𝔟 𝔠 𝔡 𝔢 𝔣 𝔤 𝔥 𝔦 𝔧 𝔨 𝔩 𝔪 𝔫 𝔬 𝔭 𝔮 𝔯 𝔰 𝔱 𝔲 𝔳 𝔴 𝔵 𝔶 𝔷

FOREWORD TO PART I

"Alles Vergängliche
Ist nur ein Gleichniss!"*

I have often before come close to writing a textbook on gas theory. I remember especially the enthusiastic request of Professor Wroblewski at the Vienna World's Fair in 1873. When I showed little inclination to write a textbook, since I did not know how soon my eyes would fail me, he answered dryly: "All the more reason to hurry up!" At present, when I no longer have this reason, the time seems less appropriate for such a textbook than it was then. For, first, gas theory has gone out of fashion in Germany, as it were; second, the second edition of O. E. Meyer's well-known text has appeared, and Kirchhoff has devoted a longer section of his lectures on the theory of heat to gas theory.† Yet Meyer's book, though acknowledged to be excellent for chemists and students of physical chemistry, has a completely different purpose. Kirchhoff's work shows the touch of a master in the selection and presentation of its topics, but it is only a posthumously published set of lecture notes on the theory of heat, which treats gas theory as an appendix, not a comprehensive textbook. Indeed, I freely confess that the interest, on the one hand, which Kirchhoff showed in gas theory, and on the other hand the many gaps in his presentation because of its brevity, have encouraged me to

* Goethe, *Faust,* Part 2, Act V, final "Chorus Mysticus." At least seven different English translations of this couplet have been published; roughly, it means that all transitory (i.e., earthly) things are only symbols or reflections (of reality). It is suggested that the reader interpret this in the light of Boltzmann's philosophy of science (see Translator's Introduction).

† O. E. Meyer, *Die kinetische Theorie der Gase* (Breslau: Maruschke und Berendt, 1877; second edition, 1899). Kirchhoff, *Vorlesungen über die Theorie der Wärme,* p. 134 *et seq.* (Leipzig: B. G. Teubner, 1894).

publish the present work, which likewise originated from lectures at the universities of Munich and Vienna.

In this book I have tried above all to make clearly comprehensible the path-breaking works of Clausius and Maxwell. The reader may not think badly of me for finding also a place for my own contributions. These were cited respectfully in Kirchhoff's lectures and in Poincaré's *Thermodynamique** at the end, but were not utilized where they would have been relevant. From this I concluded that a brief presentation, as easily understood as possible, of some of the principal results of my efforts might not be superfluous. Of great influence on the content and presentation was what I have learned at the unforgettable meeting of the British Association in Oxford and the subsequent letters of numerous English scientists, some private and some published in *Nature.*†

I intend to follow Part I by a second part, where I will treat the van der Waals theory, gases with polyatomic molecules, and dissociation. An explicit proof of Equation (110a)—which is only indicated briefly in §16, in order to avoid repetition—will also be included.

Unfortunately it was often impossible to avoid the use of long formulas to express complicated trains of thought, and I can well imagine that to many who do not read over the whole work, the results will perhaps not seem to justify the effort expended. Aside from many results of pure mathematics which, though likewise apparently fruitless at first, later become useful in practical science as soon as our mental horizon has been broadened, even the complicated formulas of Maxwell's theory of electromagnetism were often considered useless before Hertz's experiments. I hope this will not also be the general opinion concerning gas theory!

Vienna, September, 1895

Ludwig Boltzmann

* Poincaré, *Thermodynamique* (Paris: Gauthier-Villars, 1892).

† Bryan, Nature **51**, 31, 152, 176, 262, 319 (1894–1895); **52**, 244 (1895). Culverwell, Nature **50**, 617 (1894); **51**, 78, 105, 246, 581 (1894–1895); **52**, 149 (1895). Burbury, Nature **51**, 175 (1894); **52**, 316 (1895). Fitzgerald, Nature **51**, 221 (1895). Boltzmann, Nature **51**, 413, 581 (1895).

INTRODUCTION

§1. Mechanical analogy for the behavior of a gas.

Clausius has made a sharp distinction between the general theory of heat, based essentially on the two fundamental principles of thermodynamics, and the special theory of heat, which starts out by making a definite assumption—that heat is molecular motion—and then attempts to construct a more precise description of the nature of this motion.

The general theory of heat also requires hypotheses which go beyond the bare facts of nature. Nevertheless, it is obviously less dependent on special assumptions than the special theory, and it is desirable and necessary to be able to separate its exposition from that of the special theory, and to show that it is independent of the subjective assumptions of the latter. Clausius has already done this very clearly, and indeed has based the division of his book* into two parts on just this principle, and it would be useless to repeat his procedure.

Recently, the mutual relations of these two branches of the theory of heat have changed somewhat. Through the exploitation of some very interesting analogies and differences, which the properties of energy exhibit in various phenomena of physics, there has arisen the so-called Energetics, which is unfavorable to the view that heat is molecular motion. This view is in fact unnecessary for the general theory of heat and, as is well known, was not held by Robert Mayer. The further development of Energetics is certainly very important for science; however, up to now its concepts are still rather unclear, and its theorems not very precisely expressed, so that it cannot replace the older theory of heat in dealing with new special cases where the results are not already known.†

---

* Clausius, *Abhandlungen über die mechanische Wärmetheorie* (Braunschweig: F. Vieweg, Part I, 1864, Part II, 1867, Part III, 1889). More complete bibliographic information on this and other works cited may be found in the Bibliography at the end of this book.

† There is space to cite only a few of the many works on Energetics:

Likewise in the theory of electricity, the older mechanical explanations of phenomena by action at a distance, customarily employed especially in Germany, have suffered a shipwreck. Indeed, Maxwell speaks with the greatest respect of Wilhelm Weber's theory, which by determining the conversion factor of electrostatic and electromagnetic units and by discovering the relation of that factor to the velocity of light has laid the first stone in the structure of the electromagnetic theory of light; yet one accedes to the contention that Weber's mechanical hypothesis about the action of electrical force is harmful to the progress of science.

In England, views on the nature of heat and on atomistics have remained relatively unchanged. But on the continent, where earlier the assumption of central forces, useful in astronomy, had been generalized to an epistemological demand, and for this reason Maxwell's electrical theory did not receive much notice for a decade and a half (only this generalization was obnoxious), the provisional character of all hypotheses has now been generalized, and it has been concluded that the assumption that heat is motion of the smallest particles of matter will eventually be proved false and discarded.

It must however be remembered that any similarity between the kinetic theory and the doctrine of central forces is purely coincidental. Gas theory has in fact a particular kinship with Maxwell's theory of electricity; the visible motion of a gas, viscosity,

---

Georg Helm, *Die Lehre von der Energie* (Leipzig, 1887); *Grundzüge der mathematischen Chemie* (Leipzig, 1894); Ann. Physik [3] **57,** 646 (1896); *Die Energetik* (Leipzig, 1898). Wilhelm Ostwald, Leipzig Ber. **43,** 271 (1891), **44,** 211 (1892); Ann. Physik [3] **58,** 154 (1896); Verh. Ges. d. Naturf. Aerzte, Th. 1, p. 155 (1895); *Individuality and Immortality* (Boston: Houghton, Mifflin; 1906); *Die Forderung des Tages* (Leipzig, 1910); *Energetische Grundlagen der Kulturwissenschaft* (Leipzig, 1909); *Monistische Sonntagspredigten* (Leipzig, 1911); *Der energetische Imperativ* (Leipzig, 1912); *Die Energie* (Leipzig, 1912); *Die Philosophie der Werte* (Leipzig, 1913). Boltzmann, Ann. Physik [3] **57,** 39, **58,** 595 (1896), **60,** 231 (1897) **61,** 790 (1897); Wien. Ber. **106,** 83 (1897); Verh. Ges. d. Naturf. Aertze, Th. 2, p. 65 (1898); Th. 1, p. 99 (1899). Pierre Duhem, *Traité d'energetique* (Paris, 1911). Max Planck, Z. phys. Chem. **8,** 647 (1891); Ann. Physik [3] **57,** 72 (1896). Abel Rey, *La Théorie de la Physique chez les physiciens contemporains* (Paris, 1907). Jules Sageret, *La vague mystique* (Paris, 1920). René Dugas, *La Théorie Physique au sens de Boltzmann* (Neuchatel-Suisse: Editions du Griffon, 1959).

and heat, are conceived as phenomena that are essentially different in stationary or almost stationary states, while in certain transitional cases (very rapid sound waves with evolution of heat; viscosity or heat conduction in very dilute gases[1]) a sharp distinction is no longer possible between visible and thermal motion (cf. §24). Likewise in Maxwell's theory of electricity, in borderline cases the distinction between electrostatic and electrodynamic forces, etc., can no longer be maintained. It is just in this transition region that Maxwell's electrical theory has yielded new facts; likewise the gas theory has led to completely new laws in these transition regions, which appear to reduce to the usual hydrodynamic equations, corrected for viscosity and heat conduction, as purely approximate formulae (cf. §23). The completely new laws were indicated for the first time in Maxwell's 16-year-old paper, "On stresses in rarefied gases."* The phenomena which the theories based on older hydrodynamic experience can never describe are those connected with radiometer action.† Researches under many different conditions, and quantitative observations, have definitely proved that the stimulus and explanation for this previously unknown realm of experimental investigation could

---

[1] Cf. Kundt and Warburg, Ann. Physik [2] **155**, 341 (1875). [Numbered notes are by Boltzmann.—Tr.]

* Maxwell, Phil. Trans. **170**, 231 (1880).

† William Crookes, though not its original discoverer, was the first to investigate extensively and publicize the radiometer effect: see Proc. R. S. London **22**, 37 (1874), Phil. Trans. **164**, 501 (1874), and many other papers. It created a sensation in the scientific world during 1875–1877, since it apparently was a demonstration of the long-sought pressure of light. However, it was immediately suggested by Osborne Reynolds [Proc. R. S. London **22**, 401 (1874)] and others that the motion of the radiometer plates was probably due to the reaction of gas molecules leaving the heated sides. Arthur Schuster [Proc. R. S. London **24**, 391 (1876)] showed that when the entire radiometer is suspended in a larger evacuated vessel, the case turns in a direction opposite to that in which the plates turn, so that the effect must be due largely to internal rather than external causes. The light shining on the radiometer plates provides energy to heat them up, but not momentum to move them in a particular sense. While it was thus recognized that the kinetic theory must provide the explanation for the radiometer, the detailed mechanism and quantitative theory were not worked out until much later. See E. H. Kennard, *Kinetic Theory of Gases* (New York: McGraw-Hill, 1938), pp. 333–337, and L. B. Loeb, *The Kinetic Theory of Gases*, 2nd ed. (New York: McGraw-Hill, 1934), pp. 364–386.

only have come from the theory of gases; likewise, the tremendous fertility of Maxwell's theory of electricity for experimental investigations for more than twenty years has seldom been pointed out.

Although in the following exposition any qualitative difference between heat and mechanical energy will be excluded, in the treatment of collisions of molecules the old distinction between potential and kinetic energy will be retained. This does not basically affect the nature of the subject. The assumptions about the interaction of molecules during a collision have a provisory character, and will certainly be replaced by others. I have also studied a gas theory in which, instead of forces acting during collisions, one merely has conditional equations in the sense of the posthumous mechanics of Hertz,* which are more general than those of elastic collisions; I have abandoned this theory, however, since I only had to make more new arbitrary assumptions.

Experience teaches that one will be led to new discoveries almost exclusively by means of special mechanical models. Maxwell himself recognized the defect of Weber's electrical theory at first glance; on the other hand, he pursued zealously the theory of gases and the method of mechanical analogies, which goes beyond that which he calls the method of purely mathematical formulae.†

As long as a clearer and better representation is not available, we shall still need to go beyond the general theory of heat, and without impugning its importance, cultivate the old hypotheses of the special theory of heat. Indeed, since the history of science shows how often epistemological generalizations have turned out to be false, may it not turn out that the present "modern" distaste for special representations, as well as the distinction between qualitatively different forms of energy, will have been a retrogression? Who sees the future? Let us have free scope for all directions of research; away with all dogmatism, either atomistic or antiatomistic! In describing the theory of gases as a mechanical *analogy*, we have already indicated, by the choice of this word, how far removed we are from that viewpoint which would see in visible matter the true properties of the smallest particles of the body.

---

* Heinrich Hertz, *Die Prinzipien der Mechanik* (Leipzig, 1894).

† See A. E. Woodruff, Isis **53**, 439 (1962).

We shall first take the modern viewpoint of pure description, and accept the known differential equations for the internal motions of solid and fluid bodies. From these it follows in many cases, for example collisions of two solid bodies, motion of fluids in closed vessels, etc., that as soon as the form of the body deviates the least bit from a simple geometrical figure, waves must arise, which cross each other ever more randomly, so that the kinetic energy of the original visible motion must finally be dissolved into invisible wave motion. This mathematical consequence of the equations describing the phenomena leads (to a certain extent by itself) to the hypothesis that all vibrations of the smallest particles, into which the ever diminishing waves must finally be transformed, must be identical with the heat that we observe to be produced, and that heat generally is a motion in small—to us, invisible—regions.

Whence comes the ancient view, that the body does not fill space continuously in the mathematical sense, but rather it consists of discrete molecules, unobservable because of their small size. For this view there are philosophical reasons. An actual continuum must consist of an infinite number of parts; but an infinite number is undefinable. Furthermore, in assuming a continuum one must take the partial differential equations for the properties themselves as initially given. However, it is desirable to distinguish the partial differential equations, which can be subjected to empirical tests, from their mechanical foundations (as Hertz emphasized in particular for the theory of electricity). Thus the mechanical foundations of the partial differential equations, when based on the coming and going of smaller particles, with restricted average values, gain greatly in plausibility; and up to now no other mechanical explanation of natural phenomena except atomism has been successful.

A real discontinuity of bodies is moreover established by numerous, and moreover quantitatively agreeing, facts. Atomism is especially indispensable for the clarification of the facts of chemistry and crystallography. The mechanical analogy between the facts of any science and the symmetry relations of discrete particles pertains to those most essential features which will outlast all our changing ideas about them, even though the latter may themselves be regarded as established facts. Thus already today the hypothesis that the stars are huge bodies millions of miles away is similarly viewed only as a mechanical analogy for the representation of the action of the sun and the faint visual perceptions arising from the other heavenly bodies, which could also be criticized on the grounds that it replaces the world of our sense perceptions by a world of imaginary objects, and that anyone could just as well replace this imaginary world by another one without changing the observable facts.

I hope to prove in the following that the mechanical analogy between the facts on which the second law of thermodynamics is based, and the statistical laws of motion of gas molecules, is also more than a mere superficial resemblance.

The question of the utility of atomistic representations is of course completely unaffected by the fact, emphasized by Kirchhoff,* that our theories have the same relation to nature as signs to significates, for example as letters to sounds, or notes to tones. It is likewise unaffected by the question of whether it is not more useful to call theories simply descriptions, in order to remind ourselves of their relation to nature. The question is really whether bare differential equations or atomistic ideas will eventually be established as complete descriptions of phenomena.

Once one concedes that the appearance of a continuum is more clearly understood by assuming the presence of a large number of adjacent discrete particles, assumed to obey the laws of mechanics, then he is led to the further assumption that heat is a permanent motion of molecules. Then these must be held in their relative positions by forces, whose origin one can imagine if he wishes. But all forces that act on the visible body but not equally on all the molecules must produce motion of the molecules relative to each other, and because of the indestructibility of kinetic energy these motions cannot stop but must continue indefinitely.

In fact, experience teaches that as soon as the force acts equally on all parts of a body—as for example in so-called free fall—all the kinetic energy becomes visible. In all other cases, we have a loss of visible kinetic energy, and hence creation of heat. The view offers itself that there is a resulting motion of molecules among themselves, which we cannot see because we do not see individual molecules, but which however is transmitted to our

* Kirchhoff, *Vorlesungen über mathematische Physik*. I. *Mechanik* (Leipzig: B. G. Teubner, 1874.)

nerves by contact, and thus creates the sensation of heat. It always moves from bodies whose molecules move rapidly to those whose molecules move more slowly, and because of the indestructibility of kinetic energy it behaves like a substance, as long as it is not transformed into visible kinetic energy or work.

We do not know the nature of the force that holds the molecules of a solid body in their relative positions, whether it is action at a distance or is transmitted through a medium, and we do not know how it is affected by thermal motion. Since it resists compression as much as it resists dilatation, we can obviously get a rather rough picture by assuming that in a solid body each molecule has a rest position. If it approaches a neighboring molecule it is repelled by it, but if it moves farther away there is an attraction.

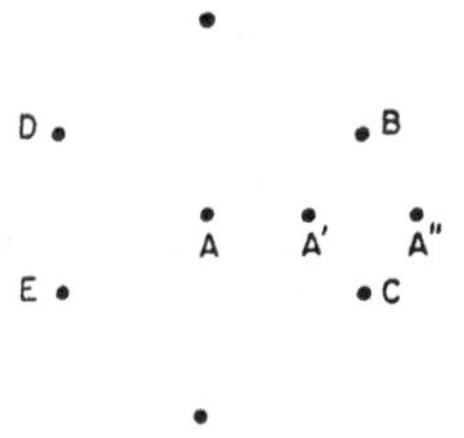

FIG. 1.

Consequently, thermal motion first sets a molecule into pendulum-like oscillations in straight or elliptical paths around its rest position A (in the symbolic Fig. 1, the centers of gravity of the molecules are indicated). If it moves to A', the neighboring molecules B and C repel it, while D and E attract it and hence bring it back to its original rest position. If each molecule vibrates around a fixed rest position, the body will have a fixed form; it is in the solid state of aggregation. The only consequence of the thermal motion is that the rest positions of the molecules will be somewhat pushed apart, and the body will expand somewhat. However, when the thermal motion becomes more rapid, one gets to the point where a molecule can squeeze between its two neighbors and move from A to A'' (Fig. 1). It will no longer then be pulled back to its old rest position, but it can instead remain where it is. When this happens to many molecules, they will crawl among each other like earthworms, and the body is molten. Although one may find this description rather crude and childish, it may be modified later and the apparent repulsive force may turn out to be a direct consequence of the motion. In any case, one will allow that when the motions of the molecules increase beyond a definite limit, individual molecules on the surface of the body can be torn off and must fly out freely into space; the body evaporates. If it is in an enclosed vessel, then this will be filled with freely moving molecules, and these can occasionally penetrate into the body again; as soon as the number of recondensing molecules is, on the average, equal to the number of evaporating ones, one says that the vessel is saturated with the vapor of the body in question.

A sufficiently large enclosed space, in which only such freely moving molecules are found, provides a picture of a gas. If no external forces act on the molecules, these move most of the time like bullets shot from guns in straight lines with constant velocity. Only when a molecule passes very near to another one, or to the wall of the vessel, does it deviate from its rectilinear path. The pressure of the gas is interpreted as the action of these molecules against the wall of the container.

§2. Calculation of the pressure of a gas.

We shall now undertake a more detailed consideration of such a gas. Since we assume that the molecules obey the general laws of mechanics, then in collisions of the molecules with each other and with the wall, the principle of conservation of kinetic energy, and of motion of the center of gravity, must be satisfied. We can make as varying pictures of the internal properties of the molecules as we like; as long as these two principles are satisfied, we shall obtain a system which shows a definite mechanical analogy with the actual gas. The simplest such picture is one in which the molecules are completely elastic, negligibly deformable spheres, and the wall of the container is a completely smooth and elastic surface. However, when it is convenient we can assume a different law of force. Such a law, provided it is in agreement with the general mechanical principles, will be no more and no less justified than the original assumption of elastic spheres.

We imagine a container of volume Ω of any arbitrary shape, filled with a gas, against whose walls the gas molecules are reflected exactly like completely elastic spheres. Let a part of the wall of the container, AB, have surface area φ. We place perpendicular to it, running from inside to outside, the positive axis of abscissas. The pressure on AB will clearly not be changed if we imagine behind this surface element a right cylinder on the base AB, in which the surface element AB is like a piston which can be displaced parallel to itself. This piston will then be pushed into the cylinder by the molecular impacts. If a force P acts from the outside in the negative direction, then its intensity can be chosen so that the molecular impacts are kept in equilibrium, and the piston makes no visible motion in either direction.

During any instant of time dt, there may be several molecules colliding with the piston AB; the first exerts a force q_1, the second a force q_2, etc., in the positive abscissa direction, on the piston. Denoting by M the mass of the piston, and by U the velocity in the positive direction, then one has for the time element dt the equation

$$M\frac{dU}{dt} = -P + q_1 + q_2 + \cdots$$

If one multiplies by dt and integrates over an arbitrary time t, then:

$$M(U_1 - U_0) = -Pt + \sum\int_0^t qdt.$$

If P is to be equal to the pressure of the gas, then the piston must not show any noticeable motion, aside from invisible fluctuations. In the above formula, U_0 is the value of its velocity in the abscissa direction at the initial time, and U_1 is its value after an interval of time t. Both quantities will be very small; indeed, one can easily choose the time t such that $U_1 = U_0$, since the piston must periodically assume the same velocity during its various small fluctuations. In any case, $U_1 - U_0$ cannot continually increase with increasing time, and therefore with increasing time the quotient $(U_1 - U_0)/t$ must approach the limit zero. Whence follows:

$$P = \frac{1}{t}\sum\int_0^t qdt \tag{1}$$

The pressure is therefore the mean value of the sum of all the small pressures that the individual colliding molecules exert on the piston at different times. We shall now calculate $\int qdt$ for any one collision which the piston experiences during the time t with a molecule. Let the mass of the molecule be m, and let the component of its velocity in the abscissa direction be u. The collision begins at time t_1 and ends at time $t_1+\tau$; before time t_1 and after time $t_1+\tau$, the molecule exerts essentially no force on the piston. Then

$$\int_0^t qdt = \int_{t_1}^{t_1+\tau} qdt$$

During the time of the collision, however, the force which the molecule exerts on the piston is equal and opposite to the force which the piston exerts on the molecule:

$$m\frac{d\dot{u}}{dt} = -q.$$

We denote by ξ the velocity component of the colliding molecule before the collision in the direction of the positive abscissa axis, and by $-\xi$ the same component after the collision, and we obtain:

$$\int_{t_1}^{t_1+\tau} qdt = 2m\xi.$$

Since the same is true for all other colliding molecules, it follows from Equation (1) that:

$$P = \frac{2}{t}\sum m\xi, \tag{2}$$

where the sum is to be extended over all molecules that strike the piston between the instants 0 and t. Only those which are in collision with the piston at the instants 0 and t are thereby omitted, which is permissible when the total time interval t is very large compared to the duration of an individual collision.

We shall see (§3) that even when only a single gas is present in the container, all the molecules can be no means have the same velocity. In order to preserve the greatest generality, we assume that in the container there are different kinds of molecules, which however are all reflected like elastic spheres from the walls. $n_1\Omega$ molecules will have mass m_1 and velocity c_1 with components ξ_1, η_1, ζ_1 in the coördinate directions. They should be uniformly distributed, on the average, throughout the interior volume Ω of the container, so that there are n_1 in unit volume. Further, there are $n_2\Omega$ molecules distributed similarly, having another velocity c_2 with components ξ_2, η_2, ζ_2 and perhaps also a different mass m_2. The quantities n_3, c_3, ξ_3, η_3, ζ_3, m_3, etc., up to n_i, c_i, ξ_i, η_i, ζ_i, m_i have similar meanings. The state of the gas in the container should remain stationary during the time t, so that when during a time τ some of the $n_1\Omega$ molecules cease to have velocity components ξ_1, η_1, ζ_1 as a result of collisions with other molecules or with the wall, an equal number of molecules on the average will acquire the same velocity components during the same time.

We must first calculate how many of our $n_1\Omega$ molecules strike the piston in the time interval t, on the average. During a short time dt, all $n_1\Omega$ molecules travel the distance c_1dt in a direction such that their projections on the coördinate axes are ξ_1dt, η_1dt, and ζ_1dt. If ξ_1 is negative, then the molecule considered cannot collide with the piston. If it is positive, we construct in the container an oblique cylinder, whose base is the piston AB, and whose side is equal to and in the same direction as the path c_1dt. Then those, and only those, of our $n_1\Omega$ molecules which at the beginning of the interval dt are in this cylinder—whose number we shall denote by $d\nu$—will collide with the piston during time dt. The $n_1\Omega$ molecules are on the average uniformly distributed in the entire container, and this uniform distribution extends even up to the wall, since molecules reflected from the wall move backwards as if it did not exist and there was instead another gas with the same properties beyond it. Then $n_1\Omega$ is to $d\nu$ as Ω is to the volume of the oblique cylinder;[1] the latter, however, is equal to $\varphi\xi_1dt$, whence it follows that:

$$d\nu = n_1\varphi\xi_1dt. \tag{3}$$

Since now the state in the container remains stationary, during any arbitrary time t, $n_1\varphi\xi_1t$ of our $n_1\Omega$ molecules collide with the piston. They all have mass m_1 and velocity component ξ in the abscissa direction before the collision, and contribute to the sum $\sum m\xi$ in Equation (2) the term:

$$\varphi tn_1m_1\xi_1^2,$$

[1] Concerning the conditions of validity of a similar proportion, cf. §3.

and since this is true for all molecules, we obtain

$$\frac{P}{\varphi} = 2 \sum n_h m_h (+\xi_h)^2,$$

where the sum is to be extended over all molecules in the container whose velocity components in the abscissa direction are positive. $P/\varphi = p$ is the pressure corresponding to unit surface. The formula will also be valid when φ is infinitesimal, so that the container wall need not be flat anywhere. Under the assumption (whose validity will be shown later, in §19) that in a stationary gas there is no preferred direction in space for the direction of motion of a molecule, there must be equally many molecules of each kind in the positive as in the negative abscissa direction, so that $\sum m_h n_h \xi_h^2$ summed over all molecules with negative ξ_h is as large as the sum over molecules with positive ξ_h, and thus one gets

$$p = \sum_{h=1}^{h=i} n_h m_h \xi_h^2, \tag{4}$$

where the summation is now over all molecules in the container, and therefore over all values of h from $h=1$ to $h=i$.

When any quantity g has the value g_1 for n_1 molecules, the value g_2 for n_2 molecules, etc., and the value g_i for n_i molecules, then we shall denote the expression:

$$\frac{\sum_{h=1}^{h=i} n_h g_h}{n}$$

by $\bar{g}$ and call it the mean value of g. Note that

$$n = \sum_{h=1}^{h=i} n_h$$

is the total number of all the molecules. Then we can write

$$p = n\overline{m\xi^2}. \tag{5}$$

If all molecules have the same mass, then

$$p = nm\overline{\xi^2}.$$

Since the gas has the same properties in all directions, we have $\overline{\xi^2} = \overline{\eta^2} = \overline{\zeta^2}$. Since furthermore, for each molecule, $c^2 = \xi^2 + \eta^2 + \zeta^2$, then also $\overline{c^2} = \overline{\xi^2} + \overline{\eta^2} + \overline{\zeta^2}$ and $\overline{\xi^2} = \frac{1}{3}\overline{c^2}$. Hence we obtain

$$p = \tfrac{1}{3}nm\overline{c^2}; \tag{6}$$

where nm is the total mass contained in unit volume of the gas, i.e., the density ρ of the gas; one has therefore

$$p = \tfrac{1}{3}\rho\overline{c^2} \tag{7}$$

Since p and ρ can be determined experimentally, one can calculate $\overline{c^2}$. One finds that at 0°C, $\sqrt{\overline{c^2}}$ is 461 m/sec for oxygen; for nitrogen it is 492 m/sec; for hydrogen it is 1,844 m/sec. It is this velocity whose square is equal to the mean square velocity of the molecule; it is also the velocity with which all molecules must move in order to produce in the gas the prevailing pressure, when all molecules have equal velocities and move equally often in all directions in space, or when one third of the molecules move back and forth in a direction perpendicular to the surface considered, the other two thirds moving in parallel directions. On the other hand, $\sqrt{\overline{c^2}}$ is of the same order of magnitude as the mean velocity of a molecule, but differs from it by a numerical factor (cf. §7).

When other gases are present in the container, let n', n'', etc. be the number of molecules in unit volume, m', m'', etc. the masses of each molecule for the different gases, and ρ', ρ'', etc. their partial densities—i.e., the density which each gas would have if it were alone in the container. It is then evident from Equations (4) and (5) that

$$p = \tfrac{1}{3}(n'm'\overline{c'^2} + n''m''\overline{c''^2} + \cdots) = \tfrac{1}{3}(\rho'\overline{c'^2} + \rho''\overline{c''^2}\cdots) \tag{8}$$

is the total pressure of the gas mixture; it is also equal to the sum of the partial pressures, i.e., the pressure which each gas would exert if it were alone in the container.

The force that two molecules exert on each other during a collision can be completely arbitrary, provided only that the sphere of action is small compared to the mean free path. On the other hand, it will be assumed that the molecules are reflected at the walls like elastic spheres. We shall make ourselves independent of the latter restrictive assumption in §20. A second general derivation of the equations of this section from the virial theorem will be given in Part II, §50.

CHAPTER I

The molecules are elastic spheres. External forces and visible mass motion are absent.

§3. Maxwell's proof of the velocity distribution law; frequency of collisions.

We shall suppose for a moment that in the container there is a single gas composed of completely identical molecules. The molecules will also from now on—unless we specify otherwise—be assumed to behave like completely elastic spheres when they collide with each other. Even if all the molecules initially had the same velocity, there would soon occur collisions in which the velocity of one colliding molecule is nearly in the direction of the line of centers, but that of the other is nearly perpendicular to it. The first molecule would thereby end up with nearly zero velocity, while the velocity of the second would become $\sqrt{2}$ times as large. In the course of further collisions it would soon happen, if the number of molecules were large enough, that all possible velocities would occur, from zero up to a velocity much larger than the original common velocity of all the molecules; it is then a question of calculating the law of distribution of velocities among the molecules in the final state thus reached, or, as one says more briefly, to find the velocity distribution law. In order to find it, we shall consider a more general case. We assume that we have two kinds of molecules in the container. Each molecule of the first kind has mass m, and each of the second has mass m_1. The velocity distribution which prevails at any arbitrary time t will be represented by drawing as many straight lines (starting from the origin of coördinates) as there are m-molecules in unit volume. Each line will be the same in length and direction as the velocity of the corresponding molecule. Its endpoint will be called the velocity point of the corresponding molecule. Now at time t let

$$f(\xi, \eta, \zeta, t)d\xi d\eta d\zeta = fd\omega \tag{9}$$

be the number of m-molecules whose velocity components in the three coördinate directions lie between the limits

(10) ξ and $\xi + d\xi$, η and $\eta + d\eta$, ζ and $\zeta + d\zeta$

and for which the velocity points thus lie in the parallelepiped having at one corner the coördinates ξ, η, ζ and having edges of length $d\xi$, $d\eta$, $d\zeta$ parallel to the coördinate axes. We shall always call this the parallelepiped $d\omega$. We shall also use the abbreviations $d\omega$ for the product $d\xi d\eta d\zeta$, and f for $f(\xi, \eta, \zeta, t)$. If $d\omega$ were a volume element of any other shape, though still infinitesimal, the number of m-molecules whose velocity points lie inside $d\omega$ would still be equal to

(11) $$f(\xi, \eta, \zeta, t)d\omega,$$

as one can see by dividing the volume element $d\omega$ into even smaller parallelepipeds. If the function f is known for one value of t, then the velocity distribution for the m-molecules is determined at time t. Similarly we can represent the velocity of each m_1-molecule by a velocity point, and denote by

(12) $$F(\xi_1, \eta_1, \zeta_1, t)d\xi_1 d\eta_1 d\xi_1 = F_1 d\omega_1$$

the number of molecules whose velocity components lie between any other limits

(13) ξ_1 and $\xi_1 + d\xi_1$, η_1 and $\eta_1 + d\eta_1$, ζ_1 and $\zeta_1 + d\zeta_1$

and for which therefore the velocity points lie in a similar parallelepiped $d\omega_1$. Likewise we write $d\omega_1$ for $d\xi_1 d\eta_1 d\zeta_1$ and F_1 for $F(\xi_1, \eta_1, \zeta_1, t)$. Moreover, we shall completely exclude any external forces, and assume that the walls are completely smooth and elastic. Then the molecules reflected from the wall will move as if they came from a gas which is the mirror image of our gas, and which is thus completely equivalent to it; the container wall is thought of as a reflecting surface. According to these assumptions, the same conditions will prevail everywhere inside the container, and if the number of molecules in a volume element whose velocity components lie between the limits (10) is initially the same everywhere in the gas, then this will also be true for all subsequent times. If we assume this, then it follows that the number of m-molecules inside any volume Φ that satisfy the conditions (10) is proportional to the volume Φ and is therefore equal to

(14) $$\Phi f d\omega;$$

likewise the number of m_1-molecules in the volume Φ that satisfy the conditions (13) is:

(14a) $$\Phi F_1 d\omega_1.$$

From these assumptions it follows that the molecules that leave any space as a result of their progressive motion will on the average be replaced by an equal number of molecules from the neighboring space or by reflection at the walls of the container, so that the velocity distribution is changed only by collisions, and not by the progressive motion of the molecules. We shall make ourselves independent of these restrictive conditions (made now to simplify the calculation) in §§15–18, where we shall take account of the effect of gravity and other external forces.

We next consider only the collisions of an m-molecule with an m_1-molecule and indeed we shall single out, from all those collisions that can occur during the interval dt, only those for which the following three conditions are satisfied:

1. The velocity components of the m-molecule lie between the limits (10) before the collision, hence its velocity point lies in the parallelepiped $d\omega$.

2. The velocity components of the m_1-molecule lie between the limits (13) before the collision, hence its velocity point lies in the parallelepiped $d\omega_1$. All m-molecules for which the first condition is fulfilled will be called "m-molecules of the specified kind," and similarly we speak of "m_1-molecules of the specified kind."

3. We construct a sphere of unit radius, whose center is at the origin of coördinates, and on it a surface element $d\lambda$. The line of centers of the colliding molecules drawn from m to m_1 must, at the moment of collision, be parallel to a line drawn from the origin to some point of the surface element $d\lambda$. The aggregate of these lines constitutes the cone $d\lambda$.

(15) Direction mm_1 in the cone $d\lambda$.

All collisions that take place in such a way that these three conditions are fulfilled will be called "collisions of the specified kind" and we have the problem of determining the number $d\nu$ of collisions of the specified kind that take place during a time interval dt in unit volume. We shall represent these collisions in Figure 2. Let O be the origin of coördinates, C and C_1 the velocity points of the two molecules before the collision, so that the lines OC and OC_1 represent these velocities in magnitude and direction, before the collision. The point C must lie inside the parallelepiped $d\omega$, and the point C_1 inside the parallelepiped $d\omega_1$. (The two parallelepipeds are not shown in the figure.) Let OK be a line of unit length which has the same direction as the line of centers of the two molecules at the instant of the collision, drawn from m to m_1. The point K must therefore lie inside the surface element $d\lambda$, which is also not shown in the figure. The line $C_1C = g$ represents in magnitude and direction the relative velocity of the m-molecule with respect to the m_1-molecule before the collision, since its projections on the coördinate axes are equal to $\xi - \xi_1$, $\eta - \eta_1$, and $\zeta - \zeta_1$, respectively. The frequency of collisions obviously depends only on the relative velocity. Hence if we wish to find the number of collisions of the specified kind, we can imagine that the specified m_1-molecule is at rest, while the m-molecule moves with velocity g. We imagine further that a sphere of radius σ (the sphere σ) is rigidly attached to each of the latter molecules, so that the center of the sphere always coincides with the center of the molecule. σ should be equal to the sum of the radii of the two molecules. Each time that the surface of such a sphere touches the center of an m_1-molecule, a collision takes place. We now draw from the center of each sphere σ a cone, similar and similarly situated to the cone $d\lambda$. A surface element of area $\sigma^2 d\lambda$ is thereby cut out from the surface of each of these spheres. Since all the spheres are rigidly attached to the corresponding molecules, all these surface elements move a distance gdt relative to the specified m_1-molecule. A collision of the specified kind occurs whenever one of these surface elements touches the center of a specified m_1-molecule, which is of course possible only if the angle ϑ between the directions of the lines C_1C and OK is acute. Each of these surface elements traverses by its relative motion toward the m_1-molecule an oblique cylinder of base $\sigma^2 d\lambda$ and height $g \cos \vartheta dt$. Since there are $fd\omega m$ molecules of the specified kind in unit volume, all the oblique cylinders traversed in this manner by all the surface elements have total volume

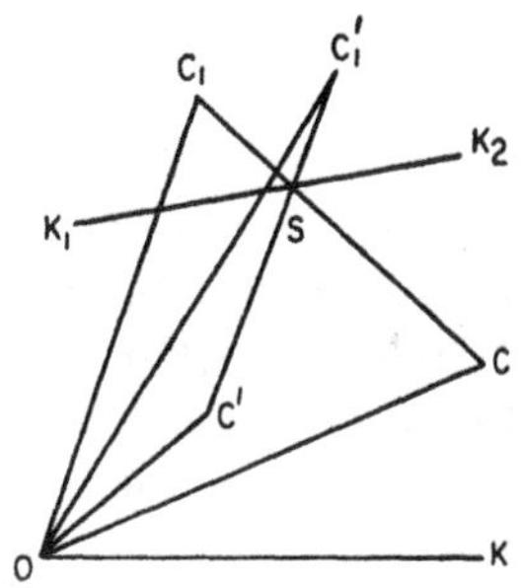

FIG. 2.

$$\Phi = fd\omega\sigma^2 g \cos \vartheta d\lambda dt. \tag{16}$$

All centers of m_1-molecules of the specified kind lying inside the volume Φ will touch during the interval dt one surface element $\sigma^2 d\lambda$, and hence the number $d\nu$ of collisions of the specified kind which occur in the volume element during time dt is equal to the number Z_Φ of centers of m_1-molecules of the specified kind that are in the volume Φ at the beginning of dt. But according to Equation (14a) this is

$$Z_\Phi = \Phi F_1 d\omega_1. \tag{17}$$

In this formula there is contained a special assumption, as Burbury[1] has clearly emphasized. From the standpoint of mechanics, any arrangement of molecules in the container is possible; in such an arrangement, the variables determining the motion of the molecules may have different average values in one part of the space filled by the gas than in another, where for example the density or mean velocity of a molecule may be larger in one half of the container than in the other, or more generally some finite part of the gas has different properties than another. Such a distribution will be called molar-ordered [*molar-geordnete*]. Equations (14) and (14a) pertain to the case of a molar-disordered distribution. If the arrangement of the molecules also exhibits no regularities that vary from one finite region to another—if it is thus molar-disordered—then nevertheless groups of two or a small number of molecules can exhibit definite regularities. A distribution that exhibits regularities of this kind will be called molecular-ordered. We have a molecular-ordered distribution if—to select only two examples from the infinite manifold of possible cases—each molecule is moving toward its nearest neighbor, or again if each molecule whose velocity lies between certain limits has ten much slower molecules as nearest neighbors. When these special groupings are not limited to particular places in the container but rather are found on the average equally often throughout the entire container, then the distribution would be called molar-disordered. Equations (14) and (14a) would then always be valid for individual molecules, but Equation (17) would not be valid, since the nearness of the m-molecule would be influenced by the probability that the m_1-molecule lies in the space Φ. The presence

[1] Burbury, Nature **51**, 78 (1894). See also Boltzmann, Weitere Bermerkungen über Wärmetheorie, Wien. Ber. **78** (June 1878), third-to-last and next-to-last pages.

of the m_1-molecule in the space Φ cannot therefore be considered in the probability calculation as an event independent of the nearness of the m-molecule. The validity of Equation (17) and the two similar equations for collisions of m- or m_1-molecules with each other can therefore be considered as defining the meaning of the expression: the distribution of states is molecular-disordered.

If the mean free path in a gas is large compared to the mean distance of two neighboring molecules, then in a short time, completely different molecules than before will be nearest neighbors to each other. A molecular-ordered but molar-disordered distribution will most probably be transformed into a molecular-disordered one in a short time. Each molecule flies from one collision to another one so far away that one can consider the occurrence of another molecule, at the place where it collides the second time, with a definite state of motion, as being an event completely independent (for statistical calculations) of the place from which the first molecule came (and similarly for the state of motion of the first molecule). However, if we choose the initial configuration on the basis of a previous calculation of the path of each molecule, so as to violate intentionally the laws of probability, then of course we can construct a persistent regularity or an almost molecular-disordered distribution which will become molecular-ordered at a particular time. Kirchhoff[2] also makes the assumption that the state is molecular-disordered in his definition of the probability concept.

That it is necessary to the rigor of the proof to specify this assumption in advance was first noticed in the discussion of my so-called H-theorem or minimum theorem. However, it would be a great error to believe that this assumption is necessary only for the proof of this theorem. Because of the impossibility of calculating the positions of all the molecules at each time, as the astronomer calculates the positions of all the planets, it would be impossible without this assumption to prove the theorems of gas theory. The assumption is made in the calculation of the viscosity, heat conductivity, etc. Also, the proof that the Maxwell velocity distribution law is a possible one—i.e., that once established it persists for an infinite time—is not possible without this assump-

[2] Kirchhoff, *Vorlesungen über Wärmetheorie*, 14th lecture, §2, p. 145, line 5.

tion. For one cannot prove that the distribution always remains molecular-disordered. In fact, when Maxwell's state has arisen from some other state, the exact recurrence of that other state will take place after a sufficiently long time (cf. the second half of §6). Thus one can have a state arbitrarily close to the Maxwellian state which finally is transformed into a completely different one. It is not a defect that the minimum theorem is tied to the assumption of disorder, rather it is a merit that this theorem has clarified our ideas so that one recognizes the necessity of this assumption.

We shall now explicitly make the assumption that the motion is molar- and molecular-disordered, and also remains so during all subsequent time. Equation (17) is then valid, and we obtain

$$d\nu = Z_\Phi = \Phi F_1 d\omega_1 = f d\omega F_1 d\omega_1 \sigma^2 g \cos \vartheta d\lambda dt. \tag{18}$$

This is the number of collisions of the specified kind in unit volume in time dt, which was to be calculated. We ignore the grazing collisions, whose number is in any case a higher-order infinitesimal, so that in each collision at least one velocity component of each molecule changes by a finite amount. In each collision of the specified kind, the number $f d\omega$ of m-molecules whose velocity components lie between the limits (10), which we call m-molecules of the specified kind, and also the number $F_1 d\omega_1$ of m_1-molecules of the specified kind, both decrease by one. In order to find the total decrease $\int d\nu$ suffered by $f d\omega$ during dt as a result of all collisions of m-molecules with m_1-molecules (without restriction on the magnitude and direction of the line of centers), we must consider ξ, η, ζ, $d\omega$ and dt as constant in Equation (18) and integrate $d\omega_1$ and $d\lambda$ over all possible values—i.e., we integrate $d\omega_1$ over all space, and $d\lambda$ over all surface elements for which the angle ϑ is acute. We shall denote the result of this integration by $\int d\nu$.

The decrease dn which the number $f d\omega$ experiences as a result of the corresponding collisions of m-molecules with each other is obviously given by a completely analogous formula; we simply denote by ξ_1, η_1, ζ_1 the velocity components of another m-molecule before the collision. All other quantities have the same meaning, except that one replaces m_1 by m and the function F by the function f, and σ by the diameter s of an m-molecule. When we have instead of $d\nu$ the expression

$$dn = ff_1 d\omega d\omega_1 s^2 g \cos \vartheta d\lambda dt, \tag{19}$$

where f_1 is an abbreviation for $f(\xi_1, \eta_1, \zeta_1, t)$. To obtain $\int dn$—i.e., the total decrease of $fd\omega$ resulting from collisions of the m-molecules with each other during dt—one must obviously consider ξ, η, ζ, $d\omega$, and dt as constant and integrate $d\omega_1$ and $d\lambda$ over all possible values. The total decrease of $fd\omega$ during dt is therefore equal to $\int d\nu + \int dn$. If the state is to be stationary, this must be exactly equal to the number of m-molecules whose velocities at the beginning of dt do not fulfill the conditions (10) but which during this time interval are changed by collisions in such a way that they now satisfy them—they obtain by collision a velocity lying between the limits (10). In other words, $\int d\nu + \int dn$ must be equal to the total increase of $fd\omega$ resulting from collisions.

§4. Continuation; values of the variables after the collision; collisions of the opposite kind.*

To find this increase, we shall next seek the velocities of the two molecules after a collision of the specified kind. Before the collision one of the colliding molecules, whose mass is m, has velocity components ξ, η, ζ; the other, of mass m_1, has components ξ_1, η_1, ζ_1. The line of centers drawn from m to m_1 forms at the instant of collision an angle ϑ with the relative velocity of the molecule m with respect to m_1. If the angle ϵ between the plane of these two lines and any other given plane—for example, that of the two velocities before the collision—is given, then the collision is completely specified. The velocity components ξ', η', ζ', and ξ_1', η_1', ζ_1' of the two molecules after the collision can thus be expressed as explicit functions of the 8 variables ξ, η, ζ, ξ_1, η_1, ζ_1, ϑ, ϵ:

$$\begin{cases} \xi' = \psi_1(\xi, \eta, \zeta, \xi_1, \eta_1, \zeta_1, \vartheta, \epsilon) \\ \eta' = \psi_2(\xi, \eta, \zeta, \xi_1, \eta_1, \zeta_1, \vartheta, \epsilon). \\ \cdots\cdots\cdots \end{cases} \tag{20}$$

We prefer however the geometric construction to the algebraic

* See Part II, §77, for corrections to this section. For further discussion of opposite collisions, see R. C. Tolman, *The Principles of Statistical Mechanics* (New York: Oxford University Press, 1938), Chapter 5.

development of the functions (20), and we return to Figure 2, p. 39. We divide the segment C_1C at the point S into two parts, such that

$$C_1S:CS = m:m_1.$$

Then the line OS represents the velocity of the common center of mass of the two molecules; for one sees that its three projections on the coördinate axes have the values:

$$\frac{m\xi + m_1\xi_1}{m + m_1}, \quad \frac{m\eta + m_1\eta_1}{m + m_1}, \quad \frac{m\zeta + m_1\zeta_1}{m + m_1} \tag{21}$$

But these are in fact just the velocity components of the common center of mass. Just as we have shown that C_1C is the relative velocity of the m-molecule with respect to m_1, it follows also that SC and SC_1 are the relative velocities of the two molecules with respect to the common center of mass before the collision. The components of these relative velocities perpendicular to the line of centers OK will not be changed by the collision. The components in the direction OK are p and p_1 before the collision, and p' and p_1' after the collision. Then according to the principle of conservation of the motion of the center of mass:

$$mp + m_1p_1 = mp' + m_1p_1' = 0,$$

and according to the principle of conservation of kinetic energy:

$$mp^2 + m_1p_1^2 = mp'^2 + m_1p_1'^2.$$

Whence it follows that

$$p' = p, \; p_1' = p_1,$$

or

$$p' = -p, \; p_1' = -p_1$$

and one sees at once that, since the molecules must separate from each other after the collision, only the latter solution is correct, and that hence the two components of relative velocity with respect to the center of mass, which fall in the direction $K_1K_2 \| OK$, will simply be reversed by the collision.

From this one obtains the following construction of the lines OC' and OC_1', which represent the velocities of the two molecules after the collision in magnitude and direction. One draws through S the line K_1K_2; then one draws in the plane of the lines

K_1K_2 and C_1C, the two lines SC' and SC_1', which are equal in length to the lines SC and SC_1, and are equally inclined on the other side from K_1K_2. The two endpoints C' and C_1' of the latter two lines are at the same time the endpoints of the required lines OC' and OC_1'. We can also call them the velocity points of the two molecules after the collision. The projections of OC' and OC_1' on the three coördinate axes are thus the velocity components ξ', η', ζ', ξ_1', η_1', ζ_1' of the two molecules after the collision. These geometrical constructions completely replace the alternative algebraic development of the functions (20). The points C_1', S, and C' obviously fall on a straight line. This line $C_1'C'$ represents the relative velocity of the molecule m with respect to the molecule m_1 after the collision, and one sees from the figure that its length is equal to C_1C, whereas the angle it forms with the line OK is $180° - \zeta$.

Up to now we have considered only one of the specified collisions, and we have constructed the velocities after the collision for it. We now consider all the specified collisions, and ask between which limits the values of the variables after the collision lie, for all these collisions—i.e., for all collisions such that the conditions (10), (13), and (15) are fulfilled before the collision. Since we assume the time duration of the collision to be infinitesimal, the direction of the line of centers is the same at the end of the collision as at the beginning, and it is only a question of finding the limits between which the velocity components ξ', η', ζ', ξ_1', η_1', ζ_1' are found after the collision. Had we calculated the function (20), then we would have considered ϑ and ϵ as constants and ξ, η, ζ, ξ_1, η_1, ζ_1 as independent variables, and we would have to express $d\xi'\, d\eta'\, d\zeta'\, d\xi_1'\, d\eta_1'\, d\zeta_1'$ in terms of $d\xi\, d\eta\, d\zeta\, d\xi_1\, d\eta_1\, d\zeta_1$ by means of the well-known Jacobian functional determinant. However, we prefer the geometric construction, and we must therefore answer the question: what volume element would be described by the points C' and C_1' when, without changing the direction of the line OK, we let the points C and C_1 describe the volume elements $d\omega$ and $d\omega_1$? First, let the position of the point C as well as the direction of the line OK remain fixed, and let C_1 sweep out the entire parallelepiped $d\omega_1$. From the complete symmetry of the figure, it follows directly that C_1' describes a congruent parallelepiped which is the mirror image of $d\omega_1$. Likewise if the point C_1 is fixed and the point C sweeps out the parallelepiped $d\omega$, then the point C' sweeps out a parallelepiped congruent to $d\omega$. For all collisions that we earlier called collisions of the specified kind, the velocity point of the m-molecule lies in the parallelepiped $d\omega'$ after the collision, and that of the m_1-molecule in $d\omega_1$, and it is always true that $d\omega' d\omega_1' = d\omega d\omega_1$. The same result would also be obtained by explicit calculation of the functions (20) and construction of the functional determinant[1]

$$\sum \pm \frac{\partial \xi'}{\partial \xi} \frac{\partial \eta'}{\partial \eta} \cdots \frac{\partial \zeta_1'}{\partial \zeta_1}$$

We shall now consider another class of collisions of an m-molecule with an m_1-molecule, which will be called the "collisions of the inverse kind." They are characterized by the following conditions:

1. The velocity point of the m-molecule lies in the volume element $d\omega'$ before the collision; the number of m-molecules in the volume element for which this condition is satisfied is, by analogy with Equation (9), $f' d\omega'$, where f' is the value of the function f obtained by replacing ξ, η, ζ by ξ', η', ζ'—i.e., it is the quantity $f(\xi', \eta', \zeta', t)$.

2. The velocity point of the m_1-molecule lies in the volume element $d\omega_1'$ before the collision. The number of m_1-molecules in unit volume for which this condition is satisfied is $F_1' d\omega_1'$, where F_1' is an abbreviation for $F(\xi_1', \eta_1', \zeta_1', t)$.

3. The line of centers of the two molecules at the instant of the collision, drawn from m_1 to m, is parallel to some line drawn from the origin of coördinates within the cone $d\lambda$. (In those integrals that refer to the collision of identical molecules, there will of course occur in place of the molecule of mass m_1 simply an m-molecule whose velocity components are ξ_1, η_1, ζ_1.)

Figure 3 represents the same collision as Figure 2, the lines having been kept fixed as far as possible. Figure 4 represents the opposite collision. The arrow directed toward the center of the molecule always represents its velocity before the collision, while

---

[1] Cf. Wien. Ber. **94**, 625 (1886); Stankevitsch, Ann. Physik [3] **29**, 153 (1886). The fact that the angles ϑ and ϵ also depend on the positions of c and c_1 does not weaken the force of the arguments in the text. One can first introduce in place of ϑ and ϵ two angles which determine the absolute position of OK in space, then transform $\xi, \eta, \cdots \zeta_1$ into $\xi', \eta', \cdots \xi_1'$ and finally again introduce ϑ and ϵ.

that directed away from the collision represents its velocity after the collision. In all collisions of the opposite kind, the relative velocity of the m-molecule with respect to that of the m_1-molecule before the collision is represented by the line $C_1'C'$ in Figure 2. Its magnitude is thus equal again to g, and it forms the angle ϑ with the line of centers drawn from m to m_1, since we have likewise reversed the direction of the line of centers. The angle ϑ must of course be acute if the collision is to be possible. The number of inverse collisions in unit volume during dt is, by analogy with Equation (18), given by

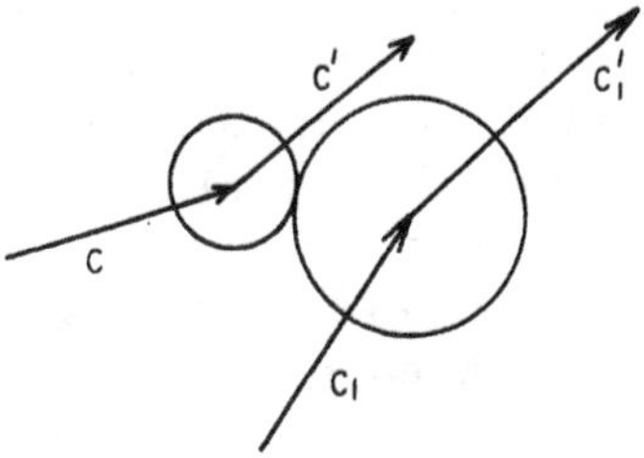

Fig. 3.

$$(22) \qquad d\nu' = f'F_1'\,d\omega'd\omega_1'\,\sigma^2 g \cos\vartheta d\lambda dt.$$

We have called these collisions those of the opposite kind, since they follow exactly the opposite course as those originally specified, so that the velocities of the two molecules *after* the collision lie between the same limits, (10) and (13), as those *before* the collision for the originally specified collisions.

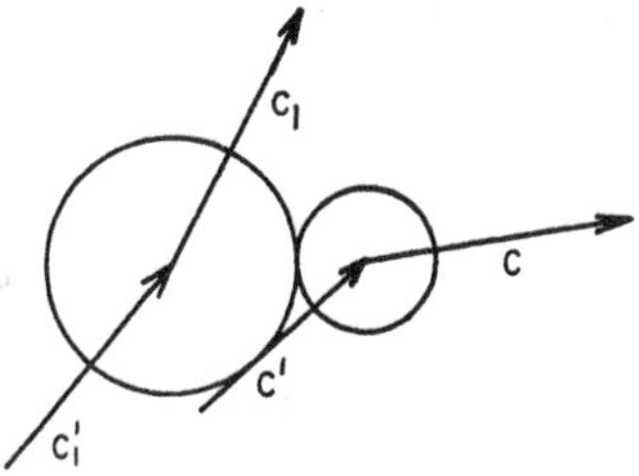

Fig. 4.

An opposite collision increases both $fd\omega$ and $F_1d\omega_1$ by one. In order to find the total increase of $fd\omega$ resulting from all collisions of m-molecules with m_1-molecules, during time dt, one must first express $\xi', \eta', \zeta', \xi_1', \eta_1', \zeta_1'$ in terms of $\xi, \eta, \zeta, \xi_1, \eta_1, \zeta_1, \vartheta$, and ϵ. Since $d\omega' d\omega_1' = d\omega d\omega_1$, we have:

$$(23) \qquad d\nu' = f'F_1'\,d\omega d\omega_1 \sigma^2 g \cos\vartheta d\lambda dt.$$

We leave the letters f', F_1' and $d\lambda$ in the equation, remembering however that one must consider their arguments ξ', η', ζ', ξ_1', η_1', ζ_1' to be functions of ξ, η, ζ, ξ_1, η_1, ζ_1, ϑ, ϵ and also $d\lambda$ through the differentials of the latter angles. As is well known, one can show that $d\lambda = \sin\vartheta d\vartheta d\epsilon$ (cf. beginning of §9). In the differential expression (23), one now considers ξ, η, ζ, $d\omega$ and dt constant and integrates over all possible values of $d\omega_1$ and $d\lambda$. Thereby all collisions will be included that can take place between an m-molecule and an m_1-molecule, such that the former has velocity components within the limits (10) before the collision, but without any other restrictions. The result of this integration, $\int d\nu'$, thus gives us the increase of $f d\omega$ resulting from all collisions of m-molecules with m_1-molecules during dt. Similarly, this quantity increases by $\int d\mathfrak{n}'$ as a result of collisions of m-molecules with each other, where

$$d\mathfrak{n}' = f' f_1' \, d\omega d\omega_1 s^2 g \cos\vartheta d\lambda dt \tag{24}$$

Here f_1' is again used as an abbreviation for $F(\xi_1', \eta_1', \zeta_1', t)$. $\xi', \eta', \zeta', \xi_1', \eta_1', \zeta_1'$ are here functions of $\xi, \eta, \zeta, \xi_1, \eta_1, \zeta_1, \vartheta$ and ϵ, inasmuch as the former represent the velocity components after a collision, and are determined by the initial conditions (10), (13), and (15), in which however both molecules have mass m.

If we subtract from the total increase of $f d\omega$ the total decrease, we obtain the net change

$$\frac{df}{dt} d\omega dt,$$

which the quantity $f d\omega$ experiences during time dt. Thus:

$$\frac{df}{dt} dt d\omega = \int d\nu' - \int d\nu + \int d\mathfrak{n}' - \int d\mathfrak{n}.$$

In the integrals $\int d\nu$ and $\int d\nu'$ the integration variables are identical, and likewise in the integrals $\int d\mathfrak{n}$ and $\int d\mathfrak{n}'$.

If we combine these integrals and divide the whole equation by $d\omega \cdot dt$, then it follows from Equations (18), (19), (23), and (24) that:

$$\left\{ \begin{aligned} \frac{\partial f}{\partial t} &= \int (f' F_1' - f F_1) \sigma^2 g \cos\vartheta d\omega_1 d\lambda + \\ &\quad + \int (f' f_1' - f f_1) s^2 g \cos\vartheta d\omega_1 d\lambda. \end{aligned} \right. \tag{25}$$

The integration is to be extended over all possible $d\omega_1$ and $d\lambda$. Likewise, one obtains for the function F the equation

$$\left\{ \begin{aligned} \frac{\partial F_1}{dt} &= \int (f' F_1' - f F_1) \sigma^2 g \cos\vartheta d\omega d\lambda + \\ &\quad + \int (F' F_1' - F F_1) s_1^2 g \cos\vartheta d\omega d\lambda. \end{aligned} \right. \tag{26}$$

Here s_1 is the diameter of an m_1-molecule; in Equation (26), ξ_1, η_1, and ζ_1 are arbitrary and should be considered constant in the integral, while ξ, η, and ζ are to be integrated over all their possible values. In the first integral, ξ', η', ζ', ξ_1', η_1', ζ_1' are the velocity components after a collision of the specified kind in the case that one of the colliding molecules has mass m, and the other has mass m_1; while in the second integral they refer to the case that both molecules have mass m_1. $\partial F_1/\partial t$, F, and F' are abbreviations for $\partial F(\xi_1, \eta_1, \zeta_1, t)/\partial t$, $F(\xi, \eta, \zeta, t)$, and $F(\xi', \eta', \zeta', t)$.

If the state is to be stationary, then the quantities $\partial f/\partial t$ and $\partial F_1/\partial t$ must vanish for all values of the variables. This certainly occurs when in all the integrals the integrand vanishes for all values of the variable of integration—when one therefore has for all possible collisions of the m-molecules with each other, the m_1-molecules with each other, and of m-molecules with m_1-molecules, the three equations

$$(27) \qquad ff_1 = f'f_1', \; FF_1 = F'F_1', \; fF_1 = f'F_1'.$$

Since the probability of the originally specified collision is given by Equation (18), and that of the opposite one by Equation (23), the general validity of the third of Equations (27) is equivalent to the statement that, however $d\omega$, $d\omega_1$, and $d\lambda_1$ may be chosen, the originally specified (or briefly, "direct") collision is as probable as the opposite one. In other words, it is equally probable that two molecules should separate from each other in a certain way as that they should collide in the opposite way. The same follows from the two other parts of the system of Equations (27) for the collisions of m-molecules with each other and of m_1-molecules with each other. However, one sees at once that a distribution of states must remain stationary when it is equally probable that two molecules separate from each other after a collision and that they collide with each other in the opposite way.

§5. Proof that Maxwell's velocity distribution is the only possible one.

We shall deal later with the solution of Equations (27), which presents no particular difficulty. It leads necessarily to the well-known Maxwell velocity distribution law. For this case the two quantities $\partial f/\partial t$ and $\partial F/\partial t$ vanish, since the integrands of all the integrals vanish identically. It still remains to be shown that the Maxwell velocity distribution, once established, is not altered by further collisions. It has still not been proved that Equations (25) and (26) cannot be satisfied by other functions that do not make the integrands vanish for all values of the variables of integration. One may give such possibilities as little weight as he wishes; I find myself inclined to dispose of them by a special proof. Since this proof has, as will appear, a not uninteresting connection with the entropy principle, I will reproduce it in the form given by H. A. Lorentz.*

We consider the same gas mixture as before, and retain the earlier notation. Furthermore, we denote by lf and lF the natural logarithms of the functions f and F. The result obtained on substituting in lf for ξ, η, ζ the velocity components of a particular gas molecule of mass m at time t, we call the value of the logarithmic function corresponding to the molecule considered at the time considered. Similarly, we obtain the value of the logarithm function corresponding to any m_1-molecule at any time when we substitute in lF_1 the velocity components ξ_1, η_1, ζ_1 of that molecule at that time. We shall now calculate the sum H of all values of the logarithm functions corresponding at a particular time to all m-molecules and m_1-molecules contained in a volume element. At time t, there are in the volume element $fd\omega$ m-molecules of the specified kind, i.e., m-molecules whose velocity components lie between the limits (10). These clearly contribute a term $f\cdot lf\cdot d\omega$ to the sum H. If we construct the analogous expression for the m_1-molecules and integrate over all possible values of the variables, then it follows that:

$$H = \int f\cdot lf\cdot d\omega + \int F_1\cdot lF_1\cdot d\omega_1. \tag{28}$$

We now seek the change experienced by H during a very small time dt. This will be due to two causes:[1]

1. Each m-molecule of the specified kind provides at time t

* H. A. Lorentz, Wien. Ber. **95**, 115 (1887).

[1] One can present the proof given in the text in the following more analytical form. We shall certainly include all required values when we integrate the integrals in Equation (28) over all variables from $-\infty$ to $+\infty$. Velocities that do not occur in the gas will not contribute to the

(*Continued on following pages*)

integral, since for these velocities either f or F must vanish. Hence the limits are constant and one obtains dH/dt by differentiating with respect to t under the integral sign, which gives:

$$\frac{dH}{dt} = \int \frac{\partial}{\partial t} d\omega + \int \frac{\partial F_1}{\partial t} d\omega_1 + \int lf \frac{\partial f}{dt} d\omega \int lF_1 \frac{\partial F_1}{dt} d\omega_1.$$

One sees at once that the first two terms represent the increase in H by what is called the first cause in the text, and that for the reasons given in the text it vanishes. The other two terms represent the increase in H due to the second cause and give, after substitution of the values of $\partial F/\partial t$ from Equations (25) and (26):

$$(29)\quad \begin{cases} \dfrac{dH}{dt} = \int lf(f'F_1' - fF_1')d\rho + \int lf(ff_1' - ff_1)dr + \int lF_1(f'F_1 - fF_1)d\rho \\ \qquad\qquad + \int lF_1(F'F_1' - FF_1)dr_1, \end{cases}$$

where $d\rho = \sigma^2 g \cos\vartheta d\omega d\omega_1 d\lambda$, $dr = s^2 g \cos\vartheta d\omega d\omega_1 d\lambda$ and $dr_1 = s_1^2 g \cos\vartheta d\omega d\omega_1 d\lambda$. All the integrations are extended over all possible values of the variables.

One sees at once that the sum $\int f'lf'd\omega' + \int F_1' lF_1' d\omega_1'$ is likewise equal to H. Its differentiation yields:

$$(30)\qquad \frac{dH}{at} = \int \frac{\partial f'}{\partial t} d\omega' + \int \frac{\partial F_1'}{\partial t} d\omega_1' + \int lf' \frac{\partial f'}{\partial t} d\omega' + \int lF_1' \frac{\partial F_1}{\partial t} d\omega_1'.$$

The quantities $\partial f'/\partial t$ and $\partial F'/\partial t$ can be found in the same way as $\partial f/\partial t$ and $\partial F/\partial t$ if one considers, instead of a collision in which the velocity components before the collision are $\xi, \eta, \zeta, \xi_1, \eta_1, \zeta_1$ and afterwards $\xi', \eta', \zeta', \xi_1', \eta_1', \zeta_1'$, a collision in which they are $\xi', \eta', \zeta', \xi_1', \eta_1', \zeta_1'$ before and $\xi, \eta, \zeta, \xi_1, \eta_1, \zeta_1$ afterwards. Then by symmetry one gets

$$\frac{\partial f'}{\partial t} = \int (fF_1 - f'F_1')\sigma^2 g \cos\vartheta d\omega_1' f d\lambda + \int (f_1 f - f'f_1')s^2 g \cos\vartheta d\omega_1' f d\lambda;$$

and similarly for $\partial F'/\partial t$. Substitution of these values in Eq. (30) yields (taking account of the fact that the first two integrals on the right hand side vanish and that $d\omega' d\omega_1' = d\omega d\omega_1$):

$$(31)\quad \begin{cases} \dfrac{dH}{dt} = \int lf'(fF_1 - f'F_1')d\rho + \int lf'(ff_1 - f'f_1')dr \\ \qquad + \int lF_1'(fF_1 - f'F_1')d\rho - \int lF_1'(FF_1 - F'F_1')dr_1. \end{cases}$$

Since in a collision of two m-molecules or two m_1-molecules, the two colliding molecules play the same role, it follows that

$$\int lf(f'f_1' - ff_1)dr = \int lf_1(f'f_1' - ff_1)dr,$$
$$\int lf'(ff_1 - f'f_1')dr = \int lf_1'(ff_1 - f'f_1)dr$$

with two similar equations for F. From this, taking the average of the two

(*Continued on following page*)

the term lf in the expression (28). After time dt, the function f experiences the increment

$$\frac{\partial f}{dt}\,dt.$$

Hence lf experiences the increment:

$$\frac{1}{f}\,\frac{\partial f}{\partial t}\,dt,$$

and each m-molecule of the specified kind provides in the expression (28) the term

$$lf + \frac{1}{f}\,\frac{\partial f}{\partial t}\,dt.$$

All m-molecules of the specified kind provide together the contribution

$$\left(lf + \frac{1}{f}\,\frac{\partial f}{\partial t}\,dt\right) f d\omega.$$

in (28). Applying the same arguments to all other m-molecules and m_1-molecules, one finds for the total increment of H, resulting from the variation of the quantities lf and lF under the integral signs in Equation (28), the value:

$$\int \frac{\partial f}{\partial t}\,dt d\omega + \int \frac{\partial F_1}{\partial t}\,dt d\omega_1.$$

But this is nothing more than the variation of the total number of

---

values (30) and (31) for dH/dt, there follows the value given in the text:

$$\frac{dH}{dt} = -\tfrac{1}{2}\int [l(f'F_1') - l(fF_1)]\cdot(f'F_1' - fF_1)d\rho$$
$$-\tfrac{1}{4}\int [l(f'f_1') - l(ff_1)]\cdot(f'f_1' - ff_1)dr$$
$$-\tfrac{1}{4}\int [l(F'F_1') - l(FF_1)]\cdot(F'F_1' - FF_1)dr_1.$$

This proof is somewhat shorter, but it appears to depend on certain mathematical conditions—permissibility of differentiation under the sign of integration, etc.—which affect only the ease of demonstrating the theorem and not its validity, since it is really a question of very large but not infinite numbers. The theorem was proved without introducing definite integrals in my paper in Wien. Ber. **66**, Oct. 1872, Section II.

molecules in unit volume, which must be equal to zero, since neither the size of the container nor the uniform distribution of molecules should undergo any change.

2. The collisions change not only lf and lF but also $fd\omega$ and $F_1 d\omega_1$—i.e., the number of molecules of the specified kind changes a little. The variation dH of H due to this second cause will, according to the above, be equal to the total variation of H during dt. In order to find it, we again denote by $d\nu$ the number of collisions of the specified kind in unit volume during dt. Each such collision decreases both $fd\omega$ and $F_1 d\omega_1$ by one.

Since each m-molecule contributes the addend lf to (28), and each m_1-molecule contributes lF_1, the total decrease of H resulting from these collisions will be

$$(lf + lF_1)d\nu$$

Each of these collisions increases $f'd\omega'$ by one, and consequently H is increased by $lf'd\nu$. Finally, each of these collisions increases $F_1' d\omega_1'$ by one, and consequently H is increased by $lF_1' d\nu$. The total increase of H during time dt is therefore:

$$\begin{aligned}&(lf' + lF_1' - lf - lF_1)d\nu \\ &\quad = (lf' + lF_1' - lf - lF_1)fF_1 d\omega d\omega_1 \sigma^2 g \cos\vartheta d\lambda dt.\end{aligned}$$

(cf. Eq. [18]).

If we keep dt constant in this expression and integrate over all the other variables (where of course $\xi', \eta', \zeta', \xi_1', \eta_1', \zeta_1'$ are to be considered as functions of $\xi, \eta, \zeta, \xi_1, \eta_1, \zeta_1$) then we obtain the total increase d_1H that H experiences as a result of all collisions of m-molecules with m_1-molecules. We write this symbolically in the form

$$\text{(31a)}\quad d_1H = dt\int (lf' + lF_1' - lf - lF_1)f\cdot F_1\cdot d\omega\cdot d\omega_1\sigma^2 g \cos\vartheta d\lambda.$$

We can also calculate the same quantity by considering inverse collisions, of which there are $d\nu'$ in the time interval dt. These collisions will decrease both $f'd\omega'$ and $F_1' d\omega_1'$ by one, and increase both $fd\omega$ and $F_1 d\omega_1$ by one. Consequently the inverse collisions will increase H by

$$\begin{aligned}&(lf + lF_1 - lf' - lF_1')d\nu' \\ &\quad = (lf + lF_1 - lf' - lF_1')f'F_1' d\omega d\omega_1 \sigma^2 g \cos\vartheta d\lambda dt\end{aligned}$$

(cf. Eq. [23]).

If we keep dt constant and integrate over all the other variables, and average the result with (31a), we then obtain for d_1H the value

$$(32)\quad \begin{cases} d_1H = \dfrac{dt}{2}\displaystyle\int [l(f'F_1') - l(fF_1)]\cdot \\ \quad\cdot [fF_1 - f'F_1']d\omega d\omega_1\sigma^2 g\cos\vartheta d\lambda. \end{cases}$$

This is the total increment experienced by H during dt as a result of all collisions of m-molecules with m_1-molecules. The increment d_2H of the same quantity resulting from collisions of the m-molecules with each other will clearly be found in a completely similar way. We simply have to replace m_1 and F by m and f, and σ by s. However, one must remember that when the two colliding molecules are identical, each collision will be counted twice in the expressions for the integrals, so that the final result must be divided by two. (The same thing happens in the calculation of the self-potential and self-induction coefficients.) Hence we find

$$d_2H = \frac{dt}{4}\int [l(f'f_1') - l(ff_1)]\cdot$$
$$\cdot [ff_1 - f'f_1']d\omega d\omega_1 s^2 g\cos\vartheta d\lambda,$$

where f_1 and f_1' have the same meaning as before. If we calculate in the same way the increment of H resulting from collisions of m_1-molecules with each other, then we obtain for the total increment dH during time dt:

$$(33)\quad \begin{cases} \dfrac{dH}{dt} = -\tfrac{1}{2}\displaystyle\int [l(f'F_1') - l(fF_1)][f'F_1' - fF_1]\sigma^2 g\cos\vartheta d\omega d\omega_1 d\lambda \\ \qquad -\tfrac{1}{4}\displaystyle\int [l(f'f_1') - l(ff_1)][f'f_1' - ff_1]s^2 g\cos\vartheta d\omega d\omega_1 d\lambda \\ \qquad -\tfrac{1}{4}\displaystyle\int [l(F'F_1') - l(FF_1)][F'F_1' - FF_1]s_1^2 g\cos\vartheta d\omega d\omega_1 d\lambda. \end{cases}$$

Since the logarithm function always increases when its argument increases, we see that in each of the three integrals the first factor in brackets has the same sign as the second. Since, moreover, g is essentially positive, and the angle ϑ is always acute, all the quantities in the integrand are essentially positive, and vanish only for glancing collisions or for collisions with zero relative velocity. The above three integrals are therefore composed purely of essentially positive terms, and the quantity that we have called H can therefore only decrease; at most it can be constant, but this is only possible when all terms of all three integrals vanish, i.e., when for all collisions Equations (27) are satisfied. Since H cannot change with time in the stationary state, we have shown that for the stationary state Equations (27) must be satisfied for all collisions. The only assumption made here is that the velocity distribution is molecular-disordered at the beginning, and remains so. With this assumption, one can prove that H can only decrease, and also that the velocity distribution must approach that of Maxwell.

§6. Mathematical meaning of the quantity H.

We shall postpone, for the moment, the solution of Equations (27), and insert a few remarks on the meaning of the quantity called H. This meaning is twofold. The first is mathematical, the second physical. We shall discuss the first only in a simple case, a single gas in a container of unit volume. Naturally we shall be able to simplify the conclusions by using this assumption, though we must also relinquish a simultaneous proof of Avogadro's law.

First we make a few preliminary remarks about the principles of the calculus of probabilities. From an urn, in which many black and an equal number of white but otherwise identical spheres are placed, let 20 purely random drawings be made. The case that only black balls are drawn is not a hair less probable than the case that on the first draw one gets a black sphere, on the second a white, on the third a black, etc. The fact that one is more likely to get 10 black spheres and 10 white spheres in 20 drawings than one is to get 20 black spheres is due to the fact that the former event can come about in many more ways than the latter. The relative probability of the former event as compared to the latter is the number $20!/10!10!$, which indicates how many permutations one can make of the terms in the series of 10 white and 10 black spheres, treating the different white spheres as identical, and the different black spheres as identical. Each one of these permutations represents an event that has the same probability as the event of all black spheres. If there were in the urn a large number of otherwise identical spheres, of which a certain number are white, an equal number black, an equal number blue, an equal number red, and so forth, then the probability of drawing a white spheres, b black spheres, c blue spheres, etc., is

$$\frac{(a+b+c\cdots)!}{a!b!c!\cdots} \tag{34}$$

times as large as the probability of drawing spheres of only one color.

Just as in this simple example, the event that all molecules in a gas have exactly the same velocity in the same direction is not a hair less probable than the event that each molecule has exactly the velocity and direction of motion that it actually has at a particular instant in the gas. But if we compare the first event with the event that the Maxwell velocity distribution holds in the gas, we find there are very many more equiprobable configurations to be counted as belonging to the latter.

In order to express the relative probabilities of these two events in terms of a permutation number, we proceed as follows: for all collisions in which the velocity point of one molecule is in some infinitesimal volume element before the collision, we saw that it will be in a volume element of exactly the same size after the collision (assuming that the other variables characterizing the collision are kept constant). We divide the total space into many (ζ) equal-sized volume elements ω (cells), so that the presence of the velocity point of a molecule in each such volume element is to be considered an event equiprobable with its presence in each other volume element—just as previously we considered the drawing of a black or white or blue sphere to be equally probable. Instead of a, the number of drawings of white spheres, we now have the number $n_1\omega$ of molecules whose velocity points lie in the first of our volume elements; instead of b, we have the number $n_2\omega$ of molecules whose velocity points lie in the second volume element ω, and so forth. Instead of Equation (34) we have

$$Z = \frac{n!}{(n_1\omega)!(n_2\omega)!(n_3\omega)!\cdots} \tag{35}$$

for the relative probability that the velocity points of $n_1\omega$ molecules lie in the first volume element, and so forth. Then

$n=(n_1+n_2+n_3+\cdots)\omega$ is the total number of all molecules in the gas. For example, the event that all molecules have equal and equally directed velocities corresponds to having all velocity points lying in the same cell. Here we would have $Z=n!/n!=1$; no other permutations are possible. It is already very much more probable that half the molecules have one particular velocity and direction, and the other half have another velocity and direction, than that all have the same velocity and direction. Then half the velocity points would be in one cell and half in another, so that

$$Z = \frac{n!}{\left(\frac{n}{2}\right)!\left(\frac{n}{2}\right)!} \text{ etc.}$$

Since now the number of molecules is extraordinarily large, $n_1\omega$, $n_2\omega$, etc., can be treated as very large numbers.

We shall use the approximation formula

$$p! = \sqrt{2p\pi}\left(\frac{p}{e}\right)^p$$

where e is the base of natural logarithms and p is an arbitrarily large number.[1]

Denoting by l the natural logarithm, we see that:

$$l[(n_1\omega)!] = (n_1\omega + \tfrac{1}{2})ln_1 + n_1\omega(l\omega - 1) + \tfrac{1}{2}(l\omega + l2\pi).$$

Omitting $\frac{1}{2}$ compared to the very large number $n_1\omega$ and forming the similar expressions for $(n_2\omega)!$, $(n_3\omega)!$, etc., one obtains:

$$lZ = -\omega(n_1ln_1 + n_2ln_2\cdots) + C,$$

where

$$C = l(n!) - n(l\omega - 1) - \frac{\zeta}{2}(l\omega + l2\pi)$$

has the same value for all velocity distributions and is therefore to be considered a constant. Then we ask for the relative probability of the distribution of different velocity points of our molecules in our cells, where of course the cell divisions, the size of a cell ω, the number of cells ζ and the total number of molecules n

[1] See Schlömilch, *Comp. der höh. Analysis*, Vol. 1, p. 437, 3d ed.

and their total kinetic energy are to be considered constant. The most probable distribution of velocity points of the molecules in our cells will be the one for which lZ is a maximum; hence the expression

$$\omega[n_1 l n_1 + n_2 l n_2 + \cdots]$$

will be a maximum. If we write $d\xi d\eta d\zeta$ for ω and $f(\xi, \eta, \zeta)$ for n_1, n_2, etc., and then transform the sum into an integral, we get

$$\omega(n_1 l n_1 + n_2 l n_2 + \cdots) = \int f(\xi, \eta, \zeta) l f(\xi, \eta, \zeta) d\xi d\eta d\zeta .$$

This expression is completely identical to the expression into which the quantity H given by Equation (28) for a single gas transforms, however. The theorem of the previous section, which states that H decreases through collisions, says simply that through collisions the velocity distribution of the gas molecules comes closer and closer to the most probable one, as soon as the state is molecular-disordered, and thus the calculus of probabilities is introduced. I must content myself here with this brief illustration, and refer the reader elsewhere[2] for more details.

In connection with the foregoing, one should mention the following point, already made long ago by Loschmidt.* Let a gas be enclosed by absolutely smooth, elastic walls. Initially there is an unlikely but molecular-disordered state—for example, all molecules have the same velocity c. After a certain time, the Maxwell velocity distribution will nearly be established. We now imagine that at time t, the direction of the velocity of each molecule is reversed, without changing its magnitude. The gas will now go through the same sequence of states backwards. We have therefore the case that a more probable distribution evolves through collisions into a less probable one, and that the quantity H increases as a result of collisions. This in no way contradicts what was proved in §5; the assumption made there that the state distribution is molecular-disordered is not fulfilled here, since after exact reversal of all velocities each molecule does not collide with others according to the laws of probability, but rather

[2] Boltzmann, Wien. Ber. **76** (Oct. 1877).

* J. Loschmidt, Wien. Ber **73,** 128, 366 (1876), **75,** 287, **76,** 209 (1877). See also E. P. Culverwell, Phil. Mag. [5] **30,** 95 (1890); Nature **51,** 581 (1895). G. H. Bryan, B. A. Rep. **61,** 85 (1891); Nature **52,** 29 (1895). G. J. Stoney, Phil. Mag. [5] **23,** 544 (1887). J. Larmor, Nature **51,** 152 (1894). H. W. Watson, Nature **51,** 105 (1894).

it must collide in the previously calculated manner. In the example cited, in which we assumed the mass of all molecules equal, all molecules at the beginning had the same velocity c. After a time in which each molecule has experienced on the average one collision, many molecules will have a velocity γ. However, ignoring the few molecules that have experienced several collisions, all these molecules come from a collision in which the other molecule left with a velocity $\sqrt{2c^2-\gamma^2}$. If we now reverse all velocities, then nearly all molecules whose velocity is γ will collide only with molecules whose velocity is $\sqrt{2c^2-\gamma^2}$, so that the characteristics of a molecular-ordered distribution are present.

The fact that H now increases does not contradict the laws of probability either; for these only predict the unlikeliness, not the impossibility, of an increase in H. Indeed, on the contrary it follows explicitly that each distribution of states, even if very unlikely, has a probability different from zero, though very small. Similarly, when one has a Maxwell distribution, the case that one molecule has the velocity that it actually has at that time, likewise for the second, third, and other molecules, is not in the least more probable than the case that all molecules have the same velocity.

It would clearly be a gross error to conclude that any motion whereby H decreases is equiprobable with one in which the velocities are reversed and H increases. Consider any motion for which H decreases from time t_0 to time t_1. When one reverses all the velocities at time t_0, he would by no means arrive at a motion for which H must increase; on the contrary, H would probably still decrease. It is only when one reverses the velocities at time t_1 that he obtains a motion for which H must increase during the time interval t_1-t_0, and even then H would probably decrease again after that, so that motions for which H continually remains very near to its minimum value are by far the most probable. Motions for which it increases to a considerably larger value, or sinks from a large value to its minimum value, are equally improbable; one knows however that H has many larger values during a definite time interval, so it must decrease with high probability.[3]

Planck has attempted to base a proof that the Maxwell

[3] Cf. Nature **51**, 413 (Feb. 1895).

velocity distribution is the only possible stationary one on this reversibility principle.* From Hamilton's principle he has not yet proved, to my knowledge, that through the reversal each stationary state distribution must transform into another one. One can still show the following: when, after a state distribution A (which is stationary to an arbitrary degree of approximation) has lasted for an arbitrarily long time, one suddenly reverses all the velocities, then he obtains a motion B which again remains stationary (with the same degree of approximation) just as long. We saw that a molecular-disordered distribution, after reversal of all velocities, can transform into a molecular-ordered one; one can therefore believe that the motion B will be molecular-ordered. Now for certain forms of containers there are certainly possible molecular-ordered motions that remain stationary for arbitrarily long times. It appears however that these could be destroyed at any time by an arbitrarily small change in the form of the container. We assume that the state distribution B cannot remain molecular-ordered during its entire duration. Moreover, for the state distribution A, we assume that each velocity is equiprobable with the opposite velocity. The state distribution B must be identical with A, since by virtue of the second assumption the magnitude and direction of each velocity in B has the same probability as in A, and by virtue of the first assumption the collisions follow the laws of probability. In B, however, each inverse collision must occur as often as the corresponding direct collision in A, since the two move in opposite directions. Hence in B, each inverse collision must be just as probable as the corresponding direct one in A. But since both distributions are identical, it follows that in each of them, each direct collision has the same probability as the corresponding inverse one, whence follow Equations (27), whose necessary consequence is the Maxwell distribution.

When one may not assume *a priori* that each velocity has the same probability as the opposite one—for example, when gravity acts—Planck's proof appears to be inapplicable, but the minimum theorem remains valid.[4]

---

* M. Planck, Mun. Ber. **24,** 391 (1895).

[4] It would be necessary to prove that the following cases are not possible: 1. Besides Maxwell's law there is another molecular-disordered stationary distribution of states in which each velocity does not have the

A remark is necessary here. The quantities previously denoted by $d\omega = d\xi d\eta d\zeta$ and now by ω are volume elements, and hence really only differentials. The number n of molecules in unit volume is indeed a very large number, but it is still finite. (When we choose the cubic centimeter as volume element, then for air under ordinary conditions this number is a few trillion.) It may therefore be surprising that we treat the expressions $n_1\omega$, $n_2\omega$, and $f(\xi, \eta, \zeta, t)d\xi d\eta d\zeta$ as very large numbers. One could also carry out the same calculations on the assumption that these are fractions; they would then simply represent probabilities. But an actual number of objects is a more perspicuous concept than a mere probability, and the considerations just carried out would have required complicated digressions and explanations since one cannot speak of the permutation number of a fraction. Such thoughts remind us, however, that we could have chosen the volume element as large as we pleased. We could have assumed so many equivalent gases to be present in the volume element that even when ω is chosen very small, the velocity points of many molecules would still always lie in it. The order of magnitude of the volume chosen as volume element is completely independent of the order of magnitude of the volume elements ω and $d\xi d\eta d\zeta$.

Even more dubious is the assumption we shall make later, that not only the number of molecules in the volume element whose velocity points lie in a differential volume, but also the number of molecules whose centers are in such a volume element, is infinitely large. The latter assumption is no longer justified as soon as one has to deal with phenomena in which finite differences in the properties of the gas are encountered in distances

---

same probability as the opposite one; and a third distribution that transforms to the second one on reversing the velocities. 2. Besides the Maxwell (most probable) distribution—which does not in general transform to a molecular-ordered state on reversing velocities, since a molecular-ordered state is as probable as a disordered state—there exists a rare molecular-disordered stationary one, which on reversal transforms into a molecular-ordered one. 3. There are also stationary, molecular-ordered state distributions. 2 and 3 are also related to the case of the presence of external forces. The impossibility of case 3 cannot be proved from the minimum theorem, and probably cannot be proved in general without special conditions. Clearly the concept "molecular-disordered" is only a limiting case, which an originally molecular-ordered motion theoretically approaches only after an infinitely long time, though actually very quickly.

that are not large compared to the mean free path (shock waves of 1/100 mm thickness, radiometer phenomena, gas viscosity in a Sprengel vacuum, etc.). All other phenomena take place in such large spaces that one can construct a volume element for which the visible motion of the gas can be taken as a differential, yet which still contains a large number of molecules. This neglect of small terms whose order of magnitude is completely independent of the order of magnitude of the terms occurring in the final result must be carefully distinguished from the omission of terms that are of the same order of magnitude as those from which the final result is derived (cf. beginning of §14). While the latter omission causes an error in the result, the former is simply a necessary consequence of the atomistic conception, which characterizes the meaning of the result obtained, and is the more permissible, the smaller the dimension of the molecule compared to that of the visible bodies. In fact from the standpoint of atomistics, the differential equations of the doctrines of elasticity and hydrodynamics are not exactly valid, but rather they are themselves approximation formulae which become more nearly exact as the space in which the visible motions occur becomes large compared to the dimensions of molecules. Likewise, the distribution law for molecular velocities is not precisely correct as long as the number of molecules is not mathematically infinite. The disadvantage of giving up the supposedly exact validity of the hydrodynamic differential equations is however compensated by the advantage of greater perspicuousness.

§7. The Boyle-Charles-Avogadro law. Expression for the heat supplied.

We now proceed to the solution of Equations (27). These are only a special case of Equations (147) which we shall treat in §18. From these equations it follows, as we shall prove explicitly there, that the functions f and F must be independent of the direction of the velocity, and can only depend on its magnitude. We could already give this proof in the same way here for a special case. In order not to repeat ourselves, we shall assume without proof that neither the form of the container nor any special circumstance influences the distribution of states. Since then all directions in space are equivalent, the functions f and F must be independent of direction, and can be functions only of the magnitudes of the relevant velocities c and c_1. If we set $f=e^{\varphi(mc^2)}$ and $F=e^{\Phi(m_1c_1^2)}$, then the last of Equations (27) becomes

$$\varphi(mc^2) + \Phi(m_1c_1^2) = \varphi(mc'^2) + \Phi(mc^2 + m_1c_1^2 - mc'^2).$$

Here the two quantities mc^2 and $m_1c_1^2$ are clearly completely independent of each other, and also the third quantity mc'^2 can take all values from zero to $mc^2+m_1c_1^2$, independently of the first two. Denoting these three quantities by x, y, and z, and differentiating the last equation with respect to x, then with respect to y, and finally with respect to z, we obtain:

$$\begin{aligned} \varphi'(x) &= \Phi'(x+y-z) \\ \Phi'(y) &= \Phi'(x+y-z) \\ 0 = \varphi'(z) &= \Phi'(x+y-z), \end{aligned}$$

whence it follows that

$$\varphi'(x) = \Phi'(y) = \varphi'(z).$$

Since the first of these expressions does not contain y or z, and the second and third must be equal to it, the second may not contain y and the third may not contain z. They do not contain any other variables; hence they must be constant; since they are equal to each other, the derivatives of the two functions φ and Φ must be equal to the same constant, $-h$, whence it follows that:

(36) $$f = ae^{-hmc^2}, \ F = Ae^{-hm_1c_1^2}.$$

The number dn_c of m-molecules in unit volume whose velocity has arbitrary direction and magnitude between c and $c+dc$ is clearly equal to the number of those for which the velocity point lies between two spherical surfaces drawn around the origin of coordinates with radii c and $c+dc$, and thus in a space of volume $d\omega=4\pi c^2dc$. Hence, according to Equation (11):

(37) $$dn_c = 4\pi a e^{-hmc^2}c^2dc.$$

The molecules whose velocity has a magnitude between c and $c+dc$ and a direction that forms an angle between ϑ and $\vartheta+d\vartheta$ with a fixed line (e.g., the axis of abscissas) are identical with those whose velocity point lies in a ring bounded by the two spherical surfaces of radii c and $c+dc$ and the two conical surfaces with apexes at the origin, whose common axis is the abscissa direction, and whose generators make angles ϑ and $\vartheta+d\vartheta$ with the axis. Since this ring has volume $2\pi c^2 \sin\vartheta \cdot dc d\vartheta$, the number $dn_{c,\vartheta}$ of such molecules is given by the following expression:

$$(38) \qquad dn_{c,\vartheta} = 2\pi a e^{-hmc^2} c^2 \sin\vartheta \cdot dc d\vartheta = \frac{dn_c \sin\vartheta \cdot d\vartheta}{2}.$$

If we integrate (37) over all possible velocities, so that c goes from 0 to ∞, we obtain the total number n of molecules in unit volume. This and the following integrations are easily performed with the aid of the two well-known integral formulae:

$$(39) \qquad \begin{cases} \displaystyle\int_0^\infty c^{2k} e^{-\lambda c^2} dc = \frac{1\cdot 3 \cdots (2k-1)\sqrt{\pi}}{2^{k+1}\sqrt{\lambda^{2k+1}}}, \\ \displaystyle\int_0^\infty c^{2k+1} e^{-\lambda c^2} dc = \frac{k!}{2\lambda^{k+1}}. \end{cases}$$

One then obtains:

$$(40) \qquad n = a\sqrt{\frac{\pi^3}{h^3 m^3}}$$

and hence instead of Equations (36) and (37) one can write:

$$(41) \qquad f = n\sqrt{\frac{h^3 m^3}{\pi^3}}\, e^{-hmc^2},$$

$$(42) \qquad F = n_1\sqrt{\frac{h^3 m_1{}^3}{\pi^3}}\, e^{-hm_1 c_1^2},$$

$$(43) \qquad dn_c = 4n\sqrt{\frac{h^3 m^3}{\pi}}\, e^{-hmc^2} c^2 dc.$$

If we multiply dn_c by c^2, the square of the velocity of those molecules whose number is equal to dn_c, integrate over all possible velocities and finally divide by the total number of molecules in unit volume, n, then we obtain the quantity that we call the mean square velocity and denote by $\overline{c^2}$. It is therefore equal to:

$$(44) \qquad \overline{c^2} = \frac{\displaystyle\int_0^\infty c^2 dn_c}{\displaystyle\int_0^\infty dn_c} = \frac{3}{2hm}.$$

Similarly one finds for the mean velocity the value:

(45) $$\bar{c} = \frac{\int_0^\infty c\,dn_c}{\int_0^\infty dn_c} = \frac{2}{\sqrt{\pi hm}}\,.$$

Therefore we have:

(46) $$\frac{\overline{c^2}}{(\bar{c})^2} = \frac{3\pi}{8} = 1{,}178 \cdots$$

We shall now put on the axis of abscissas the various values of c, and over them erect ordinates whose height is the quantity $c^2e^{-hmc^2}$, which is proportional to the probability that the velocity lies between c and $c+dc$ (where dc should have the same value for all c). We thereby obtain a curve whose largest ordinate is found at the abscissa

(47) $$c_w = \frac{1}{\sqrt{hm}}$$

This abscissa c_w* is usually called the most probable velocity.

If one now puts the velocity squared, $x=c^2$, on the axis of abscissas, and makes the ordinate proportional to the probability that c^2 lies between x and $x+dx$, where the differential dx has the same value for all x, then the ordinates will be proportional to $\sqrt{x}e^{-hmx}$. The largest ordinate is then at $x=\frac{1}{2}hm$, which does not correspond to the velocity $c=c_w^2$ but rather to $c=c_w/\sqrt{2}$. In a certain sense, $c_w^2/2$ could therefore be called the most probable velocity squared.

If one considers in the gas a surface of unit area, and looks for the mean or most probable velocities of all molecules that strike the surface in unit time, then he obtains further quantities, which differ from those previously defined as mean and most probable.

All these expressions therefore are not precisely defined; it is in no way uniquely determined what one should call the mean. Similar ambiguities will be encountered in defining the mean free path.

Since

(48) $$c^2 = \xi^2 + \eta^2 + \zeta^2, \text{ therefore } \overline{\xi^2} = \overline{\eta^2} = \overline{\zeta^2} = \frac{1}{3}\overline{c^2} \doteq \frac{1}{2hm}\,.$$

* The subscript w stands for *wahrscheinlichste*, "most probable."

In the same way, many other mean values could be calculated. For example,

$$(49)\quad \left\{\begin{aligned} \overline{\xi^4} &= \frac{\displaystyle\iiint_{-\infty}^{+\infty} \xi^4 e^{-hm(\xi^2+\eta^2+\zeta^2)} d\xi d\eta d\zeta}{\displaystyle\iiint_{-\infty}^{+\infty} e^{-hm(\xi^2+\eta^2+\zeta^2)} d\xi d\eta d\zeta} \\ &= \frac{\displaystyle\int_0^{\infty} \xi^4 e^{-hm\xi^2} d\xi}{\displaystyle\int_0^{\infty} e^{-hm\xi^2} d\xi} = \frac{3}{4h^2m^2} = 3(\overline{\xi^2})^2 \end{aligned}\right.$$

The same is of course valid for the second gas, and since h must have the same value for both gases in a mixture, it follows from Equation (44) that for two mixed gases, regardless of the density of each,

$$(50)\qquad m\overline{c^2} = m_1\overline{c_1^2}.$$

When two kinds of gas molecules are mixed in a space, in general those of one kind will impart kinetic energy to those of the other, or conversely. The above equation says that whatever happens, no matter what their density and other properties may be, in thermal equilibrium both have the Maxwellian state, and the mean kinetic energy of a molecule has the same value for either kind of gas.

In order to judge whether two gases have the same temperature, or whether one gas of higher density has the same temperature as another sample of the same gas of smaller density, we must imagine the gases to be separated by a heat-conducting wall, and ask for the thermal equilibrium in this case. The molecular processes in such a heat-conducting wall cannot be submitted to calculation according to such simple principles as those employed above, yet it seems likely—and can also be shown by calculations on the basis of certain assumptions—that the same condition of thermal equilibrium would still hold (cf., the mechanical device conceived by Bryan, discussed in §19). Experimentally, the fact that the expansion of a gas in a vacuum and the diffusion of two gases proceed without noticeable evolution of heat proves this. By the same assumption, it must in general be true that when two gases with the same properties but different densities, or with different properties, are in thermal equilibrium, then they have the same temperature, and the mean kinetic energy of a molecule must have the same value for one gas as for the other. The temperature must therefore be the same function of the mean kinetic energy for all gases. From Equation (6) it follows that for two gases at equal temperature, when the pressure at the surface is also the same, $n = n_1$, hence the number of molecules in unit volume is the same—the well-known Avogadro law. Furthermore, since m is constant for the same gas, it follows that for a gas at the same temperature but different pressures, c^2 is constant, and hence by Equation (7) the pressure p is proportional to the density ρ—the Boyle or Mariotte law.

We shall now choose a gas which is as nearly perfect as possible—say hydrogen—as the normal gas. For the normal gas, the pressure, density, mass and velocity of a molecule will be denoted by P, ρ', M, and C. For any other gas, lower-case letters will be used. We choose as our thermometric substance the normal gas at constant volume and therefore constant density—i.e., we choose the temperature scale so that the temperature T is proportional to the pressure of the normal gas on a unit surface at constant density. Then in the formula $P = \rho' \overline{C^2}/3$, the temperature T must be proportional to P at constant ρ', and therefore it must also be proportional to $\overline{C^2}$. We denote this proportionality factor by $3R$, so that for this density

$$\overline{C^2} = 3RT. \tag{51}$$

If the normal gas is at another density, then the temperature T is the same when $\overline{C^2}$ has the same value. Hence R is also independent of density and the formula $P = \rho' \overline{C^2}/3$ goes over to $P = R\rho' T$. The constant R can be chosen so that the difference between the temperature of the gas when in contact with melting ice and when in contact with boiling water is equal to 100. The absolute value of the temperature of melting ice is thereby determined. Then this must be in the same ratio to the temperature difference (100) between boiling water and melting ice as the pressure of hydrogen at the latter temperature is to the pressure difference between the two temperatures (all pressures being taken at the same density). This proportion gives the temperature of melting ice as 273.

For another gas, for which lower-case letters are used, one

obtains in the same way $p=\rho\overline{c^2}/3$, and since at equal temperatures $m\overline{c^2}=M\overline{C^2}$, it follows from Equation (51) that

$$\overline{c^2} = \frac{M\overline{C^2}}{m} = 3\frac{M}{m}RT = \frac{3R}{\mu}T = 3rT, \tag{51a}$$

where $\mu=m/M$ is the so-called molecular weight, i.e. the ratio of the mass or weight of a molecule (a freely moving particle) of the gas in question to the mass of a molecule of the normal gas. If we put this value of $\overline{c^2}$ into the equation $p=\rho\overline{c^2}/3$, then we obtain for any other gas:

$$p = \frac{R}{\mu}\rho T = r\rho T, \tag{52}$$

where r is the gas constant of the gas considered, but R is a constant which is the same for all gases. Equation (52) is the well-known expression for the combined Boyle-Charles-Avogadro law.

§8. Specific heat. Physical meaning of the quantity H.

We now imagine a simple gas in an arbitrary volume Ω. We introduce the amount of heat dQ (measured in mechanical units), which raises the temperature by dT and increases the volume by $d\Omega$. We set $dQ=dQ_1+dQ_4$, where dQ_1 represents the heat used in increasing the molecular energy, while dQ_4 represents that used in doing external work. If the gas molecules are perfectly smooth spheres, then in collisions no forces act to make them rotate. We assume that such forces in general do not exist. Then, when the molecule happens to have some rotational motion already, it will not be changed by the addition of the amount of heat qQ. Hence the total amount of heat dQ_1 will be consumed in increasing the kinetic energy with which the molecules move among each other, which we call the kinetic energy of progressive motion. We have considered only this case up to now; however, in order not to have to repeat the same calculations later, we shall now perform the following calculations for the more general case in which the molecules have another form, or consist of several particles moving with respect to each other (atoms). Then in addition to progressive motion, intramolecular motion will also be present, and work can be performed against the forces holding the atoms together (intramolecular work). In this case we set $dQ_1 = dQ_2 + dQ_3$ and denote by dQ_2 the heat consumed in increasing the kinetic energy of progressive motion, and by dQ_3 the heat consumed in increasing the kinetic energy of intramolecular motion and in performing intramolecular work. By kinetic energy of progressive motion of a molecule, we mean the kinetic energy of the total mass of the molecule, considered to be concentrated at its center of mass.

We have shown that when the volume of a gas is increased at constant temperature, the kinetic energy of progressive motion and also the law of distribution of the different progressive velocities among the molecules both remain unchanged. The molecules just move farther apart—i.e., there is a greater interval between two collisions. Although we have not investigated the internal motion, we could take it as probable that in a simple expansion at constant temperature, on the average neither the internal motion during the collision, nor that during the motion from one collision to the next, would be changed by a mere decrease in the collision frequency. The duration of a collision would still be vanishingly small compared to the time between two successive collisions. Like the kinetic energy of progressive motion, the intramolecular motion and the intramolecular potential energy can depend only on temperature. The increase of each of these energies is therefore equal to the temperature increment dT multiplied by a function of temperature, and if we set $dQ_3 = \beta dQ_4$, then β can depend only on temperature. We can return at any time to the previous case, absolutely smooth spherical molecules, simply by setting $\beta = 0$. The number of molecules in the volume Ω of our gas is $n\Omega$, and since the mean kinetic energy of progressive motion of a molecule is $m\overline{c^2}/2$, the total kinetic energy of progressive motion of all molecules is

$$\frac{n\Omega m}{2}\overline{c^2},$$

or, when one denotes the total mass of the gas by k, equal to

$$\frac{k}{2}\overline{c^2},$$

since clearly $k = \rho\Omega = nm\Omega$.

Since, further, the total mass k of the gas is not changed by the addition of heat, the increase of kinetic energy of progressive motion is

$$\frac{k}{2}\, d\overline{c^2},$$

If we measure the heat in mechanical units, then this is also equal to dQ_2. But according to Equation (51a)

$$d\overline{c^2} = \frac{3R}{\mu}\, dT,$$

and hence

$$dQ_3 = \frac{3kR}{2\mu}\, dT,$$

$$dq_1 = dq_2 + dq_3 = \frac{3(1+\beta)kR}{2\mu}\, dT.$$

The external work done by the gas is $p \cdot d\Omega$; this is therefore also equal to the quantity dQ_4, measured in mechanical units. Now the total mass of the gas remains constant, hence

$$d\Omega = kd\left(\frac{1}{\rho}\right)$$

and according to Equation (52),

$$\frac{1}{\rho} = \frac{R}{\mu}\,\frac{T}{p}\,.$$

hence

$$dQ_4 = \frac{Rkp}{\mu}\, d\left(\frac{T}{p}\right) = \frac{Rk}{\mu}\, \rho T d\left(\frac{1}{\rho}\right).$$

If one substitutes all these values he obtains, for the total heat added, the value:

$$(53)\quad \begin{cases} dQ = dQ_1 + dQ_4 = \dfrac{Rk}{\mu}\left[\dfrac{3(1+\beta)}{2}\, dT + pd\left(\dfrac{T}{p}\right)\right] \\ \qquad = \dfrac{Rk}{\mu}\left[\dfrac{3(1+\beta)}{2}\, dT + \rho T d\left(\dfrac{1}{\rho}\right)\right]. \end{cases}$$

If the volume is constant, then $d\Omega/k = d(1/\rho) = 0$, and the added heat will be

$$dQ_v = \frac{3Rk}{2\mu}(1+\beta)dT.$$

On the other hand, if the pressure is constant, then $d(T/p) = (dT)/p$ and the added heat will be

$$dQ_p = \frac{Rk}{2\mu}[3(1+\beta)+2]dT.$$

If one divides dQ by the total mass k, then he obtains the heat added per unit mass. If one divides by dT, then he obtains the amount of heat that must be added to raise the temperature by one degree, the so-called specific heat. Therefore the specific heat per unit mass of the gas at constant volume is:

$$\gamma_v = \frac{dQ_v}{k \cdot dT} = \frac{3R}{2\mu}(1+\beta). \tag{54}$$

On the other hand, the specific heat per unit mass at constant pressure is:

$$\gamma_p = \frac{R}{2\mu}[3(1+\beta)+2]. \tag{55}$$

In both expressions all quantities except β are constant. The latter can be a function of temperature. Since R refers only to the normal gas and hence has the same value for all gases, the product $\gamma_p \cdot \mu$ and also the product $\gamma_v \cdot \mu$ will have the same values for all gases for which β has the same value—e.g., in particular for all gases for which β is equal to zero. The difference of specific heats, $\gamma_p - \gamma_v$, is for all gases equal to the gas constant itself:

$$\gamma_p - \gamma_v = r = \frac{R}{\mu}. \tag{55a}$$

The product of this difference and the molecular weight μ is for all gases a constant equal to R. The ratio of specific heats is

$$\kappa = \frac{\gamma_p}{\gamma_v} = 1 + \frac{2}{3(1+\beta)}. \tag{56}$$

Conversely,

$$\beta = \frac{2}{3(\kappa - 1)} - 1. \tag{57}$$

In the case that the molecules are perfect spheres, which we have assumed previously, $\beta=0$ and hence $\kappa=1\frac{2}{3}$. This value was in fact observed by Kundt and Warburg for mercury gas,* and more recently by Ramsay for argon and helium;† for all other gases investigated up to now, κ is smaller, so that there must be intramolecular motion. We shall come back to this subject in Part II.

The general expression (53) for dQ is not a complete differential of the variables T and ρ; however if one divides it by T, then since β is a function only of T, he obtains a complete differential. If β is constant, then one has

$$\int \frac{dQ}{T} = \frac{Rk}{\mu} l[T^{(3/2)(1+\beta)}\rho^{-1}] + \text{const.}$$

This is therefore the so-called entropy of the gas.

If several gases are present in separate containers, then naturally the total added heat is equal to the sum of the amounts of heat added to the individual gases, so that whether or not they have the same temperature, their total entropy is equal to the sum of the entropies of each gas. If several gases whose masses are $k_1, k_2, \cdots$, whose partial pressures are $p_1, p_2, \cdots$, and whose partial densities are $\rho_1, \rho_2, \cdots$, are mixed in a container of volume Ω, then the total molecular energy is always equal to the sum of the molecular energies of the components. The total work is $(p_1+p_2+\cdots)d\Omega$, where

$$\Omega = k_1/\rho_1 = k_2/\rho_2 \cdots p_1 = \frac{R}{\mu_1}\rho_1 T,\ p_2 = \frac{R}{\mu_2}\rho_2 T \cdots$$

Hence it follows that the differential of the heat added to the mixture has the value:

$$dQ = R\sum \frac{k}{\mu}\left[\frac{3(1+\beta)}{2}T + \rho T d\left(\frac{1}{\rho}\right)\right]$$

From this it follows that the total entropy of several gases, when β has the same constant value for each one, is

$$R\sum \frac{k}{\mu} l[T^{(3/2)(1+\beta)}\rho^{-1}] + \text{const.} \tag{58}$$

---

* Kundt and Warburg, Ann. Physik [2] **157**, 353 (1876).

† Ramsay, Proc. R. S. London **58**, 81 (1895); Rayleigh and Ramsay, Phil. Trans. **186**, 187 (1896).

where some are in different containers, while others may be mixed; except that in the latter case ρ is the partial density, and naturally all the mixed gases must have the same temperature. Experience teaches that the constant is not changed on mixing, as long as p and ρ do not change.

Since by now we have learned the physical meaning of all the other quantities, we shall deal with the physical meaning of the quantity denoted by H in §5; for the present we shall have to limit ourselves to the case considered in §5, where the molecules are perfect spheres, and hence the ratio of specific heats is $\kappa = 1\frac{2}{3}$.

We obtain, according to Equation (28), for unit volume of a single gas, $H = \int f l f d\omega$; for the stationary state we have

$$f = ae^{-hmc^2},$$

hence

$$H = la \int f d\omega - hm \int c^2 f d\omega.$$

Now however $\int f d\omega$ is equal to the total number n of molecules, and furthermore

$$\int c^2 f d\omega = n\overline{c^2} = \frac{3n}{2hm},$$

hence

$$H = n(la - \tfrac{3}{2}).$$

Furthermore, according to Equations (44) and (51a),

$$\frac{3}{2hm} = \overline{c^2} = \frac{3RM}{m} T,$$

hence

$$h = \frac{1}{2RMT},$$

and according to Equation (40)

$$a = n \sqrt{\frac{h^2 m^3}{\pi^3}} = \rho T^{-3/2} \sqrt{\frac{m}{8\pi^3 R^3 M^3}}.$$

Hence, aside from a constant,

$$H = nl(\rho T^{-3/2}).$$

We saw that $-H$ represents, apart from a constant, the logarithm of the probability of the state of the gas considered.

The probability of simultaneous occurrence of several events is the product of the probabilities of the events; the logarithm of the former probability is therefore the sum of the logarithms of the probabilities of the individual events. Hence the logarithm of the probability of a state of a gas of doubled volume is $-2H$; of tripled volume, $-3H$; and of volume Ω, $-\Omega H$. The logarithm of the probability $\mathfrak{W}$ of the arrangement of the molecules and distribution of states among them in several gases is

$$l\mathfrak{W} = -\sum \Omega H = -\sum Qnl(\rho T^{-3/2}),$$

where the sum is to be extended over all gases present. This additive property of the logarithm of the probability is already expressed by Equation (28) for a gas mixture.

If we multiply by RM, which is constant for all gases (M is the mass of a hydrogen molecule), we obtain

$$RMl\mathfrak{W} = -\sum RM\Omega nl(\rho T^{-3/2}) = R\sum \frac{k}{\mu}\, l(\rho^{-1}T^{3/2}).$$

In nature, the tendency of transformations is always to go from less probable to more probable states. Thus if $\mathfrak{W}$ is smaller for one state than for a second, then to facilitate the transformation from the first state to the second, the action of another body may be necessary, but this transformation will still be possible without permanent changes in any other bodies. On the other hand, if $\mathfrak{W}$ is smaller for the second state, the transformation can occur only if another body takes on a more probable state. Since the quantity $RMl\mathfrak{W}$, which differs only by a constant factor and an addend from $-H$, increases and decreases with $\mathfrak{W}$, we can assert the same things about it as about $\mathfrak{W}$. But the quantity $RMl\mathfrak{W}$ is in our case, where the ratio of specific heats is equal to $1\frac{2}{3}$, in fact the total entropy of all the gases.

One sees this at once if he sets $\beta = 0$ in the empirically correct expression (58). The fact that in nature the entropy tends to a maximum shows that for all interactions (diffusion, heat conduction, etc.) of actual gases the individual molecules behave according to the laws of probability in their interactions, or at least that the actual gas behaves like the molecular-disordered gas which we have in mind.

The second law is thus found to be a probability law. We have of course proved this only in a special case, in order not to make it too difficult to understand because of too much generality. Moreover, the proof that for a gas of arbitrary volume Ω the quantity ΩH—and for several gases the quantity $\sum \Omega H$—can only decrease through collisions, and thus is to be considered as a measure of the probability of states, was only hinted at. This proof can easily be given explicitly, and will be given at the end of §19. We still need to generalize and deepen our conclusions.

Even if one concedes validity to the gas theory only as a mechanical model, I still believe that this conception of the entropy principle, to which it has led, strikes at the heart of the subject in the correct way. In one respect we have even generalized the entropy principle here, in that we have been able to define the entropy in a gas that is not in a stationary state.

§9. Number of collisions.

We shall now consider again the same mixture of two gases as in §3, and adopt all the notation used there. We proceed from the number of collisions, given by Equation (18), which occur between an m-molecule (a molecule of the first kind of gas of mass m) and an m_1-molecule (a molecule of the second kind of gas of mass m_1) in unit volume during time dt, that satisfy the three conditions (10), (13), and (15).

We consider now only the state of thermal equilibrium, for which we found in §7 the equations (41) and (42).

We ask first, how many collisions in all, without any restriction, take place between an m-molecule and an m_1-molecule in unit volume during time dt. We obtain this by dropping the three restrictive conditions to which the collisions have previously been subjected—i.e., when we integrate over the differentials. In order to find the limits of integration, we represent the velocities c and c_1 of the two molecules before collision by the lines OC and OC_1 in Figure 5. The line OG should be parallel to the relative velocity C_1C of the m-molecule with respect to the m_1-molecule before the collision, and the sphere with centre O and radius 1 (the sphere E) intersects the point G. The line OK should have the same direction as the line of centres drawn from m to m_1, and intersects the sphere E in the point K. Hence KOG is the angle denoted by ϑ. We allow the position of the line OK to vary in such a way that ϑ increases by $d\vartheta$ as the angle ϵ of the two planes KOG and COC_1 increases by $d\epsilon$. The circle shown in Figure 5 should be the intersection of the sphere E with the latter plane, which we can choose as the plane of the diagram, when we imagine the coördinate axes (of which we are now completely independent) to be lying in some oblique way. When ϑ and ϵ take all values between ϑ and $\vartheta+d\vartheta$, and ϵ and $\epsilon+d\epsilon$, then the point K describes on the sphere E a surface element of surface area $\sin\vartheta\cdot d\vartheta\cdot d\epsilon$. As indicated in §4, we can choose this surface element as the surface element $d\lambda$, so that we obtain, according to Equation (18),

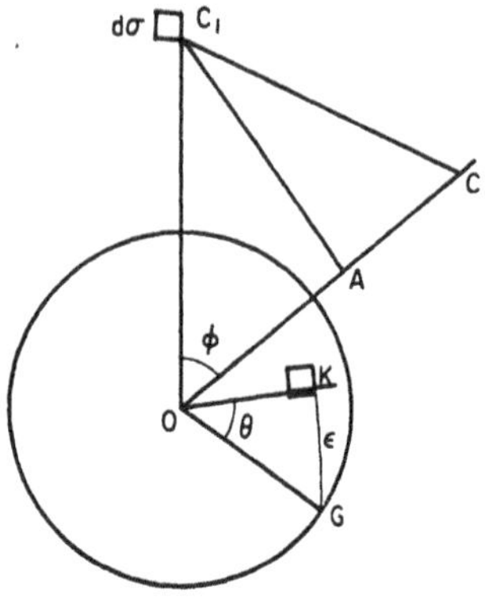

FIG. 5.

$$d\nu = fd\omega F_1 d\omega_1 g\sigma^2 \cos\vartheta \sin\vartheta d\vartheta d\epsilon dt.$$

We now leave fixed the two volume elements $d\omega$ and $d\omega_1$ within which the points C and C_1 lie, but integrate $d\nu$ with respect to ϑ and ϵ over all possible values, i.e., we integrate ϑ from 0 to $\pi/2$ and ϵ from 0 to 2π (see the conditions mentioned in §3). The result of the integration will be denoted by $d\nu_1$, and thus we obtain[1]

(59) $$d\nu_1 = fd\omega F_1 d\omega_1 g\sigma^2 \pi dt.$$

---

[1] If one substitutes in this formula the value 1 for $fd\omega$, n for $F_1 d\omega_1$, c for g, and 1 for dt, then he gets

(60) $$\nu_r = \pi n \sigma^2 c$$

for the number of collisions that a molecule will experience in unit time when it moves with constant velocity c among identical molecules at rest. σ is the sum of the radii of a moving and a stationary molecule. The path traversed by the moving molecule from one collision to the next would have on the average the length

$$\lambda_r = \frac{c}{\nu_r} = \frac{1}{\pi n \sigma^2}.$$

This is therefore the total number of collisions that occur between an m-molecule and an m_1-molecule in unit volume during time dt, such that before the collision:

1. the velocity point of the m-molecule lies in the volume element $d\omega$,
2. the velocity point of the m_1-molecule lies in the volume element $d\omega_1$.

On the other hand, condition (15) has been dropped, so that the direction of the line of centers is not subject to any restrictive condition. We shall now denote the angle COC_1 in Figure 5 by φ, the point C being held fixed while C_1 varies so that the line OC_1 takes all values between c_1 and c_1+dc_1, and the angle φ takes all values between φ and $\varphi+d\varphi$. We thereby obtain the surface element, denoted by $d\sigma$ in Figure 5, of area $c_1dc_1d\varphi$ at a distance $C_1A=c_1 \sin \varphi$ from the line OC. If we allow this surface element to rotate around the line OC as axis, it will sweep out a ring of volume $2\pi c_1^2 \sin \varphi dc_1 d\varphi$. The two integrations over φ and c_1 can always be carried out in the same way each time, so that the velocity point C_1 of the m_1-molecule always lies inside the ring R. We find the total number $d\nu$ of collisions in unit volume in time dt between an m-molecule and an m_1-molecule such that the velocity point of m lies in $d\omega$ while that of m_1 lies in the ring R by integrating the expression $d\nu_1$ with respect to $d\omega_1$ over all volume elements of the ring R. In other words, we simply set

$$d\omega_1 = 2\pi c_1^2 \sin \varphi dc_1 d\varphi$$

in $d\nu_1$, whereby we obtain

$$d\nu_2 = 2\pi^2 f d\omega F_1 c_1^2 g \sigma^2 \sin \varphi dc_1 d\varphi dt. \tag{61}$$

In order to eliminate any restrictions on the magnitude and direction of the velocity c_1, we merely have to integrate over all values of φ and c_1, keeping c constant. Thus we integrate over φ from 0 to π, and over c_1 from 0 to ∞, whence we obtain

$$d\nu_3 = 2\pi^2 \sigma^2 dt f d\omega \int_0^\infty \int_0^\pi F_1 c_1^2 g \sin \varphi dc_1 d\varphi. \tag{62}$$

Since $g^2=c^2+c_1^2-2cc_1 \cos \varphi$ and $\sin \varphi d\varphi = gdg/cc_1$, we have

$$\int_0^\pi g \sin \varphi d\varphi = \frac{g_\pi^3 - g_0^3}{3cc_1}.$$

The relative velocity g_π for $\varphi=\pi$ is $c+c_1$. The relative velocity g_0 for $\varphi=0$ is $c-c_1$ if $c_1<c$, but it is c_1-c if $c_1>c$. Hence one has

$$\int_0^\pi g \sin\varphi d\varphi = \frac{2(c_1^2+3c^2)}{3c} \text{ for } c_1 < c,$$

but

$$= \frac{2(c^2+3c_1^2)}{3c_1} \text{ for } c_1 > c.$$

One therefore must split the integration in Equation (62) into two parts:[2]

$$\text{(63)} \quad \begin{cases} d\nu_3 = \frac{4}{3}\pi^2\sigma^2 f d\omega dt \Bigg[\int_0^c F_1 c_1^2 \frac{c_1^2+3c^2}{c} dc_1 \\ \qquad + \int_c^\infty F_1 c_1^2 \frac{c^2+3c_1^2}{c_1} dc_1\Bigg]. \end{cases}$$

---

[2] If one writes in Eq. (61) the value n_1 in place of $4\pi F_1 c_1^2 dc_1$, 1 in place of dt, and 1 in place of $fd\omega$, and integrates over φ from zero to π just as in the text, then he finds the number ν' of collisions that an m-molecule would experience with m_1-molecules in unit time, if it moved with constant speed c. It is thereby assumed that there are in unit volume n_1 m_1-molecules, all of which move with the same velocity c_1 but with the velocity directions equally distributed over all possible directions in space. The integration yields

$$\text{(65)} \quad \begin{cases} \nu' = \frac{\pi\sigma^2 n_1}{3c}(c_1^2+3c^2) \text{ for } c_1 < c \\ \nu' = \frac{\pi\sigma^2 n_1}{3c_1}(c^2+3c_1^2) \text{ for } c_1 > c. \end{cases}$$

If moreover the m_1-molecules are identical with the m-molecules, there being in all n of them in unit volume, and furthermore $c=c_1$, and s is the diameter of the molecules, then one obtains for the number of collisions that one molecule experiences with identical ones of the same velocity moving in all directions, in unit time, the value

$$\text{(66)} \quad \nu'' = \frac{4}{3}\pi n s^2 c,$$

The mean path (from one collision to the next) will be

The above quantity $d\nu_3$ represents therefore the total number of collisions which the m-molecules whose velocity points lie in $d\omega$ undergo with any m_1-molecules. If we divide this number by $fd\omega$ (the number of such m-molecules) and call the quotient ν_c, then we shall obtain the probability that an m-molecule whose velocity is c collides during dt with an m_1-molecule; i.e., this quotient

$$\nu_c dt = \frac{d\nu_3}{fd\omega} \tag{64}$$

specifies what fraction of a very large number A of m-molecules, all moving with the velocity c in the gas mixture, will collide with m_1-molecules during the time interval dt.

We can also say: we imagine an m-molecule moving with constant speed c through the gas mixture. After each collision, its velocity is restored to the value c by some external cause, and also the velocity distribution in the gas mixture is not disturbed by this one molecule. Then $\nu_c dt$ will be the probability that this molecule collides with an m_1-molecule during time dt; ν_c will therefore be the number of collisions, on the average, with m_1-molecules in unit time. The two equations (63) and (64) yield, on substituting for F_1 its value from Equation (42):

$$\lambda_{\text{Claus}} = \frac{c}{\nu''} = \frac{3}{4\pi n s^2} = \frac{3}{4}\lambda_r. \tag{67}$$

This is the value calculated by Clausius for the mean free path; it differs numerically somewhat from that derived in the text, which was calculated by Maxwell. [R. Clausius, Ann. Physik [2] **105,** 239 (1859); J. C. Maxwell, Phil. Mag. [4] **19,** 19 (1860); Clausius defended his original value in a note in Phil. Mag. [4] **19,** 434 (1860). Maxwell did not bother to give a detailed proof of his own result, but this was done by W. D. Niven in Maxwell's *Scientific Papers,* Vol. I, p. 387. See S. G. Brush, Amer. J. Phys. **30,** 271 (1962); Bernstein, *Isis* **54,** 206, (1963).—Tr.]

If one has in unit volume n molecules with diameter s and n_1 with diameter $s_1 = 2\sigma - s$, and if all n of the former move with the same velocity c, while all n_1 of the latter move with a different velocity c_1, distributed uniformly in all directions in space, then one of the n molecules experiences $\nu' + \nu''$ collisions, and its mean path is:

$$\begin{cases} \lambda' = \dfrac{c}{\nu' + \nu''} = \dfrac{3c^2}{4\pi n s^2 c^2 + \pi\sigma^2 n_1(c_1^2 + 3c^2)} & \text{for } c_1 < c \\[2ex] = \dfrac{3cc_1}{4\pi n s^2 c c_1 + \pi\sigma^2 n_1(c^2 + 3c_1^2)} & \text{for } c_1 > c. \end{cases} \tag{68}$$

$$
(69)\quad \left\{
\begin{aligned}
\nu_c &= \tfrac{4}{3} n_1 \sigma^2 \sqrt{\pi h^3 m_1^3} \left[\int_0^c c_1^2 e^{-hm_1 c^2} \frac{c_1^2 + 3c^2}{c} dc_1 \right. \\
&\quad \left. + \int_c^\infty c_1^2 e^{-hmc_1^2} \frac{c^2 + 3c_1^2}{c_1} dc_1 \right] \\
&= \tfrac{4}{3} n_1 \sigma^2 \sqrt{\pi h^3 m_1^3} \left[(2hm_1 c^2 + \tfrac{3}{2}) \frac{1}{h^2 m_1^2} e^{-hm_1 c} \right. \\
&\quad \left. + \int_0^c c_1^2 e^{-hm_1 c_1^2} \frac{c_1^2 + 3c^2}{c} dc_1 \right],
\end{aligned}
\right.
$$

hence since

$$
\int c_1^{2n} e^{-\lambda c_1^2} dc_1 = -\frac{1}{2\lambda} c_1^{2n-1} e^{-\lambda c_1^2} + \frac{2n-1}{2\lambda} \int c_1^{2n-2} e^{-\lambda c_1^2} dc_1
$$

$$
(70)\quad \nu_c = n_1 \sigma^2 \sqrt{\frac{\pi}{hm_1}} \left[e^{-hm_1 c^2} + \frac{2hm_1 c^2 + 1}{c\sqrt{hm_1}} \int_0^{c\sqrt{hm_1}} e^{-x^2} dx \right].
$$

If we replace all the quantities referring to the second kind of molecule by those referring to the first—i.e., replace n_1, m_1, and σ by n, m, and s—then the above quantity ν_c becomes

$$
(71)\quad \mathfrak{n}_c = ns^2 \sqrt{\frac{\pi}{hm}} \left[e^{-hmc^2} + \frac{2mhc^2 + 1}{c\sqrt{hm}} \int_0^{c\sqrt{hm}} e^{-x^2} dx \right].
$$

The number $\mathfrak{n}_c$ is the number of times that the m-molecule moving with constant velocity c through the mixture collides on the average with another m-molecule in unit time.

The quantity dn_c given in Equation (43) specifies how many of the n m-molecules have on the average a velocity between c and $c+dc$; dn_c/n is therefore the probability that the velocity of an m-molecule lies between these limits, and when one follows an m-molecule through a sufficiently long time interval T, then the fraction of the time T during which its velocity lies between c and $c+dc$ will be equal to Tdn_c/n. During this time Tdn_c/n, the m-molecule collides $\nu_c T dn_c/n$ times with a m_1-molecule, and $\mathfrak{n}_c T dn_c/n$ times with another m-molecule. Hence each m-molecule collides a total of $(T/n)\int \nu_c dn_c$ times with an m_1-molecule and

$(T/n)\int \mathfrak{n}_c dn_c$ times with another m-molecule. Therefore in unit time each m-molecule will collide on the average $\nu = (1/n)\int \nu_c dn_c$ times with an m_1-molecule and $\mathfrak{n} = (1/n)\int \mathfrak{n}_c dn_c$ times with an m-molecule, so that there will be $(\nu + \mathfrak{n})$ collisions in all.

Integration of Equation (69) yields:

$$\nu = \frac{16}{3} n_1 s^2 h^3 \sqrt{m^3 m_1^3}(J_1 + J_2),$$

where

$$J_1 = \int_0^\infty e^{-hmc^2} c^2 dc \int_c^\infty c_1^2 e^{-hm_1c_1^2} \frac{c^2 + 3c_1^2}{c_1} dc_1$$

$$= \frac{1}{h^2 m_1^2} \int_0^\infty e^{-h(m+m_1)c^2} c^2 dc (2hm_1c^2 + \tfrac{3}{2})$$

$$= \frac{3(m + 2m_1)}{8m_1^2} \sqrt{\frac{\pi}{h^7(m + m_1)^5}},$$

$$J_2 = \int_0^\infty e^{-hmc^2} c^2 dc \int_0^c c_1^2 e^{-hm_1c_1^2} \frac{c_1^2 + 3c^2}{c} dc_1.$$

In the latter integral, c has all values from zero to infinity, but c_1 can only run through those values smaller than the given value of c. If one interchanges the order of integrations, then c_1 has all values from zero to infinity whereas c assumes all values greater than the given c_1. Hence:

$$J_2 = \int_0^\infty e^{-hm_1c_1^2} c_1^2 dc_1 \int_{c_1}^\infty c^2 e^{-hmc^2} \frac{c_1^2 + 3c^2}{c} dc_1.$$

Since one can label the variables at will in a definite integral, we can exchange the labels c and c_1. Hence we obtain for J_2 an expression that differs from the one given for J_1 only by having the symbols m and m_1 permuted. One thus obtains J_2 by permuting m and m_1 in J_1:

$$J_2 = \frac{3(m_1 + 3m)}{8m^3} \sqrt{\frac{\pi}{h^7(m + m_1)^5}},$$

hence

$$
(72)\quad \begin{cases} \nu = 2\sigma^2 n_1 \sqrt{\dfrac{\pi(m+m_1)}{hmm_1}} = \pi\sigma^2 n_1 \sqrt{\dfrac{m+m_1}{m_1}}\cdot\bar{c} \\ \quad = \pi s^2 n_1 \sqrt{(\bar{c})^2 + (\bar{c}_1)^2} = 2\sqrt{\dfrac{2\pi}{3}}\,\sigma^2 n_1 \sqrt{\overline{c^2} + \overline{c_1^2}}. \end{cases}
$$

If one writes n, m, and s in place of n_1, m_1, and σ, then it follows that:

$$
(73)\qquad \mathfrak{n} = 2ns^2 \sqrt{\frac{2\pi}{mh}} = \pi n s^2 \bar{c}\sqrt{2}.
$$

Since in unit volume there are n m-molecules, each of which collides ν times in unit time with an m_1-molecule, there are in all

$$
(74)\qquad \nu n = 2\sigma^2 n n_1 \sqrt{\pi} \sqrt{\frac{m+m_1}{hmm_1}}
$$

collisions between an m-molecule and an m_1-molecule.* But since in a collision of two m-molecules, there are always two molecules of the same kind, the number of collisions between m-molecules is:

$$
(75)\qquad \frac{\mathfrak{n}n}{2} = s^2 n^2 \sqrt{\frac{2\pi}{hm}}\,.
$$

A similar formula holds for the collisions of m_1-molecules with each other.

§10. Mean free path.

Let there be n m-molecules in unit volume; let the first have velocity c_1, the second c_2, etc. Then $\overline{c_z} = (c_1 + c_2 + \cdots)/n$ is the mean velocity. We shall call it the number-average. Since the state is stationary, $\overline{c_z}$ does not change with time. If we multiply the last equation by dt and integrate over a very long time T, then:

* An interesting application of Eq. (74) was made by Füchtbauer and Hoffmann [Ann. Phys. [4] **43,** 96 (1914)] who tested Lorentz's "collision broadening" theory of spectral lines for cesium atoms in a nitrogen discharge. They found that the Lorentz theory could not explain the observed line width except by assuming a collision rate 32 times greater than that calculated from Boltzmann's kinetic theory formula. This left the field open for Holtsmark's theory based on the Stark effect [Holtsmark, Ann. Physik [4] **58,** 577 (1919)].

$$nT\overline{c_z} = \int_0^T c_1 dt + \int_0^T c_2 dt \cdots$$

Since during a very long time all molecules behave in the same way, all the summands are equal and it follows that $\overline{c_z} = \overline{c_t}$, where

$$\overline{c_t} = \frac{1}{T}\int_0^T c dt$$

is the time-average of the velocity of any single molecule.

$$\int_0^T c dt = T\overline{c_t}$$

is the sum of all distances that it traverses during time T. But since it collides $T(\nu+\mathfrak{n})$ times with another molecule during this time, the mean distance traversed between two successive collisions (the arithmetic mean of all distances between two successive collisions) is:

$$(76) \qquad \lambda = \frac{\bar{c}}{\nu + \mathfrak{n}} = \frac{1}{\pi\left(\sigma^2 n_1\sqrt{\dfrac{m+m_1}{m}} + s^2 n\sqrt{2}\right)}.$$

The difference between time- and number-average will be ignored, since the two are equal.* The same value for λ would of course have been obtained if we had defined it as the average of all distances traversed by all the m-molecules in unit volume between two successive collisions, in unit time. For a simple gas we have:

$$(77) \qquad \lambda = \frac{\bar{c}}{\mathfrak{n}} = \frac{1}{\pi n s^2\sqrt{2}} = \frac{\lambda_{,}}{\sqrt{2}}.$$

This value is $\frac{2\sqrt{2}}{3}$ times larger than the value λ_{Claus} calculated by Clausius (cf. [60] and [67]).

The molecule earlier imagined to be moving with constant

* Boltzmann here ignores the "ergodic problem" concerning the validity of equating these averages, not because he is unaware of it but presumably because he does not wish to complicate the discussion of the elementary mean free path concept. It will be recalled that the basic postulate in this book is *not* the earlier so-called ergodic hypothesis—that the system will eventually pass through every set of values of positions and momenta consistent with its energy—but rather the assumption that the gas is always molecular-disordered (§3).

velocity c through the gas mixture would in unit time traverse a path of length c, and since during this time it would have collided $(\nu_c+\mathfrak{n}_c)$ times with other molecules, it would travel between one collision and the next the average distance:[1]

$$\lambda_c = \frac{c}{\nu_c + \mathfrak{n}_c}. \tag{78}$$

Since at any instant all molecules that have velocity c are subject to the same conditions, λ_c is also the distance that such a moleucle will travel from any instant to its next collision. When at any specified instant many m-molecules all have the same velocity c, and we take the average of all paths traversed by each of them from that instant to its next collision, we obtain again the same value λ_c. The same holds of course if we go backwards in time. At a specified time t, many m-molecules should have velocity c. If we now ask, how great is the distance which these molecules have travelled on the average from their last collision to time t, we would again obtain the value λ_c.

A false inference has been drawn from this, which Clausius* has clarified, and which deserves to be mentioned. We again consider an m-molecule which moves for a long time with the persistent velocity c. At some instant t it finds itself at the point B. We ask for the distance of the point B from the place where the molecule last collided, and take the average of all these distances for all possible locations of the point B. This would be equal to λ_c.

Likewise we could ask for the distance of B from the place where the molecule will suffer its first collision after the time t. The average of the latter distances would again be equal to λ_c. Since however the sum of the distances of B from the preceding and following collisions would be equal to the path between the two collisions, one might think that the mean distance between

[1] On substituting the values for ν_c and $\mathfrak{n}_c$, one easily sees that the quantity λ_c approaches the limit λ_r (Eq. 60) with increasing c. In fact, when the molecule considered moves with a very large velocity, the others must behave as if they were at rest. The mean free path is of course unchanged when all the velocities are increased or decreased by the same amount; λ therefore does not vary with temperature at constant density, provided the molecules can still be considered as negligibly deformable elastic bodies.

* R. Clausius, Ann. Physik [2] **115**, 1 (1862); see in particular p. 432 of the English translation in Phil. Mag. [4] **23** (1862).

two successive collisions would be equal to $2\lambda_c$. This conclusion is false, since the probability that B lies on a longer path is greater than the probability that it lies on a shorter one. If one takes the average of all paths between two successive collisions, then the shorter path-distances will be relatively more frequent than if one gives the point B all possible positions on the entire path of the m-molecule and then takes the mean of the various distances of the point B from the next collision following or preceding.

A trivial example will illustrate this much better than a long explanation. We shall make several throws with a true die; between two throws of "one" there will be on the average five others. We consider some interval J between two successive throws. Between the interval J and the next throw of "one" there will be on the average not $2\frac{1}{2}$ but rather 5 other throws. Likewise between J and the last preceding throw of "one" there will be on the average 5 throws.

Tait has defined the mean free path λ in a somewhat different manner. We saw above that at a specified time t there are in unit volume dn_c molecules whose velocity lies between c and $c+dc$, and that all these molecules travel on the average a distance λ_c from any instant of time to their next collision. If we consider all the nm-molecules that are in unit volume at any time, and take the average of all the paths that they traverse from that time to their next collision, then we obtain:

$$\lambda_T = \frac{1}{n}\int \lambda_c dn_c = \frac{1}{n}\int \frac{c\,dn_c}{\nu_c + \mathfrak{n}_c}. \tag{79}$$

This gives, after substituting the values (70) and (71) and making some easy reductions:

$$\lambda_T = \frac{1}{\pi n s^2}\int_0^\infty \frac{4x^2 e^{-x^2}dx}{\psi(x) + \dfrac{n_1\sigma^2}{ns^2}\psi\left(x\sqrt{\dfrac{m_1}{m}}\right)}, \tag{80}$$

where

$$\psi(x) = \frac{1}{x}\,e^{-x^2} + \left(2 + \frac{1}{x^2}\right)\int_0^x e^{-x^2}dx. \tag{81}$$

The expression (80) reduces, when only m-molecules are present, to:

$$\lambda_T = \frac{1}{\pi n s^2} \int_0^\infty \frac{4x^2 e^{-x^2 dx}}{\psi(x)} .$$

I found the value of the definite integral to be 0.677464.[2] Tait obtained a value that agrees to the first three decimal places.[3] Hence:

$$\lambda_T = \frac{0.677464}{\pi n s^2} . \tag{82}$$

One sees easily that the quantity λ_T must be somewhat smaller than the mean value earlier denoted by λ. Previously λ was the mean of all paths traversed by all the molecules in unit volume in unit time. Each molecule would thus contribute as many paths to the arithmetic mean as the number of collisions it experienced in unit time. By Tait's method, however, each molecule contributes only one path. Since the faster molecules collide more frequently and also travel a longer distance from one collision to the next than do the slower ones, the first method counts the longer paths relatively more often; hence the average must be larger than by the second method.

Tait points out that one could also define the mean path as the product of the mean time between two collisions and the mean velocity, which would give

$$\bar{c} \cdot \int \frac{dn_c}{\nu_c + \mathfrak{n}_c} ,$$

whence for a simple gas one gets:

$$0.734/\pi n s^2 .$$

The mean duration of time between two collisions could similarly be defined in a different way; but we have already spent too much time on these less important concepts, which can only be excused by the attempt to attain the clearest possible understanding of the fundamental concepts.

When we obtain different values for the mean free path, it is clearly not the fault of an error in the calculation. Each value is exact on the basis of its own definition. If any precisely performed

---

[2] Boltzmann, Wien. Ber. **96**, 905 (Oct. 1887).

[3] Tait, Trans. R. S. Edinburgh **33**, 74 (1886).

calculation leads to a final formula that contains the mean free path, it will be evident from the calculation itself which definition should be used. Only when the calculation by which the formula is obtained is inexact can there be any doubt.

§11. Basic equation for the transport of any quantity by the molecular motion.

We now consider a vertical cylindrical column of a simple gas, whose molecules have mass m. We draw the z-axis vertically upwards, the plane $z=z_0$ being the bottom and the plane $z=z_1$ the top of the gas column. As usual we assume that the distance between these two planes is small compared to the cross-section of the gas column, so that the effect of the wall that bounds the gas column on the side can be ignored. Let Q be any quantity which a gas molecule can have in different amounts. The top of the container shall now have the property that each molecule, however it may behave before the collision, will have after reflection the average amount G_1 of this quantity Q. Likewise each molecule will rebound from the bottom with the average amount G_0 of this quantity. For example, if the molecules were spheres of diameter s which conduct electricity, and the top and bottom were two metal plates, which had constant potentials one and zero, then each molecule would rebound from the bottom unelectrified, but would leave the top with electrical charge $s/2$. The quantity Q would then be the amount of electricity, and one would have the process of electrical conduction. If the bottom were at rest, and the top were moving in the direction of the abscissa axis in its own plane, then one would have the process of viscosity and Q would be the momentum measured in the abscissa direction. If the top and bottom were at two different temperatures, one would have heat conduction in the gas.

To be specific, we shall assume that G_1 is larger than G_0. For any z, and hence in any layer parallel to the xy plane lying between the top and the bottom, which we shall call the layer z, each molecule has on the average the amount $G(z)$ of this quantity.

We imagine in this layer a piece AB of unit surface area; a molecule that passes through AB from above to below will have experienced its last collision before its passage through AB in a higher layer.

We say briefly that it comes from a higher layer. Hence on the average it will have an amount of the quantity Q larger than $G(z)$. The molecules that pass through AB from below will carry on the average a smaller amount of this quantity with them, so that in unit time a definite amount Γ of Q will be transported from above to below, and the determination of this amount Γ is our next problem. We consider, out of all our molecules, only those whose velocities are between c and $c+dc$. There will be dn_c of these in unit volume. Of these, according to Equation (38),

$$dn_{c,\vartheta} = \frac{dn_c \sin\vartheta d\vartheta}{2}$$

move in such a way that the direction of their velocity forms an angle between ϑ and $\vartheta+d\vartheta$ with the negative z-axis. Each molecule traverses during time dt a path of length cdt, which forms an angle ϑ with the negative z-axis.

Hence there will pass through AB in time dt as many of the molecules considered as may lie at the beginning of dt in an oblique cylinder whose base is AB, and whose volume is $c \cos\vartheta dt$. This latter number is just

$$\frac{dn_c}{2} c \sin\vartheta \cos\vartheta d\vartheta dt$$

(cf. the derivation of Equation (3) in §2).

Thus during unit time, when the state is stationary,

$$d\mathfrak{N} = \tfrac{1}{2} dn_c c \sin\vartheta \cos\vartheta d\vartheta$$

molecules will pass through AB from above to below, whose velocities are between c and $c+dc$, and make an angle between ϑ and $\vartheta+d\vartheta$ with the negative z-axis. If we consider a particular one of these molecules, which passes through AB at time t, and denote the path it has traversed from its last previous collision to the instant t by λ', then clearly it comes from a layer with z-coördinate $z+\lambda' \cos\vartheta$, where each molecule possesses on the average the amount $G(z+\lambda' \cos\vartheta)$ of the quantity Q; it will therefore transport through AB this amount, which we can set equal to

$$G(z) + \lambda' \cos\vartheta \frac{\partial G}{\partial z}$$

since λ' is small.

All the $d\mathfrak{N}$ molecules considered above will therefore transport the amount

$$d\mathfrak{N}\cdot G(z) + \frac{\partial G}{\partial z}\cos\vartheta \sum \lambda'$$

through AB from above to below, where $\sum \lambda'$ is the sum of the paths of all the $d\mathfrak{N}$ molecules. We can set $\sum \lambda'$ equal to the product of the number $d\mathfrak{N}$ of these molecules and the mean free path. This mean free path, according to the remark following Equation (78) in the text, is equal to the quantity denoted by λ_c. Hence $\sum \lambda' = \lambda_c d\mathfrak{N}$ and we obtain for the amount of Q transported through unit surface from above to below by the $d\mathfrak{N}$ molecules in unit time the result:

$$d\mathfrak{N}\cdot\left[G(z) + \lambda_c \cos\vartheta \frac{\partial G}{\partial z}\right].$$

If we substitute for $d\mathfrak{N}$ its value, noting that dn_c, λ_c, G, and $\partial G/\partial x$ are not functions of ϑ, and integrate over ϑ from 0 to $\pi/2$, then we obtain for the total amount of Q transported by the molecules whose velocities are between c and $c+dc$, from above to below, the value:

$$\frac{c}{4}dn_c G(z) + \frac{c\lambda_c dn_c}{6}\frac{\partial G}{\partial x}. \tag{83}$$

Similarly we find that the molecules whose velocities lie between the same limits transport from below to above the amount

$$\frac{c}{4}dn_c G(z) - \frac{c\lambda_c dn_c}{6}\frac{\partial G}{\partial x}. \tag{84}$$

Hence all these molecules will transport the net amount

$$d\Gamma = \frac{c\lambda_c dn_c}{3}\frac{\partial G}{\partial z} \tag{85}$$

more from above to below than in the reverse direction. If we make the simplifying assumption that all molecules have the same velocity c, then the velocities of all molecules present lie between c and $c+dc$. One therefore has to replace dn_c by n, and λ_c by the mean free path of each of these molecules. Then $d\Gamma$ will be identical with the total amount Γ of Q transported in unit time through unit surface by the molecules from above to below, in excess of the amount transported in the opposite direction. Therefore we obtain

$$\Gamma = \frac{n}{3} c\lambda \frac{\partial G}{\partial z} = \frac{c}{4\pi s^2} \frac{\partial G}{\partial z}, \tag{86}$$

since here the Clausius mean free path formula is applicable.

If we do not make the simplifying assumption that all molecules have the same velocity, then we will obtain Γ by integrating the above value of $d\Gamma$ over all possible values. Equation (78) gives (since there is only one kind of gas)

$$\lambda_c = \frac{c}{\mathfrak{n}_c}.$$

If we substitute for $\mathfrak{n}_c$ and dn_c their values from Equations (71) and (43), then after some easy reductions we get:

$$I = \frac{1}{3\pi s^2} \frac{1}{\sqrt{hm}} \frac{\partial G}{\partial z} \int_0^\infty \frac{4x^3 e^{-x^2} dx}{\psi(x)}, \tag{87}$$

where $\psi(x)$ is the function defined by Equation (81).

I found the value 0.838264 for the definite integral, by mechanical quadrature.[1] Tait calculated it later to three decimal places, which agree with my value.[2]

From Equations (44), (45), and (47) it follows that:

$$\frac{1}{\sqrt{hm}} = c_w = \frac{\sqrt{\pi}}{2} \bar{c} = \sqrt{\frac{2}{3}} \sqrt{\overline{c^2}}.$$

Likewise it follows from Equations (67), (77), and (82) that:

$$\frac{1}{\pi s^2} = \lambda n \sqrt{2} = \frac{n\lambda_T}{0.677464} = \frac{4}{3} n\lambda_{\text{Claus}}.$$

When we substitute for $1/\sqrt{hm}$ and $1/\pi s^2$ any of these values, we obtain each time an equation of the form

$$\Gamma = knc\lambda \frac{\partial G}{\partial z}, \tag{88}$$

where c is either the most probable or the mean velocity or the

---

[1] Boltzmann, Wien. Ber. **84**, 45 (1881).

[2] Tait, Trans. R. S. Edinburgh **33**, 260 (1887).

square root of the mean square velocity, λ is the mean free path according to Maxwell's, Tait's, or Clausius' definition, and k is a constant which is different in each case. If we understand by c the mean velocity and by λ the Maxwell mean free path, then:

$$k = \frac{1}{3}\sqrt{\frac{\pi}{2}}\int_0^\infty \frac{4x^3}{\psi(x)} e^{-x^2} dx = 0.350271. \tag{89}$$

The coefficient therefore differs only slightly from the coefficient $\frac{1}{3}$ in Equation (86).

§12. Electrical conduction and viscosity of the gas.

We shall first consider intentionally an example in which the quantity Q is not a purely mechanical property of the molecule. Let the bottom and top of the container be two plates which are both good electrical conductors, maintained at the constant potentials 0 and 1. The distance between the bottom and top is one. The effect of the side walls will be neglected, as usual. We consider this problem simply as an exercise, and we therefore assume that the spherical gas molecule is a good conductor of electricity and that its electrical charge does not affect its molecular motion, without pretending that these conditions could be realized in nature. G is then the electricity accumulated on a molecule. For molecules reflected from the bottom it has the value $G_0=0$, and for those reflected from the top the value $G_1=s/2$. For the latter, the electrical potential in the interior and at the surface must be equal to 1. This electrical potential is equal to the quantity of electricity G_1 divided by the radius $s/2$. If the state is to be stationary, then Γ must have the same value for each cross-section. Since we assume that the molecular motion is not disturbed by the electrification, the other quantities appearing in Equation (88) have the same value for each cross-section, and it follows from this equation that $\partial G/\partial z$ is independent of z. If the distance between top and bottom is equal to 1, then:

$$\frac{\partial G}{\partial z} = \frac{s}{2}.$$

The amount of electricity transported in unit time through unit surface by molecules from above to below in excess of that transported from below to above is therefore, according to Equation (88):

$$\Gamma = \frac{k}{2} nc\lambda s. \tag{90}$$

According to our assumptions (which of course are not proved) this would be the electrical conductivity of the gas.

We shall now treat another example. The bottom will be at rest, but the top is being displaced in the abscissa direction at constant velocity. Consequently the gas molecules near the top are dragged along, while those near the bottom are held back. The mean velocity component of a molecule in the abscissa direction, i.e., the visible velocity of the gas in this direction, will therefore increase with increasing values of the z-coördinate. For the layer z it has the value u. We now understand by G the average momentum mu of a molecule in the abscissa direction, and hence we obtain:

$$\frac{\partial G}{\partial z} = m\frac{\partial u}{\partial z}, \qquad \Gamma = knc\lambda m\frac{\partial u}{\partial z} = k\rho c\lambda\frac{\partial u}{\partial z}\cdot$$

If we denote the total mass of gas between the bottom and the layer z by M, and the velocity of its center of mass in the abscissa direction by $\mathfrak{x}$, then

$$\mathfrak{x} = \frac{\sum m\xi}{M},$$

where $\sum m\xi$ is the sum of the momenta of all particles in the abscissa direction. As a result of the molecular motion of the gas, more momentum Γ will be transported downwards than upwards in unit time through unit surface. Hence during the time dt the quantity $\sum m\xi$ experiences the increment

$$\Gamma\omega dt$$

while M remains unchanged. Here ω is the surface area of the cross section of our gas cylinder. Hence $\mathfrak{x}$ increases by

$$d\mathfrak{x} = \frac{1}{M}\Gamma\omega dt$$

because of the molecular motion. The same increase would occur if the force $Md\mathfrak{x}/dt$ acted on the gas. If the state is to be stationary, an equal but opposite force must act from outside on the gas mass M. This can only come from the bottom, and since the action and reaction are equal, the gas exerts on the bottom the force

$$M \frac{d\mathfrak{x}}{dt} = \Gamma\omega = k\rho c\lambda\omega \frac{\partial u}{\partial z}$$

in the positive abscissa direction. This force is gas viscosity. It is proportional to the surface ω, and to the differential quotient of the tangential velocity u by the normal z.

The proportionality constant is the viscosity coefficient. It has the value

$$\mathfrak{R} = k\rho c\lambda. \tag{91}$$

For air at 15°C and normal barometric pressure, the experiments of Maxwell,[1] O. E. Meyer,[2] and Kundt and Warburg[3] give, nearly in agreement with each other,

$$\mathfrak{R} = 0.00019 \frac{\text{grams}}{\text{cm sec}} \cdot$$

Since oxygen and nitrogen have rather similar properties and the formula is only approximately correct in any case, we can set this equal to the viscosity constant of nitrogen. For this we find at 0°C, $\sqrt{\overline{c^2}} = 492$ m. Since $\bar{c} = 2\sqrt{(2\overline{c^2}/3\pi)}$ and since $\bar{c}$ is proportional to the square root of the absolute temperature, it follows for nitrogen at 15°C that:

$$\bar{c} = 467 \text{ m}.$$

If one understands by c in Equation (91) the mean velocity, so that one should set $k = 0.350271$, he then obtains:

$$\lambda = 0{,}00001 \text{ cm}.$$

For the number of collisions that a nitrogen molecule experiences per second at 15°C and at normal barometric pressure, one obtains:

$$\mathfrak{n} = \frac{\bar{c}}{\lambda} = 4700 \text{ Million}.$$

---

[1] Maxwell, Phil. Trans. **156,** 249 (1866); Scientific Papers **2,** 24.

[2] O. E. Meyer, Ann. Physik [2] **148,** 226 (1873).

[3] Kundt and Warburg, Ann. Physik [2] **155,** 539 (1875).

Since according to Equation (77)

$$\lambda = \frac{1}{\sqrt{2}\pi n s^2}$$

the two quantities n and s cannot be determined individually. They can however be found as soon as another relation between these quantities is known.

According to Loschmidt[4] this can be accomplished by the following arguments, whose validity he justifies by consideration of the molecular volumes of various substances. The volume of a molecule considered as a sphere is $\pi s^3/6$. If one does not have such a simple picture of a molecule, then he may consider this to be the volume of a sphere whose diameter is equal to the minimum distance of approach of the centers of gravity of two molecules in a collision. Thus $\pi n s^3/6$ is the fraction of the total gas volume (set equal to 1) that will be filled with molecules, when one regards them as spheres of the above size, while the space $1-(\pi n s^3/6)$ between them remains empty.

We assume that the gas can be liquefied and that in the liquid state the total volume is ϵ times larger than the space occupied by the molecules; then $\epsilon \pi n s^3/6$ is the volume of the liquid resulting from liquefaction of the gas, and since the volume of the gas was equal to one, we have

$$\frac{\epsilon \pi n s^2}{6} = \frac{v_f}{v_g},$$

where v_g is the volume of an arbitrary quantity of gas of density such that there are n molecules in unit volume, whereas v_f is the volume of the same amount of gas in the liquid state.* On multiplying this last equation by Equation (77), one obtains:

$$s = \frac{6\sqrt{2}}{\epsilon} \frac{v_f}{v_g} \lambda.$$

Now the volume of a liquid is not significantly changed by either pressure or temperature, and furthermore the force that two gas molecules exert on each other in a collision is probably greater than that which we can exert on liquids in our labora-

[4] Loschmidt, Wien. Ber. **52**, 395 (1865).

* The subscript f stands for *flüssig*, "liquid" or "fluid."

tories.[5] Hence we may well assume that the volume of a liquid is not more than ten times as large, and generally not smaller, than it would be if two neighboring molecules were at that distance which is in the gas their minimum distance in a collision, so that ϵ is between 1 and 10. The density of liquid nitrogen was found by Wroblewsky to be not much different from that of water.[*] Also, from the atomic volume it follows that the difference between the two densities cannot be so large that it need be taken into account in this approximate calculation. If we set the two equal, then we find for nitrogen at 15° and atmospheric pressure: $(v_g/v_f) = 813$; and we obtain, when we set $\epsilon = 1$, $s = 0.0000001$ cm = one millionth of a millimeter. Hence we may take it as probable that the mean distance of the centers of gravity of two neighboring molecules in liquid nitrogen, as well as the smallest distance to which two colliding molecules in gaseous nitrogen can approach on the average, lies between this value and the tenth part of it.

For the number $n = (1/\sqrt{2}\pi s^2 \lambda)$ of molecules in 1 cc of nitrogen at 25°C at atmospheric pressure, one obtains a number which in any case falls between $2\frac{1}{2}$ and 250 trillions.

Substitution of this value into Equation (90) gives: $\Gamma = (23 \cdot 10^9/\text{sec})$. This would be the absolute conductivity in electrostatic units. The electromagnetic specific resistance would therefore be:

$$(9 \cdot 10^{20}\ \text{cm}^2/\Gamma\ \text{sec}^2) = (4 \cdot 10^{10}\ \text{cm}^2/\text{sec}).$$

A cube of nitrogen, 1 cm on each side, has resistance $(4 \cdot 10^{10}\ \text{cm/sec}) = (40\ \text{ohm})$ while an equal cube of mercury has resistance (1/10600) ohm. Since nitrogen actually is a much poorer conductor than mercury, it follows that the hypothesis that the molecule is a conducting sphere is incorrect.

The order of magnitude of the diameter of a molecule was later calculated by Lothar Meyer,[6] Stoney,[7] Lord Kelvin,[8] Maxwell,[9] and van der Waals,[10] and they found, by several completely different methods, values in agreement with the one above.

[5] Boltzmann, Wien. Ber. **66**, 218 (July 1872).
[*] Wroblewsky, C. R. Paris **102**, 1010 (1886).
[6] L. Meyer, Ann. Chem. Pharm. **5** (Suppl.) 129 (1867).
[7] Stoney, Phil. Mag. [4] **36**, 132 (1868).
[8] Kelvin, Nature **1**, 551 (March 1870); Amer. J. Sci. **50**, 38 (1870).
[9] Maxwell, Phil. Mag. [4] **46**, 463 (1873); Scientific Papers **2**, 372.
[10] Van der Waals, *Die Continuität des Gasförmigen und Flüssigen Zustandes* (Leipzig, 1881), Chap. 10.

In order to find the dependence of the viscosity coefficient on the nature and state of the gas in question, we replace ρ by nm, and λ by its value according to Equation (77). One obtains:

$$\mathfrak{R} = \frac{km\bar{c}}{\sqrt{2}\,\pi s^2},$$

and by virtue of Equations (46) and (51a):

$$\mathfrak{R} = \frac{2k}{s^2}\sqrt{\frac{RMTm}{\pi^3}}\,.$$

Thus the viscosity coefficient is independent of the density of the gas, and proportional to the square root of its temperature. The independence of density, which is of course valid only as long as the condition of our calculation is satisfied—namely, that the mean free path is small compared to the distance between top and bottom—was verified by experiment, especially by Kundt and Warburg.* As far as the dependence on temperature is concerned, Maxwell's experiment gave a viscosity coefficient proportional to the first power of the temperature (*loc. cit.*), which is correct only for easily coercible gases, especially carbon dioxide. For less coercible gases, several later observers found close agreement with the formula developed here for the temperature coefficient of the viscosity though most values lie between the value of this calculation and the one found experimentally by Maxwell.[11]

The first remark to be made is that a more rapid increase of the viscosity with temperature than the square root of the absolute temperature cannot be attributed to the inaccuracy of our calculation, since one perceives immediately the following: when one raises the temperature without changing the density, then on the assumption of elastic, negligibly deformable molecules, the molecular motion is on the average completely unchanged except that its velocity increases in proportion to the square root of the absolute temperature. It is as though time were shortened in this ratio, and hence it follows that the amount of momentum trans-

---

* Kundt and Warburg, Ann. Physik [2] **155**, 337, 525, **156**, 177 (1875).

[11] Cf. O. E. Meyer, *Die kinetische Theorie der Gase* (Breslau: Maruschke & Berendt, 1877), p. 157 ff. [See also Hirschfelder, Curtiss, and Bird, *Molecular Theory of Gases and Liquids* (New York: Wiley, 1954), Chap. 8.—Tr.]

ported in unit time must increase by the same amount. On the other hand, according to Stefan,[12] s could decrease with increasing temperature. This would have the following meaning. The molecules are not absolutely rigid, rather they are somewhat flattened by the collision, so that their diameter appears to decrease, and indeed it decreases more the higher the temperature of the gas. Maxwell* assumes that the molecules are centers of force, which exert on each other a force that is negligible at large distances, but as they approach becomes a very rapidly increasing repulsive force, which is to be chosen as an appropriate function of the distance. In order to explain the temperature coefficient of the viscosity which he found, he sets this function equal to the inverse fifth power of the distance. I once remarked that one could obtain all the essential properties of a gas by substituting for this repulsive force a purely attractive force which is an appropriate function of the distance, and one could thereby explain dissociation phenomena and the well-known Joule-Thomson experiment.† Because of our ignorance as to the nature of molecules, all these models must of course be considered simply as mechanical analogies, which one must regard as all being on an equal footing, as long as experiment has not decided between them. In any case, however, it is probable that the diameter of a molecule is not a precisely defined quantity. Nevertheless, the neighboring molecules in the liquid state must be at such a distance that they interact strongly with each other, and the interaction of more than two of them is no longer an exceptional case. Hence they are at distances about the same as that at which gas molecules already experience a significant deviation from rectilinear paths. The quantities denoted above by s and σ represent nothing more than the order of magnitude of this distance. In order that the calculation may remain meaningful, we shall return to the assumption that the molecules are almost undeformable elastic spheres. Then it follows from the last formula for the viscosity coefficient that it is proportional to the square root of the mass of a molecule, for different gases at the same temperature, and inversely proportional to the square root of the molecular diameter.

---

[12] Stefan, Wien. Ber. **65** (2) 339 (1872).

* Maxwell, Phil. Trans. **157,** 57 (1867); Scientific Papers **2,** 36.

† Boltzmann, Wien. Ber. **89,** 714 (1884).

§13. Heat conduction and diffusion of the gas.

In order to calculate the heat conductivity from Equation (88), we have to assume that the top and bottom planes are maintained at two different temperatures. G is then the average heat contained by a molecule. The mean kinetic energy of progressive motion of a molecule is

$$\frac{m}{2}\overline{c^2}.$$

The total average energy of internal motion of a molecule will be set equal to

$$\beta\frac{m}{2}\overline{c^2},$$

hence the total average molecular motion is

$$\frac{1+\beta}{2}m\overline{c^2},$$

or, by virtue of Equation (57),

$$\frac{1}{3(\kappa-1)}m\overline{c^2}.$$

Since according to our hypothesis heat is nothing but the total energy of molecular motion, the amount of heat G pertaining to a molecule will be measured in mechanical units. If we assume that the ratio of specific heats, κ, is constant, which is probably true at least for the permanent gases, then

$$\frac{\partial G}{\partial z} = \frac{1}{3(\kappa-1)}m\frac{\partial\overline{c^2}}{\partial z}.$$

Now according to Equation (51a),

$$\overline{c^2} = \frac{3RT}{\mu},$$

where, as before, $\mu=(m/M)$ is the molecular weight of the gas. We thereby obtain

$$\frac{\partial G}{\partial z} = \frac{Rm}{(\kappa-1)\mu}\frac{\partial T}{\partial z},$$

hence according to Equation (88),

$$\Gamma = \frac{kR\rho\bar{c}\lambda}{(\kappa - 1)\mu} \frac{\partial T}{\partial z}.$$

The coefficient of $\partial T/\partial z$ is what one calls the heat conductivity $\mathfrak{L}$ of the gas. It therefore follows that

$$\mathfrak{L} = \frac{R\mathfrak{R}}{(\kappa - 1)\mu} = \frac{2k}{(\kappa - 1)s^2}\sqrt{\frac{R^3M^3T}{\pi^3 m}}. \tag{92}$$

The dependence of the heat conductivity on density and temperature is therefore, as long as κ is constant, the same as that of the viscosity. In particular, since κ depends hardly at all on density for permanent gases at constant temperature, the heat conductivity is independent of density, which was confirmed by the experiments of Stefan* and of Kundt and Warburg.†

For different gases for which κ has nearly the same value, the heat conductivity coefficient is proportional to the quotient of the viscosity coefficient divided by the molecular weight—or, as the last expression in Equation (92) shows, inversely proportional to the square of the diameter and to the square root of the molecular weight. Hence it is significantly larger for the smaller and lighter molecules than for the larger ones. This has been confirmed by experiment.

If we denote by γ_p and γ_v the specific heats of the gas referred to constant mass, at constant pressure and constant volume, respectively, where heat is again to be measured in mechanical units, then we have (Eq. 55a)

$$\frac{R}{\mu} = \gamma_p - \gamma_v = \gamma_v(\kappa - 1) = \frac{\gamma_p}{\kappa}(\kappa - 1),$$

hence

$$\mathfrak{L} = \gamma_v \mathfrak{R} = \frac{1}{\kappa}\gamma_p \mathfrak{R}. \tag{93}$$

In the last formula the unit of heat is arbitrary. If one puts, for air at 0°C and atmospheric pressure,

$$\kappa = 1.4, \quad \gamma_p = 0.2376 \frac{\text{g}\cdot\text{Calor.}}{(\text{gram mass}) \times (1°\text{ C.})}$$

* Stefan, Wien. Ber. **65,** 45 (1872); **72,** 69 (1876).

† Kundt and Warburg, Ann. Physik [2] **155,** 337, 525, **156,** 177 (1875).

and for $\mathfrak{N}$ its value given above, then it follows that:

$$\mathfrak{L} = 0.000032 \frac{\text{g}\cdot\text{Calor.}}{\text{cm/sec. } 1^\circ \text{ C.}}\,.$$

For the heat conductivity of air, different observers have found values lying between 0.000048 and 0.000058 in the above units.[1] Taking account of the fact that our calculation is only an approximate one, we consider this agreement sufficiently good.

In order to calculate the diffusion of two gases, we shall again return to the gas cylinder considered in §11. Let the gas be a mixture of two simple gases. A molecule of the first kind of gas has mass m, and diameter s; a molecule of the second kind has mass m_1 and diameter s_1. In the layer z there are (in unit volume) n molecules of the first and n_1 of the second kind of gas, where n and n_1 are to be functions of z. Also, the number dn_c of molecules of the first kind for which the magnitude of the velocity lies between c and $c+dc$ will be a function of z. One then finds by considerations similar to those in §11 that in unit time

$$d\mathfrak{N}_{c,\vartheta} = \frac{dn_c}{2} c \sin\vartheta \cos\vartheta d\vartheta$$

molecules of the first kind move through unit surface with velocities between c and $c+dc$, the angle between their direction of motion and the negative z-axis lying between ϑ and $\vartheta+d\vartheta$. These molecules come on the average from a layer whose z-coördinate has the value $z+\lambda_c \cos\vartheta$, for which therefore we can write in place of dn_c:

$$dn_c + \lambda_c \cos\vartheta \frac{\partial dn_c}{\partial z}\,.$$

If one integrates over ϑ from 0 to $\pi/2$, then the number of molecules of the first kind of gas that pass in unit time through unit surface at any angle but with velocity between c and $c+dc$ has the value:

$$\frac{cdn_c}{4} + \frac{c\lambda_c}{6}\frac{\partial dn_c}{\partial z}\,;$$

[1] O. E. Meyer, *Die Kinetische Theorie der Gase*, p. 194. From Winkelmann's experiments Kutta found, according to an improved approximate formula, the value 0.000058 (Münchn. Dissert. 1894; Ann. Physik [3] **54**, 104 [1895]).

likewise the number of molecules that pass from below to above has the value:

$$\frac{cdn_c}{4} - \frac{c\lambda_c}{6}\frac{\partial dn_c}{\partial z}$$

Hence there will be a net flow of

$$d\mathfrak{N}_c = \frac{c\lambda_c}{3}\frac{\partial dn_c}{\partial z} \tag{94}$$

molecules of the first kind of gas from above to below. On the simplifying assumption that the velocities of all molecules are the same, one would have in place of $d\mathfrak{N}_c$ simply the total number $\mathfrak{N}$ of molecules of the first kind that pass in unit time through unit surface from above to below, in excess of those that pass from below to above, and instead of dn_c one would have simply the total number n of molecules of the first kind in unit volume in the layer z. One would then have:

$$\mathfrak{N} = \frac{c\lambda}{3}\frac{\partial n}{\partial z}. \tag{95}$$

The occurrence of different velocities for molecules of the same kind will be taken account of only in the simplest case, where both kinds of gas have molecules of the same mass and diameter. In this case, which Maxwell calls self-diffusion, we assume that during the diffusion Maxwell's velocity distribution holds for the molecules of each kind of gas in each layer, so that Equation (43),

$$dn_c = 4n\sqrt{\frac{h^3m^3}{\pi}}\,c^2e^{-hmc^2}dc$$

remains unchanged, except that n is a function of z, whence one obtains:

$$\frac{\partial dn_c}{\partial z} = \frac{4\,\partial n}{\partial z}\sqrt{\frac{h^3m^3}{\pi}}\,c^2e^{-hmc^2}dc.$$

Moreover, λ_c has the same value as it would in a simple gas in which there are $n+n_1$ molecules in unit volume. λ_c is then given by Equation (78) in which $\nu_c=0$ but $\mathfrak{n}_c$ is given by Equation (71). In the latter equation, $n+n_1$ is to be substituted for n, and s

means the diameter of a molecule, which is the same for both kinds of gas. Substitution of all these values in Equation (94) and integration over c from 0 to ∞ yields for the total number of molecules of the first kind that pass through unit surface from above to below, in excess of those that pass from below to above, the value:

$$(96) \qquad \mathfrak{N} = \frac{1}{3\pi s^2 \sqrt{hm}\,(n + n_1)} \frac{\partial n}{\partial z} \int_0^\infty \frac{4x^3}{\psi(x)} e^{-x^2} dx,$$

a formula which one could have obtained directly from Equation (87) by replacing Γ and G by $\mathfrak{N}$ and $n/(n+n_1)$. Thus the probability that a molecule belongs to the first kind of gas may be treated exactly like the quantity Q, introduced in §11, pertaining to a molecule, and Γ then means the number of molecules of the first kind that pass through unit volume in unit time from above to below, in excess of those that pass in the opposite direction. Self-diffusion thus takes place, according to our approximate formulas, just as we imagined in §12 that electrical conduction might; one merely replaces the electrical charge of the molecule by the property of belonging to one or the other kind of gas. But there is an essential difference when one assumes that the electrical charges of two colliding molecules are equalized in a collision. However, since our formulas are constructed in such a way that after the collision any direction in space is equally probable for each molecule, it must follow from this that electrical conduction is just as rapid when the molecules behave like perfect insulators in collisions with each other, and like perfect conductors in collisions with the top and the bottom. Then electrical conduction would be completely analogous to diffusion.

If one introduces in Equation (96) the quantity k defined in Equation (89), then he obtains:

$$\mathfrak{N} = k\lambda\bar{c}\frac{\partial n}{\partial z} = \frac{\mathfrak{R}}{\rho}\frac{\partial n}{\partial z}\,.$$

If one multiplies on both sides by the constant m, then it follows that:

$$\mathfrak{N}m = k\lambda\bar{c}\frac{\partial(nm)}{\partial z} = \frac{\mathfrak{R}}{\rho}\frac{\partial(nm)}{\partial z}\,.$$

$\mathfrak{N}m$ is the net mass of the first gas that passes through the surface from above to below, while nm is the mass of the first gas in unit volume in the layer z, hence $\partial(nm)/\partial z$ is its gradient in the z direction. The factor multiplying this expression in the last equation is therefore what one calls the diffusion coefficient. This gives for air at 15°C and atmospheric pressure, based on the above value for $\mathfrak{R}$, the value 0.155 cm²/sec; while Loschmidt[2] has found values between 0.142 and 0.180 for different combinations of gas that behave like air. If one considers the dependence of the quantity ρ on temperature and pressure, then he finds that the diffusion coefficient is directly proportional to the $\frac{3}{2}$ power of the absolute temperature, and inversely proportional to the total pressure of the two gases. At equal temperature and equal total pressure, the diffusion constant for self-diffusion is, like the heat conductivity, inversely proportional to the quantity $s^2\sqrt{m}$, as one finds from Equation (96), since h and $n+n_1$ are constant.

In this simplest case of diffusion, where the mass and diameter of a molecule are the same for both gases, the aggregate of the two gases behaves like a single stationary gas. If we denote by $dN_{c,\vartheta}$, $dn_{c,\vartheta}$, and $dn^1_{c,\vartheta}$ the total number of molecules of both gases, the number of molecules of the first kind of gas, and of the second kind, respectively, for which the velocity lies between c and $c+dc$ and its direction forms an angle between ϑ and $\vartheta+d\vartheta$ with the positive z axis; then according to Equation (38):

$$dN_{c,\vartheta} = 2\sqrt{\frac{h^3m^3}{\pi}}(n+n_1)c^2e^{-hmc^2}dc\sin\vartheta d\vartheta.$$

One might think that consequently at least in this simple case our calculations would be exactly correct. However, we shall see that when the molecules are elastic spheres, the faster molecules diffuse more rapidly and the slower ones less rapidly.[3] Where n is small, that is at a place where the molecules of the other kind of gas predominate over the diffusing molecules, then for large values of c the quantity $dn_{c,\vartheta}$ will be larger than

$$\frac{n}{n+n_1}dN_{c,\vartheta},$$

whereas for smaller values of c it will be smaller than this. At the same place the reverse must be true for the other gas. Hence the exactness of the equation we obtained,

---

[2] Loschmidt, Wien. Ber. **61**, 367 (1870); **62**, 468 (1870).

[3] This follows from the manner in which g occurs in $\int_0^\infty gbdb\cos^2\vartheta$ (cf. §§18 and 21).

$$dn_{c,\vartheta} = \frac{n}{n + n_1} dN_{c,\vartheta}$$

is in doubt. Likewise it is doubtful whether among the molecules that collide in a layer (according to Clausius, the ones sent out from a layer) all directions of the velocity are equally probable.

§14. Two kinds of approximations; diffusion of two different gases.

One might think from what has been said up to now that Equation (87), and Equation (88), which was derived from it with the coefficient (89), are strictly correct; this would be an error. In fact, in deriving them we made the assumption that the velocity distribution would not be altered by the quantity Q associated with the molecules. In many cases, for example viscosity, when the visible motion is small compared to the mean velocity of a molecule, the velocity distribution will be altered only slightly; yet the value of the quantity dn_c in Equation (83) will always be different from its value dn_c' in Equation (84). Hence there is added to the expression (85) a term of the form

$$\frac{c}{4} G(z)(dn_c - dn_c')$$

which is of the same order of magnitude as the expression (85) itself. Also, the assumption that all directions of motion of a molecule are equally probable is doubtful.

We assumed finally that each molecule transports through the surface AB that amount $G(z+\lambda' \cos \vartheta)$ of the quantity Q which a molecule would possess on the average in the layer in which it experienced its last collision. This assumption is also arbitrary. This amount can differ for molecules that leave the layer in different directions and with different velocities; it can therefore be a function Φ of c and ϑ, so that $\partial G/\partial z$ cannot be taken out in front of the integral sign in the following integrations over ϑ and c. The amount of Q transported by a molecule through AB would then depend not only on the layer where it last collided, but also on the place where it collided the next to last time, and perhaps also on the place of the collision before that.

This is related to a situation already discussed in comparing diffusion and electrical conduction. It can happen that in a collision each of the colliding molecules retains the amount of Q that it had before the collision; however, an equalization can also take place. If we call Q electricity, then the former corresponds to the case when the molecules are conducting but are covered with an insulating layer that is penetrated in collisions with the top and bottom but not in collisions between two molecules; the latter corresponds to the case when the molecules are composed of a conducting substance right up to their surface.

In these two cases the function Φ can be different, so that the transport of Q would be unequal even though the mean value of G in a layer z would be the same in both cases, namely

$$G_0 + \frac{(G_1 - G_0)(z - z_0)}{z_1 - z_0} .$$

In fact it is more probable that a molecule will continue in nearly the same rather than the opposite direction after a collision. One sees this in the formulas (201) and (203) derived later on. Hence the transport of Q is more obstructed and hence slower when it is equalized between two colliding molecules than when it is not.

Numerous researches have been carried out to take account of the terms neglected because of all these assumptions, especially by Clausius, O. E. Meyer, and Tait.* Nevertheless, in the case of elastic spheres the perturbation of the velocity distribution by viscosity, diffusion, and heat conduction has not yet been calculated exactly, so that in all the relevant formulas terms have been neglected which are of the same order of magnitude as those which determine the result, so that they are not essentially better than the ones obtained here in a simpler way.

---

* R. Clausius, Ann. Physik [2] **115,** 1 (1862); Ann. Physik [3] **10,** 92 (1880); *Die Kinetische Theorie der Gase* (Braunschweig: F. Vieweg, 1891), chap. ii. O. E. Meyer, *Die Kinetische Theorie der Gase* (Breslau, 1877). P. G. Tait, Proc. R. S. Edinburgh **33,** 65, 251 (1886–1887); Phil. Mag. [5] **25,** 172 (1888); Trans. R. S. Edinburgh **35,** 1029 (1890); Proc. R. S. Edinburgh **15,** 225 (1889). See also J. H. Jeans, Phil. Mag. [6] **8,** 700 (1904); *The Dynamical Theory of Gases* (London: Cambridge University Press, 1904). G. Jäger, Wien. Ber. **102,** 253 (1893), **105,** 97 (1896), **108,** 447 (1899), **109,** 74 (1900), **127,** 849 (1918). M. v. Smoluchowski, Rozprawy Wydzialu mat.-przyr. Ak. Um. (Krakowie) **A46,** 129 (1906).

Such omissions, which make the results mathematically incorrect in the sense that they are not logical consequences of the assumptions made, are to be distinguished (as we explained at the end of §6) from assumptions that are physically only approximately correct, for example the assumption that the duration of a collision is small compared to the time between two collisions. As a consequence of the latter assumptions the results will also be physically inexact, i.e., their validity can only be determined by experiment. However, these physical approximations do not prevent the results from being mathematically correct, since they provide the limiting case to which the law must approach ever more closely as the physical assumptions are realized more exactly.

We shall now calculate the diffusion of two gases when the mass and diameter of a molecule are different in the two gases, but only on the simplifying assumption that the velocities of all molecules of the first kind of gas are equal to c, while those of all molecules of the second kind are equal to c_1.

Equation (95) then holds for the first kind of gas. The mean free path will then of course be calculated from Equation (68). However, since the whole calculation is only an approximate one, we shall not take account of the occurrence of different velocities here, since this simplifies the calculation, and use Equation (76). Thus we obtain for the number of molecules of the first kind of gas that pass in unit time through unit surface from above to below, in excess of those going the other way,

$$\mathfrak{N} = \mathfrak{D}_1 \frac{\partial n}{\partial z},$$

whence

$$\mathfrak{D}_1 = \frac{c}{3\pi\left[s^2 n\sqrt{2} + \left(\frac{s+s_1}{2}\right)^2 n_1 \sqrt{\frac{m+m_1}{m}}\right]}.$$

Similarly one finds for the number $\mathfrak{N}_1$ of molecules of the second kind of gas that pass downwards through the surface in excess of the number passing upwards the value:

$$\mathfrak{N}_1 = -\mathfrak{D}_2 \frac{\partial n_1}{\partial z} = +\mathfrak{D}_2 \frac{\partial n}{\partial z},$$

since $(n+n_1)$ is constant in the gas as a whole. Here we have:

$$\mathfrak{D}_2 = \frac{c_1}{3\pi \left[s_1^2 n_1 \sqrt{2} + \left(\frac{s+s_1}{2} \right)^2 n \sqrt{\frac{m+m_1}{m_1}} \right]}.$$

There now arises the difficulty that the diffusion constant does not come out the same for both gases—i.e., according to the formulas, more molecules pass through each cross section in one direction than in the other. This actually happens in diffusion through a very narrow passage or a porous wall. However in our case where we assume that the mixture is at rest and the effect of the side walls is negligible, the pressure must always be equalized, and therefore according to Avogadro's law an equal number of molecules must move in each direction.

Our formula gives a false result. Similarly, Maxwell's first formula for heat conduction* gave a visible mass motion of the heat-conducting gas. Clausius† and O. E. Meyer‡ have obtained other formulas for heat conduction, which avoid this visible mass motion, but for which the pressure is different at different places in the gas. Although this actually occurs in very dilute gases, as calculation and experiment both agree for the radiometer,** such great pressure differences as would follow from the formulas are inadmissible.[1] This is therefore a clear proof of the inaccuracy of all these calculations.

---

* Maxwell, Phil. Mag. [4] **19,** 19, **20,** 21, 33 (1860).

† Clausius, Ann. Physik [2] **115,** 1 (1862); *Die Kinetische Theorie der Gase*, chap. iv.

‡Meyer, *Die Kinetische Theorie der Gase* (Breslau, 1877).

** G. J. Stoney, Phil. Mag. [5] **1,** 177, 305 (1876); **6,** 401 (1878); G. F. Fitzgerald, Sci. Trans. Dublin **1,** 57 (1878); O. Reynolds, Proc. R. S. London **28,** 304 (1879); Phil. Trans. **170,** 727 (1880); Maxwell, Phil. Trans. **170,** 231 (1880); W. Sutherland, Phil. Mag. [5] **42,** 373, 476 (1896), **44,** 52 (1897). For summaries of later work see M. Knudsen, *Kinetic Theory of Gases* (London: Methuen, 1934); L. B. Loeb, *The Kinetic Theory of Gases*, 2d ed., (New York, 1934); E. H. Kennard, *Kinetic Theory of Gases* (New York, 1938).

[1] Kirchhoff, *Vorlesungen über die Theorie der Wärme*, ed. by Max Planck (Leipzig: B. G. Teubner, 1894, p. 210).

In the case of diffusion, with which we now concern ourselves, O. E. Meyer* has removed the contradiction by superposing on the molecular motion calculated here—in which $\mathfrak{N}-\mathfrak{N}_1$ more molecules pass through unit surface in unit time downwards than upwards—an equal but oppositely directed flow of the mixture. Since the mixture consists of $n+n_1$ molecules, n of the first kind and n_1 of the second, the flow of the mixture is imagined to be such that $n(N_1-N)/(n+n_1)$ of the first kind and $n_1(N_1-N)/(n+n_1)$ of the second kind pass downwards in excess of those going upwards. Hence according to this superposition,

$$\mathfrak{N}+\frac{n(\mathfrak{N}_1-\mathfrak{N})}{n+n_1}=\frac{n_1\mathfrak{N}+n\mathfrak{N}_1}{n+n_1}=\frac{n\mathfrak{D}_1+n_1\mathfrak{D}_2}{n+n_1}\,\frac{\partial n}{\partial z}$$

molecules of the first kind pass downwards in excess of those going upwards, and equally many molecules of the second kind pass upwards in excess of those going downwards. The diffusion coefficient is therefore now

$$\frac{n_1\mathfrak{D}_1+n\mathfrak{D}_2}{n+n_1},$$

where $\mathfrak{D}_1$ and $\mathfrak{D}_2$ have the values just found. According to this formula the diffusion coefficient would depend on the mixing ratio, and therefore would not have the same value in different layers of the gas mixture, so that for the stationary state n and n_1 would not be linear functions of z. Stefan[2] has developed another approximate theory of diffusion on other principles, and he finds that the diffusion coefficient should not depend on the mixing ratio. Experimentally the question is still open. Yet such a strong variation of the diffusion coefficient as that given by the above formula seems to be excluded.

As for the various complicated revisions undergone by these various theories of viscosity, diffusion, and heat conduction, the comparison with experimental results for different kinds of gases, and the conclusions which may be drawn about the molecular properties of various gases, I cannot go into them here. They may

---

* Meyer, *Die Kinetische Theorie der Gase.*

[2] Stefan, Wien. Ber. **65**, 323 (1872).

be found, rather exhaustively collected, in O. E. Meyer's *Kinetische Theorie der Gase*. Of the works published later, those of Tait[3] may be mentioned.*

[3] P. G. Tait, Trans. R. S. Edinburgh **33,** 65, 251 (1887); **36,** 257 (1889–1891).

* As Boltzmann remarked, a satisfactory solution of these difficulties (regarding especially the dependence of the diffusion coefficient on concentration) could not be found until the perturbation of the velocity distribution was calculated. This was first done for general types of force laws by S. Chapman, Phil. Trans. **A216,** 279 (1916), **A217,** 115 (1917) and D. Enskog, *Kinetische Theorie der Vorgänge in mässig verdünnten Gasen* (Uppsala: Almqvist and Wiksells, 1917). It was found that for elastic spheres and most other types of force law, there is a slight dependence on concentration, but not as great as that predicted by Meyer's theory. More precise values of the numerical coefficients in the expressions for viscosity and thermal conduction were also obtained, and the relation between the temperature-dependence of these coefficients and the force law was clarified (see Rayleigh, Proc. R. S. London **A66,** 68 [1900] for a deduction of the temperature dependence by dimensional arguments). But the most significant new result of the Chapman-Enskog investigations was the prediction of thermal diffusion, which was then confirmed experimentally by Chapman and Dootson, Phil. Mag. [6] **33,** 248 (1917). For details of the calculations, comparisons with experiment, and a survey of the contributions of various scientists to the theory, see the monograph by S. Chapman and T. G. Cowling, *The Mathematical Theory of Non-uniform Gases* (London: Cambridge University Press, 2d ed., 1953).

CHAPTER II

The molecules are centers of force. Consideration of external forces and visible motions of the gas.

§15. Development of partial differential equations for f and F.

We now pass to the consideration of the case when external forces act, and any arbitrary interaction may take place during collisions. In order to avoid the necessity of generalizing the formulas later on, we consider again a mixture of two gases, whose molecules have masses m and m_1 respectively. We call them again, for brevity, m-molecules and m_1-molecules respectively. Each molecule will be almost completely unaffected by the others during the greatest part of its motion; only when two molecules come unusually close to each other will there be a significant change in the magnitude and direction of their velocities. Simultaneous interactions of three molecules occur so rarely that they can be disregarded. In order to achieve a precise exposition, we think of the molecules as material points. As long as the distance r of an m-molecule from an m_1-molecule is greater than a specified very small distance σ, no interaction takes place; however, as soon as r is smaller than σ, the two molecules may exert on each other any force, whose intensity $\psi(r)$ is a function of their distance of separation, and which is sufficient to produce a significant deviation from their rectilinear paths. As soon as r becomes equal to σ we say that a collision begins. For the sake of simplicity we exclude those force laws which can cause the molecules to stick together, even though these are especially interesting since they may lead to a clarification of dissociation phenomena;* then after a short time, r will again be equal to σ, and at this instant, which we call the end of the collision, the interaction ceases. For colli-

* See Part II, chap. vi.

sions of m-molecules or of m_1-molecules with each other, we replace σ and $\psi(r)$ by s and $\Psi(r)$, s_1 and $\Psi_1(r)$, respectively. The case of elastic spheres is now a special case obtained on assuming that the functions ψ, Ψ, and Ψ_1 represent repulsive forces whose intensities increase without limit as soon as r becomes the least bit smaller than σ, s, or s_1, respectively. Everything presented up to now is therefore obtained as a special case of the equations to be developed here. In addition to these molecular forces, we now include those forces acting on the molecules that arise from external causes, called briefly external forces. We draw any fixed coördinate system in the gas. The components mX, mY, mZ of the external forces acting on any m-molecule should be independent of time and of the velocity components, and for all m-molecules they must be the same functions of the coördinates x, y, z of the molecule in question. X, Y, and Z are thus the so-called accelerating forces. The corresponding quantities for a molecule of the second kind will have the subscript 1. The external force can indeed vary from one place to another in the gas, but it should not vary markedly as long as the coördinates do not change by an amount large compared to the sphere of action (characterized by the lengths σ, s, s_1). Finally, we also do not exclude the case that the gas is in visible motion. *A priori* one can assume neither that all directions of velocity are equally probable, nor that the velocity distribution or the number of molecules in unit volume are the same in all parts of the gas, nor that they are independent of time.

We fix our attention on the parallelepiped representing all space points whose coördinates lie between the limits

$$x \text{ and } x + dx,\ y \text{ and } y + dy,\ z \text{ and } z + dz. \tag{97}$$

We set $do = dxdydz$, and we always call this parallelepiped the parallelepiped do.

We assume, following the principles mentioned earlier, that this parallelepiped can be infinitesimal yet still contain many molecules. The velocity of each m-molecule that finds itself at time t in this parallelepiped shall be represented by a line starting at the origin, and the other end point C of this line will again be called the velocity point of the molecule. Its rectangular coördinates are equal to the components ξ, η, ζ of the velocity of the molecule in those coördinate directions.

We now construct a second rectangular parallelepiped, which includes all points whose coördinates lie between the limits

(98) $$\xi \text{ and } \xi + d\xi,\ \eta \text{ and } \eta + d\eta,\ \zeta \text{ and } \zeta + d\zeta.$$

We set its volume equal to

$$d\xi d\eta d\zeta = d\omega$$

and we call it the parallelepiped $d\omega$. The m-molecules that are in do at time t and whose velocity points lie in $d\omega$ at the same time will again be called the specified molecules, or the "dn molecules." Their number is clearly proportional to the product $do \cdot d\omega$. Then all volume elements immediately adjacent to do find themselves subject to similar conditions, so that in a parallelepiped twice as large there will be twice as many molecules. We can therefore set this number equal to

(99) $$dn = f(x, y, z, \xi, \eta, \zeta, t) do d\omega.$$

Similarly the number of m_1-molecules that satisfy the conditions (97) and (98) at time t will be:

(100) $$dN = F(x, y, z, \xi, \eta, \zeta, t) do d\omega = F do d\omega.$$

The two functions f and F completely characterize the state of motion, the mixing ratio, and the velocity distribution at all places in the gas mixture. When they are given at the initial time $t=0$ for all values of their arguments, and when the external forces, the molecular forces, and the boundary conditions at the wall are also given, then the problem is completely determined, and it is completely solved as soon as one has found the values of the functions f and F for all values of t. It is always assumed here that the state is molecular-disordered. We now have to find a partial differential equation for the change of the function f during a very short time.

We shall allow a very short time dt to elapse, and during this time we keep the size and position of do and $d\omega$ completely unchanged. The number of m-molecules that satisfy the conditions (97) and (98) at time $t+dt$ is, according to Equation (99),

$$dn' = f(x, y, z, \xi, \eta, \zeta, t + dt) do d\omega$$

and the total increase experienced by dn during time dt is

(101) $$dn' - dn = \frac{\partial f}{\partial t} do d\omega dt.$$

The number dn experiences an increase as a result of four different causes.

1. All m-molecules whose velocity points lie in $d\omega$ move in the x-direction with velocity ξ, in the y-direction with velocity η, and in the z-direction with velocity ζ.

Hence through the left of the side of the parallelepiped do facing the negative abscissa direction there will enter during time dt as many molecules satisfying the condition (98) as may be found, at the beginning of dt, in a parallelepiped of base $dydz$ and height ξdt, *viz.*

$$\mathfrak{x} = \xi \cdot f(x, y, z, \xi, \eta, \zeta, t) dydzd\omega dt$$

molecules (cf. p. 33 and 88). Then, since the latter parallelepiped is infinitesimal and is infinitely near to do, the numbers $\mathfrak{x}$ and $fdod\omega$ contained in the two parallelepipeds are proportional to the volumes $\xi dydzdt$ and do of the parallelepipeds. Likewise one finds, for the number of m-molecules that satisfy the condition (98) and go out through the opposite face of do during time dt, the value:

$$\xi f(x + dx, yz, \xi, \eta, \zeta, t) dxdzd\omega dt.$$

By similar arguments for the four other sides of the parallelepiped, one finds that during time dt,

$$-\left(\xi \frac{\partial f}{\partial x} + \eta \frac{\partial f}{\partial y} + \zeta \frac{\partial f}{\partial z}\right) do \cdot d\omega dt$$

more molecules satisfying (98) enter do than leave it. This is therefore the increase V_1 which dn experiences as a result of the motion of the molecules during time dt.

2. As a result of the action of external forces, the velocity components of all the molecules change with time, and hence the velocity points of the molecules in do will move. Some velocity points will leave $d\omega$, others will come in, and since we always include in the number dn only those molecules whose velocity points lie in $d\omega$, dn will likewise be changed for this reason.

ξ, η, and ζ are the rectangular coördinates of the velocity point. Although this is only an imaginary point, still it moves like the molecule itself in space. Since X, Y, Z are the components of the accelerating force, we have:

$$\frac{d\xi}{dt} = X, \quad \frac{d\eta}{dt} = Y, \quad \frac{d\zeta}{dt} = Z.$$

Thus all the velocity points move with velocity X in the direction of the x-axis, with velocity Y in the direction of the y-axis, and with velocity Z along the z-axis, and one can employ completely similar arguments with respect to the motion of the velocity points through $d\omega$ as for the motion of the molecules themselves through *do*. One finds that, out of the velocity points belonging to m-molecules in *do*, there will enter from the left of the surface of the parallelepiped $d\omega$ parallel to the yz plane, in time dt,

$$X \cdot f(x, y, z, \xi, \eta, \zeta, t) do d\eta d\zeta dt$$

of them, while

$$X \cdot f(x, y, z, \xi + d\xi, \eta, \zeta, t) do d\eta d\zeta dt$$

of them will go out through the opposite surface. If one employs similar considerations for the other four surfaces of the parallelepiped $d\omega$, he finds that in all

$$V_2 = -\left(X \frac{\partial f}{d\xi} + Y \frac{\partial f}{\partial y} + Z \frac{\partial f}{\partial z}\right) do d\omega dt$$

more velocity points of m-molecules (in *do*) enter $d\omega$ than leave it.

Since, as noted, a molecule is included in dn when it not only lies in *do* but has its velocity point in $d\omega$, this represents the increase of dn resulting from the motion of the velocity points. But those molecules that enter *do* during the time dt, while during the same time dt their velocity points enter $d\omega$, are not taken account of, nor are those that enter *do* while their velocity points leave $d\omega$ during dt; on the other hand, those that leave *do* during dt while their velocity points enter or leave $d\omega$ are counted in V_1 as well as in V_2 and are therefore counted twice. Yet this leads to no error, since the number of all these molecules is an infinitesimal of order $(dt)^2$.

§16. Continuation. Discussion of the effects of collisions.

3. Those of our dn molecules that undergo a collision during the time dt will clearly have in general different velocity components after the collision. Their velocity points will therefore be expelled, as it were, from the parallelepiped by the collision, and

thrown into a completely different parallelepiped. The number dn will thereby be decreased. On the other hand, the velocity points of m-molecules in other parallelepipeds will be thrown into $d\omega$ by collisions, and dn will thereby increase. It is now a question of finding this total increase V_3 experienced by dn during time dt as a result of the collisions taking place between any m-molecules and any m_1-molecules.

For this purpose we shall fix our attention on a very small fraction of the total number ν_1 of collisions undergone by our dn molecules during time dt with m_1-molecules. We construct a third parallelepiped which includes all points whose coördinates lie between the limits

$$(102) \qquad \xi_1 \text{ and } \xi_1 + d\xi_1,\ \eta_1 \text{ and } \eta_1 + d\eta_1,\ \zeta_1 \text{ and } \zeta_1 + d\zeta_1.$$

Its volume is $d\omega_1 = d\xi_1 d\eta_1 d\zeta_1$; it constitutes the parallelepiped $d\omega_1$. By analogy with Equation (100), the number of m_1-molecules in do whose velocity points lie in $d\omega_1$ at time t is:

$$(103) \qquad dN_1 = F_1 do d\omega_1.$$

F_1 is an abbreviation for $F(x, y, z, \xi_1, \eta_1, \zeta_1)$.

We now ask for the number ν_2 of collisions that take place during dt between one of our dn m-molecules and an m_1-molecule, such that before the collision the velocity point C_1 of the latter molecule lies in $d\omega_1$. We again denote by C and C_1 the velocity points of the two molecules before the collision, so that the lines OC and OC_1 drawn from the origin to C and C_1 represent in magnitude and direction the velocities of the two molecules before the collision. The line $C_1C = g$ gives also the relative velocity of the m-molecule with respect to m_1, in magnitude and direction; the number of collisions clearly depends only on the relative motion. Furthermore, we assume that a collision always occurs between an m-molecule and an m_1-molecule as soon as they approach to a distance smaller than σ. The problem of finding ν_2 is thus reduced to the following purely geometrical question. In a parallelepiped do there are $dN_1 = F_1 do d\omega_1$ points. We call them again the m_1-points. Moreover, $f do d\omega$ points (the m-points) move therein with velocity g in the direction C_1C, which we call the direction g for short. ν_2 is just equal to the number of times that an m-point comes so close to an m_1-point that their distance is less than σ. Naturally we assume a molecular-disordered, i.e., completely random, distribution of the m-points and the m_1-points. In order not to have to consider those pairs of molecules that are interacting at the beginning or the end of the time dt, we assume that while dt is indeed very small, it is still large compared to the duration of a collision, just as do is very small but still contains many molecules.

In order to solve this purely geometrical problem, one can completely ignore the interaction of the molecules. The motion of the molecules during and after the collision will of course depend on the law of this interaction. However, the collision frequency can be affected by this interaction only insofar as a molecule that has already collided once during dt may collide again, with its altered velocity, during the same interval dt; yet such effects would certainly be infinitesimals of order $(dt)^2$.

We define a passage of an m-point by an m_1-point as that instant of time when the distance between the points has its smallest value; thus m would pass through the plane through m_1 perpendicular to the direction g, if no interaction took place between the two molecules. Hence ν_2 is equal to the number of passages of an m-point by an m_1-point that occur during time dt, such that the smallest distance between the two molecules is less than σ. In order to find this number, we draw through each m_1-point a plane E moving with m_1, perpendicular to the direction g, and a line G parallel to this direction. As soon as an m-point crosses E, a passage takes place between it and the m_1-point. We draw through each m_1-point a line m_1X parallel to the positive abscissa direction and similarly directed. The half-plane bounded by G, which contains the latter line, cuts E in the line m_1H, which of course again contains each m_1-point. Furthermore, we draw from each m_1-point in each of the planes E a line of length b, which forms an angle ϵ with the line m_1H. All points of the plane E for which b and ϵ lie between the limits

$$b \text{ and } b + db, \quad \epsilon \text{ and } \epsilon + d\epsilon \tag{104}$$

form a rectangle of surface area $R = bdbd\epsilon$. In Figure 6 the intersections of all these lines with a sphere circumscribed about m_1 are shown. The large circle (shown as an ellipse) lies in the plane E; the circular arc GXH lies in the half-plane defined above. In each of the planes E, an equal and identically situated rectangle will be found. We consider for the moment only those passages of

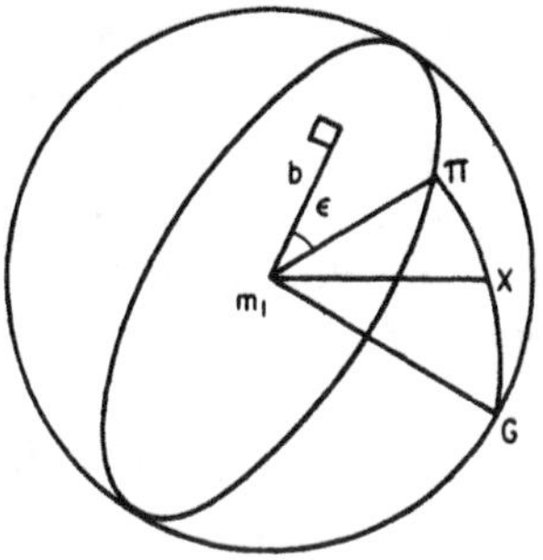

FIG. 6.

an m-point by an m_1-point in which the first point penetrates one of the rectangles R.[1] In their motion relative to m_1, each of the m-points travels the distance gdt during time dt, in a direction perpendicular to the plane of all these rectangles. Therefore during time dt all those m-points that were initially in any of the parallelepipeds whose base is one of these rectangles and whose height is equal to gdt will go through the surface of one of these rectangles. (Cf. pp. 33, 38, and 113. The state should again be molecular-disordered.) The volume of each of these parallelepipeds is therefore

$$\Pi = bdbd\epsilon gdt,$$

and since the number of m_1-points, and consequently the number of parallelepipeds, is equal to $F_1 dod\omega_1$, then the total volume of all the parallelepipeds is:

$$\sum \Pi = F_1 dod\omega_1 gbdbd\epsilon dt.$$

Since these volumes are infinitesimal, and lie infinitely close to the point with coördinates x, y, z, then by analogy with Equation (99) the number of m-points (i.e., m-molecules whose velocity points lie in $d\omega$) that are initially in the volumes $\sum \Pi$ is equal to:

$$\nu_3 = fd\omega \sum \Pi = fF_1 dod\omega d\omega_1 gbdbd\epsilon dt. \tag{105}$$

---

[1] b is the smallest possible distance of the two colliding molecules that could be attained if they moved without interaction in straight lines with the velocities they had before the collision. In other words, b is the line P_1P, where P_1 and P are the two points at which m_1 and m would be found at the moment of their closest approach if there were no interaction. Thus ϵ is the angle between the two planes through the direction of relative motion, one parallel to P_1P and the other to the abscissa axis.

This is at the same time the number of m-points that pass an m_1-point during time dt at a distance between b and $b+db$, in such a way that the angle ϵ lies between ϵ and $\epsilon+d\epsilon$.

By ν_2 we mean the number of m-points that pass an m_1-point at any distance less than σ during dt. We find ν_2 by integrating the differential expression ν_3 over ϵ from 0 to 2π, and over b from 0 to σ. Although the integration can easily be carried out, it is better for our purposes merely to indicate it. Hence we write:

$$\nu_2 = do d\omega d\omega_1 dt \int_0^\sigma \int_0^{2\pi} g \cdot b \cdot f \cdot F_1 \cdot db d\epsilon. \tag{106}$$

As we saw, ν_2 is at the same time the number of collisions of our dn molecules during dt with m_1-molecules whose velocity points lie within $d\omega_1$. The number (earlier denoted by ν_1) of all collisions of our dn molecules during dt with m_1-molecules is therefore found by integrating over the three variables ξ_1, η_1, ζ_1 whose differentials occur in $d\omega_1$, from $-\infty$ to $+\infty$; we indicate this by a single integral sign, so that we obtain:

$$\nu_1 = do \cdot d\omega \cdot dt \int \int_0^\sigma \int_0^{2\pi} f F_1 g b d\omega_1 db d\epsilon. \tag{107}$$

In each of these collisions, provided it is not merely a glancing one, the velocity point of the m-molecule involved is expelled from the parallelepiped $d\omega$, and hence dn decreases by one.

In order to find out for how many m-molecules the velocity points lie in $d\omega$ after a collision with an m_1-molecule, we simply need to ask how many collisions occur in a manner just the opposite of that considered above.

We shall consider again those collisions between m-molecules and m_1-molecules, whose number was denoted by ν_3 and is given by Equation (105). These are the collisions that occur in unit time in the volume element do in such a way that the following conditions are satisfied:

1. The velocity components of the m-molecules and the m_1-molecules lie between the limits (98) and (102), respectively, before the interaction begins.
2. We denote by b the closest distance of approach that would be attained if the molecules did not interact but retained the velocities they had before the collision. We

denote by P and P_1 the points at which they would be found at the instant when this smallest distance is realized, and by g the relative velocity before the interaction. Then b, and the angle between the two planes through g parallel to P_1P and to the abscissa axis, respectively, lie between the limits (104) (cf. footnote 1).

We call all these collisions, for short, direct collisions of the kind considered. For them the velocity components of the two molecules after the collision lie between the limits

$$(108)\quad \begin{cases} \xi' \text{ and } \xi' + d\xi',\ \eta' \text{ and } \eta' + d\eta',\ \zeta' \text{ and } \zeta' + d\zeta', \\ \xi_1' \text{ and } \xi_1' + d\xi_1',\ \eta_1' \text{ and } \eta_1' + d\eta_1,'\ \zeta_1' \text{ and } \zeta_1' + d\zeta_1'. \end{cases}$$

We denote by P_1P the smallest distance to which the two molecules would approach if they always had the same velocities as those with which they separate from each other after the collision; and by g' the relative velocity after the collision. Thus for all direct collisions the length of the segment P_1P' and the angle of the planes through g' parallel to P_1P' and to the abscissa axis, respectively, lie between the limits

$$(109)\qquad b' \text{ and } b' + db',\ \epsilon' \text{ and } \epsilon' + d\epsilon'.$$

All collisions that occur during time dt in the volume element do such that the values of the variables before the collision lie between the limits (108) and (109) will be referred to as inverse collisions. For them the direction of g' is to be reversed. They clearly follow just the opposite course as the direct collisions, and after the collision the values of the variables lie between the limits (98), (102), and (104).

Since we assume that the force law acting in the collisions is given, the values ξ', η', ζ', ξ_1', η_1', ζ_1', b' and ϵ' of all the variables after the collision can be calculated as functions of the values of the same variables ξ, η, ζ, ξ_1, η_1, ζ_1, b, and ϵ before the collision. Just as for the number of direct collisions we obtained Equation (105), so one obtains for the number of inverse collisions the value:

$$i_3 = do\,d\omega' d\omega_1' dt f' F_1' g' b' db' d\epsilon'.$$

Here we have written: $d\omega'$ for $d\xi' d\eta' d\zeta'$; $d\omega_1'$ for $d\xi_1' d\eta_1' d\zeta_1'$; and f' and F_1' for $f(x, y, z, \xi', \eta', \zeta', t)$ and $F(x, y, z, \xi_1', \eta_1', \zeta_1', t)$.

In order to be able to perform the integration, we must express all the variables as functions of ξ, η, ζ, ξ_1, η_1, ζ_1, b, and ϵ.

Later on we shall study explicitly the motion during the interaction (§21). Here we only note the following. The motion of m relative to m_1 (i.e., relative to three coördinate axes passing through m_1 and always parallel to the fixed coördinate axes, on which the quantities g, g', b, b', ϵ and ϵ' depend) we call the relative central motion. It is just the same central motion that one would obtain for the same force law by holding m_1 fixed and letting m originally have the relative velocity g in a line that has the perpendicular distance b from m_1. The latter material point must then have mass $mm_1/(m+m_1)$ instead of its actual mass. g' is nothing but the velocity of m at the end of the relative central motion; b' is the perpendicular distance of the line described by m as it recedes from m_1 at the end of the relative central motion. From the complete symmetry of any central motion it follows at once that $g'=g$, $b'=b$ (cf. Figure 7, §21). The symmetry axis of the path of m in the relative central motion, which we call the line of apses, is the line connecting m_1 with that point where m has, in its entire relative central motion, the smallest separation from m_1. It plays the same role for central motion as does the line of centers for elastic collisions. The plane of relative central motion will be called the orbital plane. It contains the four lines g, g', b, and b'. One sees that $d\epsilon=d\epsilon'$ if he introduces for $d\epsilon$ the angle of rotation $d\vartheta$ of the line of apses, so that ξ, η, ζ, ξ_1, η_1, ζ_1 transform into ξ', η', ζ', ξ_1', η_1', ζ_1' and then again introduces $d\epsilon'$ for $d\vartheta$; then the expression for $d\epsilon$ in terms of $d\vartheta$ and the values of the variables before the collision must be exactly equal to that for $d\epsilon'$ in terms of $d\vartheta$ and the values of the variables after the collision.

The proof that $d\omega=d\omega'$, $d\omega_1=d\omega_1'$ has already been given for elastic spheres. Since we used there only the theorem of kinetic energy conservation and the theorem of conservation of motion of the center of mass in constructing the proof, and since these theorems are still valid here, the proof can be constructed in the same way; in place of the line of centers one has of course to introduce the line of apses. Taking account of all these equations, one can also write:

$$i_3 = f'F_1'\, do\, d\omega\, d\omega_1\, dt\, g\, b\, db\, d\epsilon. \tag{110}$$

In Part II we shall prove a general theorem, of which the law that here

(110a) $$d\omega' d\omega_1' g'b'db'd\epsilon' = d\omega d\omega_1 gbdbd\epsilon$$

is only a special case. It is only to avoid the necessity of repeating here for a special case, as a useless digression, results that will be developed in general later on, that we have merely indicated briefly the proof of this theorem which is undoubtedly correct.

As a result of each "inverse" collision, the number dn of m-molecules in the parallelepiped do whose velocity points lie in the parallelepiped $d\omega$ increases by one. The total increment i_1 experienced by dn as a result of collisions of m-molecules with m_1-molecules is found by integrating over ϵ from 0 to 2π, over b from 0 to σ, and over ξ_1, η_1, ζ_1 from $-\infty$ to $+\infty$. We shall write the result of this integration in the form:

(111) $$i_1 = dod\omega dt \int \int_0^\sigma \int_0^{2\pi} f'F_1' gbd\omega_1 dbd\epsilon.$$

Of course we cannot perform explicitly the integration with respect to b and ϵ since the variables ξ', η', ζ', and ξ_1', η_1', ζ_1' occurring in f' and F_1' are functions of ξ, η, ζ, ξ_1, η_1, ζ_1, b, and ϵ, which cannot be computed until the force law is given. The difference $i_1 - \nu_1$ expresses the net increase of dn during time dt as a result of collisions of m-molecules with m_1-molecules. It is therefore the total increase V_3 experienced by dn as a result of these collisions, and one has

(112) $$V_3 = i_1 - \nu_1 = dod\omega dt \int \int_0^\sigma \int_0^{2\pi} (f'F_1' - fF_1)dbd\omega_1 dbd\epsilon.$$

It is to be noted that in glancing collisions it may happen that the velocity point of the m-molecule lies in the parallelepiped $d\omega$ both before and after the collision. These glancing collisions are counted in (105) and hence also in ν_1, and are subtracted from V_3, although in these collisions the velocity point of the m-molecule is not expelled from $d\omega$ but is merely pushed from one place to another inside this parallelepiped. Yet this entails no error. It is precisely because the velocity point of the m-molecule also lies in $d\omega$ after the collision that these collisions are also counted in the expression (110) for i_3 and hence also in i_1, and they are therefore added back to V_3.

These collisions are simply to be understood as collisions in which the velocity point of the m-molecule is actually expelled from $d\omega$ at the beginning of the collision, but at the end of the collision it is thrown back into the same parallelepiped. Indeed, in Equation (112) we can extend the integration over b to values greater than σ. We would thereby include in the number of passages in ν_1, and exclude from those in V_3, some passages in which no change of the velocities and directions actually occurred. But the number of these collisions will therefore be counted in the expression for i_1, and added back again to V_3. Obviously the limits of integration cannot be chosen differently in the two expressions, (107) for ν_1 and (111) for i_1. In Equation (112), where $i_1-\nu_1$ is united into a single integral, on the other hand, the integration over b can be extended as far as one wishes, since as soon as b is larger than σ, the quantities ξ', η', ζ', ξ_1', η_1', ζ_1' will be identical to ξ, η, ζ, ξ_1, η_1, ζ_1; hence $f'F_1'=fF_1$ and the integrand vanishes. This remark is important in all cases where the interaction of the molecules falls off gradually with increasing distance, so that no sharp boundary of the sphere of action can be specified. In such cases one can integrate in Equation (112) over b from 0 to ∞, and since these limits are permissible in all other cases, we shall retain them in the future. When no distance can be specified at which the interaction of two molecules drops off precisely to zero, we shall still assume of course that the interaction decreases so rapidly with increasing distance that the case where more than two molecules interact simultaneously can be neglected.

The number of molecules that collide during dt but move in such a way that even without a collision they would have left do or their velocity points would have left $d\omega$ is of course an infinitesimal quantity of order $(dt)^2$.

4. The increment V_4 experienced by dn as a result of collisions of m-molecules with each other is found from Equation (112) by a simple permutation. One now uses ξ_1, η_1, ζ_1 and ξ_1', η_1', ζ_1' for the velocity components of the other m-molecule before and after the collision, respectively, and one writes f_1 and f_1' for

$$f(x, y, z, \xi_1, \eta_1, \zeta_1, t)$$

and

$$f(x, y, z, \xi_1', \eta_1', \zeta_1', t).$$

Then:

$$V_4 = dodωdt \iint_0^\infty \int_0^{2\pi} (f'f_1' - ff_1) gbdω_1dbdε. \tag{113}$$

Since now $V_1+V_2+V_3+V_4$ is equal to the increment $dn'-dn$ of dn during time dt, and this according to Equation (101) must be equal to $(\partial f/\partial t)dodωdt$, one obtains on substituting all the appropriate values and dividing by $dodωdt$ the following partial differential equation for the function f:

$$(114)\quad \begin{cases} \dfrac{\partial f}{\partial t} + \xi\dfrac{\partial f}{\partial x} + \eta\dfrac{\partial f}{\partial y} + \zeta\dfrac{\partial f}{\partial z} + X\dfrac{\partial f}{\partial x} + Y\dfrac{\partial f}{\partial y} + Z\dfrac{\partial f}{\partial z} \\ \quad = \displaystyle\iint_0^\infty \int_0^{2\pi} (f'F_1' - fF_1) gbdω_1dbdε \\ \quad + \displaystyle\iint_0^\infty \int_0^{2\pi} (f'f_1' - ff_1) gbdω_1dbdε. \end{cases}$$

Similarly one obtains for F the partial differential equation:

$$(115)\quad \begin{cases} \dfrac{\partial F_1}{\partial t} + \xi_1\dfrac{\partial F_1}{\partial x} + \eta_1\dfrac{\partial F_1}{\partial y} + \zeta_1\dfrac{\partial F_1}{\partial z} + X_1\dfrac{\partial F_1}{\partial x} + Y_1\dfrac{\partial F_1}{\partial y} + Z_1\dfrac{\partial F_1}{\partial z} \\ \quad = \displaystyle\iint_0^\infty \int_0^{2\pi} (f'F_1' - fF_1) gbdωdbdε \\ \quad + \displaystyle\iint_0^\infty \int_0^{2\pi} (F'F_1' - FF_1) gbdω_1dbdε. \end{cases}$$

Here we have used the abbreviation F' for $F(x, y, z, \xi', \eta', \zeta', t)$.

§17. Time-derivatives of sums over all molecules in a region.

Before we go further, we shall develop some general formulas useful in gas theory. Let φ be an arbitrary function of x, y, z, ξ, η, ζ, t. The value obtained by substituting therein the actual coördinates and velocity components of a particular molecule at time t will be called the value of φ corresponding to that molecule at time t. The sum of all values of φ corresponding to all the

m-molecules that lie in the parallelepiped do and whose velocity points lie in the parallelepiped $d\omega$ at time t is obtained by multiplying φ by the number $fdod\omega$ of those moleclules. We denote it by

$$(116) \qquad \sum\nolimits_{d\omega, do}\varphi = \varphi f do d\omega.$$

Similarly we choose for the second kind of gas any other arbitrary function Φ of x, y, z, ξ, η, ζ, t and denote by

$$(117) \qquad \sum\nolimits_{d\omega_1, do}\Phi_1 = \Phi_1 F_1 do d\omega_1$$

the sum of the values of Φ corresponding to all the m_1-molecules lying in do whose velocity points lie in $d\omega_1$. Φ_1 is the abbreviation for $\Phi(x, y, z, \xi_1, \eta_1, \zeta_1, t)$.

If we keep do constant in these expressions and integrate with respect to $d\omega$ and $d\omega_1$, over all possible values, then we obtain the expressions:

$$(118) \qquad \sum\nolimits_{\omega, do}\varphi = do\int \varphi f d\omega \text{ and } \sum\nolimits_{\omega_1, do}\Phi_1 = do\int \Phi_1 F_1 d\omega_1,$$

which represent for the two gases the sums of all values of φ and Φ, respectively, that correspond to all molecules in do at time t, without any restriction on the velocities.

If we also integrate do over all volume elements of our gas, then we obtain the expressions:

$$(119) \qquad \sum\nolimits_{\omega, o}\varphi = \int\int \varphi f do d\omega \text{ and } \sum\nolimits_{\omega 1, o}\Phi_1 = \int\int \Phi_1 F_1 do d\omega_1$$

for the sums of all values of φ and Φ corresponding to all gas molecules of the first and second kinds, respectively.

We shall now calculate the increment $(\partial \sum_{d\omega, do}\varphi/\partial t)dt$ experienced by the sum $\sum_{d\omega, do}\varphi$ during an infinitesimal time dt, if there is no change in the two volume elements do and $d\omega$ in size, shape, or position. According to the latter condition, which is expressed by the symbol $\partial/\partial t$, one should differentiate only with respect to the time. Since during the time interval dt, φ changes by $(\partial\varphi/\partial t)dt$ and f by $(\partial f/\partial t)dt$, we obtain from Equation (116):

$$\frac{\partial}{\partial t}\sum\nolimits_{d\omega, do}\varphi = \left(f\frac{\partial\varphi}{\partial t} + \varphi\frac{\partial f}{\partial t}\right) do d\omega.$$

When we substitute for $\partial f/\partial t$ its value from Equation (114), the above expression appears to be a sum of five terms, each of which has its own physical meaning. Accordingly we set:

(120) $$\frac{\partial}{\partial t}\sum_{d\omega, do}\varphi = [A_1(\varphi)+A_2(\varphi)+A_3(\varphi)+A_4(\varphi)+A_5(\varphi)]do d\omega,$$

so that

(121) $$A_1(\varphi) = \frac{\partial\varphi}{\partial t}f$$

corresponds to the effect of the explicit dependence of φ on t;

(122) $$A_2(\varphi) = -\varphi\left(\xi\frac{\partial f}{\partial x}+\eta\frac{\partial f}{\partial y}+\xi\frac{\partial f}{\partial z}\right)$$

corresponds to the effect of the motion of the molecules;

(123) $$A_3(\varphi) = -\varphi\left(X\frac{\partial f}{\partial \xi}+Y\frac{\partial f}{\partial \eta}+Z\frac{\partial f}{\partial z}\right)$$

corresponds to the effect of external forces;

(124) $$A_4(\varphi) = \varphi\int\int_0^\infty\int_0^{2\pi}(f'F_1' - fF_1)gbd\omega_1 db d\epsilon$$

corresponds to the effects of collisions of m-molecules with m_1-molecules; and

(125) $$A_5(\varphi) = \varphi\int\int_0^\infty\int_0^{2\pi}(f'f_1' - ff_1)gbd\omega_1 db d\epsilon$$

corresponds to the effect of collisions of m-molecules with each other.

In order to find $(\partial/\partial t)\sum_{\omega,do}\varphi$, we have simply to integrate $(\partial/\partial t)\sum_{d\omega,do}\varphi$ over all possible values of $d\omega$. We shall again write:

(126) $$\frac{\partial}{\partial t}\sum_{\omega,do}\varphi = [B_1(\varphi)+B_2(\varphi)+B_3(\varphi)+B_4(\varphi)+B_5(\varphi)]do.$$

One obtains each B by multiplying the corresponding A by $d\omega = d\xi d\eta d\zeta$ and integrating over all these variables from $-\infty$ to $+\infty$, which we indicate by a single integral sign. Thus:

(127) $$B_1(\varphi) = \int\frac{\partial\varphi}{\partial t}fd\omega$$

(128) $$B_2(\varphi) = -\int\varphi\left(\xi\frac{\partial f}{\partial x}+\eta\frac{\partial f}{\partial y}+\zeta\frac{\partial f}{\partial z}\right)d\omega.$$

The third term B_3, which corresponds to the increase resulting from the action of external forces, can also be calculated in another way. Since we have to take account of all elements $d\omega$ sooner or later, we need not compare the $fdod\omega$ molecules whose velocity points lie in $d\omega$ at the beginning of the time interval with those whose velocity points lie in $d\omega$ at the end of the same interval. Instead we may simply follow the former $fdod\omega$ molecules in their motion during the interval dt. For each molecule the velocity components ξ, η, ζ will change by Xdt, Ydt, Zdt respectively during this time interval. Hence for each molecule the corresponding value of φ changes, as a result of the action of the external forces, by the amount:

$$\left(X\frac{\partial\varphi}{\partial\xi}+Y\frac{\partial\varphi}{\partial\eta}+Z\frac{\partial\varphi}{\partial\zeta}\right)dt. \tag{129}$$

The effect of the external forces thus amounts merely to the fact that each of these molecules provides this additional contribution to the sum $\sum_{\omega,do}\varphi$. The total increase $B_3(\varphi)dodt$ of this sum due to the action of external forces is therefore found by multiplying Equation (129) by $fdod\omega$ and integrating over all values of $d\omega$, whereby one obtains:

$$B_3(\varphi)=\int\left(X\frac{\partial\varphi}{\partial\xi}+Y\frac{\partial\varphi}{\partial\eta}+Z\frac{\partial\varphi}{\partial\zeta}\right)fd\omega. \tag{130}$$

According to the method by which Equations (127) and (128) were found—i.e., by multiplying Equation (123) by $d\omega$ and integrating over all values of this differential—one would obtain for the same quantity the value:

$$B_3(\varphi)=-\int\left(X\frac{\partial f}{\partial\xi}+Y\frac{\partial f}{\partial\eta}+Z\frac{\partial f}{\partial\zeta}\right)\varphi d\omega. \tag{131}$$

Since X, Y, Z do not contain the variables ξ, η, ζ and since moreover $d\omega$ is merely an abbreviation for $d\xi d\eta d\zeta$ and the integration signs in Equations (130) and (131) signifiy integration over ξ, η, ζ, from $-\infty$ to $+\infty$, one sees easily that if he integrates by parts in the first term with respect to ξ, in the second with respect to η, and in the third with respect to ζ, then the identity of the expressions (130) and (131) can be proved. For infinite values of ξ, η, or ζ, f must vanish, and the product $f\varphi$ must also approach the limit zero if $\sum_{\omega,do}\varphi$ is to have a meaning.

We shall also compute directly the increase $B_4(\varphi)dodt$ experienced by the quantity $\sum_{\omega,do}\varphi$ as a result of collisions of an m-molecule with an m_1-molecule.

We again use the phrase "direct collisions" for all collisions that occur between an m-molecule and an m_1-molecule during time dt in the volume element do such that before the collision the variables lie between the limits (98), (102), and (104). The total effect of each of these collisions is that an m-molecule loses the velocity components ξ, η, ζ and acquires the components ξ', η', ζ'. While it provides a term φ in $\sum_{\omega,do}\varphi$ before the collision, it provides a term φ' in the same sum after the collision (φ' is an abbreviation for $\varphi(x, y, z, \xi', \eta', \zeta', t)$).

This sum experiences an increase $\varphi' - \varphi$ because of each of these collisions, and since according to Equation (105) the number of these direct collisions is given by ν_3, one obtains the total increase $B_4(\varphi)dodt$ experienced by the sum $\sum_{\omega,do}\varphi$ as a result of collisions of an m-molecule with an m_1-molecule when he integrates the product $(\varphi' - \varphi)\nu_3$, keeping do and dt constant, over all values of all the other differentials. One thus obtains:

$$(132) \qquad B_4(\varphi) = \iiint_0^\infty \int_0^{2\pi} (\varphi' - \varphi) f F_1 g b d\omega d\omega_1 db d\epsilon.$$

One might just as well have calculated $B_4(\varphi)$ by considering those collisions between an m-molecule and an m_1-molecule in which the variables lie between the limits (108) and (109) before the collision—i.e., those collisions which we have called "inverse." For these, the value φ' would correspond to the m-molecule before the collision, and the value φ afterwards. The sum $\sum_{\omega,do}\varphi$ would therefore decrease by $(\varphi - \varphi')$ for each collision, and hence the total amount $(\varphi - \varphi')i_3$, where i_3 is the number of inverse collisions as given by Equation (110).

If we integrate over all variables except do and dt, then we must again obtain the quantity denoted by $B_4(\varphi)$ $dodt$. However, we would then have:

$$(133) \qquad B_4(\varphi) = \iiint_0^\infty \int_0^{2\pi} (\varphi - \varphi') f' F_1' g b d\omega d\omega_1 db d\epsilon.$$

We could set $B_4(\varphi)$ equal to the arithmetic mean of the two values found for it, and thus obtain:

$$(134) \quad B_4(\varphi) = \tfrac{1}{2} \iiint_0^\infty \int_0^{2\pi} (\varphi - \varphi')(f'F_1' - fF_1) g b d\omega d\omega_1 db d\epsilon.$$

Integration of Equation (124) would on the other hand lead to:

$$(134a)\quad B_4(\varphi) = \iint_0^\infty \int_0^{2\pi} \int \varphi(f'F_1' - fF_1)gbd\omega d\omega_1 dbd\epsilon.$$

One easily sees that the possibility of all these different ways of expressing $B_4(\varphi)$ follows fron the two equations:

$$\sum \varphi'\nu_3 = \sum \varphi i_3,$$
$$\sum \varphi' i_3 = \sum \varphi\nu_3,$$

where the summation signs express integration over all the differentials contained in i_3 or in ν_3 except for do and dt. These two equations follow directly; indeed, the summation over all ν_3 or all i_3 includes all collisions, and φ and φ' are interchanged when one replaces the former sum by the latter, or conversely.

If one assumes in Equation (132) or (133) that the two molecules are identical, then:

$$(135)\qquad B_5(\varphi) = \iiint_0^\infty \int_0^{2\pi} (\varphi' - \varphi)ff_1 gbd\omega d\omega_1 dbd\epsilon$$

$$(136)\qquad = \iiint_0^\infty \int_0^{2\pi} (\varphi - \varphi')f'f_1' gbd\omega d\omega_1 dbd\epsilon.$$

One may still consider the two colliding molecules to play the same role, so that one can interchange, in the last two formulae, the letters with subscript 1 and those with no subscript, without changing the value of $B_5(\varphi)$. Taking the arithmetic mean of the values of $B_5(\varphi)$ obtained by this permutation, one gets from (135):

$$(137)\quad B_5(\varphi) = \tfrac{1}{2}\iiint_0^\infty \int_0^{2\pi} (\varphi' + \varphi_1' - \varphi - \varphi_1)ff_1 gbd\omega d\omega_1 dbd\epsilon,$$

and from (136):

$$(138)\quad B_5(\varphi) = \tfrac{1}{2}\iiint_0^\infty \int_0^{2\pi} (\varphi' + \varphi_1 - \varphi - \varphi_1')f'f_1' dbd\omega d\omega_1 dbd\epsilon.$$

The arithmetic mean of these two values is

$$(139)\quad B_5(\varphi) = \tfrac{1}{4}\iiint_0^\infty \int_0^{2\pi} (\varphi + \varphi_1 - \varphi' - \varphi_1')(f'f_1' - ff_1)gbd\omega d\omega_1 db\epsilon.$$

The same result is obtained when one considers the fact that, as a result of each collision satisfying the conditions (98), (102), and (104), the value of φ for one of the colliding molecules changes from φ to φ', while for the other it changes from φ_1 to φ_1', so that for each such collision the sum $\sum_{\omega,do}\varphi$ must increase by $\varphi'+\varphi_1'-\varphi-\varphi_1$. Here φ_1 and φ_1' are abbreviations for $\varphi(x, y, z, \xi, \eta, \zeta, t)$ and $\varphi(x, y, z, \xi_1', \eta_1', \zeta_1', t)$. Now

$$ff_1gbdodwd\omega_1dbd\epsilon dt$$

of these collisions occur during dt. As a result of all these collisions, $\sum_{\omega,do}\varphi$ decreases by

$$(\varphi' + \varphi_1' - \varphi - \varphi_1)ff_1dbdodwd\omega_1dbd\epsilon dt.$$

If we integrate over $d\omega$, $d\omega_1$, db, and $d\epsilon$, then we obtain the increase of $\sum_{\omega,do}\varphi$ resulting from collisions of m-molecules with each other, i.e., the quantity $B_5(\varphi)dodt$. However, we must divide by 2, since we have counted each collision twice, and therefore we obtain Equation (137). Had we considered only the inverse collisions, then we would have likewise obtained Equation (138).

The special cases of Equation (136) obtained by assuming that φ is independent of the time and of the coordinates x, y, z will be treated in §20.

We shall now set:

$$(140) \quad \frac{d}{dt}\sum_{\omega,o}\varphi = C_1(\varphi) + C_2(\varphi) + C_3(\varphi) + C_4(\varphi) + C_5(\varphi).$$

Since in $\sum_{\omega,o}\varphi$ one has to integrate over all values of do and $d\omega$, this quantity is now a function only of time. Hence the use of the symbol $\partial/\partial t$ is unnecessary, and we can express differentiation by the usual Latin letter d.

Each C is obtained by multiplying the corresponding B by do and integrating over all volume elements, or else by multiplying the corresponding A by $dod\omega$ and integrating over all do and $d\omega$.

Since the total number of molecules remains unchanged, we can calculate the sum $[C_1(\varphi)+C_2(\varphi)+C_3(\varphi)]dt$—which includes all increments except those resulting from collisions—by simply following the $fdod\omega$ molecules through their paths during time dt. During this time their coördinates increase by ξdt, ηdt, ζdt, and their velocity components by Xdt, Ydt, Zdt. Each of these molecules thus provides in the sum $\sum_{\omega,o}\varphi$ at time t the contribution

$$\varphi(x, y, z, \xi, \eta, \zeta, t),$$

and at time $t+dt$ a contribution larger by the amount

$$dt\left(\frac{\partial\varphi}{\partial t} + \xi\frac{\partial\varphi}{\partial x} + \eta\frac{\partial\varphi}{\partial y} + \zeta\frac{\partial\varphi}{\partial z} + X\frac{\partial\varphi}{\partial\xi} + Y\frac{\partial\varphi}{\partial\eta} + Z\frac{\partial\varphi}{\partial\zeta}\right)$$

and since the number of these molecules is equal to $fdod\omega$, one has to multiply by this number and integrate over all differentials except dt. On dividing by dt, one obtains:

$$(141)\quad \begin{cases} C_1(\varphi) + C_2(\varphi) + C_3(\varphi) \\ \quad = \displaystyle\iint fdod\omega\left(\frac{\partial\varphi}{\partial t} + \xi\frac{\partial\varphi}{\partial x} + \eta\frac{\partial\varphi}{\partial y} + \zeta\frac{\partial\varphi}{\partial z}\right. \\ \quad \left. + X\frac{\partial\varphi}{\partial\xi} + Y\frac{\partial\varphi}{\partial\eta} + Z\frac{\partial\varphi}{\partial\zeta}\right). \end{cases}$$

This value represents the increase of $\sum_{\omega,o}\varphi$ (divided by dt) arising from the three causes considered, and is also still correct when the walls of the container are in motion. If one wrote instead of $C_2(\varphi)$ simply the integral of the quantity $B_2(\varphi)do$, then he would consider the position of all the volume elements do to be fixed. Therefore when the walls are in motion one must add special terms to take account of the new parts of space that come into the volume of the gas or go out of it during dt. These correspond to the surface integrals that appear on partial integration of Equation (141) with respect to the coördinates.

The two quantities $C_4(\varphi)$ and $C_5(\varphi)$ are obtained by multiplying the expressions $B_4(\varphi)$ and $B_5(\varphi)$ by do and integrating over all volume elements of the space filled by the gas, whereby one obtains:

$$(142)\quad \begin{cases} C_4(\varphi) = \frac{1}{2}\displaystyle\iiint\int_0^\infty\int_0^{2\pi}(\varphi - \varphi')(f'F_1' - fF_1) \\ \qquad \cdot\, gbdod\omega d\omega_1 dbd\epsilon \\ C_5(\varphi) = \frac{1}{4}\displaystyle\iiint\int_0^\infty\int_0^{2\pi}(\varphi + \varphi_1 - \varphi' - \varphi_1')(f'f_1' - ff_1) \\ \qquad \cdot\, gbdod\omega d\omega_1 dbd\epsilon \end{cases}$$

We have not considered here any effects due to motion of the walls of the container, since the molecules that would collide in volume elements entering the space considered because of such motion would provide terms only of order $(dt)^2$.

Since the expression for the time derivatives of the quantities denoted by $\sum_{d\omega_1,do}\Phi_1$, $\sum_{\omega_1,do}\Phi_1$ and $\sum_{\omega_1,o}\Phi$ are constructed in just the same way, we shall not write them out.

Since A, B, and C are only the increments of definite quantities resulting from specified causes, most authors express them as derivatives of those quantities. Maxwell* writes $(\partial/\partial t)\sum_{\omega,do}\varphi$, Kirchhoff† $(D/Dt)\sum_{\omega,do}\varphi$ for $B_5(\varphi)$ etc. As with all differentials, the A for a sum of two functions is equal to the A's for the addends:

$$\text{(143)} \qquad \begin{cases} A_k(\varphi+\psi) = A_k(\varphi) + A_k(\psi), \\ B_k(\varphi+\psi) = B_k(\varphi) + B_k(\psi), \\ C_k(\varphi+\psi) = C_k(\varphi) + C_k(\psi) \end{cases}$$

for any subscript k. These equations follow from the circumstance that φ occurs in all the integrals A, B, C only linearly.‡

§18. More general proof of the entropy theorem. Treatment of the equations corresponding to the stationary state.

We shall now consider that special case when $\varphi = lf$ and $\Phi = lF$, where l signifies the natural logarithm. Then

$$\sum_{\omega,o}\varphi = \sum_{\omega,o} lf = \int\int flf do d\omega,$$

$$\sum_{\omega,o}\Phi_1 = \sum_{\omega_1,o} lF_1 = \int\int F_1 lF_1 do d\omega_1,$$

and we shall set

$$\text{(144)} \quad H = \sum_{\omega,o} lf + \sum_{\omega_1,o} lF_1 = \int\int flf do d\omega + \int\int F_1 do d\omega_1.$$

One has according to Equation (141):

---

* Maxwell, Phil. Trans. **157**, 49 (1867).

† Kirchhoff, *Vorlesungen über die Theorie der Wärme.*

‡ See also Benndorff, Wien. Ber. **105**, 646 (1896) for calculations of $B_5(\varphi)$.

$$(145)\quad \begin{cases} C_1(lf) + C_2(lf) + C_3(lf) = \displaystyle\iint do d\omega \\ \left(\dfrac{\partial f}{\partial t} + \xi\dfrac{\partial f}{\partial x} + \eta\dfrac{\partial f}{\partial y} + \zeta\dfrac{\partial f}{\partial z} + X\dfrac{\partial f}{\partial \xi} + Y\dfrac{\partial f}{\partial \eta} + Z\dfrac{\partial f}{\partial \zeta}\right). \end{cases}$$

If one integrates the fifth term in the parentheses with respect to ξ, the sixth with respect to η and the last with respect to ζ, then he obtains zero each time, since X, Y, Z are not functions of ξ, η, ζ, and f vanishes at the limits $(-\infty, +\infty)$. If one integrates the second, third, and fourth terms with respect to x, y, and z, respectively, then he obtains an integral J over the total external surface of the gas. If dS is a surface element of this surface, and N is the outwardly directed velocity of an m-molecule, normal to dS, then $J = \iint dS d\omega N f$.

One sees easily that Jdt represents the total number of K molecules that leave through the whole surface S, in excess of those that enter; while the first term on the right hand side of Equation (145) multiplied by dt, *viz.*,

$$dt \iint \frac{\partial f}{\partial t} do d\omega$$

represents the total increase L experienced by the number of m-molecules lying within the surface S during time dt.

It is to be understood that we do not consider the volume element do as fixed, but rather we allow it to move with the molecules. If the gas is surrounded by a vacuum, then the surface S and the molecules of different velocities will always move together. Hence no more molecules can go out through S than come in, so that $L = K = 0$. If the gas is enclosed by walls at rest, at which the molecules are reflected like elastic spheres,[1] then in the place of each molecule that arrives at a volume element do adjacent to the wall as a result of its motion toward the wall, there appears another identical molecule which simply has the sign of its component of velocity normal to the wall reversed. Hence $L = K = 0$.

This argument will be valid as long as the wall is at rest, on account of the symmetry and equal probability of opposite motions (even though the action of the wall might be of some other

[1] One sees at once that on this assumption a completely smooth flat wall would experience no resistance to its motion in a direction lying in its own plane.

kind) so long as the wall neither adds nor subtracts kinetic energy.[2] In all these cases $(d/dt) \sum_{\omega, o} lf$ reduces to the terms

[2] We understand as usual by S an arbitrary surface completely enclosing the gas, which can everywhere be taken as close to the walls as one wishes, and the integral over do extends only over all volume elements within this surface, while that over dS extends over all surface elements. We denote by $K'dt$ the number of molecules that in time dt go out through the surface S in excess of those that come in; by $L'dt$, the decrease in the number of molecules lying within the surface S, so that we always have $K'+L'=0$. However, K' and L' are not identical with the quantities called K and L in the text, since in calculating $(d/dt) \sum_{\omega, o} lf$ we have followed each molecule along its path during time dt. The sum has therefore included the same molecules at the beginning and end of dt, and the difference of these two sums was divided by dt. We thus assume that the volume element do moves along with the molecules being considered, and that these same molecules always remain within the surface S. This is not the case as soon as S does not move with the molecules. We shall always sum over the same space element at the beginning and end of dt, so that $(d/dt) \sum_{\omega, o} lf$ is simply the expression given by Eq. (120), integrated over do and $d\omega$, in which of course lf is to be substituted for φ. Then one obtains, on substituting the values (121–125):

$$\text{(145a)} \quad \begin{cases} \dfrac{d}{dt} \sum_{\omega, o} lf = \displaystyle\int\!\!\int do\, d\omega \left[\frac{\partial f}{\partial t} - lf\left(\xi \frac{\partial f}{\partial x} + \eta \frac{\partial f}{\partial y} + \zeta \frac{\partial f}{\partial z} \right.\right. \\ \qquad\qquad \left.\left. + X \dfrac{\partial f}{\partial \xi} + Y \dfrac{\partial f}{\partial \eta} + Z \dfrac{\partial f}{\partial \zeta} \right) \right] + C_4(lf) + C_5(lf). \end{cases}$$

$C_4(lf) + C_5(lf)$ are the same quantities as before. The first term of the double integral is likewise the same as in Eq. (145), and therefore it is equal to K. Also, the last three terms can, by integrating by parts with respect to ξ, η, ζ, be put in the same form as the corresponding terms in Eq. (145). By direct integration of the fifth term with respect to ξ, the sixth with respect to η, and the seventh with respect to ζ, these can also be reduced to zero, since flf vanishes for infinite ξ, η, or ζ (since $\int_{-\infty}^{\infty} f d\xi$ must be finite). The sum of the second, third, and fourth terms of the double integral (145a) gives (since $d(flf - f) = lfdf$) on integrating over x, y, and z, the two surface integrals

$$\int\!\!\int do dS f N - \int\!\!\int do dS N flf$$

both to be extended over the surface S (which is now considered fixed). The first is the quantity previously denoted by K; the second represents, when it is multiplied by dt, the amount of the quantity H (defined by Eq. 144) that is carried into the surface S by the motion of the m-molecules, in excess of that carried out. There cannot be any creation of H inside the gas, any more than in the case considered in the text. The total amount of H contained within the surface S can only increase by an amount less than, or at most equal to, the amount of this quantity carried into the surface from without.

The quantity $-H$, which is proportional to the entropy, will never

$C_4(lf)+C_5(lf)$ furnished by the collisions, and Equations (140) and (142) yield:

$$\frac{d}{dt}\sum_{\omega,o} lf$$

$$= \tfrac{1}{4}\int\int\int\int_0^\infty\int_0^{2\pi}[l(ff_1) - l(f'f_1')](f'f_1' - ff_1)gbdod\omega d\omega_1 dbd\epsilon$$

$$+ \tfrac{1}{2}\int\int\int\int_0^\infty\int_0^{2\pi}(lf - lf')(f'F_1' - fF_1)gbdod\omega d\omega_1 dbd\epsilon.$$

Similarly:

$$\frac{d}{dt}\sum_{\omega_1,o} lF_1$$

$$= \tfrac{1}{4}\int\int\int\int_0^\infty\int_0^{2\pi}[l(FF_1) - l(F'F_1')](F'F_1' - FF_1)gbdod\omega d\omega_1 dbd\epsilon$$

$$+ \tfrac{1}{2}\int\int\int\int_0^\infty\int_0^{2\pi}(lF_1 - lF_1')(f'F_1' - fF_1)gbdod\omega d\omega_1 dbd\epsilon.$$

Hence according to Equation (144):

$$(146)\quad \begin{cases} \dfrac{dH}{dt} = -\tfrac{1}{4}\displaystyle\int\int\int\int_0^\infty\int_0^{2\pi}[l(ff_1) - l(f'f_1')](ff_1 - f'f_1') \\ \qquad gbdod\omega d\omega_1 dbd\epsilon \\ \qquad -\tfrac{1}{4}\displaystyle\int\int\int\int_0^\infty\int_0^{2\pi}[l(FF_1) - l(F'F_1')](FF_1 - F'F_1') \\ \qquad gbdod\omega d\omega_1 dbd\epsilon \\ \qquad -\tfrac{1}{2}\displaystyle\int\int\int\int_0^\infty\int_0^{2\pi}[l(fF_1) - l(f'F_1')](fF_1 - f'F_1') \\ \qquad gbdod\omega d\omega_1 dbd\epsilon. \end{cases}$$

---

be changed when visible motion is produced, or its direction is changed by external forces, or any other change occurs, as long as no molecular motion arises thereby through collisions. Even if initially one gas is in one half and another gas in the other half of the container, the entropy is not changed by their progressive motion. The mixing of course leads to a more probable state, but the velocity distribution is less probable, since each gas has an average motion in a particular direction. As soon as this average motion is annihilated by collisions (transformed into disordered molecular motion) then H decreases and the entropy increases.

Like the integrals in Equation (33), these integrals are sums of terms none of which can be negative. Hence H can never increase. We could also have constructed the proof in the same way for any number of gases in any molecular-disordered initial distribution under the action of external forces. Thus the proof of the Clausius-Gibbs theorem—that at constant volume, excluding any addition of energy, the quantity H can only decrease—which was merely indicated at the conclusion of §8, has now been completely established for monatomic gases.

The quantity dH/dt can vanish only when the integrand vanishes in all the integrals. For the stationary final state, however, which the gas mixture must reach if the walls are at rest, H cannot continue to decrease but must eventually become constant. Hence the integrands in Equation (146) must vanish for all values of the variables, i.e., for all possible collisions the following three equations must hold:

$$ff_1 = f'f_1', \quad FF_1 = F'F_1', \quad fF_1 = f'F_1'. \tag{147}$$

These functions cannot of course depend on the variable t in the equilibrium state; nevertheless, we shall not introduce this condition until later, and for the moment we look for solutions of Equations (147) that still depend on time.

We first treat the last of these equations, and consider x, y, z, t constant; for the present we look only at the dependence of the functions f and F on the variable ξ, η, and ζ. We set

$$\varphi = lf(x, y, z, \xi, \eta, \zeta, t), \qquad \varphi' = lf(x, y, z, \xi', \eta', \zeta', t).$$

$$\Phi_1 = lF(x, y, z, \xi_1, \eta_1, \zeta_1, t), \quad \Phi_1' = lF(x, y, z, \xi_1', \eta_1', \zeta_1', t);$$

so that the last of Equations (147) becomes

$$\varphi + \Phi_1 - \varphi' - \Phi_1' = 0. \tag{148}$$

In any case the equations of conservation of kinetic energy and of the motion of the center of mass must not be violated in the collisions. One has therefore:

$$\begin{cases} m(\xi^2 + \eta^2 + \zeta^2) + m_1(\xi_1^2 + \eta_1^2 + \zeta_1^2) \\ \qquad -m(\xi'^2 + \eta'^2 + \zeta'^2) - m_1(\xi_1'^2 + \eta_1'^2 + \zeta_1'^2) = 0, \\ m\xi + m_1\xi_1 - m\xi' - m_1\xi_1' = 0 \\ m\eta + m_1\eta_1 - m\eta' - m_1\eta_1' = 0 \\ m\zeta + m_1\zeta_1 - m\zeta' - m_1\zeta_1' = 0. \end{cases} \tag{149}$$

Clearly each of the eight variables ξ, η, ζ, ξ_1, η_1, ζ_1, b, and ϵ can assume any of an infinite manifold of values, independently of the others; they are the so-called independent variables. The six quantities ξ', η', ζ', ξ_1', η_1', ζ_1' will be expressed as functions of the independent variables by six equations.

All possible equations relating the twelve variables

$$\xi,\ \eta,\ \zeta,\ \xi_1,\ \eta_1,\ \zeta_1,\ \xi',\ \eta',\ \zeta',\ \xi_1',\ \eta_1',\ \zeta_1' \tag{150}$$

can arise only from these six equations by eliminating b and ϵ. However, by this elimination only four equations can be obtained. Whence it follows that these four equations (149) are the only ones that can exist relating these twelve variables. Equations (147) and (148) must therefore hold for all values of each of the twelve variables that satisfy the four conditions (149). We note that these equations are completely symmetrical with respect to the three coördinate axes, so that one can permute the coördinates cyclically in each equation that follows from them, without endangering the validity of the equation.

By the well-known method of undetermined multipliers we can make all twelve differentials of the twelve quantities (150) depend on each other, if we add to the total differential of Equation (148) the total differentials of the four equations (149) multiplied by four different factors A, B, C, D. These factors can always be chosen such that the coefficients of the total differentials vanish. Therefore we obtain:

$$d\xi\left[\frac{\partial\varphi}{\partial\xi} + 2m\,A\xi + m\,B\right]$$

$$+\ d\eta\left[\frac{\partial\varphi}{\partial\eta} + 2m\,A\eta + m\,C\right] + \cdots$$

$$+\ d\xi_1\left[\frac{\partial\Phi_1}{\partial\xi_1} + 2m_1A\xi_1 + m_1B\right] + \cdots\quad\cdots$$

$$-\ d\xi'\left[\frac{\partial\varphi'}{\partial\xi'} + 2m\,A\xi' + m\,B\right] + \cdots\quad\cdots$$

$$-\ d\xi_1'\left[\frac{\partial\Phi_1'}{\partial\xi_1'} + 2m_1A\xi_1' + m_1B\right] + \cdots\quad\cdots = 0.$$

With a suitable choice of the four factors, the coefficients of all twelve differentials will vanish; hence

$$\frac{1}{m}\frac{d\varphi}{d\xi} + 2A\xi + B = \frac{1}{m_1}\frac{d\Phi_1}{d\xi_1} + 2A\xi_1 + B = 0$$

or

$$\frac{1}{m}\frac{\partial\varphi}{\partial\xi} - \frac{1}{m_1}\frac{\partial\Phi_1}{\partial\xi_1} = 2A(\xi_1 - \xi).$$

Likewise it follows that

$$\frac{1}{m}\frac{\partial\varphi}{\partial\eta} - \frac{1}{m_1}\frac{\partial\Phi_1}{\partial\eta_1} = 2A(\eta_1 - \eta).$$

Elimination of A—which, as an undetermined factor, cannot in any case be set identically equal to zero—yields:

$$(151)\quad \left(\frac{1}{m}\frac{\partial\varphi}{\partial\xi} - \frac{1}{m_1}\frac{\partial\Phi_1}{\partial\xi_1}\right)(\eta_1 - \eta) = \left(\frac{1}{m}\frac{\partial\varphi}{\partial\eta} - \frac{1}{m_1}\frac{\partial\Phi_1}{\partial\eta_1}\right)(\xi_1 - \xi).$$

This equation contains, aside from the variables x, y, z, t, which we always consider constant, only the six completely independent variables ξ, η, ζ, ξ_1, η_1, ζ_1. If one takes the partial derivative with respect to ζ, it follows that:

$$\frac{\partial^2\varphi}{\partial\xi\partial\zeta}(\eta_1 - \eta) = \frac{\partial^2\varphi}{\partial\eta\partial\zeta}(\xi_1 - \xi).$$

Further partial differentiation of this equation with respect to η_1 gives:

$$\frac{\partial^2\varphi}{\partial\xi\partial\zeta} = 0.$$

Differentiation with respect to ξ_1 gives however:

$$\frac{\partial^2\varphi}{\partial\eta\partial\zeta} = 0,$$

and by cyclic permutation it follows that:

$$\frac{\partial^2\varphi}{\partial\xi\partial\eta} = 0.$$

These three equations express the fact that φ must break into three summands, of which the first contains only ξ, the second only η, and the third only ζ.

Similarly for the function Φ we obtain:

$$\frac{\partial^2 \Phi_1}{\partial \xi_1 \partial \eta_1} = \frac{\partial^2 \Phi_1}{\partial \xi_1 \partial \zeta_1} = \frac{\partial^2 \Phi_1}{\partial \eta_1 \partial \zeta_1} = 0. \tag{152}$$

Differentiation of Equation (151) with respect to ξ gives:

$$\frac{1}{m} \frac{\partial^2 \varphi}{\partial \xi^2} (\eta_1 - \eta) = -\frac{1}{m} \frac{\partial \varphi}{\partial \eta} + \frac{1}{m_1} \frac{\partial \Phi_1}{\partial \eta_1}, \tag{153}$$

since

$$\frac{\partial^2 \varphi}{\partial \xi \partial \eta} = 0.$$

Further differentiation of Equation (153) with respect to η_1 gives however:

$$\frac{1}{m} \frac{\partial^2 \varphi}{\partial \xi^2} = \frac{1}{m_1} \frac{\partial^2 \Phi_1}{\partial \eta_1^2}.$$

Since the two expressions on the left and right hand sides contain completely different variables, they can be equal only when both are independent of all variables and therefore are equal to a quantity independent of ξ, η, ζ, ξ_1, η_1, ζ_1.

Since the y and z axes enter the equations to be solved in exactly the same way, one can likewise prove the equation

$$\frac{1}{m} \frac{\partial^2 \varphi}{\partial \xi^2} = \frac{1}{m_1} \frac{\partial^2 \Phi_1}{\partial \zeta_1^2}$$

and also, that the last expression must be equal to

$$\frac{1}{m} \frac{\partial^2 \varphi}{\partial \eta^2}.$$

Hence all these second derivatives are equal to one and the same quantity $-2h$, independent of ξ, η, ζ, ξ_1, η_1, and ζ_1. From all these equations one easily draws the conclusion that φ must be equal to $-hm(\xi^2+\eta^2+\zeta^2)$ plus a linear function of ξ, η, ζ. The coefficients of the latter can be written in such a form that one can write, without loss of generality,

$$\varphi = -hm[(\xi - u)^2 + (\eta - v)^2 + (\xi - w)^2] + lf_0,$$

where u, v, w and f_0 are new constants, which however like h can still be functions of x, y, z, and t. Whence it follows that:

$$f = f_0 e^{-hm[(\xi-u)^2+(\eta-v)^2+(\zeta-w)^2]} \tag{154}$$

and one obtains similarly

$$F = F_0 e^{-hm[(\xi-u_1)^2+(\eta-v_1)^2+(\zeta-w_1)^2]}. \tag{155}$$

The functions f and F must in any case have this form when the three equations (147) are satisfied for all values of the variables. One sees easily that, conversely, when f and F have this form, Equations (147) are in fact satisfied, provided only that $u_1 = u$, $v_1 = v$, $w_1 = w$. The quantities f_0, F_0, u, v, w, h can be any functions of x, y, z, t.

These functions are to be determined in such a way that the two equations

$$\frac{\partial f}{\partial t} + \xi \frac{\partial f}{\partial x} + \eta \frac{\partial f}{\partial y} + \zeta \frac{\partial f}{\partial z} + X \frac{\partial f}{\partial \xi} + Y \frac{\partial f}{\partial \eta} + Z \frac{\partial f}{\partial \zeta} = 0 \tag{156}$$

and

$$\frac{\partial F}{\partial t} + \xi \frac{\partial F}{\partial x} + \eta \frac{\partial F}{\partial y} + \zeta \frac{\partial F}{\partial z} + X_1 \frac{\partial F}{\partial \xi} + Y_1 \frac{\partial F}{\partial \eta} + Z_1 \frac{\partial F_1}{\partial \zeta} = 0 \tag{157}$$

are satisfied; Equations (114) and (115) reduced to these, since their right hand sides vanish identically.

The number of molecules at time t in do whose velocity points are in $d\omega$ is:

$$f do d\omega = f_0 do e^{-hm[(\xi-u)^2+(\eta-v)^2+(\zeta-w)^2]} d\xi d\eta d\zeta.$$

If one sets

$$\xi = \mathfrak{x} + u, \quad \eta = \mathfrak{y} + v, \quad \zeta = \mathfrak{z} + w, \tag{158}$$

then he obtains just Equation (36), except that $\mathfrak{x}$, $\mathfrak{y}$, $\mathfrak{z}$ replace ξ, η, ζ.

From this one sees immediately that all the considerations

attached to Equation (36) remain valid, except that all the gas molecules have a common progressive motion through space, whose velocity components are u, v, w. When $u=u_1$, $v=v_1$, $w=w_1$, these are the components of the visible velocity with which the entire gas in do moves. If u, v, w differ from u_1, v_1, w_1, respectively, then u, v, w would be the components of the velocity with which the gas of the first kind moves through the gas of the second kind in do.

One can also see this in the following way. The number of m-molecules that lie in do at time t is:

$$dn = do\int fd\omega = dof_0\iiint_{-\infty}^{+\infty} e^{-hm[(\xi-u)^2+(\eta-v)^2+(\zeta-w)^2]}d\xi d\eta d\zeta.$$

On substituting (158) it follows that:

$$(159)\quad dn = dof_0\iiint_{-\infty}^{+\infty} e^{-hm(\mathfrak{x}^2+\mathfrak{y}^2+\mathfrak{z}^2)}d\mathfrak{x}d\mathfrak{y}d\mathfrak{z} = dof_0\sqrt{\frac{\pi^3}{h^3m^3}}\,.$$

If one multiplies this by m and divides by do, then he obtains the partial density of the first kind of gas:

$$(160)\qquad \rho = f_0\sqrt{\frac{\pi^3}{h^3m}}\,.$$

The mean value $\bar{\xi}$ of the velocity component along the abscissa direction of all m-molecules in do is:

$$(161)\qquad \bar{\xi} = \frac{\int \xi fd\omega}{\int fd\omega}\,.$$

This is clearly also the x-component of the velocity of the center of mass of the first kind of gas in do. If a surface element parallel to the yz plane moved with this velocity in the abscissa direction, an equal number of molecules of each kind would go through it, as follows directly from the concept of mean velocity. One can therefore call $\bar{\xi}$ the velocity with which the first kind of gas moves in the abscissa direction in do.

On substituting (158), the numerator of Equation (161) is transformed to:

$$f_0\iiint_{-\infty}^{+\infty} \mathfrak{x}e^{-hm(\mathfrak{x}^2+\mathfrak{y}^2+\mathfrak{z}^2)}d\mathfrak{x}d\mathfrak{y}d\mathfrak{z} + f_0u\iiint_{-\infty}^{+\infty} e^{-hm(\mathfrak{x}^2+\mathfrak{y}^2+\mathfrak{z}^2)}d\mathfrak{x}d\mathfrak{y}d\mathfrak{z}$$

One sees at once that the first term vanishes, but that the second reduces to *udn*. Hence

$$u = \bar{\xi} \tag{162}$$

Since $\mathfrak{x}$ is the relative velocity of a gas molecule with respect to a surface element moving with velocity u, and f is an even function of $\mathfrak{x}$, one sees at once that through each surface element $\perp$ to the x-axis, on the average as many molecules of the first kind go in as go out.

§19. Aerostatics. Entropy of a heavy gas whose motion does not violate Equations (147).

On substituting the values (154), Equation (156) yields many solutions each of which describes a state of the gas when it is under the influence of specified external forces in a container at rest. The walls of the container satisfy the conditions employed in the derivation of Equation (146). Thus after the cessation of all heat conduction and diffusion phenomena, there is no persistent flow of heat into or out of the gas. We shall now look for these solutions. It is clear that none of the relevant quantities here can depend on time.

Moreover, the equations $u=v=w=u_1=v_1=w_1=0$ must hold. Hence according to Equations (154) and (155):

$$f = f_0 e^{-hm(\xi^2+\eta^2+\zeta^2)}, \quad F = F_0 e^{-hm_1(\xi^2+\eta^2+\zeta^2)}, \tag{163}$$

where f_0, F_0 and h can still be functions of the coördinates. If we substitute this into Equation (156), it follows that:

$$- m(\xi^2 + \eta^2 + \zeta^2)\left(\xi \frac{\partial h}{\partial x} + \eta \frac{\partial h}{\partial y} + \zeta \frac{\partial h}{\partial z}\right)$$

$$+ \xi\left(\frac{\partial f_0}{\partial x} - 2hmf_0X\right) + \eta\left(\frac{\partial f_0}{\partial y} - 2hmf_0Y\right)$$

$$+ \zeta\left(\frac{\partial f_0}{\partial z} - 2hmf_0Z\right) = 0.$$

Since this equation must be valid for all values of ξ, η, and ζ, we see that:

$$\frac{\partial h}{\partial x} = \frac{\partial h}{\partial y} = \frac{\partial h}{\partial z} = 0.$$

Therefore h must be constant everywhere in space.

Furthermore, the coefficients of ξ, η, ζ in the following terms must vanish separately. This can happen only when X, Y, and Z are the partial derivatives of one and the same function $-\chi$ of the coördinates. If this condition is not satisfied, then the gas cannot in general be at rest. If it is satisfied, then:

$$f_0 = ae^{-2hm\chi} \tag{164}$$

where a is an absolute constant. Since in each volume element do the quantity f_0 is constant, Equation (163) must have the same form as Equation (36). The velocity distribution in each volume element is therefore exactly the same as it would be if only one kind of gas were present, and the same external forces acted on an equal partial density of it. In other words, in spite of the action of the external forces, every direction in space is equally probable for the direction of motion of a molecule. Since the problem to which the equations treated at the beginning of §7 refer is only a special case of the problem treated here, we see that the assumption made there without proof, viz., that all directions of motion of a molecule are equally probable, has now been proved. Because of the identity of the forms of the equations, the equations developed in §7 and the conclusions drawn therefrom are now applicable without change to each volume element. Hence, again corresponding to Equation (44), the mean square velocity of an m-molecule is:

$$\overline{c^2} = \frac{3}{2hm},$$

i.e., under the action of external forces the mean kinetic energy of each molecule also remains the same; likewise for the second kind of gas,

$$\overline{c_1^2} = \frac{3}{2hm_1}$$

and the constant h must have the same value for both kinds of gas. Let ρ be the partial density of the first kind of gas in do, and let p be the partial pressure which this kind of gas would exert on the wall if it were present alone in do; then according to Equations (160) and (164),

$$\rho = a\sqrt{\frac{\pi^3}{h^3 m}}\, e^{-2hm\chi}. \tag{165}$$

Since furthermore dn/do is the number of molecules in unit volume, then according to Equation (6):

$$p = \frac{m\overline{c^2}}{3}\frac{dn}{do} = \frac{\rho\overline{c^2}}{3} = \frac{\rho}{2hm}. \tag{166}$$

Hence p/ρ has the same value everywhere in the gas. Since now the gas in each volume element has exactly the same properties as it would have at the same partial density and equal energy if no external forces acted, then just as in the latter case we have $p/\rho = rT$. The gas constant r is, as earlier (§8), equal to $\frac{1}{2}hmT$. Since furthermore p/ρ has the same value everywhere and is equal to rT, then the temperature is also the same everywhere in spite of the action of the external forces.

For the second kind of gas one finds, completely independently of the presence of the first,

$$F_0 = Ae^{-2hm_1\chi_1},$$

where

$$\chi_1 = -\int (X_1dx + Y_1dy + Z_1dz).$$

The two gases therefore do not disturb each other in the equilibrium state. Thus in air completely at rest and in complete thermal equilibrium, each constituent of the air would form an atmosphere according to the laws given above, just as if the others were not present; except that for each gas, h and therefore temperature must have the same value. According to Equation (165):

$$\rho = \rho_0 e^{-2hm(\chi-\chi_0)} = \rho_0 e^{(\chi_0-\chi)/rT}, \tag{167}$$

and likewise according to (166),

$$p = p_0 e^{(\chi_0-\chi)/rT}. \tag{168}$$

Here p, ρ, and χ are the values of these quantities at any position with coördinates x, y, z, and p_0, ρ_0, χ_0 are the values of the same quantities at any other place with coördinates x_0, y_0, z_0. These are the well known formulas of aerostatics (barometric height measurements).

We shall now, following Bryan,* treat the following case

* Bryan and Boltzmann, Wien. Ber. 103, 1125 (1894).

which, although it does not occur in nature, is of theoretical interest. The two gases in the container will be separated by an arbitrary surface S_1 into two parts, a left part T_1, and a right part. To the right of the surface S_1 is a second surface S_2, which lies everywhere very close to S_1. The space between S_1 and S_2 will be called τ, and the space to the right of S_2 will be called T_2. Let χ be zero throughout T_1, and let it have positive values between S_1 and S_2, becoming infinite as one approaches S_2. Thus an m-molecule experiences no force in T_1, but in τ a force begins to act, pushing it away from S_2 toward S_1 and becoming infinite if the molecule approaches S_2. Conversely in T_2 no force acts on the m_1-molecules; however, when these enter τ, a force pushes them away from S_1 toward S_2, and becomes infinite near S_1. Thus χ_1 is zero in T_2, positive in τ, and becomes infinite near S_1.

If there are initially no m-molecules in T_2, then none can ever get into T_2; if there are initially some m-molecules in T_2, where no force acts on them, then each molecule that reaches the surface S_2 is expelled into T_1 and can never return. Hence we can assume in any case that the space T_2 contains no m-molecules, and likewise the space T_1 contains no m_1-molecules. This illustrates the formula; for one has

$$f = ae^{-hm(c^2+2\chi)}, \quad F = Ae^{-hm_1(c_1^2+2\chi_1)}.$$

In the space T_1, $\chi_1 = \infty$, hence $F=0$; in T_2, $\chi = \infty$, hence $f=0$. Also, Equation (167) gives the value zero for the partial density when χ is infinite. This in both T_1 and T_2 only pure gases are present; only in τ where both χ and χ_1 are present does one find a mixture. Our formula gives thermal equilibrium only when h has the same value for both gases, which according to Equation (44) implies that the mean kinetic energy of a molecule must have the same value for both gases. This is exactly the same condition that we also found for the thermal equilibrium of two mixed gases. The mechanical conditions just discussed are of course somewhat different from those for two gases separated by a solid heat-conducting wall; but they have a certain similarity. We can imagine a third surface S_3 to the right of S_2 and everywhere near it. By an appropriate choice of χ for three different kinds of gas, it can be arranged that molecules of the first kind are present only to the left of S_2, those of the second kind only to the right of S_2, and those of the third kind only between S_1 and S_3. This third gas thus facilitates the heat exchange between the first and the second kinds; equality of mean kinetic energy for each kind of gas is then the condition of thermal equilibrium. Since experience shows that the condition of thermal equilibrium of two bodies is independent of the nature of the body that facilitates the heat exchange, then the assumption made in §7, that equality of mean kinetic energy is the condition of thermal equilibrium when the heat exchange takes place in some other way—e.g., through a solid wall separating the gases—is seen to be very probable.

The solution of Equations (156) and (157) found in this section is the only one possible, provided that

$$u = v = w = u_1 = v_1 = w_1 = 0$$

and that everything is independent of time. However, if one assumes that these quantities are different from zero, then each of these equations has many solutions, representing motions in which H does not decrease and hence the total entropy does not increase. One can have a gas mixture moving with constant velocity in some fixed direction in space. There are also many other solutions. One sees immediately that, if the wall of the container is an absolutely smooth surface of revolution, at which the molecules are reflected like elastic spheres, then

$$\frac{d}{dt}\sum_{\omega,o} lf$$

reduces to $C_4(lf)+C_5(lf)$. Then entropy neither flows in nor out of the gas. For the stationary state we must have $dH/dt=0$, hence Equations (147) and also (156) and (157) must hold. Such a possible stationary state consists however in a uniform rotation of the entire gas mixture around the axis of rotation of the container, as if the gas were a solid body. This state must certainly be represented by Equations (154) and (155). If the z-axis is the axis of rotation, then in this case

$$u = u_1 = -by,\ v = v_1 = +bx. \quad w = w_1 = 0.$$

One can then satisfy Equations (156) and (157); f_0 and F_0 will be functions of $\sqrt{x^2+y^2}$, and will thus show the density variations created in the gas by centrifugal force. Other solutions of these equations, in which t can also occur explicitly, can also be found.[1] For example, there is one remarkable solution in which the

[1] Boltzmann, Wien. Ber. **74**, 531 (1876).

gas flies outward from a center equally in all directions in such a manner that, first, there is no viscosity; and second, while the temperature of course drops as a result of the expansion, it drops equally everywhere in space so that no heat conduction occurs. We shall no longer concern ourselves with these matters, however; we shall only ask what value the quantity H has in all these cases.

If we denote by H' the contribution to H for the first kind of gas, in Equation (144), then:

$$H' = \int \int do d\omega f l f.$$

In all cases where Equation (147) is not violated, f is given by Equation (154). If one sets, in accordance with Equation (160),

$$\rho = f_0 \sqrt{\frac{\pi^3}{h^3 m}},$$

then:

$$f = \sqrt{\frac{h^3 m}{\pi^3}} \rho e^{-hm[(\xi-u)^2+(\eta-v)^2+(\zeta-w)^2]}.$$

The integration over $d\omega = d\xi d\eta d\zeta$ can be carried out at once; it is to be extended over all values of ξ, η, ζ from $-\infty$ to $+\infty$ and one obtains, on doing this:

$$H' = \int do f_0 \sqrt{\frac{\pi^3}{h^3 m^3}} \left[l\left(\rho \sqrt{\frac{h^3 m}{\pi^3}}\right) - \frac{3}{2} \right],$$

or, using Equation (159),

$$H' = \int dn \left[l\left(\rho \sqrt{\frac{h^3 m}{\pi^3}}\right) - \frac{3}{2} \right]. \tag{169}$$

Now $mdn = dm$ is the total mass of the first gas in the volume element. If we multiply Equation (169) by the mass M of a molecule of the standard gas (hydrogen), further by the gas constant R of this gas, and finally by -1, and if we again denote by $\mu = m/M$ the molecular weight of our gas relative to the standard gas, then:

$$-MRH' = -\int \frac{Rdm}{\mu} \left[l\left(\rho \sqrt{\frac{h^3 m}{\pi^3}}\right) - \frac{3}{2} \right].$$

Since according to Equations (44) and (51a),

$$\overline{c^2} = \frac{3}{2hm} = \frac{3R}{\mu} T$$

then:

$$l\left(\rho\sqrt{\frac{h^3m}{\pi^3}}\right) = l(\rho T^{-3/2}) + l\sqrt{\frac{m}{8\pi^3M^3R^3}},$$

the last logarithm is however a constant. Further,

$$\int \frac{Rdm}{\mu} = \frac{Rm}{\mu}$$

which is a constant in any case. If one combines all the constants, then it follows that:

$$-MRH' = \int \frac{Rdm}{\mu} l(\rho^{-1}T^{3/2}) + \text{Const.} \tag{170}$$

According to Equation (58), however, this is just the sum of the entropies of all the masses in all the volume elements, and hence it is the total entropy of the first kind of gas, and one sees from Equation (144) that one simply adds the entropies of the two parts of the gas mixture. Neither a progressive motion of the gas nor the action of external forces has any effect on the entropy, as long as Equations (147) hold, and therefore the velocity distribution in each volume element is given by Equations (154) and (155). Thus we have completed the proof—which was given only incompletely in §8—that H is identical to the entropy, aside from a factor $-RM$ which is constant for all gases, and an additive constant.

§20. General form of the hydrodynamic equations.

Before we pass to the consideration of further special cases, we shall develop some general formulas. Since u, v, w are the components of the velocity with which the first gas moves as a whole, one easily sees that during time dt there will flow through the two faces perpendicular to the abscissa axis of the elementary parallelepiped $dxdydz$ the amounts $\rho udydzdt$ and

$$-\left[\rho u + \frac{\partial(\rho u)}{\partial x}dx\right]dydzdt$$

of gas, respectively.

The sum of these quantities, added to the total amount of gas that flows through the other four faces, is the total increase

$$\frac{\partial \rho}{\partial t}dxdydzdt$$

of the amount of the first kind of gas in the parallelepiped; whence it follows that:

$$\frac{\partial \rho}{\partial t} + \frac{\partial(\rho u)}{\partial x} + \frac{\partial(\rho v)}{\partial y} + \frac{\partial(\rho w)}{\partial z} = 0. \tag{171}$$

This is the so-called continuity equation. If one imagines an equal parallelepiped $do\alpha dxdydz$ moving in space at a velocity whose components are u, v, w, then during time dt the coördinates of the molecules contained therein will increase on the average by udt, vdt, and wdt. The average acceleration is therefore:

$$\frac{\partial u}{\partial t} + u\frac{\partial u}{\partial x} + v\frac{\partial u}{\partial y} + w\frac{\partial u}{\partial z}\cdot$$

If

$$\sum m = \rho dxdydz$$

is the total mass of these molecules, then their total momentum measured in the abscissa direction increases by

$$\left(\frac{\partial u}{\partial t} + u\frac{\partial u}{\partial x} + v\frac{\partial u}{\partial y} + w\frac{\partial u}{\partial z}\right)\cdot \rho dxdydz. \tag{172}$$

This increase in momentum will be in part created by the external forces acting on the total gas mass $\sum m$, whose components are

$$X\sum m, \quad Y\sum m, \quad Z\sum m.$$

If only one kind of gas is present, the total momentum will not change, because of the conservation of the motion of the center of mass in collisions; however, it will be changed by entrances and exits of molecules in do. If we denote by ξ, η, ζ the velocity components of some molecule, and set (see Eq. [158]) $\xi=u+\mathfrak{x}$, $\eta=u+\mathfrak{y}$, $\zeta=u+\mathfrak{z}$, then $\mathfrak{x}$, $\mathfrak{y}$, $\mathfrak{z}$ are the components of motion of the molecule relative to the volume element *do*. If in unit volume there are $fd\omega$ molecules whose velocity points lie within $d\omega$, then there will enter through the left of the side surface, facing the negative abscissa direction, of the parallelepiped *do*, during time *dt*,

$$\mathfrak{x} f d\omega dt dy dz$$

molecules, whose velocity points lie within $d\omega$; these transport momentum

$$m\mathfrak{x}(u + \mathfrak{x}) f d\omega dt dy dz$$

into the parallelepiped. Since $\xi = \bar{\xi} + \mathfrak{x}$,

$$\bar{\mathfrak{x}} = \frac{\int \mathfrak{x} f d\omega}{\int f d\omega} = 0$$

so that the total momentum transported through the left side of the parallelepiped *do* is

$$m dy dz dt \int \mathfrak{x}^2 f d\omega = P,$$

where the integration is to be extended over all volume elements $d\omega$.

$\int f d\omega$ is the total number of molecules in unit volume, hence

$$m \int f d\omega = \rho$$

is the density of the gas. The quantity

$$\frac{\int \mathfrak{x}^2 f d\omega}{\int f d\omega}$$

we call the mean value $\overline{\mathfrak{x}^2}$ of all $\mathfrak{x}^2$. Hence

$$P = \rho \overline{\mathfrak{x}^2} dy dz dt.$$

Through the opposite face, momentum

$$-\left[\rho\overline{\mathfrak{x}^2} + \frac{\partial(\rho\overline{\mathfrak{x}^2})}{\partial x} dx\right] dy dz dt$$

will be transported. Similarly one finds that through the sides of *do* perpendicular to the *y*-axis the amounts

$$\rho\overline{\mathfrak{x}\mathfrak{y}}\,dxdzdt$$

and

$$-\left[\rho\overline{\mathfrak{x}\mathfrak{y}}+\frac{\partial(\rho\overline{\mathfrak{x}\mathfrak{y}})}{\partial y}dy\right]dxdzdt$$

respectively of momentum in the abscissa direction will be transported. If one applies the same considerations to the last two sides, and finally sets the total increase (172) of momentum in the abscissa direction equal to the sum of the total momentum transported and the increase caused by the action of external forces, then it follows that:

$$(173)\qquad \begin{cases} \rho\left(\dfrac{\partial u}{\partial t}+u\dfrac{\partial u}{\partial x}+v\dfrac{\partial u}{\partial y}+w\dfrac{\partial u}{\partial z}\right) \\ =\rho X-\dfrac{\partial(\rho\overline{\mathfrak{x}^2})}{\partial x}-\dfrac{\partial(\rho\overline{\mathfrak{x}\mathfrak{y}})}{\partial y}-\dfrac{\partial(\rho\overline{\mathfrak{x}\mathfrak{z}})}{\partial z} \end{cases}$$

with two similar equations for the y and z axes. These equations, as well as Equation (171), are only special cases of the general equation (126) and were derived from it by Maxwell and (following him) by Kirchhoff.* One can see this in the following way:

Let ψ be an arbitrary function of x, y, z, ξ, η, ζ, t, which may be equal to the function earlier denoted by φ or may be different from it. Then the average of all values taken by ψ for all molecules in do at time t is:

$$(174)\qquad \bar{\psi}=\frac{\int\psi f d\omega}{\int f d\omega}.$$

Further,

$$m\,do\int f d\omega=\rho\,do$$

is the total mass of the first gas contained in a volume element do, so that one therefore has:

$$(175)\qquad m\int\psi f d\omega=\rho\bar{\psi}.$$

---

* Maxwell, Phil. Trans. **157**, 49 (1867); Kirchhoff, *Vorlesungen über die Theorie der Wärme* (Leipzig, 1894).

Using this kind of notation,

$$m \sum\nolimits_{\omega,do} \varphi = mdo\int \varphi f d\omega = \rho\bar{\varphi}do. \tag{176}$$

If one denotes by $\bar{\bar{\psi}}$ the mean value of ψ in all volume elements of the gas, and by $\mathfrak{m}$ the total mass of the first gas, then

$$\bar{\bar{\psi}} = \frac{\int\int \psi f do d\omega}{\int\int f do d\omega},$$

$$\mathfrak{m} = m\int\int f do d\omega$$

hence

$$\bar{\bar{\psi}} = \frac{m}{\mathfrak{m}}\int\int \psi f do d\omega.$$

One may therefore write:

$$H = \frac{\mathfrak{m}}{m}\overline{\overline{lf}} + \frac{\mathfrak{m}_1}{m_1}\overline{\overline{lF}} = \mathfrak{Z}\overline{\overline{lf}} + \mathfrak{Z}_1\overline{\overline{lF}},$$

where $\mathfrak{Z}$ and $\mathfrak{Z}_1$ are the total numbers of molecules of the first and second kinds of gas, respectively.

In the following, ψ shall be a function only of ξ, η, ζ, so that according to Equation (127):

$$B_1(\varphi) = 0.$$

Since furthermore ψ does not contain the coördinates, then according to Equations (128) and (175):

$$mB_2(\varphi) = -m\left[\frac{\partial}{\partial x}\int \xi\varphi f d\omega + \frac{\partial}{\partial y}\int \eta\varphi f d\omega + \frac{\partial}{\partial z}\int \zeta\varphi f d\omega\right]$$

$$= -\frac{\partial(\rho\overline{\xi\varphi})}{\partial x} - \frac{\partial(\rho\overline{\eta\varphi})}{\partial y} - \frac{\partial(\rho\overline{\zeta\varphi})}{\partial z}.$$

Since X, Y, Z are not functions of ξ, η, ζ, it follows from Equation (130) that:

$$mB_3(\varphi) = \rho\left[X\overline{\frac{\partial\varphi}{\partial\xi}} + Y\overline{\frac{\partial\varphi}{\partial\eta}} + Z\overline{\frac{\partial\varphi}{\partial\zeta}}\right].$$

If one collects all these terms, then Equation (126) reduces in this special case to:

$$(177)\quad \begin{cases} \dfrac{\partial(\rho\overline{\varphi})}{\partial t}+\dfrac{\partial(\rho\overline{\xi\varphi})}{\partial x}+\dfrac{\partial(\rho\overline{\eta\varphi})}{\partial y}+\dfrac{\partial(\rho\overline{\zeta\varphi})}{\partial z} \\ -\rho\left[X\overline{\dfrac{\partial\varphi}{\partial\xi}}+Y\overline{\dfrac{\partial\varphi}{\partial\eta}}+Z\overline{\dfrac{\partial\varphi}{\partial\zeta}}\right]=m[B_4(\varphi)+B_5(\varphi)]. \end{cases}$$

From this equation Maxwell calculated the viscosity, diffusion, and heat conduction, and Kirchhoff therefore calls it the basic equation of the theory.* If one sets $\varphi=1$, he obtains at once the continuity equation (171); for it follows from Equations (134) and (137) that $B_4(1)=B_5(1)=0$. Subtraction of the continuity equation, multiplied by φ, from (177) gives (using the substitution [158]):[1]

---

[1] For a better understanding of §3 of the 15th lecture of Kirchhoff's *Vorlesungen* we make the following remark:

Since φ is a function only of ξ, η, ζ, it transforms by the substitution (158) into a function of $\mathfrak{x}+u$, $\mathfrak{y}+v$, $\mathfrak{z}+w$, and therefore

$$\frac{\partial\varphi}{\partial\xi}=\frac{\partial\varphi}{\partial u}=\frac{\partial\varphi}{\partial\mathfrak{x}},$$

when in the last two derivatives φ is considered as a function of $u+\mathfrak{x}$, $v+\mathfrak{y}$, $w+\mathfrak{z}$. Therefore:

$$\overline{\frac{\partial\varphi}{\partial\xi}}=\overline{\frac{\partial\varphi}{\partial u}}=\overline{\frac{\partial\varphi}{\partial\mathfrak{x}}}.$$

Kirchhoff now denotes by $(\partial\overline{\varphi})/(\partial u)$ the derivatives obtained by leaving the quantities u, v, w, explicitly in $\varphi(u+\mathfrak{x}, v+\mathfrak{y}, w+\mathfrak{z})$ and taking partial derivatives with respect to u, holding constant the mean values of the coefficients containing $\mathfrak{x}$, $\mathfrak{y}$, $\mathfrak{z}$; these coefficients do not need to be considered as functions of u, v, w, or their derivatives with respect to the coördinates. Nor are u, v, w to be regarded as functions of x, y, z. Then

$$\overline{\frac{\partial\varphi}{\partial u}}=\frac{\partial\overline{\varphi}}{\partial u},$$

hence also

$$\overline{\frac{\partial\varphi}{\partial\xi}}=\frac{\partial\overline{\varphi}}{\partial u}.$$

The same holds of course for the other two coördinates.

$$(178)\quad \begin{cases} \rho\dfrac{\partial\overline{\varphi}}{\partial t}+\rho u\dfrac{\partial\overline{\varphi}}{\partial x}+\rho v\dfrac{\partial\overline{\varphi}}{\partial y}+\rho w\dfrac{\partial\overline{\varphi}}{\partial z}+\dfrac{\partial(\rho\overline{\mathfrak{x}\varphi})}{\partial x} \\ +\dfrac{\partial(\rho\overline{\mathfrak{y}\varphi})}{\partial y}+\dfrac{\partial(\rho\overline{\mathfrak{z}\varphi})}{\partial z}-\rho\left[X\overline{\dfrac{\partial\varphi}{\partial\xi}}+Y\overline{\dfrac{\partial\varphi}{\partial\eta}}+Z\overline{\dfrac{\partial\varphi}{\partial\zeta}}\right] \\ = m[B_4(\varphi)+B_5(\varphi)]. \end{cases}$$

If one has only one kind of gas, then $B_4(\varphi)$ always vanishes. If one substitutes in the above equation:

$$\varphi = \xi = u + \mathfrak{x},$$

then

$$\varphi + \varphi_1 = \varphi' + \varphi_1'$$

by conservation of momentum. Hence $B_5(\varphi)$ always vanishes. Further,

$$\overline{\mathfrak{x}} = \overline{\mathfrak{y}} = \overline{\mathfrak{z}} = 0, \quad \frac{\partial\varphi}{\partial\xi} = 1, \quad \frac{\partial\varphi}{\partial\eta} = \frac{\partial\varphi}{\partial\zeta} = 0$$

and one obtains precisely Equation (173).

If one denotes the six quantities

$$(179)\quad \begin{cases} \qquad \rho\overline{\mathfrak{x}^2},\ \rho\overline{\mathfrak{y}^2},\ \rho\overline{\mathfrak{z}^2},\ \rho\overline{\mathfrak{y}\mathfrak{z}},\ \rho\overline{\mathfrak{x}\mathfrak{z}},\ \rho\overline{\mathfrak{x}\mathfrak{y}} \\ \text{by} \\ \qquad X_x,\ Y_y,\ Z_z,\quad Y_z = Z_y,\quad Z_x = X_z,\quad X_y = Y_x, \end{cases}$$

then Equation (173) transforms to:

$$(180)\quad \begin{cases} \rho\left(\dfrac{\partial u}{\partial t}+u\dfrac{\partial u}{\partial x}+v\dfrac{\partial u}{\partial y}+w\dfrac{\partial u}{\partial z}\right) \\ +\dfrac{\partial X_x}{\partial x}+\dfrac{\partial X_y}{\partial y}+\dfrac{\partial X_z}{\partial z} = \rho X; \end{cases}$$

two similar equations are of course valid for the other two coördinate axes.

One would obtain exactly the same equations in a case whose mechanical conditions are completely different from those just considered. Suppose that the molecules contained in each volume element have no motion except their common motion with velocity components u, v, w, but, as in a solid elastic body, when one constructs in the gas a surface element dS perpendicular to the abscissa axis, those molecules immediately to the left of dS (facing toward the negative abscissa direction) exert on those molecules immediately to the right of dS a force whose components are $X_x dS$, $X_y dS$, $X_z dS$. Similarly, of course, there will be surface elements perpendicular to the other coördinates.

Taking account of molecular forces, one would again obtain Equation (180) together with the two analogous equations for the y and z axes. Each volume element in the gas has the same properties as if this force actually acted between the molecules on the left and those on the right of a surface element. The molecular motion produces the appearance of such a force; any force can be explained dynamically, as it were, in terms of the molecular motion in the gas. For example, when the molecules on the left of dS have larger velocities than those on the right, then the slower ones diffuse towards the left and the faster ones towards the right; the mean velocity of the molecules in a volume element to the right of dS will increase, and that of the molecules in a volume element to the left will decrease. The net effect is just the same as if the molecules on the left had exerted a force in the positive direction on the molecules on the right, and conversely.

Thus molecular motion creates the appearance of a molecular force, and in an agitated gas the pressure is not exactly the same in all directions, nor is it always exactly normal to the surface.

We now imagine that the gas is enclosed by a surface impenetrable to the molecules, and ask what force it exerts on the surface. Let dS be a surface element, and let its plane be perpendicular to the x-axis. Let the gas move with velocity components u, v, w at the place in question. If the motion of the gas itself experiences no sudden change, then during time dt there will collide with dS, $\mathfrak{x} f d\omega dt$ molecules whose velocity points lie in $d\omega$, when $\mathfrak{x}$ is positive; or they are reflected from it, when $\mathfrak{x}$ is negative.

The total momentum in the abscissa direction transported by the reflected molecules is therefore $mdSdt\int \mathfrak{x}^2 f d\omega = \rho\overline{\mathfrak{x}^2}dt\cdot dS$, and likewise that transported in the other directions is $\rho\overline{\mathfrak{x}\mathfrak{y}}dtdS$ and $\rho\overline{\mathfrak{x}\mathfrak{z}}dtdS$. X_x, Y_x, and Z_x are the components of the force corresponding to unit surface which dS exerts on the gas, and which conversely the gas exerts on dS, as long as there is no discontinuity of motion at the surface. Likewise one can also find from the kinetic theory the well-known expression for the force acting on any arbitrarily directed surface element of the wall.

In particular, when the gas is at rest in a container that is also at the rest, the pressure law follows directly from the principle of conservation of the motion of the center of gravity. If one applies this principle to a gas in a cylindrical container whose axis is parallel to the abscissa direction, then he finds that the pressure on the lateral surface of the container has no component in the abscissa direction. Application of the principle to the part of the gas between an end surface and a cross section shows that the pressure on the end surface is normal, and is equal to the momentum transported through unit cross section in the same direction; hence it must be equal to $\rho\overline{\xi^2}$, or also equal to $\frac{1}{3}\rho(\overline{\xi^2}+\overline{\eta^2}+\overline{\zeta^2})$, since in this case $\overline{\xi^2}=\overline{\eta^2}=\overline{\zeta^2}$.

Whenever Equations (147) are satisfied for all values of the variables, the number of molecules in the volume element *do* for which the components of the velocity relative to the total motion of the gas in the same volume lie between the limits $\mathfrak{x}$ and $\mathfrak{x}+d\mathfrak{x}$, $\mathfrak{y}$ and $\mathfrak{y}+d\mathfrak{y}$, $\mathfrak{z}$ and $\mathfrak{z}+d\mathfrak{z}$ is equal to

$$do f_0 e^{-hm(\mathfrak{x}^2+\mathfrak{y}^2+\mathfrak{z}^2)} d\mathfrak{x}\, d\mathfrak{y}\, d\mathfrak{z},$$

where f_0 is a function only of x, y, and z. Therefore the probability of this relative motion is given by exactly the same formula as that which applies to the absolute velocities in a stationary gas. The only difference is that the visible motion, which has components u, v, w, is included in the formula. This progressive motion of the gas as a whole obviously has no effect on its internal state, and hence no effect on the temperature and pressure of the gas, which are expressed in the same way in terms of $\mathfrak{x}$, $\mathfrak{y}$, $\mathfrak{z}$ as they are in terms of ξ, η, ζ for the stationary gas. Hence, corresponding to our previous results, we have

$$p = \rho\overline{\mathfrak{x}^2} = \rho\overline{\mathfrak{y}^2} = \rho\overline{\mathfrak{z}^2}, \quad \overline{\mathfrak{x}\mathfrak{y}} = \overline{\mathfrak{x}\mathfrak{z}} = \overline{\mathfrak{y}\mathfrak{z}} = 0. \tag{181}$$

As we shall see later, whenever there are no external forces the quantities

$$\overline{\mathfrak{x}^2} - \overline{\mathfrak{y}^2}, \quad \overline{\mathfrak{x}^2} - \overline{\mathfrak{z}^2}, \quad \overline{\mathfrak{y}^2} - \overline{\mathfrak{z}^2}, \quad \overline{\mathfrak{x}\mathfrak{y}}, \quad \overline{\mathfrak{x}\mathfrak{z}}, \quad \overline{\mathfrak{y}\mathfrak{z}} \tag{182}$$

rapidly approach zero as a result of the effect of collisions. When external forces hinder this process, these quantities still never differ significantly from zero as long as the external influences are not enormously sudden and violent. For the present, we assume as a fact of experience that in gases the normal pressure is always nearly equal in all directions, and that tangential elastic forces are very small, so that Equations (181) are approximately true. Substitution of the values given by this equation into Equation (173) yields:

$$\rho\left(\frac{\partial u}{\partial t}+u\frac{\partial u}{\partial x}+v\frac{\partial u}{\partial y}+w\frac{\partial u}{\partial z}\right)+\frac{\partial p}{\partial x}-\rho X=0 \tag{183}$$

with two similar equations for the y and z axes. These are the well-known hydrodynamic equations without viscosity or heat conduction; they are to be regarded as first approximations.

We now denote by Φ any function of x, y, z, and t. $(\partial\Phi/\partial t)dt$ is the increase experienced by the value of this function at a point A fixed in space during time dt. Now let the point A move with a velocity (u, v, w) equal to the total velocity of the first kind of gas in the volume element do. During the time dt, A becomes A'. If one now substitutes for the value of the function Φ at time $t+dt$ at A' its value at time t at A, and divides the difference by the elapsed time, then he obtains

$$\frac{\partial\Phi}{\partial t}+u\frac{\partial\Phi}{\partial x}+v\frac{\partial\Phi}{\partial y}+w\frac{\partial\Phi}{\partial z},$$

which value we shall denote by $d\Phi/dt$ for short. One can then write the continuity equation and the first of the hydrodynamic equations in the following form:

$$\frac{d\rho}{dt}+\rho\left(\frac{\partial u}{\partial x}+\frac{\partial v}{\partial y}+\frac{\partial w}{\partial z}\right)=0, \tag{184}$$

$$\rho\frac{du}{dt}+\frac{\partial(\rho\overline{\mathfrak{x}^2})}{\partial x}+\frac{\partial(\rho\overline{\mathfrak{x}\mathfrak{y}})}{\partial y}+\frac{\partial(\rho\overline{\mathfrak{x}\mathfrak{z}})}{\partial z}-\rho X=0. \tag{185}$$

The latter equation reads, in first approximation:

$$\rho\frac{du}{dt}+\frac{\partial p}{\partial x}-\rho X=0. \tag{186}$$

The completely exact Equation (178) can however be written:[2]

[2] As Poincaré (C. R. Paris **116**, 1017 [1893]) remarks, the derivatives of φ appearing in this equation may be functions only of ξ, η, ζ or of

$$
(187)\quad \begin{cases} \rho \dfrac{d\overline{\varphi}}{dt} + \dfrac{\partial(\rho\overline{\mathfrak{x}\varphi})}{\partial x} + \dfrac{\partial(\rho\overline{\mathfrak{y}\varphi})}{\partial y} + \dfrac{\partial(\rho\overline{\mathfrak{z}\varphi})}{\partial z} \\ -\rho\left(X\dfrac{\overline{\partial\varphi}}{\partial\xi} + Y\dfrac{\overline{\partial\varphi}}{\partial\eta} + Z\dfrac{\overline{\partial\varphi}}{\partial\zeta}\right) = m[B_4(\varphi) + B_5(\varphi)]. \end{cases}
$$

Now assume again that only one kind of gas is present. Then

$$
(187a)\qquad B_4(\varphi) = 0.
$$

Let φ be a complete function of ξ, η, ζ. Then

$$
(187b)\qquad \varphi(\xi,\eta,\zeta) = \mathfrak{f} = u\frac{\partial\mathfrak{f}}{\partial\mathfrak{x}} + v\frac{\partial\mathfrak{f}}{\partial\mathfrak{y}} + w\frac{\partial\mathfrak{f}}{\partial\mathfrak{z}} + Q_2,
$$

where $\mathfrak{f}$ is an abbreviation for $\varphi(\mathfrak{x}, \mathfrak{y}, \mathfrak{z})$ and Q_n denotes a function of u, v, w, containing no terms whose degree in u, v, w is lower than n. The coefficients of Q_2 are functions of $\mathfrak{x}$, $\mathfrak{y}$, $\mathfrak{z}$. According to Equation (143),

$$
(187c)\qquad B_5(\varphi) = B_5(\mathfrak{f}) + uB_5\left(\frac{\partial\mathfrak{f}}{\partial\mathfrak{x}}\right) + \cdots.
$$

Furthermore,

$$
\frac{\partial\varphi}{\partial\xi} = \frac{\partial\varphi}{\partial u} = \frac{\partial\mathfrak{f}}{\partial\mathfrak{x}} + \frac{\partial Q_2}{\partial u},
$$

hence

$$
(187d)\qquad \overline{\frac{\partial\varphi}{\partial\xi}} = \overline{\frac{\partial\mathfrak{f}}{\partial\mathfrak{x}}} + Q_1 \text{ etc.}
$$

The coefficients of Q_1 are mean values of functions of $\mathfrak{x}$, $\mathfrak{y}$ and $\mathfrak{z}$. Similar equations follow for $(\partial\varphi/\partial\eta)$ and $(\partial\varphi/\partial\zeta)$. If one

---

$u+\mathfrak{x}$, $v+\mathfrak{y}$, $w+\mathfrak{z}$; they may not be arbitrary functions of u, v, w, $\mathfrak{x}$, $\mathfrak{y}$, $\mathfrak{z}$. In the following equations, however, $\mathfrak{f}$ is a function of $\mathfrak{x}$, $\mathfrak{y}$, $\mathfrak{z}$ and $B_5(\mathfrak{f})$ is the expression obtained when one substitutes, in the expression (137) for φ, φ_1, φ', and φ_1', $\mathfrak{f}=\varphi(\mathfrak{x}, \mathfrak{y}, \mathfrak{z})$, $\mathfrak{f}_1=\varphi_1(\mathfrak{x}_1, \mathfrak{y}_1, \mathfrak{z}_1)$ and so forth. Since $\mathfrak{x}'$, $\mathfrak{y}'$, $\mathfrak{z}'$, $\mathfrak{x}_1'$, $\mathfrak{y}_1'$, $\mathfrak{z}_1'$ are given functions of $\mathfrak{x}$, $\mathfrak{y}$, $\mathfrak{z}$, $\mathfrak{x}_1$, $\mathfrak{y}_1$, $\mathfrak{z}_1$, b and ϵ, one can carry out the integrations over the latter 8 variables directly.

substitutes the values (187a–d) for φ, $\overline{\partial\varphi/\partial\xi}$, $\overline{\partial\varphi/\partial\eta}$, $\overline{\partial\varphi/\partial\zeta}$, $B_4(\varphi)$ and $B_5(\varphi)$ into Equation (187), then he obtains:

$$\rho\frac{d\overline{\mathfrak{f}}}{dt}+\frac{\partial(\rho\overline{\mathfrak{x}\mathfrak{f}})}{\partial x}+\frac{\partial(\rho\overline{\mathfrak{y}\mathfrak{f}})}{\partial y}+\frac{\partial(\rho\overline{\mathfrak{z}\mathfrak{f}})}{\partial z}-mB_5(\mathfrak{f})$$
$$+\overline{\frac{\partial\mathfrak{f}}{\partial\mathfrak{x}}}\rho\left(\frac{du}{dt}-X\right)+\rho\left(\frac{\partial u}{\partial x}\overline{\mathfrak{x}\frac{\partial\mathfrak{f}}{\partial\mathfrak{x}}}+\frac{\partial u}{\partial y}\overline{\mathfrak{y}\frac{\partial\mathfrak{f}}{\partial\mathfrak{x}}}+\frac{\partial u}{\partial z}\overline{\mathfrak{z}\frac{\partial\mathfrak{f}}{\partial\mathfrak{x}}}\right)$$
$$+\overline{\frac{\partial\mathfrak{f}}{\partial\mathfrak{y}}}\rho\left(\frac{dv}{dt}-Y\right)+\rho\left(\frac{\partial v}{\partial x}\overline{\mathfrak{x}\frac{\partial\mathfrak{f}}{\partial\mathfrak{y}}}+\frac{\partial v}{\partial y}\overline{\mathfrak{y}\frac{\partial\mathfrak{f}}{\partial\mathfrak{y}}}+\frac{\partial v}{\partial z}\overline{\mathfrak{z}\frac{\partial\mathfrak{f}}{\partial\mathfrak{y}}}\right)$$
$$+\overline{\frac{\partial\mathfrak{f}}{\partial\mathfrak{z}}}\rho\left(\frac{dw}{dt}-Z\right)+\rho\left(\frac{\partial w}{\partial x}\overline{\mathfrak{x}\frac{\partial\mathfrak{f}}{\partial\mathfrak{z}}}+\frac{\partial w}{\partial y}\overline{\mathfrak{y}\frac{\partial\mathfrak{f}}{\partial\mathfrak{z}}}+\frac{\partial w}{\partial z}\overline{\mathfrak{z}\frac{\partial\mathfrak{f}}{\partial\mathfrak{z}}}\right)=0.$$

In addition there are still terms containing first and higher powers of u, v, w. But these must vanish identically, since the internal state remains unchanged when one imparts a constant velocity through space to the gas as a whole. This velocity can always be chosen such that $u=v=w=0$. Referring to Equation (185), one can also write the last equation as:

$$(188)\quad\left\{\begin{aligned}mB_5(\mathfrak{f})&=\rho\frac{d\overline{\mathfrak{f}}}{dt}+\frac{\partial(\rho\overline{\mathfrak{x}\mathfrak{f}})}{\partial x}+\frac{\partial(\rho\overline{\mathfrak{y}\mathfrak{f}})}{\partial y}+\frac{\partial(\rho\overline{\mathfrak{z}\mathfrak{f}})}{\partial z}\\&+\rho\left(\frac{\partial u}{\partial x}\overline{\mathfrak{x}\frac{\partial\mathfrak{f}}{\partial\mathfrak{x}}}+\frac{\partial u}{\partial y}\overline{\mathfrak{y}\frac{\partial\mathfrak{f}}{\partial\mathfrak{x}}}+\frac{\partial u}{\partial z}\overline{\mathfrak{z}\frac{\partial\mathfrak{f}}{\partial\mathfrak{x}}}\right)\\&-\overline{\frac{\partial\mathfrak{f}}{\partial\mathfrak{x}}}\left(\frac{\partial(\rho\overline{\mathfrak{x}^2})}{\partial x}+\frac{\partial(\rho\overline{\mathfrak{x}\mathfrak{y}})}{\partial y}+\frac{\partial(\rho\overline{\mathfrak{x}\mathfrak{z}})}{\partial z}\right)\\&+\rho\left(\frac{\partial v}{\partial x}\overline{\mathfrak{x}\frac{\partial\mathfrak{f}}{\partial\mathfrak{y}}}+\frac{\partial v}{\partial y}\overline{\mathfrak{y}\frac{\partial\mathfrak{f}}{\partial\mathfrak{y}}}+\frac{\partial v}{\partial z}\overline{\mathfrak{z}\frac{\partial\mathfrak{f}}{\partial\mathfrak{y}}}\right)\\&-\overline{\frac{\partial\mathfrak{f}}{\partial\mathfrak{y}}}\left(\frac{\partial(\rho\overline{\mathfrak{x}\mathfrak{y}})}{\partial x}+\frac{\partial(\rho\overline{\mathfrak{y}^2})}{\partial y}+\frac{\partial(\rho\overline{\mathfrak{y}\mathfrak{z}})}{\partial z}\right)\\&+\rho\left(\frac{\partial w}{\partial x}\overline{\mathfrak{x}\frac{\partial\mathfrak{f}}{\partial\mathfrak{z}}}+\frac{\partial w}{\partial y}\overline{\mathfrak{y}\frac{\partial\mathfrak{f}}{\partial\mathfrak{z}}}+\frac{\partial w}{\partial z}\overline{\mathfrak{z}\frac{\partial\mathfrak{f}}{\partial\mathfrak{z}}}\right)\\&-\overline{\frac{\partial\mathfrak{f}}{\partial\mathfrak{z}}}\left(\frac{\partial(\rho\overline{\mathfrak{x}\mathfrak{z}})}{\partial x}+\frac{\partial(\rho\overline{\mathfrak{y}\mathfrak{z}})}{\partial y}+\frac{\partial(\rho\overline{\mathfrak{z}^2})}{\partial z}\right).\end{aligned}\right.$$

If one sets $\mathfrak{f}=\mathfrak{x}^2$ here, then, since $\overline{\mathfrak{x}}=0$,

$$(189)\quad \begin{cases} mB_5(\mathfrak{x}^2) = \rho \dfrac{d\overline{\mathfrak{x}^2}}{dt} + \dfrac{\partial(\rho\overline{\mathfrak{x}^3})}{\partial x} + \dfrac{\partial(\rho\overline{\mathfrak{x}^2\mathfrak{y}})}{\partial y} + \dfrac{\partial(\rho\overline{\mathfrak{x}^2\mathfrak{z}})}{\partial z} \\ \qquad + 2\rho\left(\overline{\mathfrak{x}^2}\dfrac{\partial u}{\partial x} + \overline{\mathfrak{x}\mathfrak{y}}\dfrac{\partial u}{\partial y} + \overline{\mathfrak{x}\mathfrak{z}}\dfrac{\partial u}{\partial z}\right) \cdot \end{cases}$$

If one sets $\mathfrak{f}=\mathfrak{x}\mathfrak{y}$, then

$$(190)\quad \begin{cases} mB_5(\mathfrak{x}\mathfrak{y}) = \rho \dfrac{d(\overline{\mathfrak{x}\mathfrak{y}})}{dt} + \dfrac{\partial(\rho\overline{\mathfrak{x}^2\mathfrak{y}})}{\partial x} + \dfrac{\partial(\rho\overline{\mathfrak{x}\mathfrak{y}^2})}{\partial y} + \dfrac{\partial(\rho\overline{\mathfrak{x}\mathfrak{y}\mathfrak{z}})}{\partial z} \\ + \rho\left\{\overline{\mathfrak{x}\mathfrak{y}}\dfrac{\partial u}{\partial x} + \overline{\mathfrak{y}^2}\dfrac{\partial u}{\partial y} + \overline{\mathfrak{y}\mathfrak{z}}\dfrac{\partial u}{\partial z} + \overline{\mathfrak{x}^2}\dfrac{\partial v}{\partial x} + \overline{\mathfrak{x}\mathfrak{y}}\dfrac{\partial v}{\partial y} + \overline{\mathfrak{x}\mathfrak{z}}\dfrac{\partial v}{\partial z}\right\}, \end{cases}$$

which is exactly correct.

If we now make the assumption that the state distribution corresponds approximately to Maxwell's law, then Equations (181) are approximately correct. In addition, $\overline{\mathfrak{x}^3}=\overline{\mathfrak{x}^2\mathfrak{y}}=\overline{\mathfrak{x}^2\mathfrak{z}}=\cdots=0$, for as a result of collisions the distribution of states always rapidly approaches the Maxwellian distribution. Any mean value that vanishes for the latter distribution will therefore be quite small, as we shall see in the next section when we consider the effect of collisions explicitly. In this approximation, Equation (189) transforms (taking account of Eq. [186]) to

$$(191)\qquad mB_5(\mathfrak{x}^2) = \rho \frac{d\left(\dfrac{p}{\rho}\right)}{dt} + 2p\frac{\partial u}{\partial x} \cdot$$

We now construct similar equations for the y and z axes, add all three equations together, and recall that

$$B_5(\mathfrak{x}^2) + B_5(\mathfrak{y}^2) + B_5(\mathfrak{z}^2) = B_5(\mathfrak{x}^2+\mathfrak{y}^2+\mathfrak{z}^2) = 0$$

since the total kinetic energy of two molecules is not changed by a collision. Thereby we obtain:

$$3\rho\frac{d\left(\dfrac{p}{\rho}\right)}{dt} + 2p\left(\frac{\partial u}{\partial x} + \frac{\partial v}{\partial y} + \frac{\partial w}{\partial z}\right) = 0,$$

or, using the continuity equation (184),

$$3\rho \frac{d\left(\frac{p}{\rho}\right)}{dt} - \frac{2p}{\rho}\frac{d\rho}{dt} = 3\frac{dp}{dt} - \frac{5p}{\rho}\frac{d\rho}{dt} = 0.$$

The integration gives, when one follows the gas mass in a volume element along its trajectory, $p\rho^{-5/3} = \text{constant}$, the well-known Poisson relation between pressure and density.* Conduction of heat is neglected here. In general we do not know anything at all about heat radiation. The ratio of specific heats is, in the case considered here, 5/3. Since the internal state of the gas is nearly the same as that of one in thermal equilibrium, moving with uniform velocity components u, v, w, the Boyle-Charles law is valid. Thus $p = r\rho T$, hence $T\rho^{-5/3} = \text{constant}$. Any compression is connected with an adiabatic temperature increase, and any expansion with a temperature decrease.

* S. D. Poisson, Ann. chim. phys. **23**, 337 (1823).

CHAPTER III

The molecules repel each other with a force inversely proportional to the fifth power of their distance.

§21. Integration of the terms resulting from collisions.

We now consider cases where Equations (147) are not satisfied; in order to be able to calculate the values ξ', η', ζ' of the variables after the collision as functions of their values before the collision, we must now consider the collision process in more detail.

We imagine a molecule of mass m (the m-molecule) to enter into collision, i.e., into interaction, with another molecule of mass m_1 (the m_1-molecule). At any time t, let x, y, z be the coördinates of the first and x_1, y_1, z_1 those of the second molecule. The force that the two molecules exert on each other will be a repulsion in the direction of their line of centers r, whose intensity $\psi(r)$ is a function of r. The equations of motion are:

$$(191a) \qquad m_1 \frac{d^2x_1}{dt^2} = \psi(r) \frac{x_1 - x}{r}, \quad m \frac{d^2x}{dt^2} = \psi(r) \frac{x - x_1}{r}$$

with four similar equations for the other coördinate axes.

In order to find the relative motion of the two molecules, we construct through m_1 a coördinate system whose axes remain parallel to the fixed coördinate axes, but move parallel to themselves in such a manner that they always go through the m_1-molecule, which is therefore at any time the origin of the second coördinate system. The coördinates of the m-molecule with respect to this second coördinate system, and thus its coördinates relative to the m_1-molecule, are:

$$\mathfrak{a} = x - x_1, \quad \mathfrak{b} = y - y_1, \quad \mathfrak{c} = z - z_1.$$

If one substitutes

$$\mathfrak{M} = \frac{mm_1}{m + m_1}, \quad \text{hence} \quad \frac{1}{\mathfrak{M}} = \frac{1}{m} + \frac{1}{m_1},$$

then he finds easily from Equations (191a)

$$\mathfrak{M}\frac{d^2\mathfrak{a}}{dt^2} = \psi(r)\frac{\mathfrak{a}}{r}$$

with two similar equations for the other two coördinate axes. Since also we have $r^2 = \mathfrak{a}^2 + \mathfrak{b}^2 + \mathfrak{c}^2$, these equations represent just the central motion that the m-molecule would perform if its mass were $\mathfrak{M}$ and it were repelled by the fixed m_1-molecule with the same force $\psi(r)$. Therefore we need to discuss only this latter central motion, which we call the relative central motion, or the central motion Z. It always takes place in the plane that includes m_1 and the initial velocity of m, which we have already called in §16 the orbital plane. The initial velocity of the m-molecule is considered to be that velocity which it has at a great distance from m_1, before the collision, and which we already denoted by g in the same section. The line g drawn from the fixed m_1-molecule in Figure 7 will represent it in magnitude and direction. Its exten-

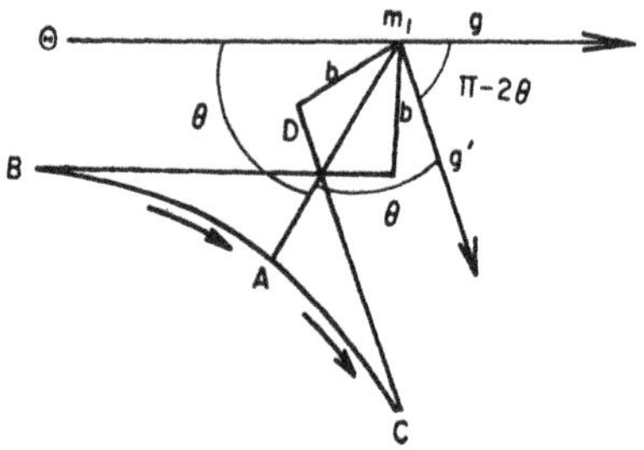

FIG. 7.

sion in the opposite direction is to be called $m\Theta$. We specify the position of m at any time t by its distance r from m_1 and by the angle β which r forms with $m\Theta$. The work performed from the beginning of the collision up to time t by the force $\psi(r)$ is:

$$\int_{\infty}^{r}\psi(r)dr = -R.$$

The integration can begin at $r = \infty$ since $\psi(r) = 0$ for distances greater than the sphere of action. For the present we consider only the central motion Z of the m-molecule around the mass $\mathfrak{M}$, since we know that the actual motion of m relative to m_1 is exactly equivalent. For this central motion Z, the kinetic energy before the collision is $\mathfrak{M}g^2/2$, but at time t it is:

$$\frac{\mathfrak{M}}{2}\left[\left(\frac{dr}{dt}\right)^2 + r^2\left(\frac{d\beta}{dt}\right)^2\right].$$

The equation of energy conservation for the central motion Z therefore reads:

$$\frac{\mathfrak{M}}{2}\left[\left(\frac{dr}{dt}\right)^2 + r^2\left(\frac{d\beta}{dt}\right)^2\right] - \frac{\mathfrak{M}g^2}{2} = -R. \tag{192}$$

As in §16 we denote by b the smallest distance from m_1 that the m-molecule would reach if there were no interaction—i.e., if both molecules moved in straight lines in the same directions as they did before the collisions. The orbit that the m-molecule describes during the central motion Z will therefore have the form of the curved line in Figure 7, which extends to infinity on both sides; both asymptotes have the same distance from m_1, namely b. Since before the collision the m-molecule has relative velocity g with respect to m_1, twice the area described by the radius vector r in unit time is equal to bg; however, at time t this must be equal to $r^2 d\beta/dt$. Hence, according to the law of areas,

$$r^2 \frac{d\beta}{dt} = bg. \tag{193}$$

From this equation and Equation (192) it follows by well-known arguments that:

$$d\beta = \frac{d\rho}{\sqrt{1 - \rho^2 - \dfrac{2R}{\mathfrak{M}g^2}}},$$

where $\rho = b/r$. Since initially β and ρ are increasing, one must choose the positive sign of the square root until it vanishes. We now specialize the function ψ in order to carry out the integration, by substituting:

$$\psi(r) = \frac{K}{r^{n+1}}. \tag{194}$$

This is the repulsion between an m-molecule and an m_1-molecule at distance r. At the same distance the repulsion between two m-molecules will be equal to K_1/r^{n+1}, while that between two m_1-molecules will be K_2/r^{n+1}.

Then:

$$R = \frac{K}{nr^n}, \quad \frac{2R}{\mathfrak{M}g^2} = \frac{2K(m + m_1)\rho^n}{nmm_1g^2b^n}.$$

If we substitute:

(195) $$b = \alpha\left[\frac{K(m + m_1)}{mm_1g^2}\right]^{1/n},$$

then:

$$d\beta = \frac{d\rho}{\sqrt{1 - \rho^2 - \frac{2}{n}\left(\frac{\rho}{\alpha}\right)^n}}.$$

In order to avoid all discussions about the value which the quantity under the square root sign may have, we assume that the force is always repulsive, so that $\psi(r)$ is always positive, and hence R and $2\rho^n/n\alpha^n$ are also positive. Since, according to Equation (193), β always increases with time, and since the square root cannot change its sign without going through zero, we see that ρ must always increase up to the point when

(196) $$1 - \rho^2 - \frac{2}{n}\left(\frac{\rho}{\alpha}\right)^n = 0.$$

The smallest positive root of this equation we denote by $\rho(\alpha)$. For given n, it can be a function only of α. When n is positive, as we assume, then moreover $\rho^2 + 2\rho^n/n\alpha^n$ can be equal to unity for only one positive value of ρ; hence Equation (196) has no other positive roots. For $\rho = \rho(\alpha)$, the moving body reaches that point A (the perihelion) of its orbit at which it is closest to m_1, and at which its velocity is perpendicular to r. Since the quantity under the square root sign would become negative if ρ increased any further, and constant ρ corresponds to a circular orbit (which is impossible for a repulsive force), then ρ must again decrease; hence the root must change its sign. Because of the complete symmetry of the problem, a congruent branch of the curve will then be described, which is the mirror image of the part described up to that point (with respect to a plane through m_1A perpendicular to the orbital plane). The angle between the radius vector $\rho(\alpha) = m_1A$ and the two asymptotic directions of the orbital curve is:

(197) $$\vartheta = \int_0^{\rho(\alpha)} \frac{d\rho}{\sqrt{1 - \rho^2 - \dfrac{2}{n}\left(\dfrac{\rho}{\alpha}\right)^n}} = \vartheta(\alpha).$$

It can also be computed as a function of α, as soon as n is given. 2ϑ is the angle between the two asymptotes of the orbital curve, and therefore between the lines along which (in its motion relative to m_1) the m-molecule approaches the m_1-molecule before the collision and recedes from it after the collision. (The former line is opposite to the direction of motion of the molecule before the collision; the latter is in the same direction as its motion after the collision.)

The angle between the two lines g and g', representing the directions of the relative velocities before and after the collision, is $\pi - 2\vartheta$. (These are the line DC and the extension of BD beyond D, in Figure 7.)

When each of the two colliding molecules is an elastic sphere, only one modification is required in Figure 7. The sum of the two radii is $m_1D = \sigma$. The m-molecule moves relative to m_1 not in the curve BAC but rather in the broken line BDC; for $b \leq \sigma$, we have:

(198) $$\vartheta = \text{arc sin}\,\frac{b}{\sigma}$$

but for large values of b, $\vartheta = \pi/2$.

We now imagine, in Figure 8, a spherical surface of center m_1 and radius 1; it is intersected at G and G' by two lines drawn

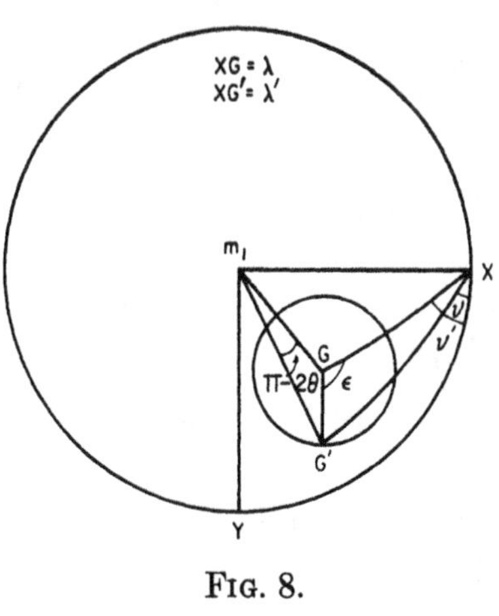

FIG. 8.

from m_1 parallel to g and g', and at X by a line through m_1 parallel to the fixed abscissa axis. Then the greatest circular arc GG' on this sphere is equal to $\pi - 2\vartheta$.

The angle ϵ has been defined in §16 in the following way. We draw through m_1 a plane E perpendicular to g. Further, we draw through m_1G two half-planes, one of which contains the line b, and the other the positive abscissa axis. The first we called the orbital plane. ϵ was then the angle of the two lines in which these two half planes intersected the plane E; therefore also the angle of the two half-planes themselves, or also the angle of the two great circle arcs GX and GG' on our sphere, where one is always supposed to understand by "great circle arc" the largest circular arc that is smaller than π.

For the spherical triangle we have:

$$\cos(G'X) = \cos(GX)\cos(GG') + \sin(GX)\sin(GG')\cos\epsilon. \tag{199}$$

Now however:

$$\sphericalangle GG' = \pi - 2\vartheta, \quad g'\cos(G'X) = \xi' - \xi_1',$$

$$g\cos(GX) = \xi - \xi_1, \qquad g\sin(GX) = \sqrt{g^2 - (\xi - \xi_1)^2},$$

where the positive sign of the root is to be taken, since $GX < \pi$.

If we multiply Equation (199) by the magnitude $g = g'$ of the relative velocity before or after the collision, then:

$$\xi' - \xi_1' = (\xi - \xi_1)\cos(\pi - 2\vartheta) + \sqrt{g^2 - (\xi - \xi_1)^2}\sin 2\vartheta\cos\epsilon.$$

If we multiply this equation by m_1 and add it to the equation

$$m\xi' + m_1\xi_1' = m\xi + m_1\xi_1 = (m + m_1)\xi + m_1\xi_1 + m_1\xi_1 - m\xi,$$

then:

$$\xi' = \xi + \frac{m_1}{m+m_1}\left[2(\xi_1 - \xi)\cos^2\vartheta + \sqrt{g^2 - (\xi - \xi_1)^2}\sin 2\vartheta\cos\epsilon\right]. \tag{200}$$

If only one kind of gas is present, then we have to put $m_1 = m$, $K = K_1$. Then:

$$\xi' = \xi + (\xi_1 - \xi)\cos^2\vartheta + \sqrt{g^2 - (\xi - \xi_1)^2}\sin\vartheta\cos\vartheta\cos\epsilon. \tag{201}$$

If we again denote $\xi - u$, $\xi' - u$, $\eta - v \cdots$ by $\mathfrak{x}$, $\mathfrak{x}'$, $\mathfrak{y}$, $\cdots$ then we obtain for $\mathfrak{x}$, $\mathfrak{y}$, $\mathfrak{z}$ an equivalent equation:

$$\mathfrak{x}' = \mathfrak{x} + (\mathfrak{x}_1 - \mathfrak{x})\cos^2\vartheta + \sqrt{g^2 - (\mathfrak{x} - \mathfrak{x}_1)^2}\sin\vartheta\cos\vartheta\cos\epsilon. \tag{202}$$

In order to find $B_5(\mathfrak{x}^2)$, we have to integrate the quantity

$$(\mathfrak{x}'^2 + \mathfrak{x}_1'^2 - \mathfrak{x}^2 - \mathfrak{x}_1) f f_1 d\omega d\omega_1 g b db dg d\epsilon$$

over ϵ from zero to 2π. The entire orbital curve thereby remains unaltered. Then we have to integrate over b, keeping $\mathfrak{x}$, $\mathfrak{y}$ $\mathfrak{z}$, $\mathfrak{x}_1$, $\mathfrak{y}_1$, $\mathfrak{z}_1$ constant. Then follows the integration over these quantities. Since the equations (201) and (202) are equivalent, we can obtain the expression for $B_5(\xi^2)$ by simply writing ξ, η, ζ for $\mathfrak{x}$, $\mathfrak{y}$, $\mathfrak{z}$ in $B_5(\mathfrak{x}^2)$.

Leaving out the terms containing the first power of $\cos\epsilon$, we get

$$\begin{aligned}\mathfrak{x}'^2 - \mathfrak{x}^2 &= 2(\mathfrak{x}_1\mathfrak{x} - \mathfrak{x}^2)\cos^2\vartheta + (\mathfrak{x}_1 - \mathfrak{x})^2\cos^4\vartheta \\ &\quad + \tfrac{1}{4}[g^2 - (\mathfrak{x} - \mathfrak{x}_1)^2]\sin^2 2\vartheta\cos^2\epsilon \\ &= (\mathfrak{x}_1^2 - \mathfrak{x}^2)\cos^2\vartheta - \mathfrak{p}^2\sin^2\vartheta\cos^2\vartheta \\ &\quad + (g^2 - \mathfrak{p}^2)\sin^2\vartheta\cos^2\vartheta\cos^2\epsilon.\end{aligned}$$

The components of the relative velocity with respect to the coördinate directions will be denoted by $\mathfrak{p}$, $\mathfrak{q}$, $\mathfrak{r}$, so that

$$(203)\qquad \begin{cases}\mathfrak{p} = \xi - \xi_1 = \mathfrak{x} - \mathfrak{x}_1 \\ \mathfrak{q} = \eta - \eta_1 = \mathfrak{y} - \mathfrak{y}_1 \\ \mathfrak{r} = \zeta - \zeta_1 = \mathfrak{z} - \mathfrak{z}_1.\end{cases}$$

If one forms the expression $\mathfrak{x}_1'^2 - \mathfrak{x}_1^2$ which arises from $\mathfrak{x}'^2 - \mathfrak{x}^2$ by simply permuting $\mathfrak{x}$ and $\mathfrak{x}_1$, it follows that:

$$\int_0^{2\pi}(\mathfrak{x}'^2 + \mathfrak{x}_1'^2 - \mathfrak{x}^2 - \mathfrak{x}_1^2)d\epsilon = 2\pi(g^2 - 3\mathfrak{p}^2)\sin^2\vartheta\cos^2\vartheta.$$

Since at the moment we are considering only one kind of gas, we must set $m_1 = m$ and $K = K_1$, and one has according to Equation (195):

$$(204)\qquad b = \left(\frac{2K_1}{m}\right)^{1/n} g^{-2/n}\alpha.$$

Since $\mathfrak{x}$, $\mathfrak{y}$, $\mathfrak{z}$, $\mathfrak{x}_1$, $\mathfrak{y}_1$, $\mathfrak{z}_1$ and hence also g are considered constant in the integration over b and ϵ, which now concerns us, it follows from this that:

$$(205)\qquad db = \left(\frac{2K_1}{m}\right)^{1/n} g^{-2/n} d\alpha.$$

Hence

$$(205a)\quad \begin{cases} \displaystyle\int_0^{\infty}\int_0^{2\pi}(\mathfrak{x}'^2+\mathfrak{x}_1'^2-\mathfrak{x}^2-\mathfrak{x}_1^2)\,b\,db\,d\epsilon \\ \displaystyle\quad = 2\pi(g^2-3\mathfrak{p}^2)\left(\frac{2K_1}{m}\right)^{2/n} g^{-4/n}\int_0^{\infty}\sin^2\vartheta\cos^2\vartheta\,\alpha\,d\alpha. \end{cases}$$

If one substitutes this in the expression $B_5(\mathfrak{x}^2)$, then he obtains under the integral sign a factor $g^{1-(4/n)}$; therefore since n must in any case be positive, there will in general be a negative or fractional power of g, which makes the integration very difficult. Only for $n=4$ does g drop out completely, and the integration becomes relatively easy. Since we have set the repulsion between two molecules equal to K/r^{n+1}, this implies that two molecules repel each other with a force inversely as the fifth power of the distance. One then obtains, as we shall see, a law of dependence of the viscosity, diffusion, and thermal conduction coefficients on temperature which for compound gases (water vapor, carbon dioxide) appears to be in good agreement with experiment, though not for the most common gases (oxygen, hydrogen, nitrogen). Other phenomena from which one could infer such a force law are not known. We certainly do not wish to assert that gas molecules actually act like point masses between which there is a repulsive force inversely proportional to the fifth power of the distance. Since it is here only a question of a mechanical model, we adopt this force law first introduced by Maxwell,* for which the calculation is easiest.[1] Moreover, with this law the repulsion increases so rapidly with decreasing distance that the motion of the molecules differs little from the motion of elastic spheres (aside from glancing collisions, which are of little importance). In order to

---

* Maxwell, Phil. Trans. **157**, 49 (1867).

[1] Likewise the assumption of an attraction proportional to the inverse fifth power of the distance produces a similar simplification (cf. Wien. Ber. **89**, 714 [1884]). However, one must then assume that for distances small compared to the distance at which the interaction becomes strong, the force follows a different law, according to which the attraction remains finite or is transformed to a repulsion, since otherwise the molecules would not separate in a finite time after collisions. In the text we always assume a repulsion inversely as the fifth power.

show this, Maxwell[2] has published a very instructive figure* in which the paths of the centers of a number of molecules are shown, projected towards a fixed molecule with the same velocity, and repelled by various force laws. In order to compare these paths with the path that would be followed by elastic spheres, we can do the following: we imagine in Maxwell's figure a marked circle whose center is S, and whose radius is Maxwell's dotted line, so that its radius is the smallest distance to which the centers of the two molecules approach for each force law. Now if the molecules are elastic spheres, whose diameter is this smallest distance, and if we hold one of them fixed and throw the others at it (not all at the same time, of course, but one after another so they won't interfere with each other) then Maxwell's figure undergoes the following modification. The center of the fixed molecule is still S. The centers of the moving ones will come from the same directions as in Maxwell's figure, but will be reflected like very small elastic spheres from the marked circle.

One sees that the resulting paths for elastic spheres are indeed quantitatively different but not essentially qualitatively different from those following from the new Maxwellian law.

In the following we set $n=4$ with Maxwell. Then it follows from Equation (205a) that:

(206) $$\int_0^{2\pi} (\mathfrak{x}'^2 + \mathfrak{x}_1'^2 - \mathfrak{x}^2 - \mathfrak{x}_1^2) g b\,db\,d\epsilon = \sqrt{\frac{K_1}{2m}} \cdot \frac{A_2}{g} \cdot (g^3 - 3\mathfrak{p}^2),$$

where

(207) $$A_2 = 4\pi \int_0^\infty \sin^2\vartheta \cos^2\vartheta\alpha \cdot d\alpha$$

is a pure number.[3]

[2] Maxwell, Phil. Mag. [4] **35,** 145 (1868); *Scientific Papers* **2,** p. 42.

* Here is Maxwell's figure:

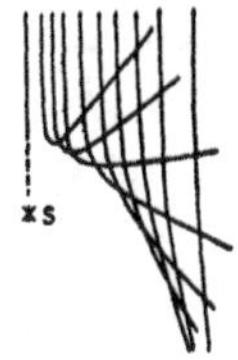

[3] Likewise one finds easily:

(208) $$2\pi \int_0^\infty g b\,db \sin^2\vartheta \cos^2\vartheta = A_2 \sqrt{\frac{K_1}{2m}}.$$

According to Equation (197):

$$\vartheta = \int_0^{\rho(\alpha)} \frac{d\rho}{\sqrt{1 - \rho^2 - \frac{1}{2}\frac{\rho^4}{\alpha^4}}}.$$

The upper limit is the only positive value for which the quantity under the square root sign vanishes. ϑ is therefore expressible in terms of a complete elliptic integral and a function of α. The integral (207) was evaluated by Maxwell by mechanical quadrature. He obtained the result:*

$$A_2 = 1{\cdot}3682 \cdots \tag{209}$$

We now have, according to Equation (137),

$$B_5(\mathfrak{x}^2) = \frac{1}{2}\iint\int_0^\infty\int_0^{2\pi} (\mathfrak{x}'^2 + \mathfrak{x}_1'^2 - \mathfrak{x}^2 - \mathfrak{x}_1^2) f f_1 g b\, d\omega\, d\omega_1\, db\, d\epsilon. \tag{210}$$

Substitution of (206) gives:

$$B_5(\mathfrak{x}^2) = \frac{1}{2}\sqrt{\frac{K_1}{2m}}\, A_2 \iint (g^2 - 3\mathfrak{p}^2) f f_1\, d\omega\, d\omega_1. \tag{211}$$

We have:

$$\begin{aligned} g^2 - 3\mathfrak{p}^2 &= \eta^2 + \eta_1^2 + \zeta^2 + \zeta_1^2 - 2\xi^2 - 2\xi_1^2 - 2\eta\eta_1 - 2\zeta\zeta_1 + 4\xi\xi_1 \\ &= \mathfrak{y}^2 + \mathfrak{y}_1^2 + \mathfrak{z}^2 + \mathfrak{z}_1^2 - 2\mathfrak{x}^2 - 2\mathfrak{x}_1^2 - 2\mathfrak{y}\mathfrak{y}_1 - 2\mathfrak{z}\mathfrak{z}_1 + 4\mathfrak{x}\mathfrak{x}_1. \end{aligned}$$

In the integration over $d\omega_1$ we can take ξ, η, ζ or $\mathfrak{x}$, $\mathfrak{y}$, $\mathfrak{z}$ outside the integral sign, and similarly in the integration over $d\omega$ we can take ξ_1, η_1, ζ_1 or $\mathfrak{x}_1$, $\mathfrak{y}_1$, $\mathfrak{z}_1$ outside the integral sign. According to Equation (175):

$$\int \eta^2 f d\omega = \frac{\rho}{m}\overline{\eta^2}, \quad \int \eta f d\omega = \frac{\rho}{m}\bar{\eta}, \quad \int \mathfrak{y}^2 f do = \frac{\rho}{m}\overline{\mathfrak{y}^2} \text{ etc.} \tag{212}$$

Since, however, the two colliding molecules are equivalent—or, if one prefers, since one can give any labels to variables over which one integrates in a definite integral—we also have:

$$\int \eta_1^2 f_1 d\omega_1 = \int \eta^2 f d\omega = \frac{\rho}{m}\overline{\eta^2} \text{ etc.}$$

---

* Later calculations of A_2 were made by K. Aichi and T. Tanukadate, reported by H. Nagaoka, Nature **69**, 79 (1903), and by S. Chapman, Mem. Proc. Manchester Lit. Phil. Soc. **66**, No. 1 (1922). The values found were 1.3704 and 1.3700, respectively.

and since $\bar{\mathfrak{x}}=\bar{\mathfrak{y}}=\mathfrak{z}=0$, we have:

$$(213)\quad \begin{cases} B_5(\mathfrak{x}^2) = \sqrt{\dfrac{K_1}{2m^5}}A_2\rho^2(\overline{\eta^2}+\overline{\zeta^2}-2\overline{\xi^2}-\bar{\eta}\cdot\bar{\eta}-\bar{\zeta}\cdot\bar{\zeta}+2\bar{\xi}\cdot\bar{\xi}) \\ \qquad = \sqrt{\dfrac{K_1}{2m^5}}A_2\rho^2(\overline{\mathfrak{y}^2}+\overline{\mathfrak{z}^2}-2\overline{\mathfrak{x}^2}) = \\ \qquad = \sqrt{\dfrac{K_1}{2m^5}}A_2\rho^2(\overline{\mathfrak{c}^2}-3\overline{\mathfrak{x}^2}). \end{cases}$$

$\mathfrak{c}=\sqrt{\mathfrak{x}^2+\mathfrak{y}^2+\mathfrak{z}^2}$ is the total velocity of a molecule relative to the average motion of all molecules in the volume element.

The quantity $B_5(\mathfrak{x}\mathfrak{y})$ is calculated by Maxwell by a coördinate transformation. We imagine new x and y axes placed in such a way that the old x and y axes are rotated by an angle λ in the xy plane. We denote the corresponding quantities in the new coördinate system by capital letters:

$$\mathfrak{x} = \mathfrak{X}\cos\lambda - \mathfrak{Y}\sin\lambda,\quad \mathfrak{y} = \mathfrak{Y}\cos\lambda + \mathfrak{X}\sin\lambda,$$
$$\mathfrak{p} = \mathfrak{P}\cos\lambda - \mathfrak{Q}\sin\lambda \text{ etc.}$$

If we substitute these values in Equation (206), we obtain the same terms with factors of $\cos^2\lambda$, $\cos\lambda\sin\lambda$, and $\sin^2\lambda$. If we set $\lambda=0$, we see that the former must separately be equal; if we set $\lambda=\pi/2$, we see that the latter must likewise separately be equal. Hence the terms multiplied by $\sin\lambda\cos\lambda$ on the right- and left-hand sides of the equality sign must separately be equal. Setting them equal to each other, we obtain:

$$\int_0^\infty\int_0^{2\pi}(\mathfrak{X}'\mathfrak{Y}'+\mathfrak{X}_1'\mathfrak{Y}_1'-\mathfrak{X}\mathfrak{Y}-\mathfrak{X}_1\mathfrak{Y}_1)gb\,db\,d\epsilon = -3\sqrt{\frac{K_1}{2m}}A_2\mathfrak{P}\mathfrak{Q}.$$

Since the new coördinate axes are just as good as the old ones, one can again use lower-case letters instead of capitals. If one carries out the further integrations exactly as for Equation (206), it follows that:

$$(214)\quad \begin{cases} B_5(\mathfrak{x}\mathfrak{y}) = \dfrac{1}{2}\displaystyle\iiint_0^\infty\int_0^{2\pi}(\mathfrak{x}'\mathfrak{y}'+\mathfrak{x}_1'\mathfrak{y}_1'-\mathfrak{x}\mathfrak{y}-\mathfrak{x}_1\mathfrak{y}_1) \\ \qquad\qquad \cdot gbff_1\,d\omega\,d\omega_1\,db\,d\epsilon \\ \qquad = -3\sqrt{\dfrac{K_1}{2m^5}}A_2\rho^2(\overline{\xi\eta}-\bar{\xi}\cdot\bar{\eta}) = -3\sqrt{\dfrac{K_1}{2m^5}}\rho^2A_2\mathfrak{x}\mathfrak{y}. \end{cases}$$

§22. Relaxation time. Hydrodynamic equations corrected for viscosity. Calculation of B_5 using spherical functions.

We now have to substitute this value into the general equation (187). We first consider a special, completely ideal case: a single kind of gas fills an infinite space. There are no external forces. The number of molecules in any one volume element do, whose velocity components lie between the limits ξ and $\xi+d\xi$, η and $\eta+d\eta$, ζ and $\zeta+d\zeta$, will at time $t=0$ be equal to $f(\xi, \eta, \zeta, 0)dod\xi d\eta d\zeta$, where the function f is the same for all volume elements. For any later time t, let this number be equal to $f(\xi, \eta, \zeta, t)dod\xi d\eta d\zeta$. Since all volume elements are subjected to the same conditions, $f(\xi, \eta, \zeta, t)$ also has the same value for all volume elements. If

$$f(\xi, \eta, \zeta, 0) = ae^{-hm[(\xi-u)^2+(\eta-v)^2+(\zeta-w)^2]},$$

where a, h, u, v, w are constants, then we would have a gas in which Maxwell's distribution holds, but which moves through space with constant velocity components u, v, w. Then we would have $\overline{(\xi-u)^2}=\overline{(\eta-v)^2}=\overline{(\zeta-w)^2}$, $\overline{(\xi-u)(\eta-v)}=\overline{(\xi-u)(\zeta-w)}$ $=\overline{(\eta-v)(\zeta-w)}=0$, and the distribution of states, seen from the viewpoint of an observer moving with the gas, would not change with time. If $f(\xi, \eta, \zeta, 0)$ is some other function of ξ, η, ζ, then at the initial time a velocity distribution differing from Maxwell's, but still the same in each volume element, holds. This distribution changes with time, but the components of the visible motion of the gas

$$u = \bar{\xi} = \frac{\int \xi f d\omega}{\int f d\omega}, \quad v = \bar{\eta} = \frac{\int \eta f d\omega}{\int f d\omega}, \quad w = \bar{\zeta} = \frac{\int \zeta f d\omega}{\int f d\omega}$$

naturally do not change with time, because of the conservation of the motion of the center of mass. If we again set $\xi-u=\mathfrak{x}$, $\eta-v=\mathfrak{y}$, $\zeta-w=\mathfrak{z}$, then in general now

$$\overline{\mathfrak{x}^2} - \overline{\mathfrak{y}^2}, \quad \overline{\mathfrak{x}^2} - \overline{\mathfrak{z}^2}, \quad \overline{\mathfrak{y}^2} - \overline{\mathfrak{z}^2}, \quad \overline{\mathfrak{x}\mathfrak{y}}, \quad \overline{\mathfrak{x}\mathfrak{z}} \text{ and } \overline{\mathfrak{y}\mathfrak{z}}$$

differ from zero, and we ask ourselves how these quantities change with time. First, since nothing depends on x, y, or z, it follows that, from (188):

$$(215) \qquad \rho \frac{\partial \bar{\mathfrak{f}}}{\partial t} = mB_5(\mathfrak{f}).$$

If one now substitutes $\mathfrak{f}=\mathfrak{x}^2$ or $\mathfrak{f}=\mathfrak{x}\mathfrak{y}$, then it follows with the help of Equations (213) and (214) that:

$$\frac{d\overline{\mathfrak{x}^2}}{dt} = \sqrt{\frac{K_1}{2m^3}}\,A_2\rho(\overline{\mathfrak{c}^2} - 3\overline{\mathfrak{x}^2}), \quad \frac{d\overline{\mathfrak{x}\mathfrak{y}}}{dt} = -3\sqrt{\frac{K_1}{2m^3}}\,A_2\rho\overline{\mathfrak{x}\mathfrak{y}}.$$

Similarly, to the first of these equations, we have:

$$\frac{d\overline{\mathfrak{y}^2}}{dt} = \sqrt{\frac{K_1}{2m^3}}\,A_2\rho(\overline{\mathfrak{c}^2} - 3\overline{\mathfrak{y}^2})$$

and consequently

$$\frac{d(\overline{\mathfrak{x}^2} - \overline{\mathfrak{y}^2})}{dt} = -3\sqrt{\frac{K_1}{2m^3}}\,A_2\rho(\overline{\mathfrak{x}^2} - \overline{\mathfrak{y}^2}).$$

Since everything is independent of x, y, z, the differential quotients with respect to t are to be taken in the usual sense. Since furthermore all volume elements are equivalent, exactly as many molecules will flow in through each lateral surface as flow out through the opposite surface. The density ρ must therefore remain constant. Hence integration of these equations, using the index 0 to characterize the values at time zero, gives the results:

$$\overline{\mathfrak{x}^2} - \overline{\mathfrak{y}^2} = (\overline{\mathfrak{x}_0^2} - \overline{\mathfrak{y}_0^2})e^{-3\sqrt{(K_1/2m^3)}A_2\rho t}, \quad \overline{\mathfrak{x}\mathfrak{y}} = (\overline{\mathfrak{x}\mathfrak{y}})_0 e^{-3\sqrt{(K_1/2m^3)}A_2\rho t}.$$

Multiplication by ρ yields (recalling the notations [179]):

$$X_x - Y_y = (\overset{0}{X}_x - \overset{0}{Y}_y)e^{-3\sqrt{(K_1/2m^3)}A_2\rho t}, \; X_y = \overset{0}{X}_y e^{-3\sqrt{(K_1/2m^3)}A_2\rho t}.$$

Similar equations follow of course for the other coördinate axes. In the simple special case now considered, the difference in the normal pressure in two different directions (e.g., $X_x - Y_y$) and likewise the tangential force (e.g., X_y) simply decreases with increasing time in geometric progression. The time after which it becomes e times smaller is the same for all these quantities, and is equal to

(216)
$$\frac{1}{3A_2\rho}\sqrt{\frac{2m^3}{K_1}} = \tau.$$

Maxwell* calls this the relaxation time. We shall see that it is very short.

---

* Maxwell, Phil. Trans. **157**, 49 (1867), esp. Eq. (130). Maxwell actually called it the "modulus of the time of relaxation."

We now return to the general case. In general we no longer have $\rho\overline{\mathfrak{x}^2}=\rho\overline{\mathfrak{y}^2}=\rho\overline{\mathfrak{z}^2}$, but these quantities are still approximately equal. We shall therefore calculate their deviations from a quantity nearly equal to one of them. For this purpose we choose their arithmetic mean. Since, according to the assumptions necessary for the validity of Equations (181), this is equal to the quantity denoted by p, we shall again denote it by p, and therefore put

$$p = \frac{\rho}{3}\left(\overline{\mathfrak{x}^2} - \overline{\mathfrak{y}^2} + \overline{\mathfrak{z}^2}\right) = \frac{\rho}{3}\overline{\mathfrak{c}^2}. \tag{217}$$

If we denote the right-hand side of Equation (189) by $\mathfrak{r}$ and substitute the value (213) on the left-hand side for $B_5(\mathfrak{x}^2)$, then it follows that:

$$\overline{\mathfrak{c}^2} - 3\overline{\mathfrak{x}^2} = \frac{1}{A_2\rho^2}\sqrt{\frac{2m^3}{K_1}}\,\mathfrak{r}. \tag{218}$$

We look for the small difference between the two quantities $\overline{\mathfrak{c}^2}=\overline{\mathfrak{x}^2}+\overline{\mathfrak{y}^2}+\overline{\mathfrak{z}^2}$ and $3\overline{\mathfrak{x}^2}$. This, and hence the right-hand side of the above Equation (218), are small first-order infinitesimal quantities; hence we need to retain on the right-hand side only the terms of the largest order of magnitude. The terms of smaller magnitude are also smaller than $\overline{\mathfrak{c}^2}-3\overline{\mathfrak{x}^2}$. In the expression for $\mathfrak{r}$ we can therefore set

$$\rho\overline{\mathfrak{x}^2} = \rho\overline{\mathfrak{y}^2} = \rho\overline{\mathfrak{z}^2} = p, \quad \overline{\mathfrak{x}\mathfrak{y}} = \overline{\mathfrak{x}\mathfrak{z}} = \overline{\mathfrak{y}\mathfrak{z}} = \overline{\mathfrak{x}^3} = \overline{\mathfrak{x}\mathfrak{y}^2} = \overline{\mathfrak{x}\mathfrak{z}^2} = 0.$$

We saw that then (see Eq. [191])

$$\mathfrak{r} = \rho\frac{d\left(\dfrac{p}{\rho}\right)}{dt} + 2p\frac{\partial u}{\partial x}.$$

We wish to find $\overline{\mathfrak{x}^2}$ and thence X_x and its dependence on the instantaneous state; we must therefore eliminate the term that contains a time-derivative. This is easy, since we find to the same degree of accuracy

$$\rho\frac{d\left(\dfrac{p}{\rho}\right)}{dt} = -\frac{2p}{3}\left(\frac{\partial u}{\partial x} + \frac{\partial v}{\partial y} + \frac{\partial w}{\partial z}\right).$$

Hence in first approximation

$$\mathfrak{r} = \frac{2p}{3}\left(2\frac{\partial u}{\partial x} - \frac{\partial v}{\partial y} - \frac{\partial w}{\partial z}\right).$$

The following terms in the expression for $\mathfrak{x}$ provide terms of smaller magnitude in $\overline{\mathfrak{c}^2}-3\overline{\mathfrak{x}^2}$, and therefore we can ignore them. Hence, according to Equation (218),

$$\overline{\mathfrak{c}^2} - 3\overline{\mathfrak{x}^2} = \frac{2p}{3A_2\rho^2}\sqrt{\frac{2m^3}{K_1}}\left(2\frac{\partial u}{\partial x} - \frac{\partial v}{\partial y} - \frac{\partial w}{\partial z}\right),$$

therefore, since we set $\rho\overline{\mathfrak{c}^2}=3p$,

$$X_x = \rho\overline{\mathfrak{x}^2} = p - \frac{2p}{9A_2\rho}\sqrt{\frac{2m^3}{K_1}}\left(2\frac{\partial u}{\partial x} - \frac{\partial v}{\partial y} - \frac{\partial w}{\partial z}\right).$$

We now wish to substitute the value (214) for $B_5(\mathfrak{x}\mathfrak{y})$ into Equation (190). On the right-hand side of this equation we can set $\rho\overline{\mathfrak{x}^2}=\rho\overline{\mathfrak{y}^2}=\rho\overline{\mathfrak{z}^2}=p$ for the same reason as before, and set the terms containing odd powers of $\mathfrak{x}$, $\mathfrak{y}$, or $\mathfrak{z}$ under the bar symbol equal to zero. We thereby obtain:

$$\overline{\mathfrak{x}\mathfrak{y}} = -\frac{p}{3A_3\rho^2}\sqrt{\frac{2m^3}{K_1}}\left(\frac{\partial v}{\partial x} + \frac{\partial u}{\partial y}\right). \tag{218a}$$

If one substitutes the abbreviation

$$\frac{p}{3A_3\rho}\sqrt{\frac{2m^3}{K_1}} = p\tau = \mathfrak{R}, \tag{219}$$

then he obtains the following values:

$$\left\{\begin{aligned}
X_x &= \rho\overline{\gamma^2} = p - \frac{2\mathfrak{R}}{3}\left(2\frac{\partial u}{\partial x} - \frac{\partial v}{\partial y} - \frac{\partial w}{\partial z}\right),\\
Y_y &= \rho\overline{\mathfrak{y}^2} = p - \frac{2\mathfrak{R}}{3}\left(2\frac{\partial v}{\partial y} - \frac{\partial u}{\partial x} - \frac{\partial w}{\partial z}\right),\\
Z_z &= \rho\overline{\mathfrak{z}^2} = p - \frac{2\mathfrak{R}}{3}\left(2\frac{\partial w}{\partial z} - \frac{\partial u}{\partial x} - \frac{\partial v}{\partial y}\right),\\
X_y &= Y_x = \rho\overline{\mathfrak{x}\mathfrak{y}} = -\mathfrak{R}\left(\frac{\partial v}{\partial x} + \frac{\partial u}{\partial y}\right),\\
X_z &= Z_x = \rho\overline{\mathfrak{x}\mathfrak{z}} = -\mathfrak{R}\left(\frac{\partial w}{\partial x} + \frac{\partial u}{\partial z}\right),\\
Y_z &= Z_y = \rho\overline{\mathfrak{y}\mathfrak{z}} = -\mathfrak{R}\left(\frac{\partial v}{\partial z} + \frac{\partial w}{\partial z}\right).
\end{aligned}\right. \tag{220}$$

These equations are not completely exact, of course;

however, they are one degree more exact than the equations $X_x = Y_y = Z_z = p$, $X_y = Y_x = X_z = Z_x = Y_z = Z_y = 0$. Substitution of these values into the equations of motion (185) yields

$$
(221)\quad \begin{cases}
\rho \dfrac{du}{dt} + \dfrac{\partial p}{\partial x} - \Re\left[\Delta u + \dfrac{1}{3}\dfrac{\partial}{\partial x}\left(\dfrac{\partial u}{\partial x} + \dfrac{\partial v}{\partial y} + \dfrac{\partial w}{\partial z}\right)\right] - \rho X = 0 \\[2ex]
\rho \dfrac{dv}{dt} + \dfrac{\partial p}{\partial y} - \Re\left[\Delta v + \dfrac{1}{3}\dfrac{\partial}{\partial y}\left(\dfrac{\partial u}{\partial x} + \dfrac{\partial v}{\partial y} + \dfrac{\partial w}{\partial z}\right)\right] - \rho Y = 0 \\[2ex]
\rho \dfrac{dw}{dt} + \dfrac{\partial p}{\partial z} - \Re\left[\Delta w + \dfrac{1}{3}\dfrac{\partial}{\partial z}\left(\dfrac{\partial u}{\partial x} + \dfrac{\partial v}{\partial y} + \dfrac{\partial w}{\partial y}\right)\right] - \rho Z = 0.
\end{cases}
$$

Here $\Re$ is considered constant, which is also not strictly correct, since $\Re$ is a function of temperature, and temperature changes during compression or rarefaction. However, since the actual temperature-dependence of $\Re$ is still in doubt, and since the gas moves, in the case of less violent motions, almost like an incompressible fluid, so that there is no significant compression or rarefaction, this error is not important. Equations (221) are the well-known hydrodynamic equations corrected for viscosity.* These equations are satisfied and one obtains therefore a possible motion, when he sets p equal to a constant and $X = Y = Z = 0$, $v = w = 0$, and $u = ay$. Each layer of gas parallel to the xy plane moves with velocity ay in the x-direction. a is the velocity difference between two such layers at unit distance from each other. One of these layers must clearly be artificially held fixed, and another must be artificially maintained in a state of constant motion. The tangential force on unit surface of these layers has the value $a\Re$ according to Equations (220), and $\Re$ is therefore the quantity that we already called the viscosity coefficient in §12. From Equation (219) it follows that it is proportional to p/ρ and therefore to the absolute temperature, but at a given temperature it is independent of pressure and density. The latter statement is also true when the molecules are elastic spheres, but then $\Re$ is proportional to the square root of the absolute temperature. Of course the mean free path cannot now be calculated from the numerical value of $\Re$, since the beginning and end of a collision

* Now usually called the Navier-Stokes equations; however Boltzmann, following Maxwell (*op. cit.*) has suppressed the second viscosity coefficient (bulk or dilatational viscosity). This point is discussed in §23.

are not sharply defined; this value can only provide a relation between the mass m of the molecule and the constant K_1 in the force law. It also permits the calculation of the relaxation time $\tau = \mathfrak{R}/p$. From the value of $\mathfrak{R}$ for nitrogen used in §12, we find a relaxation time of about $\tau = 2 \cdot 10^{-10}$ sec at atmospheric pressure and 15°C.

We now proceed to the calculation of $B_5(\mathfrak{x}^3)$, $B_5(\mathfrak{x}\mathfrak{y}^2)$, etc. There is no difficulty in raising the expression (201) to the third power and then performing the integration, as we have done for the calculation of $B_5(\mathfrak{x}^2)$. The same coördinate transformation as before gives the values of $B_5(\mathfrak{x}\mathfrak{y}^2)$ and $B_5(\mathfrak{x}\mathfrak{z}^2)$ and the other B_5's, which contain as arguments of the function terms of third order in $\mathfrak{x}$, $\mathfrak{y}$, and $\mathfrak{z}$. $B_5(\mathfrak{x}\mathfrak{y}\mathfrak{z})$ must be found by a spatial (3-dimensional) coördinate transformation. We shall adopt here another method which Maxwell, in his paper "On stresses in rarified gases,"[1] indicated in bracketed notes added in the last months of his life.

Any function p of the n'th degree in x, y, z, which satisfies the equation

$$\frac{\partial^2 p}{\partial x^2} + \frac{\partial^2 p}{\partial y^2} + \frac{\partial^2 p}{\partial z^2} = 0$$

we call a (solid) spherical function of the n'th degree. If we substitute $x = \cos \lambda$, $y = \sin \lambda \cos \nu$, $c = \sin \lambda \cos \nu$, then it is transformed to a spherical surface function of the n'th degree: $p^{(n)}(\lambda, \nu)$. Further, we denote the coefficient of x^n in the power series that arises from the expansion of

$$(1 - 2\mu x + x^2)^{-1/2} \tag{222}$$

by $P^{(n)}(\mu)$ (zonal spherical function or spherical function of one argument). Now let G and G' be any two points on a spherical surface with polar coördinates λ, ν and λ', ν' and let G_i be the symbol representing $n+1$ arbitrary other points on the same spherical surface. Let the polar coördinates of G_i be λ_i and ν_i. Then[2]

$$p^{(n)}(\lambda', \nu') = \sum_{i=1}^{i=2n+1} c_i P^{(n)}(s_i'), \tag{223}$$

[1] Maxwell, Phil. Trans. **170,** 231 (1879); *Scientific Papers* **2,** p. 681.
[2] Heine, *Handbuch der Kugelfunctionen* (2d ed.), p. 322.

where s_i' is the cosine of the spherical angle $G'G_i$. The c_i are constant coefficients which can be determined. Now hold G and G_i constant, while G' describes a circle such that the spherical angle GG' always remains constant. Its cosine is called μ. Finally, denote by ϵ the angle between the great circle GG' and a fixed circle drawn through G. Then:

$$\frac{1}{2\pi}\int_0^{2\pi} p^{(n)}(\lambda', \nu')d\epsilon = \sum_{i=1}^{i=2n+1} \frac{c_i}{2\pi}\int_0^{2\pi} P^{(n)}(s_i')d\epsilon.$$

Furthermore:[3]

$$\int_0^{2\pi} P^{(n)}(s_i')d\epsilon = 2\pi P^{(n)}(\mu)\cdot P^{(n)}(s_i),$$

where s_i is the cosine of the spherical angle GG_i. One has therefore:

$$\int_0^{2\pi} p^{(n)}(\lambda', \nu')d\epsilon = 2\pi P^{(n)}(\mu)\cdot \sum_{i=1}^{i=2n+1} c_i P^{(n)}(s_i).$$

As in Equation (223), the latter sum has the value $p^{(n)}(\lambda, \nu)$. One therefore obtains the formula:[4]

$$(224) \qquad \int_0^{2\pi} p^{(n)}(\lambda', \nu')d\epsilon = 2\pi P^{(n)}(\mu)\cdot p^{(n)}(\lambda, \nu).$$

We wish to show the application of this theorem to the calculation of B_5 in a special case, and in particular to calculate $B_5(\mathfrak{x}\mathfrak{y})$.

As before let $\xi, \eta, \zeta, \xi_1, \eta_1, \zeta_1, \xi', \eta', \zeta', \xi_1', \eta_1', \zeta_1'$ be the velocity components of the two molecules before and after the collision; let $\mathfrak{x}, \mathfrak{y}, \mathfrak{z}, \mathfrak{x}_1, \mathfrak{y}_1, \mathfrak{z}_1, \mathfrak{x}', \mathfrak{y}', \mathfrak{z}', \mathfrak{x}_1', \mathfrak{y}_1', \mathfrak{z}_1'$ be the same velocities relative to the average motion of all the m-molecules in the volume element, so that therefore $\xi - \mathfrak{x} = u$, $\eta - \mathfrak{y} = v \cdots$ etc. where u, v, w are the components of the mean velocity of all m-molecules in the volume element. Further, let

$$\mathfrak{p} = \xi - \xi_1 = \mathfrak{x} - \mathfrak{x}_1,\quad \mathfrak{q} = \eta - \eta_1 = \mathfrak{y} - \mathfrak{y}_1,\quad \mathfrak{r} = \zeta - \zeta_1 = \mathfrak{z} - \mathfrak{z}_1,$$
$$\mathfrak{p}' = \xi' - \xi'_1 = \mathfrak{x}' - \mathfrak{x}'_1,\quad \mathfrak{q}' = \eta' - \eta'_1 = \mathfrak{y}' - \mathfrak{y}_1',\quad \mathfrak{r}' = \zeta' - \zeta'_1 = \mathfrak{z}' - \mathfrak{z}_1'.$$

be the components before and after the collision, respectively, of the velocities g and g' respectively of the molecule that has before the collision velocity components ξ, η, ζ relative to the other one

[3] Heine, *op. cit.*, p. 313.

[4] I thank Prof. Gegenbauer for this proof of Maxwell's theorem.

that has velocity components ξ_1, η_1, ζ_1. The latter will again be called the m_1-molecule, even though it actually has mass m. Finally, we denote by

$$\mathfrak{u}=\mathfrak{x}+\mathfrak{x}_1=\mathfrak{x}'+\mathfrak{x}_1', \quad \mathfrak{v}=\mathfrak{y}+\mathfrak{y}_1=\mathfrak{y}'+\mathfrak{y}_1', \quad \mathfrak{w}=\mathfrak{z}+\mathfrak{z}_1=\mathfrak{z}'+\mathfrak{z}_1'$$

twice the velocity components of the center of gravity of the system formed by the two colliding molecules, in its motion relative to the mean motion of all m-molecules in the volume element. These are equal before and after the collision. Then

$$\begin{aligned}
4\mathfrak{x}\,\mathfrak{y} &= \mathfrak{p}\,\mathfrak{q} + \mathfrak{u}\mathfrak{q} + \mathfrak{v}\mathfrak{p} + \mathfrak{u}\mathfrak{v}\\
4\mathfrak{x}_1\mathfrak{y}_1 &= \mathfrak{p}\,\mathfrak{q} - \mathfrak{u}\mathfrak{q} - \mathfrak{v}\mathfrak{p} + \mathfrak{u}\mathfrak{v}\\
4\mathfrak{x}'\mathfrak{y}' &= \mathfrak{p}'\mathfrak{q}' + \mathfrak{u}\mathfrak{q}' + \mathfrak{v}\mathfrak{p}' + \mathfrak{u}\mathfrak{v}\\
4\mathfrak{x}_1'\mathfrak{y}_1' &= \mathfrak{p}'\mathfrak{q}' - \mathfrak{u}\mathfrak{q}' - \mathfrak{v}\mathfrak{p}' + \mathfrak{u}\mathfrak{v}
\end{aligned}$$

hence

$$(225) \qquad 2(\mathfrak{x}'\mathfrak{y}' + \mathfrak{x}'_1\mathfrak{y}_1' - \mathfrak{x}\mathfrak{y} - \mathfrak{x}_1\mathfrak{y}_1) = \mathfrak{p}'\mathfrak{q}' - \mathfrak{p}\mathfrak{q}.$$

We now construct around m_1 a sphere of radius 1. The lines through m_1 parallel to the abscissa axis, representing the relative velocities g and g', should intersect this sphere in the points X, G, and G' respectively (Fig. 8). λ, ν and λ', ν' are the polar coördinates of the points G and G' (i.e., λ and λ' are the angles Xm_1G and Xm_1G', while ν and ν' are the angles made by the planes GmX and $G'mX$ with the xy plane). Since $\mathfrak{p}$, $\mathfrak{q}$, $\mathfrak{r}$ and $\mathfrak{p}'$, $\mathfrak{q}'$, $\mathfrak{r}'$ are the projections of g and g' on the coördinate directions, we have

$$\begin{aligned}
\mathfrak{p} &= g\cos\lambda, \quad \mathfrak{q} = g\sin\lambda\cos\nu, \quad \mathfrak{r} = g\sin\lambda\sin\nu\,,\\
\mathfrak{p}' &= g\cos\lambda', \quad \mathfrak{q}' = g\sin\lambda'\cos\nu', \quad \mathfrak{r}' = g\sin\lambda'\sin\nu',
\end{aligned}$$

hence

$$\mathfrak{p}\mathfrak{q} = g^2 p^{(2)}(\lambda, \nu), \quad \mathfrak{p}'\mathfrak{q}' = g^2 p^{(2)}(\lambda', \nu'),$$

where $p^{(2)}(\lambda, \nu)$ is the spherical function $\cos\lambda\sin\lambda\cos\nu$. Earlier we denoted by ϵ the spherical-triangle angle XGG', and by $\pi-2\vartheta$ the angle Gm_1G'. Then, according to the theorem cited on spherical functions,

$$(226) \qquad \int_0^{2\pi} p^{(2)}(\lambda', \nu')d\epsilon = 2\pi p^{(2)}(\lambda, \nu)\cdot P^{(2)}(\mu),$$

where $\mu=\cos(\pi-2\vartheta)$. On expanding Equation (222) one finds:

$$P^{(2)}(\mu) = \tfrac{3}{2}\mu^2 - \tfrac{1}{2} = \tfrac{3}{2}\cos^2(2\vartheta) - \tfrac{1}{2} = 1 - 6\sin^2\vartheta\cos^2\vartheta.$$

Hence

$$\int_0^{2\pi}(\mathfrak{x}'\mathfrak{y}' + \mathfrak{x}_1'\mathfrak{y}_1' - \mathfrak{x}\mathfrak{y} - \mathfrak{x}_1\mathfrak{y}_1)d\epsilon = -\pi g^2 p^{(2)}(\lambda, \nu)\cdot 6\sin^2\vartheta\cos^2\vartheta$$
$$= -6\pi\mathfrak{p}\mathfrak{q}\sin^2\vartheta\cos^2\vartheta.$$

Whence follows, by comparison with Equation (208),

$$\int_0^\infty gbdb\int_0^{2\pi}(\mathfrak{x}'\mathfrak{y}' + \mathfrak{x}_1'\mathfrak{y}_1' - \mathfrak{x}\mathfrak{y} - \mathfrak{x}_1\mathfrak{y}_1)d\epsilon = -3A_2\sqrt{\frac{K_1}{2m}}\,\mathfrak{p}\mathfrak{q}.$$

$$B_5(\mathfrak{x}\mathfrak{y}) = \frac{1}{2}\iint\int_0^\infty\int_0^{2\pi}(\mathfrak{x}'\mathfrak{y}' + \mathfrak{x}_1'\mathfrak{y}_1' - \mathfrak{x}\mathfrak{y} - \mathfrak{x}_1\mathfrak{y}_1)gbff_1d\omega d\omega_1 dbd\epsilon$$
$$= -\frac{3}{2}A_2\sqrt{\frac{K_1}{2m}}\iint\mathfrak{p}\mathfrak{q}ff_1d\omega d\omega_1,$$

whence finally, according to Equation (212), we obtain:

$$B_5(\mathfrak{x}\mathfrak{y}) = -3A_2\rho^2\sqrt{\frac{K_1}{2m^5}}\,\overline{\mathfrak{x}\mathfrak{y}}.$$

Since Equation (226) is valid for any spherical function of the second degree, it follows in general that:

$$B_5[p^{(2)}(\mathfrak{x}, \mathfrak{y})] = -3A_2\rho^2\sqrt{\frac{K_1}{2m^5}}\,\overline{p^{(2)}(\mathfrak{x}, \mathfrak{y})}$$

$$\text{e.g. } B_5(\mathfrak{x}^2 - \mathfrak{y}^2) = -3A_2\rho^2\sqrt{\frac{K_1}{2m^5}}\,(\overline{\mathfrak{x}^2} - \overline{\mathfrak{y}^2}).$$

When f is not a function of x, y, z, and $X = Y = Z = 0$ (and the effect of the wall vanishes) it follows from Equation (188) that

$$\rho\frac{d\overline{\mathfrak{f}}}{dt} = mB_5(\mathfrak{f}). \tag{227}$$

Hence if $\mathfrak{f}$ is any spherical function of the second degree, it follows in general that

$$\overline{\mathfrak{f}} = \overline{\mathfrak{f}}_0 e^{-3A_2\rho\sqrt{(K_1/2m^3)}\,t}. \tag{228}$$

Therefore

$$\frac{1}{\tau} = \frac{\Re}{p} = 3A_2\rho\sqrt{\frac{K_1}{2m^3}} \tag{229}$$

is the reciprocal of the relaxation time for all spherical functions of the second degree in $\mathfrak{x}$, $\mathfrak{y}$, and $\mathfrak{z}$—i.e., the time in which, by the action of collisions, the mean value of that spherical function decreases to $1/e$ of its original value. This confirms our earlier result.

We now pass to spherical functions of the third degree, e.g., $\mathfrak{x}^3-3\mathfrak{x}\mathfrak{y}^2$. By analogy with Equation (225) we find

$$4[\mathfrak{x}'^3 + \mathfrak{x}_1'^3 - \mathfrak{x}^3 - \mathfrak{x}_1^3 - 3(\mathfrak{x}'\mathfrak{y}'^2 + \mathfrak{x}_1'\mathfrak{y}_1'^2 - \mathfrak{x}\mathfrak{y}^2 - \mathfrak{x}_1\mathfrak{y}_1^2)]$$
$$= 3\mathfrak{u}(\mathfrak{p}'^2 - \mathfrak{q}'^2 - \mathfrak{p}^2 + \mathfrak{q}^2) - 6\mathfrak{v}(\mathfrak{p}'\mathfrak{q}' - \mathfrak{p}\mathfrak{q}).$$

If we denote the expression in brackets by Φ, then according to the theorem on spherical functions,

$$\int_0^{2\pi} \Phi d\epsilon = \frac{3\pi}{2}(\mathfrak{u}\mathfrak{p}^2 - \mathfrak{u}\mathfrak{q}^2 - 2\mathfrak{v}\mathfrak{p}\mathfrak{q})\,\frac{3}{2}\,(\mu^2 - 1).$$

Now recall that $\mu^2-1=-4\sin^2\vartheta\cos^2\vartheta$. If one substitutes $\mathfrak{u}=\mathfrak{x}+\mathfrak{x}_1$, $\mathfrak{v}=\mathfrak{y}+\mathfrak{y}_1$, $\mathfrak{p}=\mathfrak{x}-\mathfrak{x}_1$, $\mathfrak{q}=\mathfrak{z}-\mathfrak{z}_1$, applies Equation (212) and assumes that $\bar{\mathfrak{x}}=\bar{\mathfrak{y}}=\bar{\mathfrak{z}}=0$, then it follows, on taking account of Equation (208), that

$$(230)\quad \begin{cases} B_5(\mathfrak{x}^3 - 3\mathfrak{x}\mathfrak{y}^2) = \dfrac{1}{2}\displaystyle\int\int\int_0^\infty\int_0^{2\pi} \Phi f f_1 g b d\omega d\omega_1 db d\epsilon \\ = -\dfrac{9}{2}A_2\rho^2\sqrt{\dfrac{K_1}{2m^5}}\,(\overline{\mathfrak{x}^3} - \overline{3\mathfrak{x}\mathfrak{y}^2}) = -\dfrac{3p\rho}{2m\mathfrak{R}}\,(\overline{\mathfrak{x}^3} - \overline{3\mathfrak{x}\mathfrak{y}^2}). \end{cases}$$

The equality is valid for any spherical function of the third degree. In general,

$$(231)\qquad B_5[p^{(3)}(\mathfrak{x}, \mathfrak{y}, \mathfrak{z})] = -\frac{3p\rho}{2m\mathfrak{R}}\,\overline{p^{(3)}(\mathfrak{x}, \mathfrak{y}, \mathfrak{z})}.$$

The reciprocal of the relaxation time of a spherical function of the third degree is therefore

$$\frac{3}{2}\,\frac{p}{\mathfrak{R}}\,.$$

Any function of the third degree in $\mathfrak{x}$, $\mathfrak{y}$, $\mathfrak{z}$ can be represented as a sum of spherical functions of the third degree and of the three functions $\mathfrak{x}(\mathfrak{x}^2+\mathfrak{y}^2+\mathfrak{z}^2)$, $\mathfrak{y}(\mathfrak{x}^2+\mathfrak{y}^2+\mathfrak{z}^2)$, $\mathfrak{z}(\mathfrak{x}^2+\mathfrak{y}^2+\mathfrak{z}^2)$ each multiplied by some constant. The latter three functions are the products of the spherical functions of the first degree by the expression

$(\mathfrak{x}^2+\mathfrak{y}^2+\mathfrak{z}^2)$. The relaxation time of these latter products still has to be found.

We have

$$2[\mathfrak{x}'(\mathfrak{x}'^2+\mathfrak{y}'^2+\mathfrak{z}'^2)+\mathfrak{x}_1'(\mathfrak{x}_1'^2+\mathfrak{y}_1'^2+\mathfrak{z}_1'^2)-\mathfrak{x}(\mathfrak{x}^2+\mathfrak{y}^2+\mathfrak{z}^2)$$
$$-\mathfrak{x}_1(\mathfrak{x}_1^2+\mathfrak{y}_1^2+\mathfrak{z}_1^2)]=\mathfrak{u}(\mathfrak{p}'^2-\mathfrak{p}^2)+\mathfrak{v}(\mathfrak{p}'\mathfrak{q}'-\mathfrak{p}\mathfrak{q})+\mathfrak{w}(\mathfrak{p}'\mathfrak{r}'-\mathfrak{p}\mathfrak{r}).$$

If we denote the expression in brackets by Ψ then we have

$$\int_0^{2\pi}\Psi d\epsilon=+\left[\frac{\mathfrak{u}}{6}(2\mathfrak{p}^2-\mathfrak{q}^2-\mathfrak{r}^2)+\frac{\mathfrak{v}}{2}\mathfrak{p}\mathfrak{q}+\frac{\mathfrak{w}}{2}\mathfrak{p}\mathfrak{r}\right]3\pi(\mu^2-1).$$

$$\text{(231a)}\quad\begin{cases}\displaystyle\int_0^\infty gbdb\int_0^{2\pi}d\epsilon\Psi=-\frac{1}{2}[\mathfrak{u}(2\mathfrak{p}^2-\mathfrak{q}^2-\mathfrak{r}^2)\\ \displaystyle\qquad+3\mathfrak{v}\mathfrak{p}\mathfrak{q}+3\mathfrak{w}\mathfrak{p}\mathfrak{r}]A_2\sqrt{\frac{2K_1}{m}}\end{cases}$$

hence

$$\text{(232)}\quad\begin{cases}\displaystyle B_5[\mathfrak{x}(\mathfrak{x}^2+\mathfrak{y}^2+\mathfrak{z}^2)]=\frac{1}{2}\int\int\int_0^\infty\int_0^{2\pi}\Psi ff_1gbd\omega d\omega_1dbd\epsilon\\ \displaystyle=-2A_2\rho^2\sqrt{\frac{K_1}{2m^5}}(\overline{\mathfrak{x}^3}+\overline{\mathfrak{x}\mathfrak{y}^2}+\overline{\mathfrak{x}\mathfrak{z}^2})=-\frac{2p\rho}{3m\mathfrak{R}}(\overline{\mathfrak{x}^3}+\overline{\mathfrak{x}\mathfrak{y}^2}+\overline{\mathfrak{x}\mathfrak{z}^2}).\end{cases}$$

Therefore:

$$\text{(233)}\quad B_5[(\mathfrak{x}^2+\mathfrak{y}^2+\mathfrak{z}^2)p^{(1)}(\mathfrak{x},\mathfrak{y},\mathfrak{z})]=-\frac{2p\rho}{3m\mathfrak{R}}\overline{(\mathfrak{x}^2+\mathfrak{y}^2+\mathfrak{z}^2)p^{(1)}(\mathfrak{x},\mathfrak{y},\mathfrak{z})}.$$

The reciprocal relaxation time of the product of $(\mathfrak{x}^2+\mathfrak{y}^2+\mathfrak{z}^2)$ by a spherical function of the first degree is

$$\frac{2}{3}\frac{p}{\mathfrak{R}}.$$

§23. Heat conduction. Second method of approximate calculations.

We now wish to set $\mathfrak{f}=\mathfrak{x}^3$ in Equation (188) and retain only terms of the largest order of magnitude, thus neglecting the deviation of the state distribution from that which holds for a gas moving with constant velocity, so that $\overline{\mathfrak{x}^3}=\overline{\mathfrak{y}\mathfrak{x}^3}=\overline{\mathfrak{x}^2\mathfrak{y}}=0$. Thereby we obtain from Equation (188):

$$mB_5(\mathfrak{x}^3) = \frac{\partial(\rho\overline{\mathfrak{x}^4})}{\partial x} - 3\overline{\mathfrak{x}^2}\cdot\frac{\partial(\rho\overline{\mathfrak{x}^2})}{\partial x}.$$

Since the present approximate calculation is based on calculating the terms as if the Maxwell distribution were valid, if one writes $\mathfrak{x}$, $\mathfrak{y}$, $\mathfrak{z}$ for ξ, η, ζ then he can apply Equation (49) provided he writes therein $\mathfrak{x}$, $\mathfrak{y}$, $\mathfrak{z}$ in place of ξ, η, ζ. Hence

$$\rho\overline{\mathfrak{x}^4} = 3\rho(\overline{\mathfrak{x}^2})^2 = 3\frac{p^2}{\rho}, \quad \rho\overline{\mathfrak{x}^2} = p.$$

Therefore

$$mB_5(\mathfrak{x}^3) = 3p\frac{\partial\left(\dfrac{p}{\rho}\right)}{\partial x}.$$

If one substitutes $\mathfrak{f}=\mathfrak{x}\mathfrak{y}^2$, then it follows using the same approximations that

$$mB_5(\mathfrak{x}\mathfrak{y}^2) = \frac{\partial(\rho\overline{\mathfrak{x}^2\mathfrak{y}^2})}{\partial x} - \overline{\mathfrak{y}^2}\frac{\partial(\rho\overline{\mathfrak{x}^2})}{\partial x}.$$

Since now

$$\overline{\mathfrak{x}^2\mathfrak{y}^2} = \overline{\mathfrak{x}^2}\cdot\overline{\mathfrak{y}^2} = \frac{p^2}{\rho^2},$$

then

$$mB_5(\mathfrak{x}\mathfrak{y}^2) = p\frac{\partial\left(\dfrac{p}{\rho}\right)}{\partial x}.$$

Likewise

$$mB_5(\mathfrak{x}\mathfrak{z}^2) = p\frac{\partial\left(\dfrac{p}{\rho}\right)}{\partial x},$$

hence

$$mB_5(\mathfrak{x}^3 - 3\mathfrak{x}\mathfrak{y}^2) = 0$$

$$mB_5(\mathfrak{x}^3 + \mathfrak{x}\mathfrak{y}^2 + \mathfrak{x}\mathfrak{z}^2) = 5p\frac{\partial\left(\dfrac{p}{\rho}\right)}{\partial x}$$

and according to Equations (230) and (232):

$$(234)\quad \begin{cases} \overline{\mathfrak{x}^3} - 3\overline{\mathfrak{x}\mathfrak{y}^2} = \overline{\mathfrak{x}^3} - 3\overline{\mathfrak{x}\mathfrak{z}^2} = 0 \\ \rho(\overline{\mathfrak{x}^3} + \overline{\mathfrak{x}\mathfrak{y}^2} + \overline{\mathfrak{x}\mathfrak{z}^2}) = -\dfrac{15\mathfrak{R}}{2}\dfrac{\partial\left(\dfrac{p}{\rho}\right)}{\partial x}. \end{cases}$$

Whence it follows that

$$(235)\quad \begin{cases} \overline{\mathfrak{x}^3} = -\dfrac{9}{2}\dfrac{\mathfrak{R}}{\rho}\dfrac{\partial\left(\dfrac{p}{\rho}\right)}{\partial x}, \quad \overline{\mathfrak{x}\mathfrak{y}^2} = \overline{\mathfrak{x}\mathfrak{z}^2} = -\dfrac{3}{2}\dfrac{\mathfrak{R}}{\rho}\dfrac{\partial\left(\dfrac{p}{\rho}\right)}{\partial x}; \\ \text{similarly it follows that} \\ \overline{\mathfrak{y}^3} = -\dfrac{9}{2}\dfrac{\mathfrak{R}}{\rho}\dfrac{\partial\left(\dfrac{p}{\rho}\right)}{\partial y}, \quad \overline{\mathfrak{x}^2\mathfrak{y}} = \overline{\mathfrak{y}\mathfrak{z}^2} = -\dfrac{3}{2}\dfrac{\mathfrak{R}}{\rho}\dfrac{\partial\left(\dfrac{p}{\rho}\right)}{\partial y} \\ \overline{\mathfrak{z}^3} = -\dfrac{9}{2}\dfrac{\mathfrak{R}}{\rho}\dfrac{\partial\left(\dfrac{p}{\rho}\right)}{\partial z}, \quad \overline{\mathfrak{x}\mathfrak{z}^2} = \overline{\mathfrak{y}\mathfrak{z}^2} = -\dfrac{3}{2}\dfrac{\mathfrak{R}}{\rho}\dfrac{\partial\left(\dfrac{p}{\rho}\right)}{\partial z}. \end{cases}$$

These values can be used to carry the approximate solution of Equations (189) and (190) one step further than has been done up to now.

We next add to Equation (189) the analogous equations for the y and z axes. Now we have $B_5(\mathfrak{x}^2)+B_5(\mathfrak{y}^2)+B_5(\mathfrak{z}^2)=0$. If one takes account of Equation (234) and the two equations obtained from it by cyclic permutation, as well as the continuity equation (184), and if one substitutes finally for $\rho\overline{\mathfrak{x}^2}=X_x$, $\rho\overline{\mathfrak{x}\mathfrak{y}}=X_y$ etc. the values given in Equation (220), it follows that:

$$(236)\quad \begin{cases} \dfrac{3\rho}{2}\dfrac{d\left(\dfrac{p}{\rho}\right)}{dt} = \dfrac{p}{\rho}\dfrac{d\rho}{dt} + \dfrac{15}{4}\left[\dfrac{\partial}{\partial x}\left(\mathfrak{R}\dfrac{\partial\left(\dfrac{p}{\rho}\right)}{\partial x}\right) + \dfrac{\partial}{\partial y}\left(\mathfrak{R}\dfrac{\partial\left(\dfrac{p}{\rho}\right)}{\partial y}\right)\right. \\ \left. + \dfrac{\partial}{\partial z}\left(\mathfrak{R}\dfrac{\partial\left(\dfrac{p}{\rho}\right)}{\partial z}\right)\right] + \mathfrak{R}\left[2\left(\dfrac{\partial u}{\partial x}\right)^2 + 2\left(\dfrac{\partial v}{\partial y}\right)^2 + 2\left(\dfrac{\partial w}{\partial z}\right)^2\right. \\ -\dfrac{2}{3}\left(\dfrac{\partial u}{\partial x} + \dfrac{\partial v}{\partial y} + \dfrac{\partial w}{\partial z}\right)^2 + \left(\dfrac{\partial v}{\partial z} + \dfrac{\partial w}{\partial y}\right)^2 + \left(\dfrac{\partial u}{\partial z} + \dfrac{\partial w}{\partial x}\right)^2 \\ \left. + \left(\dfrac{\partial u}{\partial y} + \dfrac{\partial v}{\partial x}\right)^2\right]. \end{cases}$$

Here $3p/\rho=\overline{\mathfrak{x}^2}+\overline{\mathfrak{y}^2}+\overline{\mathfrak{z}^2}$ is the mean square velocity of thermal motion of a molecule in the volume element *do*. By thermal motion, we mean the motion of the molecule relative to the visible motion of the gas in *do*, the latter having velocity components *u*, *v*, *w*. ρdo is the mass of all molecules in *do*.

$$\frac{3}{2}\rho do\cdot\frac{d\left(\frac{p}{\rho}\right)}{dt}dt$$

is therefore the increment of heat measured in mechanical units, i.e., the increase of the kinetic energy of thermal motion of all molecules contained in *do* during time *dt*. However, in this case the volume element *do* does not remain fixed in space; rather, during time *dt* it must experience that deformation and progressive motion which is required in order that each point in it may move with velocity components *u*, *v*, *w*. The same molecules therefore remain in *do*, aside from the exchanges effected by molecular motion. The amount of heat supplied by the latter will then be included in the calculation as that conducted and created by viscosity.

From §8 we find that the amount of compressional work added to a gas during time *dt* is $-pd\Omega=-pkd(1/\rho)$. In our case, $k=\rho do$ and $d(1/\rho)=-(1/\rho^2)(d\rho/dt)dt$. Hence the term

$$\frac{p}{\rho}\frac{d\rho}{dt}dtdo$$

in Equation (236) represents the work done by the external pressure *p* during *dt* on *do*, and hence the heat of compression produced by the pressure *p*. If one applies the same considerations by which the work of deformation of an elastic body is calculated, then he finds that the last term with the factor $\mathfrak{R}$ outside the differential sign in Equation (236), when it is multiplied by *dodt*, expresses the total work done by additional forces that must be added to the pressure in order to obtain the forces X_x, $X_y\cdots$ given by Equations (220).[1] This term therefore corresponds to the heat developed by viscosity. The next to last term, multiplied by the factor 15/4, must therefore represent (if one multiplies it by *dodt*) the heat introduced by heat conduction into the volume element. If we imagine the volume element to be a parallelepiped with edges *dx*, *dy*, *dz*, and draw the *x*-axis from left to right, the

[1] Cf. Kirchhoff, *Vorlesungen über die Theorie der Wärme* (Teubner, 1894), p. 118.

y-axis from back to front, and the z-axis from below to above, and denote by T the temperature, by $\mathfrak{L}$ the coefficient of heat conductivity—then, according to the old Fourier theory of heat conduction (which is established by experiment, at least approximately)

$$\mathfrak{L}\frac{\partial T}{\partial x}dydzdt, \quad \mathfrak{L}\frac{\partial T}{\partial y}dxdzdt \text{ and } \mathfrak{L}\frac{\partial T}{\partial z}dxdydt$$

are the quantities of heat that leave the parallelepiped on the left, in back, and below, respectively;

$$\left[\mathfrak{L}\frac{\partial T}{\partial x}+\frac{\partial}{\partial x}\left(\mathfrak{L}\frac{\partial T}{\partial x}\right)dx\right]dydzdt,$$

$$\left[\mathfrak{L}\frac{\partial T}{\partial y}+\frac{\partial}{\partial y}\left(\mathfrak{L}\frac{\partial T}{\partial y}\right)dy\right]dxdzdt$$

and

$$\left[\mathfrak{L}\frac{\partial T}{\partial z}+\frac{\partial}{\partial z}\left(\mathfrak{L}\frac{\partial T}{\partial z}\right)dz\right]dxdydt$$

are the quantities of heat that come in on the opposite sides. The total increment of heat caused by heat conduction into the parallelepiped do during time dt is therefore

$$\left[\frac{\partial}{\partial x}\left(\mathfrak{L}\frac{\partial T}{\partial x}\right)+\frac{\partial}{\partial y}\left(\mathfrak{L}\frac{\partial T}{\partial y}\right)+\frac{\partial}{\partial z}\left(\mathfrak{L}\frac{\partial T}{\partial z}\right)\right]dodt. \tag{237}$$

The term multiplied by 15/4 in Equation (236) is small. We can therefore neglect higher powers of this quantity, and treat the gas as if u, v, w were constant, while $\mathfrak{x}$, $\mathfrak{y}$, $\mathfrak{z}$ are given by the Maxwell velocity distribution. Its internal state will then be determined only by $\mathfrak{x}$, $\mathfrak{y}$, $\mathfrak{z}$, and we can apply the formulae of §7 and §8 as if it were a gas at rest. If r is the gas constant of our gas, and R that of the normal gas, while m/μ is the mass of a molecule of the latter, then according to Equation (52) we have

$$\frac{p}{\rho}=rT=\frac{R}{\mu}T.$$

Hence the term with the factor 15/4 in Equation (236), multiplied by $dodt$, has the following form:

$$\frac{15}{4}\frac{R}{\mu}\left[\frac{\partial}{\partial x}\left(\mathfrak{R}\frac{\partial T}{\partial x}\right)+\frac{\partial}{\partial y}\left(\mathfrak{R}\frac{\partial T}{\partial y}\right)+\frac{\partial}{\partial z}\left(\mathfrak{R}\frac{\partial T}{\partial z}\right)\right] do dt.$$

This agrees completely with the empirical expression (237), if one sets

$$\mathfrak{L} = \frac{15}{4}\frac{R\mathfrak{R}}{\mu}. \tag{238}$$

In order to make this independent of the thermal units used, we introduce instead of R the specific heat. Since we assume there is no intramolecular motion, the quantity β in Equation (54) is zero; this equation then gives:

$$\gamma_v = \frac{3R}{2\mu},$$

hence[2]

$$\mathfrak{L} = \frac{5}{2}\gamma_v\mathfrak{R}. \tag{239}$$

This value is 5/2 times as large as that given by Equation (93) and is about as much larger than the experimental values as the latter value is smaller than them. One should not expect quantitative agreement in cases where the assumptions made (e.g., $\beta=0$) are clearly not satisfied. Since R, μ, and hence also γ_v are constants, $\mathfrak{L}$ depends on temperature and pressure in the same way that $\mathfrak{R}$ does.

We have thus obtained all the formulae accepted by the so-called descriptive theory, except that a coefficient in the terms representing viscosity, which remains arbitrary in the descriptive theory, has here a particular value. In the descriptive theory $(p-X_x)\cdot(3/2\mathfrak{R})$ is equal to

$$3\frac{\partial u}{\partial x}-\epsilon\left(\frac{\partial u}{\partial x}+\frac{\partial v}{\partial y}+\frac{\partial w}{\partial z}\right),$$

---

[2] As a result of an error in calculation, Maxwell (Phil. Mag. [4] **35**, 216 (1868), *Scientific Papers* **2**, 77, Eq. [149]) found for $\mathfrak{L}$ only $\frac{2}{3}$ of the above value, as I noted in Wien. Ber. **66**, 332 (1872). Poincaré made the same remark: C. R. Paris **116**, 1020 (1893).

while here it is equal to

$$3\frac{\partial u}{\partial x}-\left(\frac{\partial u}{\partial x}+\frac{\partial v}{\partial y}+\frac{\partial w}{\partial z}\right).$$

Therefore in the descriptive theory, in the expression for $X_x - p$, the compression-dependent expression

$$\frac{\partial u}{\partial x}+\frac{\partial v}{\partial y}+\frac{\partial w}{\partial z}$$

is multiplied by a coefficient that is independent of the coefficient of $\partial u/\partial x$, while in our theory the latter coefficient is exactly three times as large as the former. The same holds for Y_y and Z_z. The latter coefficient must indeed here, as also in the descriptive theory, be twice the coefficient of

$$\frac{\partial w}{\partial y}+\frac{\partial v}{\partial z}$$

in the expression for Y_z, and thus it is twice the experimentally measurable viscosity coefficient.*

In the light of our theory, all these formulas are approximate ones. There is no difficulty in carrying the approximation further. The equations thus extended to a higher degree of approximation would not necessarily agree with experiment everywhere, since many of our hypotheses are still arbitrary, but they will probably be useful guidelines when experiments are initiated. It will be difficult, but not completely hopeless, to test them by experiment, and it is to be expected that they will teach us new facts going beyond the old hydrodynamic equations. In order to indicate briefly how the approximation is to be carried further, we shall substitute in Equations (189) and (190) the values just found. From the latter, it follows by Equations (214, 235, 220, 52, 238) that

---

* The result of Maxwell and Boltzmann, that a gas does not have two independent viscosity coefficients, is certainly not true in general for real gases, nor is it a strict consequence of the kinetic theory except perhaps for certain idealized models. For a list of recent papers on "bulk viscosity" see S. G. Brush, Chem. Revs. **62**, 513 (1962).

$$
(239a) \quad \left\{
\begin{aligned}
X_y = \rho\overline{\mathfrak{x}\mathfrak{y}} = -\frac{\Re}{p}\Big[\rho\frac{d\overline{\mathfrak{x}\mathfrak{y}}}{dt} + X_y\frac{\partial u}{\partial x} + Y_y\frac{\partial u}{\partial z} + Y_z\frac{\partial u}{\partial z} \\
+ X_x\frac{\partial v}{\partial x} + X_y\frac{\partial v}{\partial y} + X_z\frac{\partial v}{\partial z} \\
-\frac{2}{5}\frac{\partial}{\partial x}\left(\mathfrak{L}\frac{\partial T}{\partial y}\right) - \frac{2}{5}\frac{\partial}{\partial y}\left(\mathfrak{L}\frac{\partial T}{\partial x}\right) + \frac{\partial(\rho\overline{\mathfrak{x}\mathfrak{y}\mathfrak{z}})}{\partial z}\Big].
\end{aligned}
\right.
$$

If one makes the substitution $\mathfrak{f}=\mathfrak{x}\mathfrak{y}\mathfrak{z}$ in Equation (188), he obtains only terms that vanish to the present degree of accuracy. One can therefore set

$$
mB_5(\mathfrak{x}\mathfrak{y}\mathfrak{z}) = -\frac{3p}{2\Re}\rho\overline{\mathfrak{x}\mathfrak{y}\mathfrak{z}}
$$

equal to zero, and therefore also

$$
\frac{\partial(\rho\overline{\mathfrak{x}\mathfrak{y}\mathfrak{z}})}{\partial z} = 0.
$$

For X_x, $X_y \cdots$ one is to substitute the values on the right side of Equation (220). Further, according to (218a),

$$
\frac{d\overline{\mathfrak{x}\mathfrak{y}}}{dt} = -\frac{d}{dt}\left[\frac{\Re}{\rho}\left(\frac{\partial v}{\partial x}+\frac{\partial u}{\partial y}\right)\right],
$$

and since one uses here only the terms of largest magnitude,

$$
\frac{d\overline{\mathfrak{x}\mathfrak{y}}}{dt} = -\frac{\Re}{\rho}\left(\frac{\partial v}{\partial x}+\frac{\partial u}{\partial y}\right)\left(\frac{\partial u}{\partial x}+\frac{\partial v}{\partial y}+\frac{\partial w}{\partial z}\right)
$$

$$
+\frac{\Re}{\rho}\frac{\partial}{\partial x}\left(\frac{1}{\rho}\frac{\partial p}{\partial y}-X\right)+\frac{\Re}{\rho}\frac{\partial}{\partial y}\left(\frac{1}{\rho}\frac{\partial p}{\partial x}-Y\right)-\frac{1}{\rho}\left(\frac{\partial v}{\partial x}+\frac{\partial u}{\partial y}\right)\frac{d\Re}{dt}.
$$

Similarly X_x, $Y_z \cdots$ can be calculated. One would thus obtain some very complicated expressions, which would appear to Continental physicists as strange as did at first those in Maxwell's electrical theory. Yet who knows whether many terms of these equations may not some day play an important role? Here we shall indicate only a special case already considered by Maxwell. 1. Let there be neither mass motion nor external forces acting on the gas; hence $u=v=w=X=Y=Z=0$. 2. Suppose that some heat flow is taking place. Then the derivatives with respect to t will vanish, hence, according to Equation (239a),

$$X_y = Y_x = \frac{2}{5}\frac{\mathfrak{R}}{p}\left[\frac{\partial}{\partial x}\left(\mathfrak{L}\frac{\partial T}{\partial y}\right)+\frac{\partial}{\partial y}\left(\mathfrak{L}\frac{\partial T}{\partial x}\right)\right].$$

In this special case, Equation (189) gives:

$$Y_y + Z_z - 2X_x = \frac{3\mathfrak{R}}{p}\left[\frac{\partial(\rho\overline{\mathfrak{x}^3})}{\partial x}+\frac{\partial(\rho\overline{\mathfrak{x}^2\mathfrak{y}})}{\partial y}+\frac{\partial(\rho\overline{\mathfrak{x}^2\mathfrak{z}})}{\partial z}\right].$$

Therefore, taking account of Equations (235),

$$2X_x - Y_y - Z_z$$
$$= \frac{6\mathfrak{R}}{5p}\left[3\frac{\partial}{\partial x}\left(\mathfrak{L}\frac{\partial T}{\partial x}\right)+\frac{\partial}{\partial y}\left(\mathfrak{L}\frac{\partial T}{\partial y}\right)+\frac{\partial}{\partial z}\left(\mathfrak{L}\frac{\partial T}{\partial z}\right)\right],$$

hence since $X_x+Y_y+Z_z=3p$,

$$X_x = p+\frac{2\mathfrak{R}}{5p}\left[3\frac{\partial}{\partial x}\left(\mathfrak{L}\frac{\partial T}{\partial x}\right)+\frac{\partial}{\partial y}\left(\mathfrak{L}\frac{\partial T}{\partial y}\right)+\frac{\partial}{\partial z}\left(\mathfrak{L}\frac{\partial T}{\partial z}\right)\right]$$
$$= p+\frac{4\mathfrak{R}}{5}\frac{\partial}{\partial x}\left(\mathfrak{L}\frac{\partial T}{\partial x}\right),$$

since for stationary heat flow

$$\frac{\partial}{\partial x}\left(\mathfrak{L}\frac{\partial T}{\partial x}\right)+\frac{\partial}{\partial y}\mathfrak{L}\left(\frac{\partial T}{\partial y}\right)+\frac{\partial}{\partial z}\left(\mathfrak{L}\frac{\partial T}{\partial z}\right) = 0.$$

In this case we also have

$$\frac{\partial X_x}{\partial x}+\frac{\partial Y_x}{\partial y}+\frac{\partial Z_x}{\partial z} = 0;$$

therefore the volume elements in the interior of the gas are in equilibrium. But the customary view (cf. the last page of Kirchhoff's lectures on heat theory, already cited, where what is said otherwise about the old heat conduction theory is quite true)—that in stationary heat flow the pressure can be everywhere equal—is shown to be false. The pressure varies from place to place, and at a given point it may be different directions, and not exactly normal to the surface.

Hence, if a solid body is completely surrounded by a heat-conducting gas, it will in general start to move, since the pressure is not equal everywhere. Maxwell was quite correct in attributing the cause of radiometer phenomena to this effect.* Moreover, a gas in contact with a solid wall cannot remain at rest if the wall cannot exert a finite tangential force on the gas. These motions, created by pressure-differences inside the gas, are not to be confused with those which arise through the action of gravity, in consequence of the different density of warmer and colder gases. The latter motions canot play any role in the radiometer, since the axis of rotation is vertical. Also, our formulas do not apply to the latter motions, since we set $X=Y=Z=0$.

Up to now we have followed the ingenious methods devised by Maxwell, and applied by Kirchhoff and others. These methods permit one to avoid calculating the velocity distribution function $f(x, y, z, \xi, \eta, \zeta, t)$. There is another method which proceeds on the contrary from a calculation of this function. Although the latter method has not been used, I will say a few words about it here, since to calculate the entropy we need to know f.

The starting point is the general equation (114) in which, since we consider only one kind of gas, the next to last term vanishes. If we write, instead of the previously used constants $a, h, u, v, w,$

$$e^a, \frac{k}{m}, u_0, v_0, w_0,$$

then we know that the equation will be satisfied if we put

$$f = e^{a-k[(\xi-u_0)^2+(\eta-v_0)^2+(\zeta-w_0)^2]} \tag{240}$$

as long as a, k, u_0, v_0, w_0 are constants. Then u_0, v_0, w_0 are the velocity components of the gas as a whole.

Now let k, a, u_0, v_0, w_0 be functions of x, y, z, t; their variations (i.e., their derivatives with respect to these variables) are assumed to be so small that only small correction terms need to be added to the expression (240) in order to satisfy Equation (114). We shall represent these corrections in the form of a power series. Since a, k, u_0, v_0, w_0 are arbitrary, we can choose their values in such a way that the terms multiplied by ξ, η, and ζ in

* Maxwell, Phil. Trans. **170**, 231 (1879).

the power series will vanish. This can therefore be done without any loss of generality. Moreover, we can choose the coefficients of ξ^2, η^2, and ζ^2 to be such that their sum is equal to zero. We introduce the variables

$$(241) \qquad \mathfrak{x}_0 = \xi - u_0, \quad \mathfrak{y}_0 = \eta - v_0, \quad \mathfrak{z}_0 = \zeta - w_0$$

and put

$$(242) \qquad \begin{cases} f = f^{(0)}(1 + b_{11}\mathfrak{x}_0^2 + b_{22}\mathfrak{y}_0^2 + b_{33}\mathfrak{z}_0^2 + b_{12}\mathfrak{x}_0\mathfrak{y}_0 + b_{13}\mathfrak{x}_0\mathfrak{z}_0 \\ \quad + b_{23}\mathfrak{y}_0\mathfrak{z}_0 + c_1\mathfrak{x}_0\mathfrak{c}_0^2 + c_2\mathfrak{y}_0\mathfrak{c}_0^2 + c_3\mathfrak{z}_0\mathfrak{c}_0^2), \end{cases}$$

where

$$(243) \qquad f^{(0)} = e^{a-k(\mathfrak{x}_0^2+\mathfrak{y}_0^2+\mathfrak{z}_0^2)}$$

and

$$(244) \qquad b_{11} + b_{22} + b_{33} = 0.$$

The left side of Equation (114) is now transformed to

$$\mathfrak{I} = \frac{\partial f}{\partial t} + (\mathfrak{x}_0 + u_0)\frac{\partial f}{\partial x} + (\mathfrak{y}_0 + v_0)\frac{\partial f}{\partial y} + (\mathfrak{z}_0 + w_0)\frac{\partial f}{\partial z}$$
$$+ X\frac{\partial f}{\partial \mathfrak{x}_0} + Y\frac{\partial f}{\partial \mathfrak{y}_0} + Z\frac{\partial f}{\partial \mathfrak{z}_0}\,\cdot$$

Since all the differential quotients are small, we can replace f by $f^{(0)}$ in them. If we write $\mathfrak{c}_0^2$ for $\mathfrak{x}_0^2+\mathfrak{y}_0^2+\mathfrak{z}_0^2$ and d_0/dt for $\partial/\partial t+u_0\partial/\partial x+v_0\partial/\partial y+w_0\partial/\partial z$, then we find:

$$(245) \qquad \begin{cases} \dfrac{1}{f^{(0)}}\mathfrak{I} = \dfrac{d_0 a}{dt} - \mathfrak{c}_0^2\dfrac{d_0 k}{dt} + \mathfrak{x}_0\left[\dfrac{\partial a}{\partial x} - 2k\left(\dfrac{d_0 u_0}{dt} - X\right)\right] \\ + \mathfrak{y}_0\left[\dfrac{\partial a}{\partial y} + 2k\left(\dfrac{d_0 v_0}{dt} - Y\right)\right] + \mathfrak{z}_0\left[\dfrac{\partial a}{\partial z} + 2k\left(\dfrac{d_0 w_0}{dt} - Z\right)\right] \\ + 2k\left[\mathfrak{x}_0^2\dfrac{\partial u_0}{\partial x} + \mathfrak{y}_0^2\dfrac{\partial v_0}{\partial y} + \mathfrak{z}_0^2\dfrac{\partial w_0}{\partial z} + \mathfrak{y}_0\mathfrak{z}_0\left(\dfrac{\partial v_0}{\partial z} + \dfrac{\partial w_0}{\partial y}\right)\right. \\ \left. + \mathfrak{x}_0\mathfrak{z}_0\left(\dfrac{\partial w_0}{\partial x} + \dfrac{\partial u_0}{\partial z}\right) + \mathfrak{x}_0\mathfrak{y}_0\left(\dfrac{\partial u_0}{\partial y} + \dfrac{\partial v_0}{\partial x}\right)\right] \\ - \mathfrak{c}_0^2\left(\mathfrak{x}_0\dfrac{\partial k}{\partial x} + \mathfrak{y}_0\dfrac{\partial k}{\partial y} + \mathfrak{z}_0\dfrac{\partial k}{\partial z}\right). \end{cases}$$

If one considers the coefficients b to be small, so that their products and squares may be ignored, then the right side of Equation (114) becomes

$$\mathfrak{r} = \int\int_0^\infty \int_0^{2\pi} f^{(0)} f_1^{(0)} d\omega_1 g b db d\epsilon [b_{11}(\mathfrak{x}'^2 + \mathfrak{x}_1'^2 - \mathfrak{x}^2 - \mathfrak{x}_1^2)$$

$$+ b_{22}(\mathfrak{y}'^2 + \mathfrak{y}_1'^2 - \mathfrak{y}^2 - \mathfrak{y}_1^2) \cdots].$$

In order to avoid an accumulation of too many indices, we shall drop the index zero from the quantities $\mathfrak{x}$, $\mathfrak{y}$, $\mathfrak{z}$ until we get to Equation (246)—i.e., it will not be explicitly indicated that they arise by subtracting the quantities u_0, v_0, w_0 rather than u, v, w from the corresponding quantities ξ, η, ζ. Since in $f^{(0)}$ and $f_1^{(0)}$ it is u_0, v_0, w_0 and not u, v, w that are subtracted from ξ, η, ζ, we find just as before that:

$$U = \int_0^\infty g b db \int_0^{2\pi} d\epsilon (\mathfrak{x}'\mathfrak{y}' + \mathfrak{x}_1'\mathfrak{y}_1' - \mathfrak{x}\mathfrak{y} - \mathfrak{x}_1\mathfrak{y}_1)$$

$$= -3A_2 \sqrt{\frac{K_1}{2m}} (\mathfrak{x}\mathfrak{y} - \mathfrak{x}\mathfrak{y}_1 - \mathfrak{x}_1\mathfrak{y} + \mathfrak{x}_1\mathfrak{y}_1).$$

Hence:

$$\int f_1^{(0)} d\omega_1 U = -3A_2 \sqrt{\frac{K_1}{2m^2}} \rho \mathfrak{x}\mathfrak{y}.$$

The same result holds for the products $\mathfrak{x}\mathfrak{z}$ and $\mathfrak{y}\mathfrak{z}$. Since now

$$\int \mathfrak{x}_1^2 f_1^{(0)} d\omega_1 = \int \mathfrak{y}_1^2 f_1^{(0)} d\omega_1 = \int \mathfrak{z}_1^2 f_1^{(0)} d\omega_1 \text{ and } b_{11} + b_{22} + b_{33} = 0$$

it follows that $b_{11}\mathfrak{x}^2 + b_{22}\mathfrak{y}^2 + b_{33}\mathfrak{z}^2$ can be represented as a sum of spherical functions of the second degree, and we have

$$\int\int_0^\infty \int_0^{2\pi} f_1^{(0)} g b d\omega_1 db d\epsilon (b_{11}\mathfrak{X} + b_{22}\mathfrak{Y} + b_{33}\mathfrak{Z})$$

$$= -\frac{3}{2} A_2 \rho \sqrt{\frac{2K_1}{m^2}} (b_{11}\mathfrak{x}^2 + b_{22}\mathfrak{y}^2 + b_{33}\mathfrak{z}^2),$$

where the abbreviation $\mathfrak{X}$ has been used for $\mathfrak{x}'^2 + \mathfrak{x}_1'^2 - \mathfrak{x}^2 - \mathfrak{x}_1^2$. $\mathfrak{Y}$ and $\mathfrak{Z}$ have similar meanings.

If one sets

$$\mathfrak{X}_1 = \mathfrak{x}'\mathfrak{c}'^2 + \mathfrak{x}_1'\mathfrak{c}_1'^2 - \mathfrak{x}\mathfrak{c}^2 - \mathfrak{x}_1\mathfrak{c}_1^2$$

$$\mathfrak{Y}_1 = \mathfrak{y}'\mathfrak{c}'^2 + \mathfrak{y}_1'\mathfrak{c}_1'^2 - \mathfrak{y}\mathfrak{c}^2 - \mathfrak{y}_1\mathfrak{c}_1^2$$

$$\mathfrak{Z}_1 = \mathfrak{z}'\mathfrak{c}'^2 + \mathfrak{z}_1'\mathfrak{c}_1'^2 - \mathfrak{z}\mathfrak{c}^2 - \mathfrak{z}_1\mathfrak{c}_1^2,$$

then he finds likewise, according to the principles of the previous section (cf. Eq. [231**a**]):

$$\int_0^\infty gbdb \int_0^{2\pi} d\epsilon \mathfrak{X}_1 = -A_2\sqrt{\frac{K_1}{2m}}\cdot[2(\mathfrak{x}^2 - \mathfrak{x}_1^2)(\mathfrak{x} - \mathfrak{x}_1) - (\mathfrak{x} + \mathfrak{x}_1)(\mathfrak{y} - \mathfrak{y}_1)^2 - (\mathfrak{x} + \mathfrak{x}_1)(\mathfrak{z} - \mathfrak{z}_1)^2 + 3(\mathfrak{y}^2 - \mathfrak{y}_1^2)(\mathfrak{x} - \mathfrak{x}_1) + 3(\mathfrak{z}^2 - \mathfrak{z}_1^2)(\mathfrak{x} + \mathfrak{x}_1)].$$

$$\iint_0^\infty \int_0^{2\pi} f_1^{(0)} d\omega_1 gbdbd\epsilon(c_1\mathfrak{X}_1 + c_2\mathfrak{Y}_1 + c_3\mathfrak{Z}_1) =$$

$$-2A_2\rho\sqrt{\frac{K_1}{2m^3}}\left[(c_1\mathfrak{x}_1 + c_2\mathfrak{y}_2 + c_3\mathfrak{z})\mathfrak{c}^2 - \frac{5}{2k}(c_1\mathfrak{x} + c_2\mathfrak{y} + c_3\mathfrak{z})\right],$$

hence finally

$$(246)\quad \begin{cases} \dfrac{\mathfrak{r}}{f^{(0)}} = -3A_2\rho\sqrt{\dfrac{K_1}{2m^3}}\Big\{b_{11}\mathfrak{x}_0^2 + b_{22}\mathfrak{y}_0^2 + b_{33}\mathfrak{z}_0^2 \\ + b_{23}\mathfrak{y}_0\mathfrak{z}_0 + b_{13}\mathfrak{x}_0\mathfrak{z}_0 + b_{12}\mathfrak{x}_0\mathfrak{y}_0 + \dfrac{2}{3}\mathfrak{c}_0^2(c_1\mathfrak{x}_0 + c_2\mathfrak{y}_0 + c_3\mathfrak{z}_0) \\ - \dfrac{5}{3k}(c_1\mathfrak{x}_0 + c_2\mathfrak{y}_2 + c_3\mathfrak{z}_0)\Big\}. \end{cases}$$

Equation (114) must be satisfied identically. Hence expressions (245) and (246) must be equal for all values of x_0, y_0, and z_0. The terms independent of x_0, y_0, z_0 must be equal; hence:

$$(247)\qquad \frac{d_0 a}{dt} = 0.$$

Since $b_{11}+b_{22}+b_{33}+0$, the terms of second order in $\mathfrak{x}_0$, $\mathfrak{y}_0$, $\mathfrak{z}_0$ give:

$$
(248)\quad \begin{cases}
\dfrac{d_0 k}{dt} + \dfrac{2k}{3}\left(\dfrac{\partial u_0}{\partial x} + \dfrac{\partial v_0}{\partial y} + \dfrac{\partial w_0}{\partial z}\right) = 0, \\[2ex]
b_{11} = \dfrac{2k}{9A_{2\rho}}\sqrt{\dfrac{2m^3}{K_1}}\left(\dfrac{\partial v_0}{\partial y} + \dfrac{\partial w_0}{\partial z} - 2\dfrac{\partial u_0}{\partial x}\right), \\[2ex]
b_{22} = \dfrac{2k}{9A_{2\rho}}\sqrt{\dfrac{2m^3}{K_1}}\left(\dfrac{\partial u_0}{\partial x} + \dfrac{\partial w_0}{\partial z} - 2\dfrac{\partial v_0}{\partial y}\right), \\[2ex]
b_{33} = \dfrac{2k}{9A_{2\rho}}\sqrt{\dfrac{2m^3}{K_1}}\left(\dfrac{\partial u_0}{\partial x} + \dfrac{\partial v_0}{\partial y} - 2\dfrac{\partial w_0}{\partial z}\right), \\[2ex]
b_{23} = -\dfrac{2k}{3A_{2\rho}}\sqrt{\dfrac{2m^3}{K_1}}\left(\dfrac{\partial v_0}{\partial z} + \dfrac{\partial w_0}{\partial y}\right), \\[2ex]
b_{13} = -\dfrac{2k}{3A_{2\rho}}\sqrt{\dfrac{2m^3}{K_1}}\left(\dfrac{\partial w_0}{\partial x} + \dfrac{\partial u_0}{\partial z}\right), \\[2ex]
b_{12} = -\dfrac{2k}{3A_{2\rho}}\sqrt{\dfrac{2m^3}{K_1}}\left(\dfrac{\partial u_0}{\partial y} + \dfrac{\partial v_0}{\partial x}\right), \\[2ex]
c_1 = \dfrac{1}{2A_{2\rho}}\sqrt{\dfrac{2m^3}{K_1}}\,\dfrac{\partial k}{dx}, \\[2ex]
c_2 = \dfrac{1}{2A_{2\rho}}\sqrt{\dfrac{2m^3}{K_1}}\,\dfrac{\partial k}{\partial y}, \\[2ex]
c_3 = \dfrac{1}{2A_{2\rho}}\sqrt{\dfrac{2m^3}{K_1}}\,\dfrac{\partial k}{\partial z}.
\end{cases}
$$

Setting equal the terms containing first powers of $\mathfrak{x}_0$, $\mathfrak{y}_0$, $\mathfrak{z}_0$ gives finally (taking account of the values found for c_1, c_2 and c_3):

$$
(249)\quad \begin{cases}
\dfrac{d_0 u_0}{dt} - X + \dfrac{1}{2k}\dfrac{\partial a}{\partial x} - \dfrac{5}{4k^2}\dfrac{\partial k}{\partial x} \\[2ex]
= \dfrac{d_0 v_0}{dt} - Y + \dfrac{1}{2k}\dfrac{\partial a}{\partial y} - \dfrac{5}{4k^2}\dfrac{\partial k}{\partial y} \\[2ex]
= \dfrac{d_0 w_0}{dt} - Z + \dfrac{1}{2k}\dfrac{\partial a}{\partial z} - \dfrac{5}{4k^2}\dfrac{\partial k}{\partial z} = 0.
\end{cases}
$$

Since $b_{11}+b_{22}+b_{33}=0$, and each term containing an odd power of $\mathfrak{x}_0$, $\mathfrak{y}_0$ or $\mathfrak{z}_0$ vanishes on integration, it follows that (if one writes $d\omega$ and $d\omega_0$ for $d\xi d\eta d\zeta$ and $d\mathfrak{x}_0 d\mathfrak{y}_0 d\mathfrak{z}_0$):

$$\iiint_{-\infty}^{+\infty} f d\omega = \iiint_{-\infty}^{+\infty} f^{(0)} d\omega_0.$$

Hence

$$\rho = m\sqrt{\frac{\pi^3}{k^3}}\, e^a,$$

if we do not carry the approximation far enough to produce any correction to the density of the gas. Likewise:

$$\int (\mathfrak{x}_0^2 + \mathfrak{y}_0^2 + \mathfrak{z}_0^2) f d\omega = \int (\mathfrak{x}_0^2 + \mathfrak{y}_0^2 + \mathfrak{z}_0^2) f^{(0)} d\omega_0.$$

Hence the mean square velocity of motion of the molecule relative to a point moving with velocity u_0, v_0, w_0 is equal to $3/2k$.

On the other hand, u_0, v_0, w_0 are only approximately equal to the components of visible velocity of the gas in the volume element do. These components are actually defined as $\bar{\xi}$, $\bar{\eta}$, $\bar{\zeta}$. Now we have $\bar{\xi}=u_0+\bar{\mathfrak{x}}_0$, and furthermore

$$\overline{\mathfrak{x}_0} = \frac{\int \mathfrak{x}_0 f d\omega}{\int f d\omega} = c_1 \frac{\int \mathfrak{x}_0^2 c_0^2 f^{(0)} d\omega_0}{\int f^{(0)} d\omega_0} = \frac{5c_1}{2k}.$$

If we denote the exact components $\bar{\xi}$, $\bar{\eta}$, $\bar{\zeta}$ of the visible motion of the gas by u, v, w, and those of the motion of a molecule relative to the visible motion by $\mathfrak{x}$, $\mathfrak{y}$, $\mathfrak{z}$, then we obtain in this approximation

$$u = u_0 + \frac{5c_1}{2k}, \quad v = v_0 + \frac{5c_2}{2k}, \quad w = w_0 + \frac{5c_3}{2k}$$

$$\mathfrak{x} = \mathfrak{x}_0 - \frac{5c_1}{2k}, \quad \mathfrak{y} = \mathfrak{y}_0 - \frac{5c_2}{2k}, \quad \mathfrak{z} = \mathfrak{z}_0 - \frac{5c_3}{2k}.$$

Furthermore,

$$p = \frac{\rho}{3}(\overline{\mathfrak{x}^2} + \overline{\mathfrak{y}^2} + \overline{\mathfrak{z}^2}) = \frac{\rho}{3}\left(\overline{\mathfrak{x}_0^2} + \overline{\mathfrak{y}_0^2} + \overline{\mathfrak{z}_0^2} - \frac{25}{4}\frac{c_1^2 + c_2^2 + c_3^2}{k^2}\right)$$

$$= \rho\left(\frac{1}{2k} - \frac{25}{12}\frac{c_1^2 + c_2^2 + c_3^2}{k^2}\right).$$

One has therefore as a first approximation

$$u = u_0,\ v = v_0,\ w = w_0,\ \frac{d_0}{dt} = \frac{d}{dt},\ k = \frac{\rho}{2p} = \frac{1}{2rT},$$

$$a = l\left(\frac{\rho}{m}\sqrt{\frac{k^3}{\pi^3}}\right) = l\left(\frac{\rho^{5/2}p^{-3/2}}{m\sqrt{8\pi^3}}\right) = l\left(\frac{\rho T^{-3/2}}{m\sqrt{8\pi^3 r^3}}\right).$$

Hence, according to Equation (247),

$$p\rho^{-5/2} = \text{const. or } \rho T^{-3/2} = \text{const.}$$

which is Poisson's law. Furthermore,

$$\frac{1}{2k} = \frac{p}{\rho},\ \frac{\partial a}{\partial x} = \frac{5}{2\rho}\frac{\partial \rho}{\partial x} - \frac{3}{2p}\frac{\partial p}{\partial x},\ \frac{1}{k}\frac{\partial k}{\partial x} = \frac{1}{\rho}\frac{\partial \rho}{\partial x} - \frac{1}{p}\frac{\partial p}{\partial x},$$

hence

$$\frac{1}{2k}\left(\frac{\partial a}{\partial x} - \frac{5}{2k}\frac{\partial k}{\partial x}\right) = \frac{1}{\rho}\frac{\partial p}{\partial x}.$$

Therefore Equations (249) yield:

$$\frac{du}{dt} - X + \frac{1}{\rho}\frac{\partial p}{\partial x} = \frac{dv}{dt} - Y + \frac{1}{\rho}\frac{\partial p}{\partial y} = \frac{dw}{dt} - Z + \frac{\partial p}{\partial z} = 0.$$

If we wish to carry the approximation one step further, we can make the above substitutions in the terms that are small, and thereby find:

$$X_y = \rho\overline{\mathfrak{x}\mathfrak{y}} = \rho\frac{\int \mathfrak{x}_0\mathfrak{y}_0 f d\omega_0}{\int f^{(0)}d\omega_0} = \rho b_{12}\frac{\int \mathfrak{x}_0^2\mathfrak{y}_0^2 f^{(0)} d\omega_0}{\int f^{(0)}d\omega_0} = \frac{\rho b_{12}}{4k^2}$$

$$= \frac{p b_{12}}{2k} = -\frac{p}{3A_2\rho}\sqrt{\frac{2m^3}{K_1}}\left(\frac{\partial v}{\partial z} + \frac{\partial w}{\partial y}\right) = -\Re\left(\frac{\partial v}{\partial z} + \frac{\partial w}{\partial y}\right).$$

Likewise the rest of Equations (220) follow, and there is again no difficulty in extending the degree of approximation.

§24. Entropy for the case when Equations (147) are not satisfied. Diffusion.

Up to now we have calculated H only under the restrictive assumption that Equations (147) are satisfied. We now wish to calculate it under the general assumption that f is given by

Equation (242), so that viscosity and heat conduction are present. We assume a simple gas. Then

$$H = \int\int f l f do d\omega.$$

Since f is given by Equation (242), it will be approximated by

$$lf = a - k(\mathfrak{x}_0^2 + \mathfrak{y}_0^2 + \mathfrak{z}_0^2) + A - \frac{A^2}{2}\cdot$$

where the expression in parenthesis in Equation (242) has been denoted by $1+A$.

We now wish to construct the expression H for the gas contained in the volume element do. The value thus found will be multiplied by $-RM$ and divided by do. Let this quantity be

$$J = -RM\int f l f d\omega.$$

Jdo is then the entropy of the gas contained in do.

If we now substitute the above values for f and lf, we obtain first a term independent of the coefficients b and c. This is the entropy (divided by do) that the gas in do would have if it had the same energy (heat) content and the same progressive motion in space, and obeyed the Maxwell velocity distribution law. It can be calculated just as in §19, and has, as shown there, the value

$$\frac{R\rho}{\mu}\, l(T^{3/2}\rho^{-1})$$

apart from a constant. Second, we obtain terms linear in the coefficients b and c. These all vanish. Since

$$\int \mathfrak{x}_0^a \mathfrak{y}_0^b \mathfrak{z}_0^c \exp\{-k(\mathfrak{x}_0^2 + \mathfrak{y}_0^2 + \mathfrak{z}_0^2)\} d\omega_0 = 0$$

if one of the numbers a, b, c is an odd integer, the coefficients of b_{12}, b_{13}, b_{23}, c_1, c_2 and c_3 all vanish. However, if all three numbers a, b, c are even integers, then the integral does not change its value under cyclic permutations of $\mathfrak{x}_3$, $\mathfrak{y}_3$ and $\mathfrak{z}_0$. Hence b_{11}, b_{22} and b_{33} have the same coefficients, and the sum of the terms in question vanishes in any case, since

$$b_{11} + b_{22} + b_{33} = 0.$$

Since we are omitting higher order terms, there still remain in the expression for J only terms of second order in the coefficients b and c. Their sum is

$$J_1 = -\frac{R\rho}{2\mu}\Big(b_{11}^2\overline{\mathfrak{x}_0^4} + b_{22}^2\overline{\mathfrak{y}_0^4} + b_{33}^2\overline{\mathfrak{z}_0^4} + 2b_{11}b_{22}\overline{\mathfrak{x}_0^2}\,\overline{\mathfrak{y}_0^2}$$

$$+ 2b_{11}b_{33}\overline{\mathfrak{x}_0^2}\,\overline{\mathfrak{z}_0^2} + 2b_{22}b_{33}\overline{\mathfrak{y}_0^2}\,\overline{\mathfrak{z}_0^2} + b_{12}^2\overline{\mathfrak{x}_0^2}\,\overline{\mathfrak{y}_0^2} + b_{13}^2\overline{\mathfrak{x}_0^2}\,\overline{\mathfrak{z}_0^2}$$

$$+ b_{23}^2\overline{\mathfrak{y}_0^2}\,\overline{\mathfrak{z}_0^2} + c_1^2\overline{\mathfrak{x}_0^2\mathfrak{c}_0^4} + c_2^2\overline{\mathfrak{y}_0^2\mathfrak{c}_0^4} + c_3^2\overline{\mathfrak{z}_0^2\mathfrak{c}_0^4}\Big).$$

The next terms to be added in Equation (242) (which we have not calculated) will of course be of the same order of magnitude as these, but it is not improbable that they would also vanish on integration

We now find:

$$\overline{\mathfrak{x}_0^4} = \overline{\mathfrak{y}_0^4} = \overline{\mathfrak{z}_0^4} = \frac{3}{4k^2}, \quad \overline{\mathfrak{x}_0^2} = \overline{\mathfrak{y}_0^2} = \overline{\mathfrak{z}_0^2} = \frac{1}{2k}$$

and one finds easily:

$$\overline{\mathfrak{x}_0^2\mathfrak{c}_0^4} = \overline{\mathfrak{y}_0^2\mathfrak{c}_0^4} = \overline{\mathfrak{z}_0^2\mathfrak{c}_0^4} = \frac{1}{3}\overline{\mathfrak{c}_0^6} = \frac{35}{8k^3}.$$

Since

$$\frac{1}{2k} = \frac{RT}{\mu}$$

we have therefore

$$J_1 = -\frac{R^3T^2\rho}{2\mu^3}\Big\{3(b_{11}^2 + b_{22}^2 + b_{33}^2) + 2(b_{11}b_{22} + b_{11}b_{33} + b_{22}b_{33})$$

$$+ b_{12}^2 + b_{13}^2 + b_{23}^2$$

$$+ \frac{5\cdot7\cdot9}{16}\,\frac{\mathfrak{R}^2\mu}{R\rho^2T^3}\Big[\Big(\frac{\partial T}{\partial x}\Big)^2 + \Big(\frac{\partial T}{\partial y}\Big)^2 + \Big(\frac{\partial T}{\partial z}\Big)^2\Big]\Big\}.$$

On substituting the value of b, one finds, writing θ for

$$\frac{\partial u}{\partial x} + \frac{\partial v}{\partial y} + \frac{\partial w}{\partial z}$$

the following value for the total entropy of the gas contained in the volume element do:

$$
(250)\quad \left\{
\begin{aligned}
&Jdo = \frac{R\rho do}{2\mu} l(T^{3/2}\rho^{-1}) - \frac{4\Re^2R^3T^2\rho do}{p^2\mu^3}\left\{2\left(\frac{\partial u}{\partial x} - \frac{1}{3}\theta\right)^2\right.\\
&+ 2\left(\frac{\partial v}{\partial y} - \frac{1}{3}\theta\right)^2 + 2\left(\frac{\partial w}{\partial z} - \frac{1}{3}\theta\right)^2 + \left(\frac{\partial v}{\partial z} + \frac{\partial w}{\partial y}\right)^2\\
&+ \left(\frac{\partial w}{\partial x} + \frac{\partial u}{\partial z}\right)^2 + \left(\frac{\partial v}{\partial x} + \frac{\partial u}{\partial y}\right)^2 + \frac{5\cdot 7\cdot 9}{64}\frac{\mu}{RT^3}\left[\left(\frac{\partial T}{\partial x}\right)^2\right.\\
&\left.\left.+ \left(\frac{\partial T}{\partial y}\right)^2 + \left(\frac{\partial T}{\partial z}\right)^2\right]\right\} = \frac{R\rho do}{2\mu} l(T^{3/2}\rho^{-1})\\
&- \frac{4\Re^2R^3T^2\rho do}{p^2\mu^3}\left\{2\left[\left(\frac{\partial u}{\partial x}\right)^2 + \left(\frac{\partial v}{\partial y}\right)^2 + \left(\frac{\partial w}{\partial z}\right)^2\right]\right.\\
&- \frac{2}{3}\left(\frac{\partial u}{\partial x} + \frac{\partial v}{\partial y} + \frac{\partial w}{\partial z}\right)^2 + \left(\frac{\partial v}{\partial z} + \frac{\partial w}{\partial y}\right)^2\\
&+ \left(\frac{\partial w}{\partial x} + \frac{\partial u}{\partial z}\right)^2 + \left(\frac{\partial u}{\partial y} + \frac{\partial v}{\partial x}\right)^2\\
&\left.+ \frac{5\cdot 7\cdot 9}{64}\frac{\mu}{RT^3}\left[\left(\frac{\partial T}{\partial x}\right)^2 + \left(\frac{\partial T}{\partial y}\right)^2 + \left(\frac{\partial T}{\partial z}\right)^2\right]\right\}.
\end{aligned}
\right.
$$

The sum of all the terms containing derivatives of u, v, w with respect to x, y, z is what Lord Rayleigh calls the dissipation function of viscosity.* The sum of the last three terms has been called by Ladislaus Natanson the dissipation function of heat conduction.†

The energeticist holds that the different forms of energy are qualitatively different; to him, an energy halfway between kinetic energy and heat is very strange. Hence the oft-emphasized principle of the superposition of properties of different energies contained in a body. The principle is valid for the static state, and for completely stationary visible motion, where to a certain extent the forms of energy can be separated. On the other hand, if the above equation is correct, then in the presence of viscosity and

---

* Rayleigh, Proc. London Math. Soc. **4**, 357 (1873); Phil. Mag. [5] **36**, 354 (1893).

† Natanson, Rozprawy Krakow **7**, 273, **9**, 171 (1895); Phil. Mag. [5] **39**, 455, 501 (1895).

heat conduction the entropy of the gas is not the same as it would be at the same temperature and velocity, when there is no dissipation. Thus we have to deal with a kinetic energy that is, so to speak, half visible kinetic energy and half transformed into thermal motion, so that in the expression for the entropy it appears in a form that would not have been foreseen from the laws of static phenomena. If we deform a completely elastic body by means of an external force, then we get back all the energy we put into it, in the form of work, when it returns to its initial state. If we produce viscosity in a gas by means of an external force, then the work done is transformed into heat energy. After the removal of external forces, this transformation becomes complete after a time considerably greater than the relaxation time has elapsed. While the external forces are acting, our equations predict that the entropy at each instant is somewhat smaller than it would be if the energy lost from visible motion were completely transformed into heat. Instead, this energy is in a state intermediate between ordinary heat and visible energy, and part of it can still be transformed back into work, since the Maxwell velocity distribution does not yet hold exactly. This description of the dissipation of energy, based on a purely mechanical model, seems to me especially remarkable.

Now suppose that two kinds of gas are present. Let m be the mass of a molecule of the first kind, and m_1 the mass of one of the second kind. The mean value u of the velocity components ξ of all molecules of the first kind found in a volume element will be called the x-component of the total velocity of the first kind of gas in this volume element. It need not be equal to the mean value u_1 of the velocity components ξ_1 of all molecules of the other kind of gas in the same volume element. u_1 will be called the x-component of the total motion of the second kind of gas in the volume element do. v, w, v_1 and w_1 have similar meanings. Let ρ and ρ_1 be the partial densities of the two kinds of gas—i.e., ρ is the total mass of all molecules of the first kind contained in do, divided by do, and similarly for ρ_1. Let p and p_1 be the partial pressures—i.e., the pressure that each kind of gas would exert on unit surface if the other were not present. Let $P=p+p_1$ be the total pressure. Finally, let $\mathfrak{x}$, $\mathfrak{y}$, $\mathfrak{z}$ and $\mathfrak{x}_1$, $\mathfrak{y}_1$, $\mathfrak{z}_1$ be the excess of the velocity components over the total velocity components of the corresponding kind of gas:

$$\xi = u + \mathfrak{x}, \quad \eta = v + \mathfrak{y}, \quad \zeta = w + \mathfrak{z}$$
$$\xi_1 = u_1 + \mathfrak{x}_1, \quad \eta_1 = v_1 + \mathfrak{y}_1, \quad \zeta_1 = w_1 + \mathfrak{z}_1.$$

Then the continuity equation is valid for each kind of gas, as we already proved before we made the assumption that only one kind of gas is present. Hence:

$$(251) \quad \begin{cases} \dfrac{\partial \rho}{\partial t} + \dfrac{\partial(\rho u)}{\partial x} + \dfrac{\partial(\rho v)}{\partial y} + \dfrac{\partial(\rho w)}{\partial z} = 0 \\ \dfrac{\partial \rho_1}{\partial t} + \dfrac{\partial(\rho_1 u_1)}{\partial x} + \dfrac{\partial(\rho_1 v_1)}{\partial y} + \dfrac{\partial(\rho_1 w_1)}{\partial z} = 0. \end{cases}$$

We shall now imagine that the volume element do moves during time dt with the velocity components u, v, w of the first kind of gas in this volume element. The difference between the values of any quantity Φ at time $t+dt$ in the volume element in its new position, and at time t in the volume element in its old position, divided by dt, we denote by $d\Phi/dt$, so that:

$$\frac{d\Phi}{dt} = \frac{\partial \Phi}{\partial t} + u\frac{\partial \Phi}{\partial x} + v\frac{\partial \Phi}{\partial y} + w\frac{\partial \Phi}{\partial z}.$$

A similar meaning is given to:

$$\frac{d_1\Phi}{dt} = \frac{\partial \Phi}{\partial t} + u_1\frac{\partial \Phi}{\partial x} + v_1\frac{\partial \Phi}{\partial y} + w_1\frac{\partial \Phi}{\partial z}.$$

In constructing the latter quantity, one imagines that the volume element is moving with velocity components u_1, v_1, w_1. Then the two continuity equations can also be written:

$$(252) \quad \begin{cases} \dfrac{d\rho}{dt} + \rho\left(\dfrac{du}{\partial x} + \dfrac{\partial v}{\partial y} + \dfrac{\partial w}{\partial z}\right) = 0 \\ \dfrac{d_1\rho_1}{dt} + \rho_1\left(\dfrac{\partial u_1}{\partial x} + \dfrac{\partial v_1}{\partial y} + \dfrac{\partial w_1}{\partial z}\right) = 0. \end{cases}$$

We ignore the deviations from the Maxwell velocity distribution law. Then:

$$p = \rho\,\overline{\mathfrak{x}^2} = \rho\,\overline{\mathfrak{y}^2} = \rho\,\overline{\mathfrak{z}^2}, \quad \overline{\mathfrak{x}\,\mathfrak{y}} = \overline{\mathfrak{x}\,\mathfrak{z}} = \overline{\mathfrak{y}\,\mathfrak{z}} = 0.$$
$$p_1 = \rho_1\overline{\mathfrak{x}_1^2} = \rho_1\overline{\mathfrak{y}_1^2} = \rho_1\overline{\mathfrak{z}_1^2}, \quad \overline{\mathfrak{x}_1\mathfrak{y}_1} = \overline{\mathfrak{x}_1\mathfrak{z}_1} = \overline{\mathfrak{y}_1\mathfrak{z}_1} = 0.$$

The mean kinetic energy of a molecule cannot in any case be very much different for the two kinds of gas. One has approximately:

$$\frac{m}{2}(\overline{\xi^2}+\overline{\eta^2}+\overline{\zeta^2}) = \frac{m_1}{2}(\overline{\xi_1^2}+\overline{\eta_1^2}+\overline{\zeta_1^2}).$$

Since to the present degree of approximation we can ignore the squares of the small velocity components u, v, w with which the gases diffuse through each other, compared to ξ^2, η^2, $\cdots$ we also have:

$$m(\overline{\mathfrak{x}^2}+\overline{\mathfrak{y}^2}+\overline{\mathfrak{z}^2}) = m_1(\overline{\mathfrak{x}_1^2}+\overline{\mathfrak{y}_1^2}+\overline{\mathfrak{z}_1^2}).$$

We set these quantities again (cf. Eq. [51a]) equal to $3RMT$, and we call T the temperature in *do*. Here M is the mass of a molecule of some third gas (the normal gas) and R is a constant corresponding to the temperature scale to be chosen (the gas constant of the normal gas). Since each of the original two gases behaves like a gas at rest,

$$p = r\rho T = \frac{R}{\mu}\rho T, \quad p_1 = r_1\rho_1 T = \frac{R}{\mu_1}\rho_1 T_1, \tag{253}$$

where r and r_1 are the gas constants of the original two gases, and $\mu = m/M$, $\mu_1 = m_1/M$.

We shall now set $\varphi = \xi = u + \mathfrak{x}$ in Equation (187). Then:

$$\overline{\varphi} = u,\ \rho\overline{\mathfrak{x}\varphi} = \rho\overline{\mathfrak{x}^2} = p,\ \overline{\mathfrak{y}\varphi} = \overline{\mathfrak{z}\varphi} = 0,\ \overline{\frac{\partial\varphi}{\partial\xi}} = 1,\ \overline{\frac{\partial\varphi}{\partial\eta}} = \overline{\frac{\partial\varphi}{\partial\zeta}} = 0.$$

$B_5(\varphi) = 0$. One therefore obtains:

$$\rho\frac{du}{dt} + \frac{\partial p}{\partial x} - \rho X = mB_4(\xi), \tag{254}$$

where, according to Equation (132),

$$B_4(\xi) = \iiint_0^\infty\int_0^{2\pi}(\xi' - \xi)fF_1 d\omega d\omega_1 gbdbd\epsilon.$$

We now have (see Eq. [200]):

$$\xi' - \xi = \frac{m_1}{m+m_1}\left[2(\xi_1 - \xi)\cos^2\vartheta + \sqrt{g^2 - (\xi - \xi_1)^2}\sin 2\vartheta\cos\epsilon\right],$$

hence

$$\int_0^{2\pi}(\xi' - \xi)d\epsilon = \frac{4\pi m_1}{m + m_1}(\xi_1 - \xi)\cos^2\vartheta$$

$$\int_0^{\infty} gbdb \int_0^{2\pi}(\xi' - \xi)d\epsilon = \frac{m_1}{m + m_1}(\xi_1 - \xi)g\int_0^{\infty} 4\pi\cos^2\vartheta bdb.$$

We also set (Eq. [195]):

$$b = \left[\frac{K(m + m_1)}{mm_1}\right]^{1/n} g^{-2/n}\cdot\alpha$$

$$db = \left[\frac{K(m + m_1)}{mm_1}\right]^{1/n} g^{-2/n}d\alpha$$

and subsequently put $n=4$. Therefore:

$$\int_0^{\infty}\int_0^{2\pi}(\xi' - \xi)gbdbd\epsilon$$

$$= m_1(\xi_1 - \xi)\sqrt{\frac{K}{mm_1(m + m_1)}}\int_0^{\infty} 4\pi\cos^2\vartheta\alpha d\alpha.$$

Maxwell calls this definite integral A_1 and finds*

(255) $$A_1 = 2\cdot 6595.$$

We set:

(256) $$A_3 = A_1\sqrt{\frac{K}{mm_1(m + m_1)}}$$

and obtain

$$\int_0^{\infty}\int_0^{2\pi}(\xi' - \xi)gbdbd\epsilon = m_1A_3(\xi_1 - \xi).$$

Whence it follows, furthermore, that:

$$mB_4(\xi) = A_3\left[m\int fd\omega\cdot m_1\int \xi_1F_1d\omega_1 - m\int \xi fd\omega\cdot m_1\int F_1d\omega_1\right].$$

Now according to Equation (175):

$$m\int fd\omega = \rho, \quad m\int \xi fd\omega = \rho\bar{\xi} = \rho u$$

* Maxwell, Phil. Trans. **157,** 49 (1867). According to the calculations of Aichi and Tanukadate, reported by Nagaoka, Nature **69,** 79 (1903), $A_1=2.6512$; according to Chapman, Mem. Proc. Manchester Lit. Phil. Soc. **66,** No. 1 (1922), $A_1=2.6514$.

and since clearly the same result holds for the second kind of gas,

$$m_1\int F_1 d\omega_1 = \rho_1, \quad m_1\int \xi_1 F_1 d\omega_1 = \rho_1 u_1,$$

we have

$$mB_4(\xi) = A_3\rho\rho_1(u_1 - u)$$

and Equation (254) reduces to

$$\rho\frac{du}{dt} + \frac{\partial p}{\partial x} - \rho X + A_3\rho\rho_1(u - u_1) = 0. \tag{257}$$

Likewise one obtains for the second kind of gas,

$$\rho_1\frac{du_1}{dt} + \frac{\partial p_1}{\partial x} - \rho_1 X_1 + A_3\rho\rho_1(u_1 - u) = 0. \tag{257a}$$

These are the familiar hydrodynamic equations. According to our present assumptions, viscosity and heat conduction cannot be important. Only the last term describes the interaction of the two kinds of gas. On these assumptions, therefore, this interaction has exactly the same effect as if one added to the force $X\rho do$, which acts from outside on the gas of the first kind in do, the contribution $-A_3\rho\rho_1(u-u_1)do$. We can arrange things so that this gas is unaffected by the other forces that act on it, and encounters only this resistance to its motion through the second kind of gas. Since the same holds for the y and z axes, this resistance is equal to the product of the partial densities of the two gases, their relative velocity $\sqrt{(u-u_1)^2+(v-v_1)^2+(w-w_1)^2}$, the volume do of the volume element, and the constant A_3. It has the direction of this relative motion, and acts on each kind of gas against the relative motion. If we set $\varphi=\xi^2+\eta^2+\zeta^2$ in Equation (187), then we find that, according to the present approximations,

$$\frac{d}{dt}(\overline{\mathfrak{x}^2} + \overline{\mathfrak{y}^2} + \overline{\mathfrak{z}^2}) = 0$$

as soon as initially $m(\overline{x^2+y^2+z^2}) = m_1(\overline{x_1^2+y_1^2+z_1^2})$. Therefore the temperature is not changed as a result of the diffusion process.

We shall apply these equations only to the gas-diffusion experiments of Prof. Loschmidt.* These experiments were set up as follows: a vertical cylindrical container is divided into two parts by a thin partition. The lower space is filled with the heavier

* Loschmidt, Wien. Ber. **61**, 367, **62**, 468 (1870).

gas, and the upper with the lighter gas. The pressure and temperature are made equal in the two gases, and when all mass motion has stopped, the partition is suddenly withdrawn as smoothly as possible. After the gas has diffused for a certain time, the partition is again inserted and the contents of both parts of the container are analyzed. Here the effect of gravity can be ignored, hence we shall set $X = Y = Z$. Furthermore, the motion takes place only in the axis of the cylinder. If we choose this as the abscissa axis, then

$$v = w = \frac{\partial}{\partial y} = \frac{\partial}{\partial z} = 0.$$

Finally, the motion takes place so slowly that it can be considered as stationary at each point, so that du/dt can be neglected.

We can also justify this as follows: we have for the reciprocal relaxation time:

$$\frac{1}{\tau} = 3A_2\rho\sqrt{\frac{K_1}{2m^3}},$$

and according to Equation (256):

$$A_3\rho_1 = A_1\rho_1\sqrt{\frac{K_1}{mm_1(m + m_1)}}.$$

A_1 is a number less than twice as large as A_2. ρ is of the same order of magnitude as ρ_1, and m is of the same order as m_1. We also assume that the two constants K_1 and K appearing in the force laws for the interaction of m- and m_1-molecules are of the same order of magnitude. Then in Equation (257) the magnitude of the first term is to that of the last as du/dt to $(u-u_1)/\tau$. This ratio can be set equal to zero since, because of the slowness of the diffusion process, the time τ_1 within which u can experience the increase $u-u_1$ must be enormously large compared to the relaxation time τ. We can therefore neglect the first term in Equation (257) also, and we obtain:

(258) $$\frac{\partial p}{\partial x} = A_3\rho\rho_1(u - u_1).$$

Likewise:

(259) $$\frac{\partial p_1}{\partial x} = A_3\rho\rho_1(u_1 - u).$$

From the two continuity equations, however, it follows that:

$$\frac{\partial \rho}{\sigma t} + \frac{\partial(\rho u)}{\partial x} = \frac{\partial \rho_1}{\partial t} + \frac{\partial(\rho_1 u_1)}{\partial x} = 0. \tag{260}$$

The temperature T should be held constant during the entire experiment. Hence, according to Equation (253), p is proportional to ρ and p_1 to ρ_1, and one can therefore write Equation (260) in the form:

$$\frac{\partial p}{\partial t} + \frac{\partial(pu)}{\partial x} = \frac{\partial p_1}{\partial t} + \frac{\partial(p_1 u_1)}{\partial x} = 0. \tag{261}$$

If we set $p+p_1=P$, so that P is the total pressure, then it follows from (258) and (259) that:

$$\frac{\partial P}{\partial x} = 0.$$

Furthermore, from (261):

$$\frac{\partial P}{\partial t} + \frac{\partial(pu + p_1 u_1)}{\partial x} = 0$$

and on differentiating the last equation with respect to x:

$$\frac{\partial^2(pu + p_1 u_1)}{\partial x^2} = 0,$$

therefore

$$pu + p_1 u_1 = C_1 x + C_2.$$

Now no gas can flow in or out of the container, either at the top or bottom. Hence at the abscissa values corresponding to the top and bottom we have $u=u_1=0$, hence also $pu+p_1u_1=0$.

Whence it follows that $C_1=C_2=0$ and

$$pu + p_1 u_1 = 0. \tag{262}$$

If one uses this equation to eliminate u_1 from Equation (258), then it follows that

$$\frac{\partial p}{\partial x} = -A_3 \frac{\rho \rho_1}{p p_1} P \cdot pu,$$

hence, according to (253):

$$\frac{\partial p}{\partial x} = -\frac{A_3 \mu \mu_1 P}{R^2 T^2} pu. \tag{263}$$

If one differentiates again with respect to x, and refers to Equation (261), then he obtains:

$$\frac{\partial p}{\partial t} = \mathfrak{D} \frac{\partial^2 p}{\partial x^2},$$

where

$$\mathfrak{D} = \frac{R^2 T^2}{A_3 \mu \mu_1 P}.$$

This equation has the same form as the one established by Fourier for heat conduction. Both natural processes therefore follow the same law. In our special case, the diffusion takes place just as if instead of a cylindrical mass of gas one had a homogeneous metal cylinder, whose upper half is maintained initially at a temperature of 100°C, and whose lower half initially has zero temperature; through the entire surface of the metal cylinder, no heat may enter or leave, either by conduction or radiation.

$\mathfrak{D}$ is the diffusion constant. It is directly proportional to the square of the absolute temperature T, and inversely proportional to the total pressure P. It is independent of the mixing ratio, so that it is constant at all times during the diffusion process for all layers of the container. If the molecules behaved like elastic spheres, $\mathfrak{D}$ would be proportional to the $\frac{3}{2}$ power of T, and would depend on the mixing ratio. The dependence on P remains the same in both cases.

A simple definition of the diffusion constant $\mathfrak{D}$ can be obtained in the following way. We multiply Equation (263) by $-\mu\mathfrak{D}/RT$ and obtain:

$$\rho u = - \frac{R^2 T^2}{A_3 \mu \mu_1 P} \frac{\partial \rho}{\partial x} = - \mathfrak{D} \frac{\partial \rho}{\partial x}.$$

ρu is clearly the total amount of gas that goes through unit cross section in unit time. It is proportional to the gradient $\partial \rho / \partial x$ of the partial density of the gas in the direction of the axis of the container. The proportionality constant is just the diffusion constant.

If we retain the assumption of inverse fifth power forces, then we cannot draw any conclusion about K from the force constants K_1 and K_2. Thus from the properties of two gases by themselves, we can draw no conclusion about their interaction with each other. However, we may be able to draw such conclusions if we imagine that the repulsive force is transmitted by means of compressible ether-shells. We can then ascribe the value s to the diameter of the ether-shell of an m-molecule, and the diameter s_1 to the ether-shell of the m_1-molecule. In a collision, the centers of two m-molecules will approach up to a minimum distance s, on the average. If we imagine one such molecule held fixed and the other one moving toward it with mean kinetic energy $\mathfrak{l}$, then the latter will have zero velocity at the distance s. Then:

$$\mathfrak{l} = \int_s^\infty \frac{K_1 dr}{r^5} = \frac{K_1}{4s^4}. \tag{264}$$

Likewise it follows that

$$\mathfrak{l} = \frac{K_2}{4s_1^4}.$$

An m_1-molecule will, however, approach an m-molecule up to a minimum distance equal to the sum of the radii $(s+s_1)/2$ on the average. If we again fix one molecule and let the other approach it with the mean kinetic energy of all the molecules, then its velocity will be annihilated at the distance $(s+s_1)/2$, which gives:

$$\mathfrak{l} = \frac{4K}{(s+s_1)^4}.$$

From these equations it follows that:

$$2\sqrt[4]{K} = \sqrt[4]{K_1} + \sqrt[4]{K_2}.$$

Now we found (Eq. [256]):

$$A_3 = A_1 \sqrt{\frac{K}{mm_1(m+m_1)}} = \frac{A_1}{M^{3/2}} \sqrt{\frac{K}{\mu\mu_1(\mu+\mu_1)}}$$
$$= \frac{A_1}{4M^{3/2}} \frac{(\sqrt[4]{K_1}+\sqrt[4]{K_2})^2}{\sqrt{\mu\mu_1(\mu+\mu_1)}}.$$

The viscosity coefficient of the first gas was (Eq. [219]):

$$\mathfrak{R} = \frac{p}{3A_2\rho}\sqrt{\frac{2m^3}{K_1}} = \frac{RTM^{3/2}}{3A_2}\sqrt{\frac{2\mu}{K_1}}.$$

Likewise the viscosity coefficient of the second gas was:

$$\mathfrak{R}_1 = \frac{RTM^{3/2}}{3A_2}\sqrt{\frac{2\mu_1}{K_2}},$$

hence

$$\sqrt{K_1} = \frac{RTM^{3/2}}{3A_2}\frac{\sqrt{2\mu}}{\mathfrak{R}}, \quad \sqrt{K_2} = \frac{RTM^{3/2}}{3A_1}\frac{\sqrt{2\mu_1}}{\mathfrak{R}_1}$$

(265) $$A_3 = \frac{A_1RT}{6\sqrt{2}A_2\sqrt{\mu\mu_1}(\mu+\mu_1)}\left(\frac{\sqrt[4]{\mu}}{\sqrt{\mathfrak{R}}} + \frac{\sqrt[4]{\mu_1}}{\sqrt{\mathfrak{R}_1}}\right)^2$$

$$\mathfrak{D} = \frac{6\sqrt{2}A_2RT}{A_1P}\sqrt{\frac{\mu+\mu_1}{\mu\mu_1}}\cdot\frac{1}{\left(\frac{\sqrt[4]{\mu}}{\sqrt{\mathfrak{R}}} + \frac{\sqrt[4]{\mu_1}}{\sqrt{\mathfrak{R}_1}}\right)^2}.$$

This equation permits one to calculate the diffusion constant of two gases from their molecular weights and viscosity coefficients. It agrees approximately with experiment. However, one should not suppose that it is exactly correct. Nevertheless, it seems to have a more rational basis than any other formula yet proposed for this purpose.

If in Equation (264) we set:

$$\mathfrak{l} = \frac{m}{2}\overline{c^2},$$

then

$$K_1 = 2ms^4\overline{c^2},$$

hence

$$\mathfrak{R} = \frac{pm}{3A_2\rho s^2\sqrt{\overline{c^2}}}.$$

Now we have

$$\frac{p}{\rho} = \frac{1}{3}\overline{c^2},$$

hence

$$\mathfrak{R} = \frac{m\sqrt{\overline{c^2}}}{9A_2s^2} = 0.0812\frac{m\sqrt{\overline{c^2}}}{s^2}.$$

According to Equation (91),

$$\mathfrak{R} = knmc\lambda,$$

$$\lambda = \frac{1}{\pi ns^2\sqrt{2}}.$$

Further, according to Equation (89):

$$k = 0.350271,$$

if

$$c = \bar{c} = \sqrt{\frac{8}{3\pi}}\sqrt{\overline{c^2}}.$$

Therefore:

$$\mathfrak{R} = 0.350271 \frac{2}{\pi\sqrt{3\pi}} \frac{m\sqrt{\overline{c^2}}}{s^2} = 0.0726 \frac{m\sqrt{\overline{c^2}}}{s^2}.$$

One sees that the numerical coefficient is only insignificantly different.

The concepts of mean free path and number of collisions are not suited to the theory of repulsive forces proportional to the inverse fifth power of the distance. In order to define them, one must make a new arbitrary assumption. One must, for example, decide that an encounter of two molecules is to be considered a collision if the relative velocity is deflected through an angle greater than 1°.

It would be of the greatest interest to carry the approximation further in the calculation of diffusion, as well as to calculate the entropy of the two diffusing gases. In the first case there would probably be fluctuations of temperature during the diffusion, whose calculation according to the principles established would not be difficult; likewise, it would be easy to calculate a new dissipation function, that of diffusion, by determining the entropy of the two diffusing gases. However, we shall not pursue this matter any further.*

* Modern research on the properties of gases is reviewed by J. S. Rowlinson, J. E. Mayer, H. Grad, L. Waldmann and others in *Handbuch der Physik*, Vol. XII (Berlin: Springer, 1958). Older work is surveyed by J. R. Partington, *Advanced Treatise on Physical Chemistry*, Vol. I (London: Longmans, Green, 1949).

PART II

Van der Waals' theory; Gases with compound molecules; Gas dissociation; Concluding remarks.

FOREWORD TO PART II

> "The impossibility of an incompensated decrease of entropy seems to be reduced to an improbability."[1]

As the first part of *Gas Theory* was being printed, I had already almost completed the present second and last part, in which the more difficult parts of the subject were not to have been treated. It was just at this time that attacks on the theory of gases began to increase.* I am convinced that these attacks are merely based on a misunderstanding, and that the role of gas theory in science has not yet been played out. The abundance of results agreeing with experiment, which van der Waals has derived from it purely deductively, I have tried to make clear in this book. More recently, gas theory has also provided suggestions that one could not obtain in any other way. From the theory of

[1] Gibbs, Trans. Conn. Acad. **3**, 229 (1875); p. 198 in Ostwald's German edition.

* In addition to the works cited in the footnote on p. 24, see: R. Mayer, *Mechanik der Wärme* (Stuttgart, 1867), p. 9. E. Mach, "Die ökonomische Natur der physikalischen Forschung," Wien. Almanach 293 (1882); *Über die Erhaltung der Arbeit* (Prague, 1872); Monist **1**, 48, 393 (1891), **5**, 167 (1894); *Die Prinzipien der Wärmelehre* (Leipzig: J. A. Barth, 1896, 2nd ed., 1900), pp. 362–364, 429–431. K. Pearson, *The Grammar of Science* (London: Scott, 1895), pp. 200, 214, 311. J. B. Stallo, *The Concepts and Theories of Modern Physics* (New York: Appleton, 1882), chap. viii. P. Duhem, *La Théorie Physique, son Objet et sa Structure* (Paris: Chevalier et Riviera, 1906). J. Ward, *Naturalism and Agnosticism* (New York: Macmillan, 1899). A. Aliotta, *La reazione idealistica contro la scienza* (Palermo, 1912). G. Hirn, Mem. Acad. Sci. Bruxelles **43** (1881), **46** (1886); C. R. Paris **107**, 166 (1888). W. Ostwald, Lecture at the 64th meeting of the Deutscher Naturforscher und Ärzte, Halle, 1891, in his *Abhandlungen und Vorträge*, p. 34. H. Poincaré, Nature **45**, 485 (1892); *Thermodynamique* (Paris: Gauthier-Villars, 1892), p. xviii. F. Wald, *Die Energie und ihre Entwerthung* (Leipzig: Engelmann, 1889). V. Lenin, *Materialism and Empirio-Criticism* (Moscow, 1908), chap. 5.

the ratio of specific heats, Ramsay inferred the atomic weight of argon and thereby its place in the system of chemical elements—which he subsequently proved, by the discovery of neon, was in fact correct.* Likewise, Smoluchowski deduced from the kinetic theory of heat conduction the existence and magnitude of a temperature discontinuity in the case of heat conduction in a very dilute gas.†

In my opinion it would be a great tragedy for science if the theory of gases were temporarily thrown into oblivion because of a momentary hostile attitude toward it, as was for example the wave theory because of Newton's authority.

I am conscious of being only an individual struggling weakly against the stream of time. But it still remains in my power to contribute in such a way that, when the theory of gases is again revived, not too much will have to be rediscovered. Thus in this book [this Part] I will now include the parts that are the most difficult and most subject to misunderstanding, and give (at least in outline) the most easily understood exposition of them. When consequently parts of the argument become somewhat complicated, I must of course plead that a precise presentation of these theories is not possible without a corresponding formal apparatus.

I thank especially Dr. Hans Benndorf for collecting numerous literature citations during my absence from Vienna.

Volosca, Villa Irenea, August, 1898

Ludwig Boltzmann

---

* Ramsay and Travers, Proc. R. S. London **62,** 316, **63,** 437 (1898). Ramsay, Mem. Proc. Manchester Lit. Phil. Soc. **43,** No. 4 (1900); *The Gases of the Atmosphere* (London: Macmillan, 1896), pp. 207–232.

† Smoluchowski, Wien. Ber. **107,** 304 (1898), **108,** 5 (1899); Ann. Phys. [3] **64,** 101 (1898); Prace Mat.-Fiz. **10,** 33 (1898).

CHAPTER I

Foundations of van der Waals' theory.

§1. General viewpoint of van der Waals.

When the distance at which two gas molecules interact with each other noticeably is vanishingly small relative to the average distance between a molecule and its nearest neighbor—or, as one can also say, when the space occupied by the molecules (or their spheres of action) is negligible compared to the space filled by the gas—then the fraction of the path of each molecule during which it is affected by its interaction with other molecules is vanishingly small compared to the fraction that is rectilinear, or simply determined by external forces. Then the Boyle-Charles law holds for the gas in question, whether the molecules are simply material points or solid bodies, or when they are compound aggregates. The gas is "ideal" in all these cases.

Gases found in nature only partly satisfy these conditions of the ideal gas state, and hence a theory that takes account of the finite extension of the spheres of action of the molecules is very much to be desired.

Such a theory was given by van der Waals,* who considered the molecules to be negligibly deformable elastic spheres, as we did at the beginning of Part I. He generalized the theory in two ways:

1. he does not assume that the space actually occupied by the elastic spheres representing the molecules is vanishingly small compared to the total volume of the gas;

2. he assumes that, in addition to the instantaneous elastic forces that act during collisions, there is also an attractive force between the molecules, which acts in the line of their centers, and

* J. D. van der Waals, *Over de continuiteit van den gas- en vloeistoftoestand* (Dissertation, Leiden, 1873); Amsterdam Verslagen [2] **10**, 321, 337 (1876), and many other papers (see the *Royal Society Catalogue of Scientific Papers* for a list up to 1900).

whose intensity is a function of the distance of centers. We call this attractive force the van der Waals cohesion force.

The necessity of assuming an attractive force between molecules follows directly from the possibility—now demonstrated for all gases—of liquefaction, since the simultaneous existence of a liquid and a gaseous phase of the same substance at the same temperature and pressure in the same container is understandable only if there is a force between molecules that causes them to rebound at collisions, and also an attractive force.

This attractive force can be demonstrated directly by the following experiment. One suddenly brings a container filled with a compressed gas into communication with another container filled with the same gas in a dilute state. As it flows out, the gas in the first container does work against the pressure and cools itself; in the latter container there is first a visible streaming, which turns into heat as a result of viscosity. If there were only a repulsive force between the molecules, then the heat finally created must be completely equivalent to the cooling in the first container. If there is also an attractive force, then this equivalence is not complete; rather there is a net loss of heat, since the average distance of the molecules will be larger, and hence a definite amount of heat must be used to overcome the attraction.

The experiments conducted by this method by Gay-Lussac[1] and later by Joule and Lord Kelvin[2] did not give a definite answer to the question of the presence of this attractive force, but the latter two scientists have proved experimentally, by a more indirect method, the existence of this attractive force, by expansion experiments with gases.[3] They showed that a gas which (without any addition of heat from outside) is pushed by pressure through a porous stopper experiences thereby a small cooling, while calculation shows that a completely ideal gas would not change its temperature.

The simultaneous existence of an attractive force and an elastic core for a molecule has of course a certain improbability.

---

[1] Gay-Lussac, Mem. soc. d'Arcueil **1,** 180 (1807); cf. Appendix to Mach's *Prinzipien der Wärmelehre.*

[2] Joule and Thomson, Phil. Mag. [3] **26,** 369 (1845); Joule's *Scientific Papers,* 171; German translation by Sprengel. [Boltzmann is incorrect in listing Thomson as a co-author of this paper.—TR.]

[3] Joule and Thomson, Phil. Trans. **144,** 321 (1854); **152,** 579 (1862).

In particular, it appears to be diametrically opposed to the assumption made in Part I, Chapter III, that two molecules repel each other with a force inversely proportional to the 5th power of their distance. Nevertheless, both assumptions can provide a certain approximation to the truth, if the molecules actually exert a weak attraction at great distance and repel each other at very small distances with an inverse fifth power force. The attraction must then turn into repulsion as the distance decreases, so that the former does not come into consideration in collisions, because of the greatly predominating repulsion at very small distances.

We shall have to leave a more precise formulation of the possible assumptions to the future, and in the following we shall not be concerned with the exact relation between the hypotheses discussed in Part I and the assumptions of van der Waals. From the viewpoint of our theory, the latter assumptions are considered to form a picture that is correct in many but not all respects. Indeed, up to now, realizing our ignorance about the actual properties of molecules, we have not pretended that our assumptions are precisely realized in nature. On the other hand, we have laid the greatest weight on the requirement that the calculations must be exactly correct—i.e., that the results must be logical consequences of the assumptions. The resulting development of mathematical methods was our principal purpose. If we know the consequences of various kinds of assumptions, it should be easier to find experiments that will test them, and at the same time it should be made possible that, as our knowledge progresses, mathematical methods for the investigation of newly discovered laws should be readily available.

Unfortunately, van der Waals had to abandon mathematical rigor at a certain point in order to carry out his calculations. Nevertheless, his theory has proved to be of great practical value, for the resulting formula gives in general a sufficiently good description of the behavior of a gas up to its point of liquefaction, even if it does not achieve complete quantitative agreement with experiment. From this one is justified in concluding that its foundations could hardly be replaced by completely different ones.

In this chapter I will derive the equation of van der Waals in the simplest and shortest possible way, leaving further refinements to Chapter V.

§2. External and internal pressure.

A container of volume V contains n identical molecules, which are completely elastic, negligibly deformable spheres of diameter σ. The space occupied by the spheres themselves will be fairly small but not completely negligible compared to the total volume of the container. It will be shown that the formulas obtained are also approximately applicable to the state of the substance when it is no longer a gas but rather a liquid. Therefore in the following we shall call it simply a substance, not a gas, although we shall still always have in mind such cases when its state approximates that of a gas.

Between the centers of two molecules there acts an attractive force (the van der Waals cohesion force), which vanishes at macroscopic distances but decreases so slowly with increasing distance that it may be considered constant within distances large compared to the average separation of two neighboring molecules.* Consequently the van der Waals cohesion forces exerted on each molecule in the interior of the container by the surrounding molecules are very nearly equal in all directions in space, and they balance each other in such a way that the motion of the individual molecule is like that of the usual gas molecule, and is not noticeably modified by the cohesion forces. Hence even though we did not treat such forces in Part I, we can still calculate the molecular motions by the same principles established there.

The van der Waals cohesion force has an appreciable effect only on the molecules that are very near to the surface of the substance. These molecules will thus be acted on by two forces. The first is the counter-pressure of the wall on the gas; the second is the cohesion force. The intensity with which the first force acts on the molecules lying on unit surface we call p, while that of the latter we call p_i, so that the total force on these molecules is

(1) $$p_g = p + p_i.$$

Now suppose that a part DE of the wall of the container has surface area Ω. The total force

$$\Omega p_g = \Omega(p + p_i),$$

* See the remarks by van der Waals on this point, mentioned at the end of §61.

which acts on the molecules at the surface DE and which they exert back on the wall (in the equilibrium state) is, according to §1 of Part I, equal to the total momentum (in the direction of the normal N to the surface DE) that the molecules would carry through this surface in unit time, if it were placed in the interior of the gas, plus the momentum corresponding to the velocities with which these molecules are reflected from the surface back into the interior of the gas.

§3. Number of collisions against the wall.

We first pick out of all the molecules only those for which the magnitude of their velocity, c, lies between c and $c+dc$, and the angle ϑ formed by the direction of the velocity with the outwardly directed normal N at the surface DE lies between ϑ and $\vartheta+d\vartheta$; furthermore, the angle ϵ between a plane normal to DE containing the velocity direction and a fixed plane normal to DE must lie between ϵ and $\epsilon+d\epsilon$. We call the set of these conditions:

"the conditions (2)."

All molecules that satisfy (2) will be called molecules of the specified kind, and we ask first: how many molecules of the specified kind collide with the surface DE during a very short time dt?

Each molecule is to be considered a sphere of diameter σ, so that it collides with DE when this sphere touches it. During the time interval dt, the centers of all the specified molecules travel nearly the same distance, cdt, in nearly the same direction. We shall find the number of collisions between molecules of the specified kind and the plane DE during dt as follows:

We allow DE to touch, at each of its points, a sphere whose diameter is equal to the molecular diameter σ. The centers of all these spheres lie in a second plane of surface area Ω. Through each point of this second plane, we draw a line that is equal in length and direction to the path travelled by the specified molecules during dt. All these lines fill up an oblique cylinder γ of base Ω and height

$$dh = cdt\cos\vartheta, \tag{3}$$

and hence of volume Ωdh. One sees easily that the molecules that collide with DE during dt are just the ones whose centers lie in γ at the beginning of the time interval dt.

§4. Relation between molecular extension and collision number.

In order to find the number dz of these latter molecules, we first determine quite generally the probability that, for a given configuration of the other molecules, the center of a particular molecule lies within the cylinder γ. This molecule cannot be less than a distance σ from the center of any other molecule. We find the volume available to the center of this molecule, when the positions of the other are fixed, as follows: we construct a sphere of radius σ, which we call the covering sphere, around the center of each of the $n-1$ other molecules. Its volume is 8 times the volume of the molecule, if it were considered as an elastic sphere. The total volume, $4\pi(n-1)\sigma^3/3$, of all these $n-1$ covering spheres is to be subtracted from the total volume V of the gas. We may also write n for $n-1$, since n is a very large number.

In order to find dz, we compare this space $V-4\pi n\sigma^3/3$ with the space available in the cylinder γ. The latter is found by subtracting from the total volume Ωdh of γ the volume of that part of it which lies within the covering sphere of any of the $n-1$ other molecules. The covering spheres of these $n-1$ molecules will clearly be equally distributed throughout the entire volume V of the gas, except for those regions very close to the wall of the container. If γ were somewhere in the interior of the container, then that part A of the total volume $4\pi n\sigma^3/3$ of the covering spheres of all molecules which lies within γ would be in the same ratio to the total volume of these covering spheres as the volume Ωdh of the cylinder γ is to the total volume V of the gas. Hence we would have

$$A = \frac{4\pi n\sigma^3}{3V}\,\Omega dh.$$

Of all the molecules whose covering spheres penetrate γ, one can ignore those whose centers lie within γ, since the height dh of this cylinder is infinitesimal. The centers of all molecules whose covering spheres penetrate γ would lie equally often on both sides of γ, if it were inside the container.

Since γ is not inside the container but is rather at a distance $\frac{1}{2}\sigma$ from the wall, the centers of the $n-1$ molecules can only lie on one side of it. Hence one must omit half of the molecules enumerated above, so that the part of γ occupied by any of the $n-1$ molecules is only:[1]

$$\frac{A}{2} = \frac{2\pi n\sigma^3}{3V}\,\Omega dh.$$

The total remaining volume,

$$\Omega dh\left(1 - \frac{2\pi n\sigma^3}{3V}\right)$$

of γ is available to the center of the specified molecule, in case we wish to know the probability that it lies in the cylinder γ.

[1] This formula can also be derived in a more complicated way. We call the end-surface of the cylinder γ that faces the wall of the container its base. A center of one of the covering spheres can naturally lie only on the side of the base away from the wall of the container. We now construct on this side two planes parallel to the base of γ, both of surface area Ω, at distances ξ and $\xi + d\xi$ from the base. The spacc between these two planes will be called the cylinder γ_1; its volume is $\gamma_1 = \Omega d\xi$. The number of covering-spheres whose centers lie in γ at any given time is:

$$\frac{\gamma_1(n-1)}{V - \dfrac{4\pi(n-1)\sigma^3}{3}}\,.$$

Since the term we are now calculating is only a small correction, we can write this as

$$\frac{n\gamma_1}{V} = \frac{n\Omega d\xi}{V}\,.$$

Each covering sphere cuts a circle of area $\pi(\sigma^2 - \xi^2)$, hence a space of volume $\pi(\sigma^2 - \xi^2)dh$ from the cylinder γ. If we multiply this by the number of covering spheres, $n\Omega d\xi/V$, and integrate over all possible values of ξ—i.e., over ξ from 0 to σ—then we obtain for the total space cut out by the covering spheres from the cylinder γ—and thus for the space not available for the center of the specified molecule—the value:

$$\frac{n\Omega\pi dh}{V}\int_0^\sigma (\sigma^2 - \xi^2)d\xi = \frac{2\pi n\sigma^3\Omega dh}{3V}$$

in agreement with the formula in the text.

This probability is the quotient of the available space in the cylinder γ divided by the available space in the entire gas volume, namely:

$$\frac{\Omega dh}{V} \frac{1 - \dfrac{2\pi n\sigma^3}{3V}}{1 - \dfrac{4\pi n\sigma^3}{3V}}, \tag{4}$$

for which we can write (since the quantities subtracted in the numerator and denominator are very small)

$$\frac{\Omega dh}{V - B}, \tag{5}$$

where

$$B = \frac{2\pi n\sigma^3}{3} \tag{6}$$

is half of the space filled by the covering spheres of all the molecules, hence four times the volume of all the molecules.

§5. Determination of the impulse imparted to the molecules.

Since there are actually n molecules in the gas and not just the one specified molecule, the total number of molecules whose centers lie in γ is:

$$\frac{n\Omega dh}{V - B} = \nu. \tag{7}$$

Of these,

$$\nu\varphi(c)dc = \nu_1$$

have a velocity that lies between c and $c+dc$, where

$$\varphi(c)dc = 4\sqrt{\frac{h^3m^3}{\pi}}\, c^2 e^{-hmc^2} dc \tag{8}$$

is the probability that the velocity of a molecule lies between c and $c+dc$, i.e., the number of molecules whose velocities satisfy this condition divided by the total number of molecules n. Among the ν_1 molecules we shall find

$$\nu_2 = \frac{\nu}{2}\varphi(c)dc\sin\vartheta d\vartheta$$

for which the angle ϑ lies between the limits ϑ and $\vartheta+d\vartheta$,[1] and among these

$$\frac{\nu_2 d\epsilon}{2\pi} = \frac{\nu}{4\pi}\varphi(c)dc\sin\vartheta d\vartheta d\epsilon,$$

for which ϵ lies between the limits ϵ and $\epsilon+d\epsilon$. This is therefore the number of molecules (previously denoted by dz) that lie in a cylinder of volume

$$\Omega dh = \Omega c\cos\vartheta dt \tag{9}$$

and whose velocities satisfy the conditions (2) (see §3). These are the same molecules that collide with the part DE of the wall of the container, of surface area Ω, during time dt, in such a way that their velocities again satisfy the conditions (2). On substituting the values (7) and (9), the expression for the number of these molecules becomes:

$$dz = \frac{n\Omega c\cos\vartheta\sin\vartheta}{4\pi(V-B)}\varphi(c)dcd\vartheta d\epsilon dt. \tag{10}$$

We now assume that the state is stationary. During any time t_2-t_1, there will collide with the surface DE a total of $(t_2-t_1)dz/dt$ molecules of the specified kind. Each of these has momentum $mc\cos\vartheta$ in the direction N before the collision, and acquires on the average the same momentum in the opposite direction after the collision, so that momentum $2mc\cos\vartheta$ in the direction of the normal inwards from DE must be imparted to it. This momentum is part of the total impulse $\Omega p_g(t_2-t_1)$ supplied by the force Ωp_g. All molecules of the specified kind therefore provide a contribution

$$2mc\cos\vartheta(t_2-t_1)dz/dt \tag{11}$$

to the impulse $\Omega p_g(t_2-t_1)$. If one substitutes for dz the value (10) and integrates over all possible values—i.e., over ϵ from 0 to ∞—then he obtains the total impulse $\Omega p_g(t_2-t_1)$. If we divide through

[1] Cf. Part I, Eqs. (38) and (43).

by $\Omega(t_2-t_1)$ in this equation, and perform the integration over ϵ, we find:

$$p_g = \frac{nm}{V-B}\int_0^\infty c^2\varphi(c)dc\int_0^{\pi/2}\cos^2\vartheta\sin\vartheta d\vartheta. \tag{12}$$

The integral over ϑ has the value $\frac{1}{3}$, as is well known. Furthermore, $\int_0^\infty c^2\varphi(c)dc$ is equal to the mean square velocity $\overline{c^2}$ of a molecule. One thus obtains:

$$p_g = \frac{nm\overline{c^2}}{3(V-B)} \tag{13}$$

If the attractive force of the gas molecules (van der Waals' cohesion force) were not present, then p_g would be just the external pressure of the gas. Because of this cohesion force, however, the total pressure p_g consists of two parts: first, the pressure p exerted by the walls; and second, the attractive force exerted on those molecules near the wall by the other molecules. If we denote the total intensity of the attractive force acting on each molecule lying at the surface by p_i, as before, then we obtain the equation already denoted by (1):

$$p_g = p + p_i.$$

§6. Limits of validity of the approximations made in §4.

In deriving Equations (5) and (13) we neglected all terms of order B^2/V^2. Hence we cannot expect that these equations will be valid for values of V comparable to B. In fact, Equation (10) predicts an infinite pressure for $V=B$. Yet this volume is still 4 times larger than the space actually filled by the molecules, so that the corresponding pressure is certainly not yet infinite. On the other hand, the pressure can be infinitely large when the molecules are so densely packed that one cannot put any more spheres into the space.

One of the densest arrangements of many equal spheres can be obtained by piling them up in a pyramid like cannonballs. An easy calculation shows that the total volume that they occupy, including the small spaces in between the spheres, is then $3\sqrt{2}/\pi$ of the volume occupied by the spheres themselves.

If the gas molecules were arranged in this way, then

$$V = \frac{3\sqrt{2}}{4\pi} B = 0.33762B. \tag{14}$$

Thus p_g would first become finite when V is about $\frac{1}{3}B$,[1] whereas according to Equation (13) it is already infinite when $V = B$.

We shall see in Chapter V, §58, that Equation (13) does not give the correct coefficients even for the terms of order B^2/V^2.

From this viewpoint it appears that van der Waals has substituted for the exact formula another one that is certainly wrong as soon as V is not large compared to B. Although there may be an important quantitative difference between an expression that vanishes for $V = B$ and one that vanishes for $V = \frac{1}{3}B$, the correct formula must still follow a course qualitatively similar to the one that van der Waals put in its place.* This fact helps to explain the beautiful qualitative agreement of the van der Waals equation with the actual properties of gases and liquids; however, it also clarifies the essential quantitative difference. Inasmuch as the calculation of the exact formula still encounters insuperable mathematical difficulties, one must be satisfied with van der Waals' approximation.

One has therefore to distinguish between the properties of a substance that exactly conforms to van der Waals' assumptions, and one that is represented by van der Waals' equation; in the following we shall always discuss the latter.

§7. Determination of internal pressure.

In order to calculate p_i, van der Waals assumes that the attraction between two molecules acts only at short distances, which are still large compared to the average distance between

[1] Then we have the relation $v = \frac{1}{3}b$ between the quantities appearing in Eq. (19), and the quantity ω introduced in Eq. (32) would be equal to $\frac{1}{3}$.

* The recent work of Alder and Wainwright (J. Chem. Phys. **27**, 1208 (1957), **31**, 459 (1959); Phys. Rev. **127**, 359 (1962)) on phase transitions in small systems of elastic spheres throws doubt on the validity of this assertion. See also Brush, Am. J. Phys. **29**, 603 (1961).

neighboring molecules of the substance. We now find the van der Waals cohesion force p_i that acts on unit surface in the following way: we choose any surface element ds on the boundary surface of the substance, and construct inside the substance a right cylinder which has this surface element as its base. We also construct two cross sections of this cylinder at distances ν and $\nu+d\nu$ from the base ds. The volume between these two cross sections is $\zeta = dsd\nu$; the mass of the substance found in this volume is therefore $\rho dsd\nu$, if ρ is its density.[1] Since m is the mass of a molecule, we see that ζ contains

$$\frac{\rho}{m} dsd\nu \tag{15}$$

molecules. Each of these molecules is at nearly the same distance from the boundary surface, and hence subjected to nearly the same conditions. It will be drawn toward the surface by those molecules nearer the surface, and away from it by those molecules farther from the surface. Since there are more of the latter, there remains a net resultant force away from the surface.

If we select a particular molecule inside the cylinder ζ and any volume element ω near it, then all molecules in ω will exert nearly the same force on our specified molecule. Their total attractive force—and hence also its component normal to ds—will be proportional to the number of molecules in ω, and hence to the density ρ of the gas. The proportionality factor can depend only on the size of ω and its position relative to the specified mole-

---

[1] The van der Waals cohesion force will of course produce a variation in density near the wall of the container, which we neglect in the text, following van der Waals. We would still be led to the same formula if we assumed only that when the pressure or temperature changes, the densities at different distances from the boundary surface change in the same proportion. Then there would appear in Eqs. (15) and (16) a factor F independent of ρ and T, and we could set FC equal to $f(\nu)$, instead of $C = f(\nu)$ as in the text.

Furthermore, the calculation of the collision number performed in the text is certainly correct when the cohesion force is not present. When it is present, however, one might doubt whether it is permissible to leave this calculation unchanged and simply add the cohesion force to the external pressure, as we have actually done, following van der Waals, in §4. Finally, we have considered the wall of the container to be impenetrable, without any adhesive force with respect to the substance within.

In chap. v, I will give a derivation of van der Waals' formula against which none of these objections can be raised.

cule. In particular, according to our assumptions, it is independent of temperature at a given density. Temperature merely determines the rapidity of the molecular motion, whereas according to our assumptions the attractive force between the molecules must be independent of their motion. All these conclusions hold equally well for all the other volume elements $\omega_1, \omega_2 \cdots$ lying in the neighborhood of our specified molecule, and the sum of the components normal to ds of all the forces exerted on it by the surrounding molecules must be proportional to the density but independent of the temperature. We write this sum as ρC, where C depends only on the distance of the molecule from the boundary surface. Since all the molecules in the infinitesimal cylinder ζ are subjected to the same conditions, and the number of these molecules is $\rho ds d\nu/m$, according to Equation (15), we see that the total force that is exerted on all these molecules, normal to ds, is equal to

$$\frac{\rho^2 C ds d\nu}{m}. \tag{16}$$

Since moreover the value of C does not depend on temperature or density but only on the distance ν of the cylinder ζ from the surface, we shall denote it by $f(\nu)$. The total force on all the molecules in the cylinder Z is

$$\frac{\rho^2 ds}{m}\int_0^\infty f(\nu)d\nu.$$

Since the value of the expression $(1/m)\int_0^\infty f(\nu)d\nu$ depends neither on density nor temperature but is a constant characteristic of the substance, we denote it by a. The total force on all molecules in the cylinder Z is then $a\rho^2 ds$. It is proportional to ds. The force that pulls in the molecules lying on unit surface—which we have called p_i—is therefore[2] $a\rho^2$, and we obtain, according to Equations (1) and (13):

$$p + a\rho^2 = \frac{nm\overline{c^2}}{3(V - B)}, \tag{17}$$

[2] This expression still needs to be corrected if the wall is curved. Indeed it is by this correction that van der Waals explains capillary phenomena in a manner similar to that of Laplace and Poisson (cf. §23). According to van der Waals, therefore, the pressure of a gas is not absolutely independent of the curvature of the wall. Nevertheless, this correction becomes negligible when the sphere of action of the cohesion force becomes larger compared to the diameter of a molecule.

where nm is the total mass of the substance. $V/nm = v$ is therefore the volume of unit mass at the specified temperature and pressure, the so-called specific volume. Since the total mass is $nm = \rho V$, it follows that

$$\rho = \frac{1}{v} \tag{18}$$

and we can write Equation (17) as follows:

$$p + \frac{a}{v^2} = \frac{\overline{c^2}}{3(v - b)} \tag{19}$$

where

$$b = \frac{B}{nm} = \frac{2\pi\sigma^2}{3m}. \tag{20}$$

This is also a constant of the gas; it is equal to half the volume of the covering spheres in unit mass of the gas, or 4 times the volume of the molecules in unit mass.

§8. An ideal gas as a thermometric substance.

We shall now choose as a measure of the temperature, the pressure that an ideal gas (the normal gas) would exert at various temperatures, at constant volume. By an ideal gas we mean a gas of the kind considered in Part I, and defined in Part II at the beginning of §1: its molecules exert an appreciable force on each other only at distances that are vanishingly small compared to the average distance of two neighboring molecules.

For a particular ideal gas we denote the mass of a molecule by M, the mean square velocity of the center of mass of a molecule by $\overline{C^2}$, and the number of molecules in unit volume by N. Then, according to §1, Part I, the pressure on unit surface is $p = \frac{1}{3}NM\overline{C^2}$. At constant volume, N is also constant. According to our choice of temperature units, the absolute temperature is therefore proportional to the quantity $\overline{C^2}$. In agreement with Equation (51) of Part I, we shall set $\overline{C^2} = RT$, where R is a constant determined by the temperature unit.

In Chapter III, §35, and Chapter IV, §42, we shall give good arguments in favor of the proposition that, at equal temperatures, the mean kinetic energy of the motion of the center of mass of a molecule is generally the same for all bodies. We have already proved in Part I that this is the condition for thermal equilibrium of two ideal gases with monatomic molecules. The validity of that conclusion is not affected by the attractive force assumed by van der Waals, since this acts at distances large compared to the distance between two neighboring moelcules, so that it does not perturb the motion of molecules during collisions. Hence the condition for thermal equilibrium between the normal gas and another gas characterized by Equation (19)—assuming that the former is monatomic—will be the equality of mean kinetic energy of the molecules in each gas. Hence at equal temperatures, $m\overline{c^2}=M\overline{C^2}$. Since the latter quantity is equal to $3RMT$, then at the same temperature we also have $m\overline{c^2}=3RMT$ for the other gas. We shall now compare the molecular weight of the other gas with that of the normal gas, and denote the quantity m/M by μ and the quantity R/μ by r; then

$$\overline{c^2} = 3rT = \frac{3R}{\mu}T \tag{21}$$

and hence, according to Equation (19),

$$p + \frac{a}{v^2} = \frac{rT}{v-b} = \frac{RT}{\mu(v-b)}. \tag{22}$$

This is the van der Waals relation between pressure, temperature, and volume of a gas. r, a, and b are constants characteristic of the gas; R is a constant referring to the normal gas, and is completely independent of the nature of the other gas.

In chemistry one usually understands by the molecular weight of a gas the ratio of the mass of a molecule of that gas to the mass of a single hydrogen atom. Thus for ordinary hydrogen, whose molecules are diatomic, $\mu=2$, and the gas constant is $r_H=\frac{1}{2}R$, where R would be the gas constant of hydrogen if its molecules were dissociated into single atoms. If we do not wish to base our definition on such a dissociated gas, we have to define R empirically as being twice the gas constant of ordinary hydrogen gas.

§9. Temperature-pressure coefficient. Determination of the constants of van der Waals' equation.

We now consider a gas for which the relation between pressure, density, and temperature is sufficiently well expressed by van der Waals' equation (22).

We first determine for this gas the temperature coefficient of the pressure at constant volume—i.e., we shall heat it from T_1 to T_2 at constant volume, denote the pressure on unit surface at these two temperatures by p_1 and p_2, and determine the quotient $(p_2-p_1)/(T_2-T_1)$. We find, from Equation (22):

$$p_1 + \frac{a}{v^2} = \frac{rT_1}{v-b}, \quad p_2 + \frac{a}{v^2} = \frac{rT_2}{v-b}, \tag{23}$$

whence it follows that:

$$\frac{p_2 - p_1}{T_2 - T_1} = \frac{r}{v-b}\,. \tag{24}$$

Thus the pressure difference is proportional to the temperature difference, and the proportionality factor is simply a function of the volume of unit mass. The pressure difference of a gas following van der Waals' law, at constant volume, is therefore always a measure of the temperature difference. If we denote by p_3 the pressure at a third temperature T_3—keeping the same volume per unit mass, v—then:

$$p_3 - p_1 : p_2 - p_1 = T_3 - T_1 : T_2 - T_1. \tag{25}$$

We next assume that we have a second gas, for example hydrogen, which can be considered ideal with sufficient accuracy. For this latter gas, therefore,

$$p_3 : p_2 : p_1 = T_3 : T_2 : T_1.$$

One can therefore determine absolute temperatures directly, if he establishes a unit for the temperature degree—e.g., if he sets the difference between the temperature of boiling water and that of melting ice (both at atmospheric pressure) equal to 100.

One can next test how well Equation (25) is satisfied for the first gas; in other words, one can determine how well van der Waals' law predicts the actual dependence of the pressure on the temperature. If one calculates the temperature coefficient of the pressure, $r/(v-b)$, from Equation (24) for two different densities, then he can determine r and b for the gas. If one knows the chemical constitution of the molecules of the gas, then he can test whether the equation $\mu r = R$ is accurately satisfied. One can also determine μ from the empirically determined van der Waals constant r, instead of from the vapor density. If one determines the temperature coefficient (24) of the pressure at constant volume for more than two values of r, then he can test how well it is represented as a function of v by the van der Waals formula.

One remark should be made here. According to §6, the expression $r/(v-b)$ was found by an approximation that is no longer permissible when v approaches the value b. For the smaller values of v, one must replace b by $(b/3)$. In fact, experiment shows that when b is determined in the above way for different values of v, it is not actually constant, but decreases with decreasing v. It does not by any means follow from this that the basic assumptions of van der Waals are wrong for such substances. Indeed, if the equation of state could be derived exactly from these basic assumptions, similar consequences would follow. Unfortunately it has not yet been possible to find what function of v must replace $r/(v-b)$ in the exact equation of state, on the basis of the assumptions of van der Waals.* In the following we must therefore limit ourselves to the discussion of Equation (22), and keep in mind the fact that we cannot expect it to achieve any more than qualitative agreement for small values of v. From Equation (23) it follows that

$$a = v^2 \frac{p_2 T_1 - p_1 T_2}{T_2 - T_1}, \tag{26}$$

from which one may also determine the value of the constant a. If one calculates this value for several values of v, he can find out how well the term added to p on the left side of van der Waals' equation (22) agrees with experience. Thus one can test the validity of van der Waals' assumption that the cohesion force

---

* One way to obtain a more accurate function is to calculate the available space taking account of the possible configurations of groups of two, three, or more other spheres, to obtain what are now called the third, fourth, and higher virial coefficients. Boltzmann discusses this approach in chap. v, but does not claim that the exact high-density equation of state can be obtained in this way. Van der Waals himself later adopted the position that the observed variation of b with density cannot be explained in this way, but is rather due to an actual compression of the molecule itself: see *Boltzmann-Festschrift*, p. 305 (1904); Amsterdam Verslagen [4] **21**, 800, 1074 (1912–1913).

extends to distances large compared to the average distance of two neighboring molecules.

§10. Absolute temperature. Compression coefficient.

We can never actually determine the temperature by means of an ideal gas, since no known gas, not even hydrogen, possesses exactly the properties that we ascribe to an ideal gas. The most rational definition of temperature is of course that based on Lord Kelvin's temperature scale. As is well known, this scale is derived from the maximum work that can be gained by transforming heat from one temperature to a lower one. However, since the direct experimental determination of this work is always performed very inexactly, one is forced to calculate it from the equation of state of some body. Now the deviations of hydrogen from the ideal gas state are rather small, so that if one treats these deviations on the basis of van der Waals' assumptions, one ought to be able to construct the absolute Kelvin temperature scale with an accuracy that can hardly be surpassed at the present time.[1] One can then use the equations developed above for the determination of the absolute temperature, except that one cannot assume that T_1, T_2, and T_3 can be determined by comparison with another more ideal gas. One can first express the temperature differences in terms of numbers, by means of the proportion (25), if he merely chooses any arbitrary temperature unit, for example as indicated above. As a control, one can determine the temperature at several different densities. If one finds pressures p_1, p_2, p_3 at temperatures T_1, T_2, T_3 for specific volume v, and pressures p_1', p_2', p_3' at the same temperatures for specific volume v', then the pressure must satisfy the relation

$$p_3 - p_1 : p_2 - p_1 = p_3' - p_1' : p_2' - p_1'$$

if the gas is to satisfy van der Waals' equation with sufficient accuracy.

If p_1' and p_2' are the pressures corresponding to temperatures

[1] For all temperatures that are not too low, the properties of air satisfy van der Waals' assumptions with greater accuracy. One can therefore also use air (which is more convenient for experiments) instead of hydrogen.

T_1 and T_2 at a specific volume v', then one can write Equation (26) as follows:

$$a = v^2\left[(p_2 - p_1)\frac{T_1}{T_2 - T_1} - p_1\right]$$
$$= v'^2\left[(p_2' - p_1')\frac{T_1}{T_2 - T_1} - p_1'\right].$$

If one defines T_1 to be the temperature of melting ice, and T_2 to be the temperature of boiling water, and sets $T_2 - T_1 = 100$, then all the other quantities in the last two expressions in this equation are accessible to observation, and one can therefore calculate T_1. Moreover, one can determine the value of the constant a for hydrogen.

Since one now knows the absolute temperature, he can immediately determine the values of r and b for hydrogen, according to the method given earlier. Here the following remarks must be made: when we consider the van der Waals equation (22) as simply a given fact of experience, then we must write on its right-hand side, instead of T, some function $f(T)$ of Kelvin's absolute temperature. The absolute temperature itself would not then be determinable without some empirical information about specific heats or Joule-Thomson cooling, etc.[2]

In order to test the relation between p and v at constant temperature T—i.e., the pressure coefficient of density—we write the van der Waals equation in the form

$$pv = \frac{rT}{1 - \dfrac{b}{v}} - \frac{a}{v} = rT - \frac{a - rbT}{v} \cdots$$

As long as v is large compared to b, and also to a/rT, Boyle's law is nearly valid; pv is almost constant at constant temperature. The gas is far from the region of liquefaction. As long as $a > rbT$, the correction to Boyle's law resulting from the van der Waals cohesion force will dominate that due to the finite extension of the molecular cores, and $pv = p/\rho$ will increase with increasing volume. The pressure coefficient of density, $d\rho/dp$, decreases with decreasing pressure. For any gas at very high temperatures, $a < rbT$, hence the latter correction must dominate the former, and there-

[2] Boltzmann, Mun. Ber. **23,** 321 (1894); Ann. Phys. [3] **53,** 948 (1894).

for pv will decrease with increasing v. This is already the case for hydrogen at ordinary temperatures.

In addition, the differential quotient of specific volume, v, at constant pressure with respect to temperature, which we shall call the temperature coefficient of the volume, is, as our formula shows, not constant.

§11. Critical temperature, critical pressure, and critical volume.

We shall now examine more closely the relation between pressure, temperature, and specific volume represented by Equation (22). We see that the pressure is infinite when $v=b$, for any temperature. For a substance that exactly conforms to van der Waals' assumption, we saw that the pressure would not actually be infinite until the volume decreases to about $\frac{1}{3}b$. We shall not pursue this matter further, since the object of our present investigations is not the original assumptions of van der Waals but rather Equation (22), insofar as that equation is approximately correct for large volumes and gives at least a qualitatively correct expression for small volumes.

The volume $v=b$ is therefore impossible; and likewise any smaller volume, since for a stable equilibrium state the pressure must increase as the volume decreases, and it would therefore have to be even greater than infinity.

We now seek the isotherms—i.e., the relation between pressure and volume at constant temperature. Since T is constant, it follows from Equation (22) that

$$\frac{dp}{dv} = \frac{2a}{v^3} - \frac{rT}{(v-b)^2}. \tag{27}$$

Here the right-hand side is negative for the smallest possible values of v (those which are only slightly larger than b) and likewise for very large values of v. Furthermore, it changes in such a way that its derivative with respect to v is continuous for all values of v considered. The right-hand side of (27) can vanish only for

$$T = \frac{2a(v-b)^2}{rv^3}.$$

In this equation, the expression on the right side has a very small positive value for volumes slightly greater than b, and also for very large volumes. It is continuous in between, and has a single maximum value: $T_k=8a/27rb$ for $v=3b$. Hence when $T>T_k$, dp/dv cannot vanish, and hence cannot become positive; the isotherm falls off with increasing v. When $T<T_k$, dp/dv goes through zero to a negative value, and then again goes through zero to a positive value. The ordinate of the isotherm has a minimum and a maximum. For $T=T_k$, dp/dv is always negative; it is only once equal to zero, when $v=3b$. Hence p decreases with increasing v, but at this point only by an amount that is an infinitesimal of higher order than the infinitesimal increase in volume. This point is called the critical point. We give the values of v, p, and T corresponding to this point the index k and call them the critical values,* thus:

$$v_k = 3b, \quad T_k = 8a/27rb. \tag{28}$$

For the corresponding value of p, the critical pressure, one finds from Equation (22):

$$p_k = a/27b^2. \tag{29}$$

v_k, T_k and p_k are therefore three real positive values. The former is larger than the minimum volume b of the substance. From Equation (27) one finds (with T constant):

$$\frac{d^2p}{dv^2} = 2\left(\frac{rT}{(v-b)^3} - \frac{3a}{v^4}\right)$$

and one easily sees that for the critical values, d^2p/dv^2 vanishes, as was to be expected, since we have seen that for the critical values the isotherm has a completely regular maximum-minimum [point of inflection].

In addition I shall describe an algebraic property of the critical quantities. If we bring all the quantities to the same side of the equality sign in Equation (22), eliminate the fractions, and collect powers of v, then this equation becomes

$$pv^3 - (bp + rT)v^2 + av - ab = 0. \tag{30}$$

For given values of p and T, this is a third-degree equation for v. We shall denote its left-hand side by $f(v)$. If not only $f(v)$ but also

* German *kritische* = critical, hence the subscript "k."

$f'(v)$ and $f''(v)$ vanish for a particular set of values of p, T, and v, then this third-degree equation has three equal roots for v; here $f'(v)$ is the first derivative, and $f''(v)$ the second derivative of $f(v)$ at constant p and T.

If one keeps only T constant, then it follows from Equation (30) that

$$(v^3 - bv^2)\frac{dp}{dv} = -f'(v),$$

$$(v^3 - bv^2)\frac{d^2p}{dv^2} + (3v^2 - 2bv)\frac{dp}{dv} = -f''(v).$$

The quantities dp/dv and d^2p/dv^2 are the same ones mentioned above, which we have proved vanish for the critical values of p, v, and T. For these values, therefore, not only $f(v)$ but also $f'(v)$ and $f''(v)$ vanish—i.e., Equation (30) has three equal roots for v, if one substitutes the critical values for p and T. Now the coefficient of v^2, taken negative and divided by p, is equal to the sum of the roots. The coefficient of the term that does not contain v, likewise taken negative and divided by p, is equal to the product of the roots. Finally, the coefficient of v, taken positive and divided by p, is equal to the sum of the products of each pair of roots.* Hence one finds for the values p_k and T_k (for which Eq. [30] has three equal roots, whose common value we denote by v_k) the three equations

$$3v_k = b + \frac{rT_k}{p_k}, \quad 3v_k^2 = \frac{a}{p_k}, \quad v_k^3 = \frac{ab}{p_k},$$

from which follow the values of v_k, p_k, and T_k already found. For those values of the temperature for which the ordinate of the isotherm has no minimum, there corresponds to each value of p only one value of v, so that Equation (30) has only one real root, which is greater than b; however, for any temperature for which the ordinate of the isotherm has a minimum p_1 and a maximum p_2, Equation (30) has three real roots for v, greater than b, when p lies between p_1 and p_2, as one can perceive at once from the form of the isotherms.

Up to now we have made no particular stipulation about the units of pressure and volume. The formula will become especially

* For proof of these three statements see any algebra textbook.

simple if we choose the critical volume v_k and the critical pressure p_k of each substance as the units of volume and pressure, in discussing the properties of that substance. We shall also ignore the empirical unit of temperature derived from the freezing and boiling points of water, and choose for each gas its absolute critical temperature T_k as the unit of absolute temperature. We therefore set:

$$(31) \qquad \begin{cases} v = v_k\omega = 3b\omega, \\ p = p_k\pi = \dfrac{a}{27b^2}\pi, \quad T = T_k\tau = \dfrac{8a}{27rb}\tau. \end{cases}$$

Thus we measure the volume by ω (the ratio of the volume to the critical volume) and likewise the pressure and temperature by π and τ.

These three quantities ω, π, and τ we call the reduced volume, reduced pressure, and reduced temperature, or—when there is no question of comparison with another system of units—simply the volume, pressure, and temperature of the substance.

We have of course introduced different units for each gas—which we shall call the van der Waals units—but this disadvantage is outweighed by the advantage that the equations become much simpler. Since we can calculate a, b, and r, and hence also v_k, p_k, and T_k for each gas from its empirical properties, we can transform from van der Waals' units to any other units whenever we wish. If in Equation (22) we replace p, v, and T by π, ω, and τ, then we obtain (after dividing through by a factor that can never be zero)

$$(32) \qquad \pi = \frac{8\tau}{3\omega - 1} - \frac{3}{\omega^2}\,.$$

All the constants characterizing the gas have dropped out of this equation. If one bases measurements on the van der Waals units, then he obtains the same equation of state for all gases. Van der Waals believes that this equation is valid up to the liquefaction of the gas, and indeed even into the liquid region. Only the values of the critical volume, pressure, and temperature depend on the nature of the particular substance; the numbers that express the actual volume, pressure, and temperature as multiples of the critical values satisfy the same equation for all substances. In other words, the same equation relates the reduced volume, reduced pressure, and reduced temperature for all substances.

Obviously such a broad general relation is unlikely to be exactly correct; nevertheless, the fact that one can obtain from it an essentially correct description of actual phenomena is very remarkable.

§12. Geometric discussion of the isotherms.

In order to gain some insight into the relation represented by Equation (32), we shall draw on the positive abscissa axis $O\Omega$ from the origin of coördinates O the reduced volume ω as the abscissa OM. Over the point M we erect the ordinate MP, representing the reduced pressure π, parallel to the ordinate axis $O\Pi$. Then each state of the gas, characterized by its pressure and volume, is represented by a point P in the plane. The corresponding reduced temperature is the value of τ that would be obtained from Equation (32) for the assumed values of ω and π. If one assumes than van der Waals' equation is correct, the reduced pressure would become infinite at $\omega = \frac{1}{3}$ for any positive τ. As we pointed out earlier, one can compress the substance to the volume $\omega = \frac{1}{3}$ only by exerting an infinite pressure, and since the pressure must increase as the volume decreases, the smaller volumes for which the formula gives negative pressures are impossible.

We must therefore limit our considerations to abscissas $\geq \frac{1}{3}$.

By an isotherm we mean the locus of all points representing those states of our substance for which the temperature has a fixed value. The equation of an isotherm is any relation between π and ω that follows from Equation (32) if we substitute any arbitrary constant value for τ. We obtain the set of all possible isotherms by letting τ take all possible values from a very small positive value to $+\infty$. From Equation (32) it follows that for each τ, π has a very large positive value when ω is slightly larger than $\frac{1}{3}$. On the other hand, π has a very small positive value when ω is very large. Moreover, at constant τ it follows that

$$\frac{d\pi}{d\omega} = 6\left[\frac{1}{\omega^3} - \frac{4\tau}{(3\omega - 1)^2}\right]. \tag{33}$$

Since this expression is finite for $\frac{1}{3} < \omega < \infty$, all isotherms between $\omega = \frac{1}{3}$ and $\omega = \infty$ must be continuous curves. As ω approaches the limit $\frac{1}{3}$, they approach asymptotically the line AB, which is parallel to the ordinate axis at a distance $\frac{1}{3}$ from it, on the side of positive ordinates. As ω becomes very large, the isotherms likewise approach the abscissa axis on the side of positive ordinates. In the former case π has a very large positive value, while $d\pi/d\omega$ has a very large negative value; in the latter case, π is small and positive, while $d\pi/d\omega$ is small and negative. All isotherms have two branches, going to infinity on the positive side of the abscissa axis. On the other hand, it is possible for π to be negative between $\omega = \frac{1}{3}$ and $\omega = \infty$; the curve representing Equation (32) for constant τ can dip down below the abscissa axis.

In order to form a picture of this behavior, we note first that, as a glance at Equation (32) shows, smaller values of τ will always correspond to smaller values of π, when ω is the same. Therefore each isotherm must lie below the isotherms corresponding to higher temperatures, so that for each abscissa ω the isotherm at the higher temperature has a larger ordinate than the isotherm at the lower temperature; two isotherms can never intersect.

We now discuss the expression for $d\pi/d\omega$, Equation (33). It is a continuous function of ω for ω between $\frac{1}{3}$ and $+\infty$. For very large values of ω, and also when ω is slightly larger than $\frac{1}{3}$, the second term dominates, so that $d\pi/d\omega$ is negative, as we mentioned before. $d\pi/d\omega$ cannot become positive within this interval without going through zero. The latter event can happen only for

$$\tau = \frac{(3\omega - 1)^2}{4\omega^3} \tag{34}$$

according to Equation (33). Not only when ω is slightly larger than $\frac{1}{3}$ but also when ω is very large, the right hand side of this equation has a very small positive value. Its value changes continuously with ω in this interval; as one can find by well-known methods, it has a single maximum value, 1, for $\omega = 1$. Thus there are three cases to be distinguished:

1. For $\tau > 1$, Equation (34) cannot be satisfied, and $d\pi/d\omega$ cannot vanish but must be negative in the entire region considered, so that the isotherm (marked **0** in Fig. 1) sinks continuously toward the abscissa axis as ω increases.

2. Let $\tau = 1$, so that the isotherm corresponds exactly to the critical temperature. Then, according to what we have said about the right-hand side of Equation (34), $d\pi/d\omega$ vanishes only for

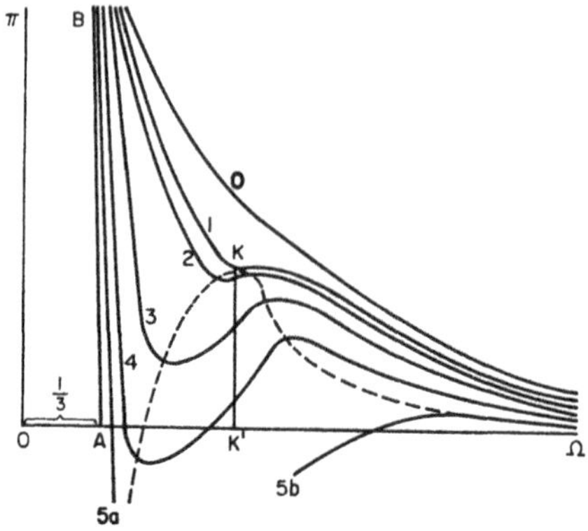

FIG. 1.

$\omega=1$. According to Equation (32), then $\pi=1$ also. The substance therefore has its critical temperature, volume, and pressure. This state (the critical state) is represented by the point K of Figure 1, whose abscissa and ordinate are both equal to 1. On taking the derivatives at constant τ, it follows from Equation (33) that:

$$(35)\quad \begin{cases} \dfrac{d^2\pi}{d\omega^2} = 18\left[\dfrac{8\tau}{(3\omega-1)^3} - \dfrac{1}{\omega^4}\right] = \dfrac{6}{3\omega-1}\left[\dfrac{3(1-\omega)}{\omega^4} - \dfrac{d\pi}{d\omega}\right], \\ \dfrac{d^3\pi}{d\omega^3} = 72\left[\dfrac{1}{\omega^5} - \dfrac{18\tau}{(3\omega-1)^4}\right]. \end{cases}$$

For the critical state we therefore have also $d^2\pi/d\omega^2=0$ (as was to be expected since we know that d^2p/dv^2 vanishes for the critical state). However, $d^3p/d\omega^3$ is negative. The isotherm therefore has an inflection point. Its tangent is parallel to the abscissa axis, but the ordinate decreases with increasing ω on both sides. The same isotherm has a second point of inflection at $\omega=1.87$; it therefore turns its concave side downwards between this value and the abscissa 1, but for other abscissas it turns its convex side downwards. Curve 1 in Figure 1 represents the critical isotherm.

The two inflection points first appear for isotherms corresponding to a reduced temperature $\tau=3^7\cdot2^{-11}=1.06787$, where they both occur at $\omega=\frac{4}{3}$.[1] For smaller τ they separate from each

[1] $d^2\pi/d\omega^2$ vanishes for $\tau=(3\omega-1)^3/8\omega^4$. The expression on the right side is positive and vanishingly small for very large values of ω as well as for values slightly larger than $\frac{1}{3}$. It is continuous between these limits and

other. For larger τ, the isotherms fall to the positive abscissa axis without inflection points, and $d\pi/d\omega$ decreases steadily.

3. Suppose the temperature is below the critical temperature: $0<\tau<1$. Then, as one sees from Equation (33), $d\pi/d\omega$ is positive for $\omega=1$, while for large values of ω and for values near $\frac{1}{3}$ it is negative. Hence $d\pi/d\omega$ must vanish for a value of ω greater than 1, and also for a value between 1 and $\frac{1}{3}$. For none of these values can $d^2\pi/d\omega^2$ also vanish, since from Equation (35) it follows that $d\pi/d\omega$ and $d^2\pi/d\omega^2$ can simultaneously vanish only when $\omega=1$. This follows also from the fact that Equation (30), which differs from the present equation only in the choice of units, must then have three equal roots for v, and as we saw this is possible only at the critical pressure, volume, and temperature. For the value of ω between $\frac{1}{3}$ and 1, at which $d\pi/d\omega$ changes from a negative to a positive value as ω increases, and $d^2\pi/d\omega^2$ is positive according to Equation (35), π therefore has a minimum, while for the other value of ω it has a maximum. $d\pi/d\omega$ cannot vanish for a third value of ω, since Equation (34), which gives the condition for this, can be written in the form

$$4\tau\omega^3 - (3\omega - 1)^2 = 0. \tag{36}$$

The polynomial is negative for $\omega=0$ and positive for $\omega=\frac{1}{3}$, so that its third root must lie somewhere between these two values of ω, and hence in the interval that we do not consider. For all isotherms corresponding to temperatures below the critical temperature, the ordinate π has a minimum for one abscissa ω between $\frac{1}{3}$ and 1, and a maximum for one abscissa greater than 1. Curve 3, Figure 1, shows its general form.

§13. Special cases.

We now consider two special cases of the third case.

3a. Let τ be slightly smaller than 1, say $1-\epsilon$. The two ordinates for which π has an extremal value are then close to 1, and we can write them in the form $1+\xi$. Substitution of $\tau=1-\epsilon$ and $\omega=1+\xi$ into Equation (36) yields (retaining only terms of the first order of magnitude) $\xi=\pm\sqrt{4\epsilon/3}$, while Equation (32) yields

has a single maximum $3^7\cdot2^{-11}$ for $\omega=\frac{4}{3}$. For $\tau>3^7\cdot2^{-11}$, therefore, $d^2\pi/d\omega^2$ cannot vanish.

$\pi = 1 - 4\epsilon$. Therefore τ and π differ from 1 by an infinitesimal of order ϵ, while the difference between ω and 1 is an infinitesimal of order $\sqrt{\epsilon}$. The isotherms just below the critical isotherm are therefore almost horizontal in the neighborhood of the critical point K, as would also follow from the fact that they have both a minimum and a maximum near this point. The locus of the maxima and minima of all the isotherms (the dotted curve in Fig. 1) therefore has its maximum at K where it touches the critical point.

3b. Let τ be very small, so that the temperature is near absolute zero. Then one easily finds, from Equation (36), the following value for the root near $\frac{1}{3}$:

$$\omega_1 = \frac{1}{3} + \frac{2}{9}\sqrt{\frac{\tau}{3}}\,.$$

The other root in which we are interested is very large, and one finds for it

$$\omega_2 = \frac{9}{4\tau}\,.$$

The minimum of π therefore corresponds to an abscissa slightly greater than $OA = \frac{1}{3}$. The value of this smallest ordinate is

$$\pi_1 = -27 + 12\sqrt{3\tau}.$$

The isotherm in question (5a in Fig. 1) thus sinks infinitely close to the downwards extension of the line BA, to the ordinate -27 below the abscissa axis. It comes back again (Curve 5b, Fig. 1, where the branch going upwards is shown increasing much too steeply at first, and is drawn too close to the origin of coördinates) and cuts the abscissa axis once again. If we denote by ω_3 the abscissa for which it cuts the abscissa axis again, then:

$$\frac{8\tau}{3\omega_3 - 1} = \frac{3}{\omega_3^2}, \qquad -\frac{3}{\omega_3^2} + \frac{9}{\omega_3} = 8\tau.$$

The transformation to the latter equation is permissible since we are not interested in values of ω_3 lying near $\frac{1}{3}$; indeed we already know that there is such a solution of the equation, corresponding to the intersection of the descending branch 5a with the abscissa axis. Instead we now want the solution for which ω_3 is large, and we see that in this case ω_3 will be approximately

$9/8\tau$. Thus for very low temperatures, the ordinates of the curve 5b do not become positive until the volume is very large. At $\omega = \omega_2 = 9/4\tau$ (in other words, for an abscissa twice as far from the origin) this ordinate attains its maximum, which is only a very small value: $\pi_2 = 16\tau^2/27$. After this, the curve again approaches the abscissa axis.

Between this extreme isotherm and the isotherm 3 of Figure 1, which has only positive values, there are of course numerous isotherms that go under the abscissa axis; curve 4 of Figure 1 is an example. One can understand the situation more clearly by drawing a third coördinate axis perpendicular to $O\Omega$ and $O\Pi$. In the various planes parallel to the plane $\Omega O \Pi$ one constructs the various isotherms corresponding to a sequence of increasing temperatures, and one models in gypsum the surface formed of all these isotherms.

CHAPTER II

Physical discussion of the van der Waals' theory.

§14. Stable and unstable states.

Let us now consider the physical meaning of the diagrams in Figure 1. Each point P of the quadrant bounded by the two infinite lines $A\Omega$ and AB represents a certain volume and a certain pressure, and thus a certain state of the substance, since Equation (32) provides the corresponding pressure. We call this state simply the state P. Each curve PQ lying in this quadrant (which, like the point P, is not shown in the figure) therefore represents a variation of state, in particular the sequence of different states corresponding to different points on the curve. We say that the substance experiences the state variation PQ as it passes through all the states represented by the different points on this curve.

The isotherm **0** in Figure 1 represents, for example, a compression of the substance at a constant temperature $\tau_0 > 1$, starting from a very large volume. The pressure increases continuously as the volume decreases, and for large volumes it is nearly inversely proportional to the volume, since then the quantities b and a/v^2 in Equation (22) are relatively very small. The substance then behaves almost like an ideal gas. On the other hand, if ω is slightly larger than 1/3, then the volume is nearly equal to 1/3 of the critical volume; then the isotherm rises very rapidly and approaches the line AB asymptotically. In this case, a very small decrease in volume results in a very large increase in pressure; the substance is almost incompressible, and behaves like a liquid. The transformation from the gaseous to the liquid state takes place very gradually; no break in the continuity of the transition is ever noticeable. This is also true for the isotherm 1 corresponding to the critical temperature, in Figure 1, except that at the critical point the tangent of the isotherm is parallel to the abscissa axis, so that an infinitesimal isothermal change in volume

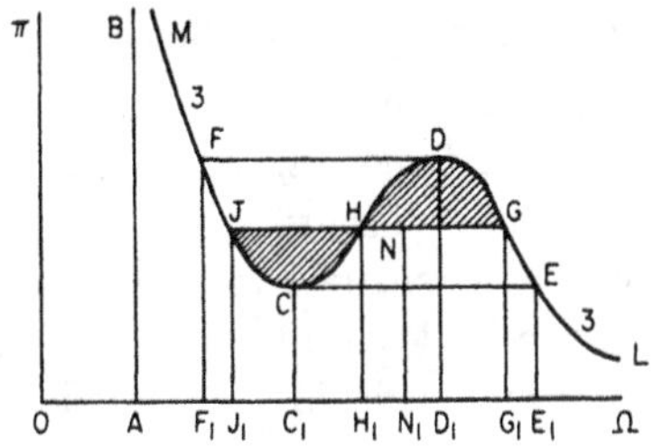

FIG. 2.

corresponds then to a higher order infinitesimal change in pressure.

Now suppose that we compress the substance at a temperature below the critical temperature. Then we follow an isotherm for which $\tau < 1$, for example isotherm 3 in Figure 1, which corresponds to the temperature τ_3. We draw this isotherm again in Figure 2, and denote by C and D the points whose ordinates CC_1 and DD_1 have the minimum and maximum values, respectively. The lines drawn from C and D parallel to the abscissa axis intersect the isotherm again at E and F, respectively. Let the projections of the latter two points on the abscissa axis be E_1 and F_1, respectively. As long as the pressure is less than EE_1, we find ourselves on the branch LE of the isotherm. At a given temperature, only one state of the substance is possible for each pressure, and Boyle's law is obeyed with greater and greater accuracy as the pressure decreases. On the other hand, as soon as the pressure becomes equal to EE_1, two completely different states (phases of the substance) are possible at the same temperature and pressure, represented by the two points E and C on the isotherm, which have equal ordinates. The phase represented by the point E (or simply, the phase E) corresponds to a larger specific volume, hence to a smaller density, while the phase C corresponds to a larger density. The former phase is the vapor, the latter the liquid. If the isotherm corresponds to a temperature that is only slightly below the critical temperature, then the two points C and E will be very close together, and the substance will have only slightly different properties in these two states. Far below the critical temperature, however, the liquid and vapor phases are completely different.

As soon as the pressure becomes equal to DD_1, there are

again two points D and F on the isotherm, each of which represents a possible phase at that temperature and pressure. However, if the pressure is between EE_1 and DD_1, say GG_1, then we have three points on the isotherm: G, H, and J, to which this pressure corresponds. One easily convinces himself that the state corresponding to the middle point H is unstable.

Suppose the substance is in a cylindrical container closed with an easily displaceable piston, and it is initially in the state H, so that at equilibrium the pressure HH_1 is exerted in the piston. Now if we push the piston in a little bit, without changing the external pressure, then the volume will decrease somewhat. We assume that the substance is surrounded by a good heat conductor, so that its temperature remains constantly equal to that of its surroundings. Then, as the nature of the isotherm near H shows, the pressure of the substance on the piston will decrease; the piston will therefore be pushed in by the external pressure until the volume becomes equal to OC_1. Since the same thing would happen for an infinitesimal isothermal expansion, the least motion of the piston will cause the volume to change by a finite amount from its original value. Hence for any pressure between EE_1 and DD_1, only two stable phases are possible at the temperature τ_3 corresponding to isotherm 3.

§15. Undercooling. Delayed evaporation.

What would happen if the critical volume of unit mass of the substance were between OC_1 and OD_1, at temperature τ_3? Such a volume does not correspond to any stable state of the substance at temperature τ_3, yet it must be a possible volume since there must be some kind of transition between the smaller volume of the liquid phase and the larger volume of the vapor. This paradox is resolved by the possibility that one part of the substance may be in the liquid phase, while simultaneously another part is in the vapor phase, so that the points representing the two coexisting phases must lie on the same isotherm if there is to be thermal and mechanical equilibrium, and must have the same distance from the abscissa axis, since the temperature and pressure must be equal for the two phases. Under the influence of gravity, the heavier liquid phase will of course collect at the bottom of the container and the vapor will stay on top.

Coexistence of the liquid and vapor phases can also occur if the volume is between OF_1 and OC_1, or between OD_1 and OE_1. If the substance first undergoes the state variation LE and then its volume decreases still further isothermally, then there are two possibilities, according to our present considerations. The further state variation can be represented by the curve ED, so that all of the substance is in the same state at every instant. However, at any moment it can happen that, while one part of the substance continues along the curve DE, another part can go from a state G over to a state J having the same temperature and pressure. Further decrease in the volume can then cause a larger amount of the substance to pass from G to J, instead of continuing along the curve GD. In fact, if there is no dust or other substance present that can initiate condensation, a vapor can be compressed without condensation, even though under other circumstances—especially in the presence of a small amount of the same substance in the liquid state—it would already have started to condense. Since this state is more often reached by cooling than by compression, it is called the undercooled vapor. When condensation finally occurs, a larger amount of it suddenly liquefies irreversibly (i.e., the liquid-vapor mixture thereby produced cannot be transformed back into undercooled vapor in such a way that it goes through the same sequence of states in reverse).

The same thing happens of course with evaporation. The substance first passes through the states on the curve MF, so that it is initially a highly compressed liquid. When the volume has increased isothermally to the state F, it can continue along the curve FJC; or, at any point, the coexistence of two phases can begin, so that from then on a part of the substance is transformed into the vapor state. On further expansion, one part of the substance will remain at the state J on the branch FC, while the other will go over to the state G, which has the same height, on the curve DE. Evaporation may be delayed when the liquid and the wall of the container are free of air. However, on further expansion or heating, a large amount will suddenly evaporate (evaporation- or boiling-delay). The latter process is likewise not reversible.

If we find ourselves on an isotherm which, like isotherm 4 in Figure 1, dips below the abscissa axis, then the pressure may even become negative. Mercury extracted from a barometer tube provides a good example of this. If one gradually draws the mercury out of a barometer tube (which is sealed at the top) then the pressure at the upper end continually decreases; the mercury itself goes through the sequence of states $MFJC$ in Figure 2. The mercury column still does not break apart, even after it has become higher than the barometer position, which shows that the point C, where the ordinate of the isotherm has its minimum, lies below the abscissa axis, like isotherm 4 in Figure 1. Finally the column breaks and the mercury quickly evaporates; of course this is barely noticeable because of the small tension of mercury vapor. However, if distilled water is placed in the barometer tube above the mercury, then the copious evolution of vapor in the water is visible to the eye.

A negative pressure can also occur in distilled water at room temperature. Hence for water the isotherm corresponding to room temperature also dips below the abscissa axis, like isotherm 4 in Figure 1. On the other hand, for ether the isotherms do not go below the abscissa axis for easily observable temperatures. If one had put ether above the mercury in the experiment described above, then he could make the mercury column so long that the pressure in the ether would be smaller than the saturation pressure of ether vapor, though not so low that it would be negative.

In processes that one usually calls boiling-delay (superheating), the substance is in contact with its own vapor. The state is not then an equilibrium state, but on the contrary violent evaporation takes place on the upper surface of the liquid; the temperature is equal to the boiling temperature corresponding to the pressure of the vapor standing over it. It is hotter in the interior, where the same state as in evaporation-delay is found, and this state can be maintained only by continuous evaporation at the surface and heat conduction through the interior.

§16. Stable coexistence of both phases.

It appears that our description of the nature of state variations is not uniquely determined. However, it is to be expected that a unique determination can be achieved if we exclude irreversible transitions, such as undercooling and delayed evaporation.

For this purpose we consider a certain sample q of our substance, whose mass shall be equal to 1. It is initially in the state L (Fig. 2), and it will be compressed isothermally. As soon as it reaches E, it will from time to time be brought into contact with liquid at the same temperature and pressure. Initially, when q is still near the state E, this liquid will be in the state of delayed evaporation. It will explosively evaporate into q. Nevertheless we shall restore the previous state of q, compress it again a little bit, and again bring it into contact with the liquid at the same temperature and pressure. We repeat this process until eventually we come to a state in which, when q is brought into contact with the liquid at the same temperature and pressure, none of the liquid evaporates into q, and none of q condenses; thus liquid and vapor are in equilibrium. At this temperature the liquid is in the state J in Figure 2, and the vapor is in the state G.

This equilibrium state cannot depend on the mixing ratio of the two phases, but only on the state at the surface of contact, for the molecules at the surface of contact are the only ones that are in equilibrium with the other phase. However, the size of the surface of contact cannot be important, since each part is subjected to similar conditions.[1] Hence any arbitrary quantity of the substance in phase J can be in equilibrium with any arbitrary quantity of the same substance in phase G, as soon as equilibrium between the two phases is possible at all.

The state G then forms the boundary between the normal and the undercooled vapor, while J forms the boundary between the normal and superheated liquid.

If one compresses q, excluding bodies that can effect condensation, then it can pass through states along the curve GD. In each such state, however, if it is brought into contact with the liquid at the same temperature and pressure, it will suddenly con-

[1] The curvature of the surface of contact will have an effect, if this is very small. Over a concave upper surface (as at the meniscus in a capillary tube) the vapor pressure is less than the hydrostatic pressure of the vapor column that lies between the level of the meniscus and that of the plane liquid-surface outside the capillary tube. Over a surface that is convex to the same degree, it would be larger by the same amount. (Cf. end of §23.)

dense in an irreversible way. But if it is compressed while in contact with the liquid, starting from the state G, then it will condense more and more until it is completely transformed into the liquid state, and this process is reversible, since the substance can be evaporated again in the same way, if one increases the volume at constant temperature.

The positions of the points G and J, which form the boundary between the normal and the superheated or undercooled states for a given isotherm, were found by Maxwell in the following way, by introducing a hypothesis.[2] As is well known, for any reversible cyclic process one has $\int dQ/T=0$, where dQ is the added heat and T the absolute temperature. The heat is to be measured in mechanical units. Maxwell assumes that this equation still remains valid even when unstable states intervene between the initial and final states, for example those states represented by the branch CHD in Figure 2.

If the cyclic process takes place at constant temperature, then the factor $1/T$ can be taken outside the integral sign, and one is left with $\int dQ=0$. dQ is equal to the excess dJ of the internal energy of the substance over the externally performed work. The latter is equal to pdv when, as assumed here, the external force is just the normal pressure force whose intensity on unit surface is equal to p for all surface elements.

Since $\int dJ$ vanishes for any cyclic process, one has therefore $\int pdv=0$, for which one can also write $\int \pi d\omega=0$, since the choice of units is completely arbitrary. We now consider unit mass of the substance. It undergoes the following cyclic process at constant temperature. Initially all its parts are in the phase J, then it is transformed bit by bit into phase G. The pressure remains constant and equal to JJ_1, but the volume increases from OJ_1 to OG_1. The external work thereby performed, $\int \pi d\omega$, is equal to the product of the pressure and the volume increase, hence it is equal to the area of the rectangle $JJ_1G_1G=R$. Now we imagine that we return to the original state by following the curve $GDHCJ$. The volume will decrease, so that external work is done on the substance. This work is equal to the negative of the integral $\int \pi d\omega$ taken along the entire state variation. Since ω is the abscissa and

[2] Maxwell, Nature **11**, 357, 374 (1875); *Scientific Papers* **2**, 424. See also Clausius, *Die Kinetische Theorie der Gase* (Braunschweig, 1889–1891), p. 201.

π the ordinate of the curve, the integral is equal to the area $J_1JCHDGG_1J_1 = \Phi$, which is bounded above by the curve $JCHDG$, below by the abscissa axis, and on the right and left by the two ordinates JJ_1 and GG_1. At the end of the process the substance returns to its original state K; $\int \pi d\omega$ extended over the entire state variation is therefore equal to the difference $R-\Phi$, which is equal to the difference $JCH-HDG$ of the two shaded areas in Figure 2. If one makes the Maxwell hypothesis that the second law must be valid even though part of the imaginary path goes through unstable states, then he finds the following result: the line GHJ, which connects the two phases G and J, which are in contact at equilibrium (the two-phase line) must be drawn in such a way that the two shaded areas in Figure 2 turn out to be equal. If this condition is not satisfied, then the two phases G and J cannot be in equilibrium, even though they lie on the same isotherm and are at the same distance from the abscissa axis. (Concerning the equation that expresses this condition, cf. §60.)

§17. Geometric representation of the states in which two phases coexist.

If in the future we always mean by GHJ a line parallel to the abscissa axis, for which the two shaded areas are equal, then the results concerning the behavior of a substance under isothermal compression at a temperature τ_3 can be expressed as follows. As long as the volume is greater than OE_1, it is in the vapor state. If the volume is between OE_1 and OG_1, the liquid phase still cannot coexist with the vapor phase. Condensation can occur only when a salt, or rather a body whose particles attract the particles of the substance more strongly than they attract each other, is present. The liquid thereby formed will dissolve the salt or cover the body, and, if an infinite amount is not present, the vapor pressure drops as this process continues (premature condensation). If no such body is present, the substance will remain gaseous until its volume becomes equal to OG_1. Here, if it is brought into contact with the least amount of the same substance in the liquid state, further isothermal compression causes condensation, and the vapor pressure does not increase until all of the substance has liquefied, since vapor of a higher pressure cannot exist over the liquid phase (normal condensation). If no body is present to facilitate normal condensation, then the substance can be still further compressed without condensation, so that its states are represented by the curved line GD (undercooled vapor). However, if condensation sets in—which must happen in any case if the volume becomes less than OD_1—then a finite amount of the substance suddenly liquefies, and the pressure drops to the value GG_1 if the temperature is held constant. The substance behaves in a similar way if it is initially liquid and is gradually expanded; instead of bringing a small amount of liquid in contact with the vapor, one now creates in the liquid an empty or vapor-filled cavity.

We must omit from each isotherm the section CHD, since it corresponds to states that cannot be physically realized. Moreover, we shall consider neither delayed evaporation (superheating) nor undercooled or prematurely condensed vapor, but only normal condensation, which represents the directly reversible transition from the liquid to the vapor state. We then have to retain only the parts MJ and GL for each isotherm, in Figure 2. In the intermediate region one part of the substance will be liquid, in a state represented by the point J, while the other will be gaseous, in the state G, both parts being at the same temperature and pressure. Each such intermediate state can be represented by the two points G and J simultaneously, each point having a weight corresponding to the fraction of the substance in that state. It is preferable to represent these states by different points on the line JG (the two-phase line). The ordinate NN_1 (Fig. 2) of an arbitrary point N on this line represents the pressure, which is the same for both coexisting phases. The abscissa ON_1 should be chosen so that it is equal to the total volume of the substance, i.e., the sum of the liquid and gaseous parts. The larger the liquid fraction, the closer to J will lie the point representing the state, and conversely. If we denote by x the mass of the liquid, and by $1-x$ the mass of the vapor, in the state N, then x has the following property: since OJ_1 is the specific volume of the liquid, and OG_1 that of the vapor, then $x \cdot OJ_1$ is the volume of the liquid and $(1-x) \cdot OG_1$ that of the gaseous part of the state represented by N. Since the sum of the volumes is equal to the abscissa ON_1, one has the equation

$$x \cdot OJ_1 + (1 - x) \cdot OG_1 = ON_1,$$

from which it follows that:

$$(37)\quad \begin{cases} x = \dfrac{N_1G_1}{J_1G_1} = \dfrac{NG}{JG}, \quad 1 - x = \dfrac{J_1N_1}{J_1G_1} = \dfrac{JN}{JG}, \\ \dfrac{x}{1-x} = \dfrac{N_1G_1}{J_1N_1} = \dfrac{NG}{JN}. \end{cases}$$

If one imagines that the mass x of the liquid is concentrated at the point J, and that of the vapor, $1-x$, at the point G, then N is the center of gravity of the system formed from the two masses. The rule that the reciprocal of the abscissa always represents the density does not of course hold for points on the two-phase line. On the contrary, if ρ_1 is the density of the liquid, and ρ_2 that of the gas, then the abscissa is

$$ON_1 = \frac{x}{\rho_1} + \frac{1-x}{\rho_2}.$$

If we represent the states where liquid and gas coexist in this way, and ignore the undercooled vapor as well as the superheated liquid, then the isotherms will have the form shown in Figure 3 instead of that in Figure 1. The isotherm labeled 3 in Figures 1 and 2 is likewise marked as 3 in Figure 3. The part JG is a

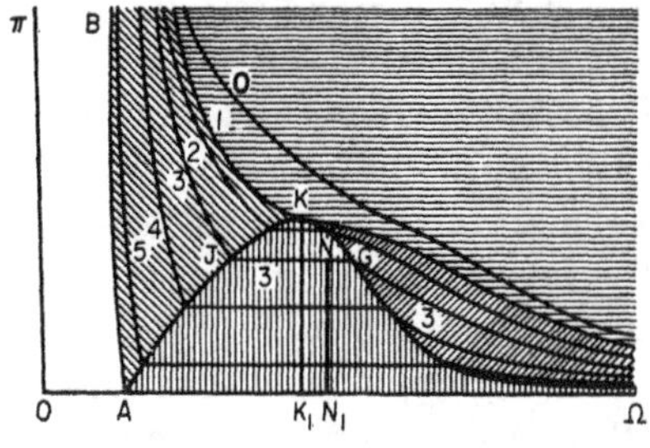

FIG. 3.

straight line, and a point N on this line represents a state in which a part x of the substance is liquid, while the remainder $1-x$ is gaseous, and the two parts are both at a pressure NN_1, while the sum of their volumes is ON_1. x and $1-x$ are then given by Equations (37). Also, for those isotherms for which the part representing delayed evaporation sinks below the abscissa axis, there is a line JG, always lying above the abscissa axis, for which the two shaded areas in Figure 2 are equal; for the area between the abscissa axis and the part of the isotherm that goes below the axis is always finite, while the area between the part of the isotherm corresponding to larger abscissas and the abscissa axis becomes logarithmically infinite as the abscissa goes to infinity. Hence equating the two shaded areas in Figure 2 always leads to a two-phase line lying above the abscissa axis.

§18. Definition of the concepts gas, vapor, and liquid.

We may denote the region lying above the critical isotherm 1, shaded horizontally in Figure 3, as the gas region. The points in this region that have large abscissas will in fact represent states close to the ideal gas state. The smaller abscissas, corresponding to points lying near the line AB, represent of course states in which the substance behaves like a liquid, but since these states can be transformed isothermally without discontinuity into states that are undoubtedly gaseous, we shall count them as belonging to the gas region. An example is provided by compressed air at ordinary temperatures.

The region filled by two-phase lines will be called the two-phase region; it is shaded vertically in Figure 3. Since the dotted curve of Figure 1, which is the geometrical locus of all maxima and minima of the isotherms, lies entirely within the two phase region in any case, the curve bounding this region has a smaller curvature than the dotted curve at the critical point, and in any case it has no cusp.

Beneath the gas region and to the right of the two-phase region lies the vapor region, in which the substance behaves like a gas; on the left of the two-phase region is the liquid region (bounded by the line AB) in which we call the substance a "condensable fluid." These two last-named regions are shaded with slanted lines in Figure 3. They are characterized by the property that no state of one can be transformed isothermally into a state of the other without condensation.

A typical isotherm going through these regions is curve 3 of Figure 3. If one starts from the low-density region, he can make the following statements about the properties of the substance under isothermal compression: in the vapor region, as long as the volume is large, Boyle's law is obeyed approximately. If one decreases the volume, deviations from this law become more and more appreciable. As soon as one reaches the two-phase region, the pressure remains constant on further compression, and an increasing proportion of the substance becomes liquid. After all the substance has been liquefied, the pressure increases rapidly on further compression.

Two extreme cases are illustrated by isotherms 2 and 5. The first lies near the critical isotherm. Here the liquid phase differs but little from the vapor. Condensation lasts only a short time, so that it has only the character of a temporary irregularity in the compressibility. Isotherm 5, on the other hand, corresponds to a temperature far below the critical; in general the vapor exerts only a vanishingly small pressure, and indeed it is hardly noticeable at all. As soon as the pressure reaches an appreciable value, and no other substance is mixed with the one considered, then it can exist only as a liquid, which has a definite volume that can be altered only slightly by pressure. The compressibility of the liquid at temperatures far below the critical temperature is therefore very small. Of course, according to our diagram, the vapor pressure can only approach asymptotically to zero. A trace of vapor must always be present even at the lowest temperatures.

§19. Arbitrariness of the definitions of the preceding section.

We have based our definitions of the concepts gas, vapor, and liquid on isothermal state variations. This is of course an arbitrary procedure, which can at best be justified by saying that in practice one usually tries to keep the temperature equal to that of the surroundings, and thus as nearly constant as possible. We could also consider state variations such that the substance is in a cylindrical container closed by an airtight, easily movable piston, subject to a constant pressure. Suppose the temperature is initially very high. As heat is lost, the volume decreases. We call such a state variation an isobaric one; it is represented by a line parallel to the abscissa axis (the isobar). If the constant pressure acting on the substance is greater than the critical pressure, then it goes over from an approximately gaseous state to an approximately liquid state, without discontinuity. If, on the other hand, the pressure is smaller than the critical pressure, then the temperature drops with increasing volume only until the two-phase region is reached. Since the isobar then coincides with the isotherm, the temperature remains constant until all of the substance is liquefied. If one wanted to base the definition of the concepts "vapor" and "gas" on isobaric compression, then the separating line between the two states would be a line parallel to the abscissa axis passing through the critical point, since above this line, isobaric transformation of any state into any other state never involves any condensation, whereas below it, isobaric compression always goes through the two-phase region.

One would obtain yet another distinction between vapor and gas if he used adiabatic variations, i.e., variations not involving addition or loss of heat. In order to compute these, one must set the differential of the added heat, dQ, equal to zero (cf. §21). We shall not pursue this further, and we remark only that if one does not adopt some definite criterion—isothermal, isobaric, or adiabatic—for state variations, then a distinction between states that can or cannot be continuously transformed into other states is not in general possible. For one can go from any state into any other state without passing through the two-phase region, and thus without condensation or evaporation. One can see this in the following way: let points in the two-phase region have the significance that their ordinates are equal to the pressure, and their abscissas are equal to the total volume of the liquid and vapor states combined, so that any point of the quadrant $BA\Omega$ in Figure 3 represents a possible state of the substance. Let two arbitrary states of the substance be given. In each of these states, let the entire substance be in the same phase, so that the two points, called P and Q, are both outside the two-phase region. We can always connect them with a curve that goes above K from P to Q without going through the two-phase region. This curve represents a continuous sequence of states by which the substance can be transformed from the state P to the state Q without any part of the substance ever being in a phase different from the rest. However, we can also draw a curve from P toward the left boundary line AJK of the two-phase region, then across this region from left to right, and finally above it to Q. This curve would represent a state variation in which the substance starts out in the liquid phase, gradually evaporates, and then when it has completely vaporized, passes continuously to the final state Q. Conversely, a curve might go from P, above the two-phase region to a point lying on the right-hand boundary curve $KG\Omega$ of the two-phase region, then across this region to a point on the curve AJK, and then above the two-phase region to Q. Such a curve would represent a state variation in which the substance is transformed from the state P continuously to a vapor state, then condenses, and finally ends up in the state Q.

A substance can even be transformed from a liquid state to a vapor state, not by evaporation, but rather by condensation. One merely needs to start with a small volume, heat the substance above the critical temperature, then expand it to a large volume, cool it below the critical temperature, then condense it, then again heat the liquid above the critical temperature, and then transform it to the desired final state. Likewise one can transform a vapor to a liquid by evaporation.

Obviously one would call a substance a liquid, vapor, or gas, when its state is represented by a point near the lower part of the line AB, or near the curve $KG\Omega$, or above the critical isotherm far from AB, respectively. In the intermediate region, however, these states can gradually transform into each other, so that if one desired a sharp boundary he would have to establish it by some arbitrary definition.

§20. Isopycnic changes of state.

If we enclose a fixed amount of a substance (say unit mass) in a tube closed on both sides, and gradually warm it, then we produce very nearly a change of state at constant volume (isopycnic change of state). If the volume is exactly equal to one-third of the critical volume (in general it cannot be any smaller than that) then the pressure is always infinite, except at absolute zero temperature. At any other volume, we would always find ourselves in the two-phase region at sufficiently low temperatures. Hence some of the substance will be found as a liquid in the lower part of the tube. Above it stands the vapor of the same substance, whose pressure is nearly zero at very small temperatures, and increases with temperature. The boundary between liquid and vapor is called the meniscus. Since we assume that the volume is constant, the change of state must be represented by a line parallel to the ordinate axis, passing through the two-phase region—for example, the line N_1N in Figure 3. In the state represented by the point N, the mass of the liquid part of the substance is, according to Equation (37),

$$x = \frac{NG}{JG}$$

and that of the vapor part is

$$1 - x = \frac{JN}{JG} \cdot$$

We now have to distinguish three cases: 1. The line N_1N representing our state variation lies to the right of the line KK_1, which is parallel to the ordinate axis and goes through the critical point. The chosen constant volume is then larger than the critical volume. Then, as the temperature increases, NG becomes smaller and smaller relative to JG. The amount of liquid in the tube decreases with increasing temperature, and the meniscus drops. Finally, when the line N_1N has reached the boundary of the two-phase region, the entire substance becomes vapor. 2. The line NN_1 lies to the left of KK_1. Then JN will instead decrease relative to NG, the meniscus will rise, and the entire substance becomes liquid at the moment when the boundary of the two-phase region is reached. 3. When the substance has precisely the critical volume OK_1, the ratio of JN to NG always remains finite as the substance is heated at constant volume, until the critical point is reached. The meniscus always stays between the upper and lower ends of the tube, until finally it vanishes at the critical temperature, since the liquid and vapor both have the same properties then.

We saw that the boundary of the two-phase region becomes nearly horizontal in the neighborhood of the critical point. Therefore the meniscus almost vanishes if NN_1 lies in the neighborhood of KK_1. Theoretically it must remain in the interior of the tube until the temperature is almost equal to the critical temperature, and then it moves very quickly to the upper or lower end of the tube.* This cannot be observed, since it has previously

* See van der Waals' remark on this point, mentioned at the end of §61.

already become so indistinct that one can no longer see it. Moreover, small impurities in the substance cause significant perturbations at the critical point.

§21. Calorimetry of a substance following van der Waals' law.

Since we have adopted a definite mechanical model, there is no difficulty in determining an expression for the differential of the added heat, dQ. As in Equation (19) and in Part I, §8, let $\overline{c^2}$ be the mean square velocity of the center of mass of a molecule (progressive motion), so that $\frac{1}{2}\overline{c^2}$ is the mean kinetic energy of the motion of the center of mass of the molecules found in unit mass of the substance. If the temperature of unit mass is raised by dT, then the part of the added heat used to increase this kinetic energy is (in mechanical units)

$$dQ_2 = \tfrac{1}{2}d(\overline{c^2}) = \tfrac{3}{2}rdT.$$

The latter relation follows from Equation (21).

Although in the derivation of van der Waals' law we assumed that the molecules behave almost like elastic spheres in collisions, in general we shall not exclude intramolecular motion; we set (as in Part I, §8) the work done against intermolecular forces by the applied heat equal to

$$dQ_3 = \beta dQ_2. \tag{37a}$$

Since the van der Waals cohesion force acts on each molecule almost equally in all directions, it will not influence the internal motion. The same results that we shall derive in §§42–44 for an ideal gas of compound molecules will therefore be valid for this internal motion. It does not depend on how often the molecules collide with each other, but only on the temperature, so that β can only be a function of temperature. If the molecules are rigid solids of revolution, then, as we shall see, $\beta = \frac{2}{3}$; but if they are rigid solids of some other shape, then β is equal to 1. If intramolecular motions are present, and if f is the number of degrees of freedom of a molecule, then the corresponding fraction of the average kinetic energy will be $\frac{1}{3}f - 1$. One may also add a contribution for the work done by intramolecular forces, which can be a function of temperature.

At present we shall not go into these details, but simply consider β to be some function of temperature in Equation (37a).

The specific heat of unit mass at constant volume is therefore:

$$\gamma_v = \frac{dQ_2 + dQ_3}{dT} = \frac{3}{2} r(1 + \beta).$$

If the volume simultaneously increases by dv, then the work done to overcome the external pressure is pdv, while that done to overcome the internal molecular pressure is adv/v^2. Hence the total heat added to unit mass is

$$(38)\quad dQ = \frac{3r(1+\beta)}{2} dT + \left(p + \frac{a}{v^2}\right) dv = \frac{3r(1+\beta)}{2} dT + \frac{rT}{v-b} dv.$$

Since van der Waals' kinetic hypothesis not only determines the equation of state but also permits one to make statements about the specific heat, it is clear how one can determine the specific heat by using a substance that satisfies the conditions of the hypothesis, but not for a substance that merely satisfies empirically the van der Waals equation of state. The entropy is:

$$S = \int \frac{dQ}{T} = r\left[l(v - b) + \frac{3}{2}\int (1 + \beta) \frac{dT}{T}\right],$$

which, if β is constant, reduces to:

$$rl[(v - b)T^{3(1+\beta)/2}] + \text{const.}$$

where l means the natural logarithm. Setting this quantity constant yields the equation for adiabatic changes of state. The specific heat of unit mass at constant pressure is

$$\gamma_p = \frac{3r(1 + \beta)}{2} + \frac{r}{1 - \dfrac{2a(v - b)^2}{rTv^3}} = \frac{3r(1 + \beta)}{2} + \frac{r}{1 - \dfrac{2a(v - b)}{v(pv^2 + a)}}$$

and the ratio of specific heats is

$$\kappa = 1 + \frac{2}{3(1 + \beta)\left[1 - \dfrac{2a(v - b)^2}{rTv^3}\right]}$$

If the gas suddenly expands into a vacuum (as in Gay-Lussac's experiment) so that the specific volume and density have initially the values v and ρ, and afterwards the values v' and ρ', then the work done against molecular attraction, per unit mass of the gas,

$$\int \frac{adv}{v^2} = \frac{a}{v} - \frac{a}{v'} = a(\rho - \rho'),$$

is therefore independent of temperature.

However, if the gas expands adiabatically in a reversible manner, then it follows from Equation (38) that

$$\left(\frac{dT}{dv}\right)_S = -\frac{2T}{3(1+\beta)(v-b)} = -\frac{T}{\gamma_v\left(\frac{dT}{dp}\right)_v}.$$

Let unit mass of the substance be originally in the liquid state, and then evaporate at constant temperature T, at the saturation pressure p of the vapor, corresponding to this temperature. Then one has to set $dT=0$ in Equation (38). The total heat of vaporization is therefore:

$$\int dQ = \frac{a}{v} - \frac{a}{v'} + \int pdv = a(\rho - \rho') + p(v' - v).$$

where v and ρ are the specific volume and density of the liquid, and v' and ρ are the same quantities for the vapor at the same temperature.

The last term of this equation represents the work needed to overcome the external pressure on the vapor. If one neglects the density of the vapor compared to that of the liquid, then:

$$\mathfrak{T} = a\rho$$

is the work of separation of the liquid particles.

One can now calculate the constant a from the deviations of the dilute heated vapor from the Boyle-Charles law. From this, one obtains for the liquid state of the same substance the co-called internal or molecular pressure (i.e., the difference between the pressure in the interior of the liquid near the surface, and that on the outside of the surface): $a\rho^2$. The heat of vaporization of the liquid, measured in mechanical units (more precisely, the heat of separation of its particles) is $a\rho$, a quantity that can be compared with experiment.

The latter result is independent of the assumption from which we have obtained van der Waals' equation, and depends only on the form of that equation, so that it would remain correct if one one were simply given this equation as an empirical law.

§22. Size of the molecule.

By calculating the constant b from the deviations of a gas from the Boyle-Charles law, one can improve Loschmidt's determination of the size of a molecule (see §12 of Part I). We can now calculate the amount of space actually occupied by the molecules in unit mass, since this is equal to $1/4b$, whereas in Part I we had to use the assumption that the volume of the molecules in the liquid state cannot be smaller than the total volume of these molecules themselves, and should not be more than ten times that volume.

From Equations (77) and (91) of Part I, one obtains for the viscosity coefficient of a gas the value $\mathfrak{R} = k\rho c/\pi n\sigma^2\sqrt{2}$, where, according to Equation (89) of Part I, k can differ only slightly from $\frac{1}{3}$, if one understands by c the mean velocity of progressive motion of the gas molecule.* σ is the diameter of a molecule, ρ is the mass in unit volume, n is the number of molecules in unit volume, so that $\rho/n = m$ is the mass of a molecule. One can therefore write also

$$\mathfrak{R} = \frac{kmc}{\pi\sigma^2\sqrt{2}};$$

further, we have:

$$b = \frac{2\pi\sigma^2}{3m},$$

hence

$$\sigma = \frac{3\mathfrak{R}b}{\sqrt{2}kc},$$

where the mean velocity c can be calculated with sufficient accuracy from Equations (7) and (46) of Part I.

I will abstain from giving numerical values, since the choice of the most reliable numbers would not be possible without an analysis of the experimental data, and this would be completely out of place in the present book.

---

* According to the Chapman-Enskog theory, the correct value of k in this formula is 0.499.

§23. Relations to capillarity.

Van der Waals obtained the term a/v^2 in his formula by a route somewhat longer than the one we used in §7; he employed the arguments by which Laplace and Poisson had derived the basic equations of capillarity.* Since this connection with capillarity is of some importance, I will summarize here the original derivation of van der Waals.

The following considerations are applicable to both liquids and gases, but more especially to the former, wherefore we shall call the substance under consideration "the fluid" for short. We assume that the attraction of two mass particles m and m' acts in the direction of their line of centers, and is a function of their distance f, which we call $mm'F(f)$. We set:

$$\int_f^\infty F(f)df = \chi(f), \quad \int_f^\infty f\chi(f)df = \psi(f),$$

so that $mm'\chi(f)$ is the work required to bring the two particles m and m' from the distance f to a much larger distance. In any case $F(f)$ will have a value different from zero at molecular distances. We assume that it decreases more rapidly than the inverse third power of f as f increases, so that not only $F(f)$ but also $\chi(f)$ and $\psi(f)$ always vanish when f does not have a very small value. From our hypothesis it follows that the force acting between m and m' is equal to $-mm'd\chi(f)/df$.

We now construct a sphere K of radius b in the fluid,[1] denote by do a surface element of K, and further construct the right cylinder Z, situated outside this sphere on the surface element do, which has a very great length $B-b$. The center O of the sphere is chosen as origin of coördinates, and we place the positive abscissa axis along the axis of the cylinder.

We now pose the problem of finding the attraction dA exerted by the fluid in K on the fluid contained in the cylinder Z.

For this purpose, we construct in Z the volume element dZ, which lies between the cross sections with abscissas x and $x+dx$.

---

* Laplace, *Traité de mécanique célèste* (Paris: J. B. M. Duprat, 1798–1825), Supplément au Xe Livre. Poisson, *Nouvelle Théorie de l'action capillaire* (Paris: Bachelier, 1831).

[1] This is not to be confused with the constant earlier denoted by b.

Its volume is $dodx$, hence it contains mass $\rho dodx$ of fluid (the density ρ is assumed to be constant everywhere).

We then cut out from the sphere K a concentric spherical shell S, which lies between spherical surfaces of radii u and $u+du$, and from this shell we cut out a ring R, for which the line connecting it with the origin of coördinates makes an angle between ϑ and $\vartheta+d\vartheta$ with the positive abscissa axis. The volume of R is $2\pi u^2 \sin \vartheta du d\vartheta$. When multiplied by ρ, this is the mass contained in the ring.

Any fluid particle of mass m' that lies in R and has abscissa x' exerts on any other fluid particle m that lies in the cylinder Z and has abscissa x the attraction

$$-mm' \frac{d\chi(f)}{df}$$

whose component in the direction of the negative abscissa axis is

$$-mm' \frac{d\chi(f)}{df} \frac{x-x'}{f} = -mm' \frac{d\chi(f)}{dx}$$

The total fluid mass contained in the ring R therefore exerts an attraction

$$-2\pi\rho^2 dodxu^2 \sin \vartheta du d\vartheta \frac{d\chi(f)}{dx}$$

on the mass in dZ. The total attraction of the fluid in the sphere K on the fluid in the cylinder Z can be found by choosing the following order of integration, which is the most convenient one for this calculation:

$$dA = -2\pi\rho^2 do \int_0^b udu \int_b^B dx \frac{d}{dx} \int_0^\pi \chi(f) u \sin \vartheta d\vartheta.$$

In carrying out the integration over ϑ we consider just the spherical shell S, so that we can consider u constant. Substituting f for u as the variable of integration, we therefore obtain

$$ux \sin \vartheta d\vartheta = fdf$$

and the limits for f will be $x-u$ and $x+u$. Hence:

$$\int_0^\pi \chi(f) u \sin \vartheta d\vartheta = \frac{1}{x} [\psi(x-u) - \psi(x+u)],$$

where ψ is the function defined at the beginning of this section. The integration over x will now be performed by setting $x=B$ in this expression and subtracting from this the value obtained by setting $x=b$ in the same expression.

It is to be recalled that the function ψ always vanishes if the argument of the function is not very small. B and b are very large compared to molecular dimensions, and therefore also larger than all values of u that come into consideration. Hence $\psi(B+u)$, $\psi(B-u)$ and $\psi(b+u)$ vanish. Only $\psi(b-u)$ can take a value different from zero, and one obtains:

$$\int_b^B dx \frac{d}{dx} \int_0^\pi \chi(f) u \sin\vartheta d\vartheta = -\frac{1}{b}\psi(b-u),$$

hence

$$dA = \frac{2\pi\rho^2 do}{b} \int_0^b u du \psi(b-u).$$

If one introduces the variable $z=b-u$ in the definite integral, then it is transformed to

$$b\int_0^b \psi(z)dz - \int_0^b z\psi(z)dz.$$

Since $\psi(z)$ vanishes for large values of z, one can write ∞ instead of b for the upper limit of the integral. If one sets

$$(39) \qquad a = 2\pi \int_0^\infty \psi(z)dz, \quad \alpha = \pi \int_0^\infty z\psi(z)dz,$$

then

$$\frac{dA}{do} = a\rho^2 - \frac{2\alpha\rho^2}{b}.$$

If the cylinder Z is surrounded by fluid on all sides, then clearly all the forces acting on it are neutralized. Hence the fluid surrounding the sphere K must exert a force on the fluid in the cylinder exactly equal and opposite to that which the fluid in K exerts. Now if the fluid in K is taken away, then there remains only a fluid mass whose outer surface is the spherical surface K. dA is then the traction exerted by this fluid mass on the cylinder Z placed over one of its surface elements do, in a direction toward the inside, which was denoted in §2 by $p_i do$.

If the surface is plane, or if its radius of curvature is so large that $1/b$ can be neglected, we obtain the value already found in §7: $p_i=a\rho^2$. However, according to Equation (39) the constant a can be expressed in terms of the law of attraction of the molecules.

The term $-2\alpha\rho^2/b$ shows that this expression requires a small correction if the surface is curved. This correction gives rise to capillary phenomena, as is well known; when the surface is not spherical, it takes the form

$$- \alpha\rho^2\left(\frac{1}{\mathfrak{R}_1} + \frac{1}{\mathfrak{R}_2}\right), \tag{40}$$

where $\mathfrak{R}_1$ and $\mathfrak{R}_2$ are the two principal radii of curvature of the surface.

§24. Work of separation of the molecules.

We shall now express the heat of vaporization in terms of the function χ. We construct around a fluid particle of mass m a spherical shell S, which is bounded by two spherical surfaces of radii f and $f+df$, in which therefore the amount of fluid mass is $4\pi\rho f^2 df$. Since $mm'\chi(f)$ is the work that must be done in order to bring the fluid particles m and m' to infinite separation, if they are initially at a distance f, then the work required to bring m from the midpoint of the spherical shell S to a great distance is:

$$4\pi\rho m f^2 \chi(f) df.$$

The total work needed in order to bring m from the interior of the fluid to a point far away from all other fluid particles is therefore:

$$B = 4\pi\rho m \int_0^\infty f^2 \chi(f) df.$$

If unit mass of the fluid contains n particles, we have $mn=1$. If one assumes that in the vapor each particle is already far away from the sphere of action of all the others, then the work done in evaporating unit mass of the fluid, to overcome the cohesive force, is

$$\mathfrak{T} = \frac{nB}{2} = 2\pi\rho \int_0^\infty f^2 \chi(f) df.$$

To this we must of course add the work $\int p dv$ done against the external pressure in evaporation.

$\mathfrak{T}$ is therefore only half of nB, since in the expression for nB the work of separation of each particle from each other one is counted twice. In §21 we found the value $a\rho$ for the work of separation $\mathfrak{T}$, where a is given by the first of Equations (39). Partial integration of the right hand side of this equation gives in fact:

$$\int_0^\infty \psi(z)dz = -\int_0^\infty z\frac{d(\psi)z}{dz}dz = \int_0^\infty z^2\chi(z)dz.$$

The value of $\mathfrak{T}$ obtained previously therefore agrees with the one found here.

One would obtain the work of separation directly by integration of the first of Equations (39), if he first calculated the work of separation of a particle at a distance h from another one in a plane layer of thickness dh in the fluid. This would be:

$$(41)\qquad 2\pi\rho m dh\int_0^\infty rdr\chi(\sqrt{h^2+r^2}) = 2\pi\rho m dh\int_h^\infty f\chi(f)df.$$

When multiplied by n and integrated over h from zero to infinity, this gives the total work of separation $\mathfrak{T}$.

By means of a similar formula we can solve the following problem. Let there be given a cylinder of cross-section 1; we draw through it anywhere a cross-section AB and ask for the work necessary to separate the fluid on one side from the fluid on the other side.

We first calculate the work of separation of a layer of thickness dx at a distance x from the bottom of the container, lying below AB, from another layer of thickness dh above AB. We set $m=\rho dx$ in Equation (41). Then the work of separation is equal to

$$2\pi\rho^2 dh dx\int_h^\infty f\chi(f)df.$$

Here h is the distance between the two layers. We keep this fixed for the moment, and integrate over all allowed values of x. If c is the distance between AB and the bottom, then when h is constant we have to integrate from $x=c-h$ to $x=c$, which gives:

$$2\pi\rho^2 dh\int_h^\infty f\chi(f)df\int_{c-h}^{c} dx = 2\pi\rho^2 h dh\int_h^\infty f\chi(f)df.$$

If one integrates h over all possible values from zero to ∞, he obtains for the total work of separation of the fluid above AB from that below AB the value:

$$2\pi\rho^2\int_0^\infty hdh\int_h^\infty f\chi(f)df = 2a\rho^2.$$

Since in this process of separation the surface area of the fluid increases by 2 units, the work required to increase the surface area of the fluid by 1 is just half of this, and is therefore equal to $\alpha\rho^2$.[1] However, this quantity is at the same time the coefficient of

$$\frac{1}{\Re_1}+\frac{1}{\Re_2}$$

in the basic equation of capillarity, (40). In fact it is well known that this coefficient represents the work required to increase the fluid surface by 1 unit.

These equations would be considerably more complicated if one introduced the improvement that Poisson made in the old Laplace theory of capillarity, by taking account of the variation of density during the transition from the interior to the surface of the fluid. Yet the form of the equations obtained remains the same; only the expression for the constant in terms of a definite integral is different. Since capillarity theory is only of incidental interest here, I will not go into this further, and merely refer to Stefan's treatment.[2]

[1] α is the quantity given by the second of Eqs. (39).

[2] Stefan, Wien. Ber. **94**, 4 (1886); Ann. Phys. [3] **29**, 655 (1886).

CHAPTER III

Principles of general mechanics needed for gas theory.

§25. Conception of the molecule as a mechanical system characterized by generalized coördinates.

In all calculations up to now (except for the specific heat) we have treated the gas molecules as completely elastic spheres or as centers of force, without any internal structure. Many circumstances show that this assumption cannot exactly conform to reality.

All gases can be made luminous, and their light then sometimes provides a wonderfully complicated spectrum. This would be impossible for simple material points; moreover, the vibrations of elastic spheres could scarcely produce the observed spectral phenomena, even if one took account in the calculations of the internal motions of the elastic substance, which we have so far ignored.

Furthermore, the facts of chemistry compel the assumption that, in chemically compound gases, the molecules consist of many heterogeneous parts. One can show that even the molecules of the chemically most simple gases must consist of at least two separate parts. For example, if Cl and H are bound into ClH, then the ClH gas occupies at equal temperature and pressure exactly the same space that the Cl and H gases did together. Since, according to Avogadro's law (Part I, §7), there are the same number of molecules in all gases at equal temperatures and equal pressures, a molecule of chlorine and a molecule of hydrogen must combine to form two molecules of ClH; hence both the chlorine and the hydrogen molecules must have been composed of two parts. One half of the chlorine molecule combines with one half of the hydrogen molecule to form a ClH molecule; the other two halves of the two original molecules combine to form another ClH molecule.

In order to take into account this undubitable composite structure of the gas molecule, one will have to consider it as an aggregate of a definite number of material points, held together by central forces. One does not obtain very good agreement with experiment in this way; on the contrary, for many gases the thermal phenomena, at least, are better interpreted by assuming that the molecules are rigid nonspherical bodies. Thus it appears that the connection of the parts of these molecules is so intimate that they behave like rigid solids with respect to thermal phenomena, even though in other cases the constituents appear to vibrate against each other.

In view of this circumstance it would be best to make our assumptions about the properties of molecules so general that all these possibilities can be included as special cases. We shall therefore obtain a mechanical model that will have the greatest possible capacity to explain new experimental results.

We shall consider the molecule as a system of whose nature we know no more than that its changes of configuration are determined by the general mechanical equations of Lagrange and Hamilton. It is then a question of studying those properties of a mechanical system that will later be needed in the most general way.

Let the state of an arbitrary mechanical system by given. The positions of all its parts will be uniquely determined by μ independently variable quantities $p_1, p_2, \cdots, p_\mu$, which we call the generalized coördinates. Since the geometric nature of the system, and the masses of all its parts, are given, we also know the kinetic energy L of the system as a function of the velocity-changes of its coördinates. This is a homogeneous quadratic function of the derivatives $p_1', p_2', \cdots, p_\mu'$ of the coördinates with respect to the time, whose coefficients can be any functions of the coördinates. The partial derivatives of the function L with respect to p' are the momenta q, so that one therefore has for each value of i:

$$q_i = \frac{\partial L(p, p')}{\partial p_i'} .$$

The q are therefore linear functions of the p', and the coefficients in these functions can again be functions of the p. Conversely, one can express the p' as functions of the q. If one substitutes the appropriate values into $L(p, p')$, then he obtains L as a function of the p and q. This function $L(p, q)$ is therefore also determined by the geometric nature of the system.*

The forces acting on the various parts of the system should likewise be exactly specified. They should be derived from a potential function V, which is a function only of the p, and whose negative partial derivatives with respect to the coördinates give the force, so that for each arbitrary displacement of the system, the increase dV of this function represents the work done by the system. If the kinetic energy of the system increases by dL at the same time, then according to the conservation of energy, we have $dV+dL=0$.

Not only the geometric nature of the system in question but also the forces acting on it are given. The equations of motion of the system are thereby determined. If one wishes to calculate the actual values of all the coördinates and momenta at some time t, then the initial state of the system must also be given. One might be given the values of the coördinates and their time-derivatives at the initial time (time zero). However, one might equally well be given the values of the coördinates and momenta at time zero, since the momenta are given as functions of the p'. We shall denote the values of the coördinates and momenta at time zero by $P_1, P_2, \cdots, P_\mu, Q_1, Q_2, \cdots, Q_\mu$. The values $p_1, p_2, \cdots, p_\mu, q_1, q_2, \cdots, q_\mu$ of the coördinates and momenta at time t are to be considered as given functions of these values and of the elapsed time t.

Since L and V are given as functions of the p and q, we can calculate L and V for each instant of time as functions of P, Q, and t. When we put these into the integral

$$W = \int_0^t (L - V)dt$$

then it is also a function of the initial values P, Q, and of the elapsed time t, since the entire motion is determined by P and Q, and the integral can be calculated as soon as we are given the limits of integration.

We saw just now that the 2μ quantities p, q are given as functions of P, Q, and t—i.e., there are 2μ equations between

* The reader accustomed to modern notation should note that Boltzmann uses p for position and q for momentum, not the other way around.

$4\mu+1$ quantities p, q, P, Q, and t. From these 2μ equations we have to determine the 2μ quantities p, q as functions of the others. However, we can also imagine that we have solved these equations for the 2μ quantities q, Q, so that the q and Q are expressed as functions of the $2\mu+1$ other quantities p, P, t. For the moment we shall indicate a function of these latter variables by putting a bar above it. Thus $\bar{q}_i$ means that the quantity q_i is to be considered expressed as a function of p, P, and t. Since we can find W as a function of P, Q, and t, we can also express therein the Q as functions of p, P, and t, so that W itself (now denoted by $\overline{W}$) becomes a function of p, P, and t. It is well known[1] that

$$\frac{\partial \overline{W}}{\partial P_i} = \bar{q}_i, \quad \frac{\partial \overline{W}}{\partial P_i} = -\overline{Q_i}.$$

Whence it follows at once that

$$\frac{\partial \overline{q_i}}{\partial P_j} = -\frac{\partial \overline{Q_i}}{\partial p_j}, \tag{42}$$

where the bar means that in the differentiation with respect to any one p, all the other p, all the P and the time are to be considered constant; similarly in the differentiation with respect to any one of the P. i and j can take any integer values from 1 to μ, independently of each other.

§26. Liouville's Theorem.*

When one wishes to discuss any curve whose equation contains an arbitrary parameter, it is customary to consider simultaneously all the curves obtained by giving this parameter all its possible values. We are now dealing with a mechanical system (characterized by given equations of motion) whose motion depends on the values of the 2μ parameters P, Q. Just as one can represent a curve infinitely many times, each time with a different value of a parameter, so we can represent our mechanical system infinitely often, so that we obtain infinitely many mechanical systems, all of the same nature and subject to the same

[1] Jacobi, *Vorlesung. üb. Dynamik*, 19th lecture, Equation 4, p. 146.

* For references see the footnotes for §29.

equations of motion, but with different initial conditions. Among this infinite number, or family, of mechanical systems, we are given certain ones for which the initial values of the coördinates and momenta are specified within infinitesimally close limits, e.g.,

$$(43)\quad \begin{cases} P_1 \text{ and } P_1 + dP_1,\ P_2 \text{ and } P_2 + dP_2 \cdots P_\mu \text{ and } P_\mu + dP_\mu \\ Q_1 \text{ and } Q_1 + dQ_1,\ Q_2 \text{ and } Q_2 + dQ_2 \cdots Q_\mu \text{ and } Q_\mu + dQ_\mu. \end{cases}$$

After these systems have moved for the same time t, according to their equations of motion, the coördinates and momenta will lie between the limits

$$(44)\quad \begin{cases} p_1 \text{ and } p_1 + dp_1,\ p_2 \text{ and } p_2 + dp_2 \cdots p_\mu \text{ and } p_\mu + dp_\mu \\ q_1 \text{ and } q_1 + dq_1,\ q_2 \text{ and } q_2 + dq_2 \cdots q_\mu \text{ and } q_\mu + dq_\mu. \end{cases}$$

Our problem is to express the product

$$(45)\qquad dp_1 dp_2 \cdots dp_\mu dq_1 dq_2 \cdots dq_\mu$$

in terms of the product

$$(46)\qquad dP_1 dP_2 \cdots dP_\mu dQ_1 dQ_2 \cdots dQ_\mu.$$

We know that we can express the q as functions of the p, P, and t. We can therefore introduce the variables p and P instead of p and q in the differential expressions (45). The time t is considered a constant. Thence follows next, according to the well-known theorem of Jacobi on the so-called functional determinants:

$$(47)\quad \begin{cases} dp_1 dp_2 \cdots dp_\mu dq_1 dq_2 \cdots dq_\mu = \\ D dp_1 dp_2 \cdots dp_\mu dP_1 dP_2 \cdots dP_\mu, \end{cases}$$

where

$$(48)\qquad D = \begin{vmatrix} 100 \cdots 0, & 0 \cdots & \\ 010 \cdots 0, & 0 \cdots & \\ \cdot\ \cdot\ \cdot\ \cdot\ \cdot\ \cdot & & \\ 000 \cdots \dfrac{\partial q_1}{\partial P_1}, & \dfrac{\partial q_2}{\partial P_2} & \cdots \\ 000 \cdots \dfrac{\partial q_1}{\partial P_1}, & \dfrac{\partial q_2}{\partial P_2} & \cdots \\ \cdot\ \cdot\ \cdot\ \cdot\ \cdot\ \cdot & & \end{vmatrix} = \begin{vmatrix} \dfrac{\partial q_1}{\partial P_1}, & \dfrac{\partial q_2}{\partial P_1} & \cdots \\ \dfrac{\partial q_1}{\partial P_2}, & \dfrac{\partial q_2}{\partial P_2} & \cdots \\ \cdot\ \cdot\ \cdot & \cdot\ \cdot\ \cdot & \end{vmatrix}.$$

Likewise, we can also introduce the variables p and P in the expression (46) by expressing the Q as functions of the P, p, and t, whereby we then obtain:

$$(49)\qquad \begin{cases} dP_1dP_2\cdots dP_\mu dQ_1dQ_2\cdots dQ_\mu = \\ \Delta dP_1dP_2\cdots dP_\mu dp_1dp_2\cdots dp_\mu, \end{cases}$$

where

$$(50)\qquad \Delta = \begin{vmatrix} \dfrac{\partial Q_1}{\partial p_1}, & \dfrac{\partial Q_2}{\partial p_1} & \cdots \\ \dfrac{\partial Q_1}{\partial p_2}, & \dfrac{\partial Q_2}{\partial p_2} & \cdots \\ \cdots & \cdots & \cdots \end{vmatrix}.$$

The partial derivatives in the functional determinant D are to be understood in the same sense as the quantities previously denoted by a bar; the q are considered to be functions of the p, P, and t. Similarly for the partial derivatives in the functional determinant Δ, in which the Q are to be considered as functions of the same variables p, P, and t. However, we can apply Equations (42), and since a determinant does not change its value when one permutes the horizontal and vertical rows, and conversely, we have

$$(51)\qquad \Delta = \begin{vmatrix} -\dfrac{\partial q_1}{\partial P_1}, & -\dfrac{\partial q_2}{\partial P_1} & \cdots \\ -\dfrac{\partial q_1}{\partial P_2}, & -\dfrac{\partial q_2}{\partial P_2} & \cdots \\ \cdots & \cdots & \cdots \end{vmatrix} = (-1)^\mu D.$$

Since only the magnitude and not the sign is important, it follows from Equations (47), (49), and (51) that

$$(52)\qquad \begin{cases} dp_1dp_2\cdots dp_\mu dq_1dq_2\cdots dq_\mu = \\ dP_1dP_2\cdots dP_\mu dQ_1dQ_2\cdots dQ_\mu \end{cases}$$

which is the desired relation. Indeed, we obtain in general the correct sign if we take account of the changes of sign resulting from the changes in the orders of differentials in going from Equation (47) to Equation (49).

Suppose one introduces, in place of 2μ arbitrary variables

$x_1, x_2, \cdots, x_{2\mu}$ in some differential expression, 2μ other variables $\xi_1, \xi_2, \cdots, \xi_{2\mu}$ which are related to the former set by the following equations:

$$\xi_1 = x_{\mu+1},\ \xi_2 = x_{\mu+2} \cdots \xi_\mu = x_{2u},\ \xi_{\mu+1} = x_1 \cdots \xi_{2\mu} = x_\mu,$$

then it follows from the theorem on functional determinants that:

$$dx_1 dx_2 \cdots dx_{2\mu} = \Theta d\xi_1 d\xi_2 \cdots d\xi_{2\mu},$$

where

$$\Theta = \begin{vmatrix} \frac{\partial \xi_1}{\partial x_1}, & \frac{\partial \xi_2}{\partial x_1} & \cdot \\ \frac{\partial \xi_1}{\partial x_2}, & \frac{\partial \xi_2}{\partial x_2} & \cdot\ \cdot \\ \cdot\ \cdot\ \cdot & \cdot\ \cdot\ \cdot & \cdot \end{vmatrix} = \begin{vmatrix} 000 \cdots 100 \cdots \\ 000 \cdots 010 \cdots \\ \cdot\ \cdot\ \cdot\ \cdot\ \cdot\ \cdot\ \cdot\ \cdot \\ 100 \cdots 000 \cdots \\ 010 \cdots 000 \cdots \\ \cdot\ \cdot\ \cdot\ \cdot\ \cdot\ \cdot\ \cdot\ \cdot \end{vmatrix}.$$

Now since permutation of two horizontal rows changes the sign of the determinant, we have

$$\Theta = (-1)^\mu \begin{vmatrix} 100 \cdots \\ 010 \cdots \\ \cdot\ \cdot\ \cdot\ \cdot \end{vmatrix} = (-1)^\mu.$$

If one sets

$$x_1 = p_1,\ x_2 = p_2 \cdots x_{u+1} = P_1,\ x_{\mu+2} = P_2 \cdots,$$

then

$$\xi_1 = P_1,\ \xi_2 = P_2 \cdots \xi_{\mu+1} = p_1,\ \xi_{\mu+2} = p_2 \cdots,$$

hence

$$dp_1 dp_2 \cdots dp_\mu dP_1 dP_2 \cdots dP_\mu$$
$$= (-1)^\mu dP_1 dP_2 \cdots dP_\mu dp_1 dp_3 \cdots dp_\mu,$$

and it follows from (47) that

$$dp_1 dp_2 \cdots dp_\mu dq_1 dq_2 \cdots dq_\mu$$
$$= (-1)^\mu D dP_1 dP_2 \cdots dP_\mu dp_1 dp_2 \cdots dp_\mu,$$

hence, according to Equation (51),

$$dp_1 dp_2 \cdots dp_\mu dq_1 dq_2 \cdots dq_\mu$$
$$= \Delta dP_1 dP_2 \cdots dP_\mu dp_1 dp_2 \cdots dp_\mu,$$

which, in conjunction with Equation (49), gives Equation (52) with the correct sign.

§27. On the introduction of new variables in a product of differentials.

Equation (52) is the fundamental equation for the following. Before I proceed to its application, I shall mention a difficulty that often arises in the theory of definite integrals, and is not always completely clarified.

I consider the most general case. Let there be given n arbitrary functions $\xi_1, \xi_2, \cdots, \xi_n$ of the n independent variables $x_1, x_2, \cdots, x_n$, the functions being single-valued and continuous in a specified region. Conversely, let the x be single-valued continuous functions of the ξ. If we set

$$D = \begin{vmatrix} \frac{\partial \xi_1}{\partial x_1}, & \frac{\partial \xi_2}{\partial x_1} & \cdots & \frac{\partial \xi_n}{\partial x_1} \\ \frac{\partial \xi_1}{\partial x_2}, & \frac{\partial \xi_2}{\partial x_2} & \cdots & \frac{\partial \xi_n}{\partial x_2} \\ \cdot & \cdot & \cdot & \cdot \\ \frac{\partial \xi_1}{\partial x_n}, & \frac{\partial \xi_2}{\partial x_n} & \cdots & \frac{\partial \xi_n}{\partial x_n} \end{vmatrix},$$

then the differentials are related by the equation

$$dx_1 dx_2 \cdots dx_n = \frac{1}{D} d\xi_1 d\xi_2 \cdots d\xi_n. \tag{53}$$

The meaning of this equation is perfectly clear if ξ_1 is only a function of x_1, ξ_2 is only a function of x_2, etc. If then only x_1 changes by dx_1 and all the other x remain constant, then only ξ_1 will change by $d\xi_1$, and all the other ξ will remain constant. Likewise a definite increment $d\xi_2$ of ξ_2 will correspond to a definite increment dx_2 of x_2, and so forth. Equation (53) will then give the relation between the increments of x and the increments of ξ.

If x_1 runs through all its possible values between x_1 and x_1+dx_1, while x_2 has a constant value somewhere between x_2 and x_2+dx_2, and likewise all the other x have constant values, then in general not only ξ_1 but also all the other ξ will change at the same time. Likewise, in general all the ξ will change when x_2 runs through all values between x_2 and x_2+dx_2, while all the other x remain constant; and each ξ will experience in general a completely different increment in the second case. We have therefore to consider n different increments of ξ_1; none of them is equal to the quantity denoted by $d\xi_1$ in Equation (53). Neither does one obtain Equation (53) by assuming that $d\xi_h$ means the largest increase that ξ_h can experience when each variation of x_1 between x_1 and x_1+dx_1 is combined with each variation of x_2 between x_2 and x_2+dx_2, and each such pair of values of x_1 and x_2 is combined with each variation of x_3 between x_3 and x_3+dx_3, and so forth.

In order to present clearly the meaning of Equation (53) in the general case, we must examine the matter more closely. This equation in general has a meaning only when it refers to the transformation of a definite integral which is to be extended over a certain set of values of all the x, to another integral in which the variables ξ replace the x. We shall use the following notation. Let the value of each of the variables x be given; the corresponding values of all the ξ are thereby determined. We call them the values of ξ corresponding to the given x. A region G of values of the x means the aggregate of the set of values of these variables bounded in the following way: we first include all values of x_1 that lie between two arbitrary given limits x_1^0 and x_1^1. Those values of x_1 that lie between these limits are to be associated with all values of the second variable x_2 that lie between arbitrary given limits x_2^0 and x_2^1, where however x_2^0 and x_2^1 can be continuous functions of the values of x_1 that are to be associated with the value of x_2. Likewise, each pair of values of x_1 and x_2 satisfying the above conditions will be associated with all values of x_3 that lie between x_3^0 and x_3^1, where x_3^0 and x_3^1 can be continuous functions of x_1 and x_2, and so forth.[1] It is then well known what one means by the definite integral

$$\int\int\cdots f(x_1, x_2 \cdots x_n)dx_1dx_2\cdots dx_n$$

extended over the entire region G. This region of values is then said to be of infinitesimal extent with respect to all n dimensions

[1] Exceptions to the continuity must be limited to individual points.

—or briefly, n-fold infinitesimal—when the difference $x_1^1 - x_1^0$ is infinitesimal, and also the difference $x_2^1 - x_2^0$ is infinitesimal for all values of x_1, the difference $x_3^1 - x_3^0$ is infinitesimal for all pairs of values of x_1 and x_2, and so forth. When $n = 2$, x_1 and x_2 can be represented as the coördinates of a point in the plane; each region of values then corresponds to a bounded part of the surface in the plane; for $n = 3$, each region of values can be represented by a bounded volume in space.

Each set of values of the x that lies in the region G corresponds to a set of values of the ξ. The region g of the ξ, which corresponds to the region G of the x, means the aggregate of all sets of values of the ξ that correspond to all sets of values of the x lying in the region G.

According to the way we have formulated our definitions, Jacobi's theorem on functional determinants can be expressed in the following completely unambiguous way.

Let there be given an arbitrary single-valued continuous function of the independent variables $x_1, x_2, \cdots, x_n$. We denote it by $f(x_1, x_2, \cdots, x_n)$. If we express therein $x_1, x_2, \cdots, x_n$ in terms of $\xi_1, \xi_2, \cdots, \xi_n$, then the function $f(x_1, x_2, \cdots, x_n)$ is transformed to $F(\xi_1, \xi_2, \cdots, \xi_n)$ so that we have identically

$$f(x_1, x_2 \cdots x_n) = F(\xi_1, \xi_2 \cdots \xi_n).$$

However, it is by no means the case that

$$\begin{aligned}&\int\int \cdots f(x_1, x_2 \cdots x_n) dx_1 dx_2 \cdots dx_n \\ = &\int\int \cdots F(\xi_1, \xi_2 \cdots \xi_n) d\xi_1 d\xi_2 \cdots d\xi_n,\end{aligned}$$

when the former integral is extended over an arbitrary region G of the x, and the latter over the corresponding region g of the ξ. If we again denote the functional determinant

$$\begin{vmatrix} \dfrac{\partial \xi_1}{\partial x_1}, & \dfrac{\partial \xi_2}{\partial x_1} & \cdots \\ \dfrac{\partial \xi_1}{\partial x_2}, & \dfrac{\partial \xi_2}{\partial x_2} & \cdots \\ \cdots & \cdots & \cdots \end{vmatrix}$$

by D, then the definite integral extended over the region G,

$$\iint \cdots f(x_1, x_2 \cdots x_n)dx_1dx_2 \cdots dx_n,$$

is always equal to the definite integral extended over the corresponding region g:[2]

$$\iint \cdots F(\xi_1, \xi_2 \cdots \xi_n) \frac{1}{D} d\xi_1 d\xi_2 \cdots d\xi_n.$$

If G is infinitesimal, then, if the values of ξ in question are continuous functions of the x, the region g is also infinitesimal, and the values of the functions f and F, as well as the functional determinant, will be considered constant throughout the region. Since the values of the two functions are the same, one can divide through by these values and it follows that:[3]

$$(54) \quad \frac{1}{D}\iint \cdots d\xi_1 d\xi_2 \cdots d\xi_n = \iint \cdots dx_1dx_2 \cdots dx_n.$$

---

[2] Naturally we have similarly

$$\iint \cdots F(\xi_1, \xi_2 \cdots \xi_n)d\xi_1 d\xi_2 \cdots d\xi_n$$
$$= \iint \cdots f(x_1, x_2 \cdots x_n) D \cdot dx_1dx_2 \cdots dx_n$$

$$\iint \cdots f(x_1, x_2 \cdots x_n)dx_1dx_2 \cdots dx_n$$
$$= \iint \cdots F(\xi_1, \xi_2 \cdots \xi_n)\Delta \cdot d\xi_1 d\xi_2 \cdots d\xi_n,$$

$$\iint \cdots f(x_1, x_2 \cdots x_n) \frac{1}{\Delta} dx_1dx_2 \cdots dx_n$$
$$= \iint \cdots F(\xi_1, \xi_2 \cdots \xi_n)d\xi_1 d\xi_2 \cdots d\xi_n,$$

where

$$\Delta = \begin{vmatrix} \frac{\partial x_1}{\partial \xi_1}, & \frac{\partial x_2}{\partial \xi_1} & \cdots \\ \frac{\partial x_1}{\partial \xi_2}, & \frac{\partial x_2}{\partial \xi_2} & \cdots \\ \cdots & \cdots & \cdots \end{vmatrix}.$$

All the integrals over x are to be extended over an arbitrary region G, and those over ξ are to be extended over the corresponding region g.

[3] Or

$$\Delta \iint \cdots d\xi_1 d\xi_2 \cdots d\xi_n = \iint \cdots dx_1dx_2 \cdots dx_n.$$

Indeed, Kirchhoff writes the equation in this form.[4] It is customary to write (as we did in the previous paragraph) simply

$$\frac{1}{D} d\xi_1 d\xi_2 \cdots d\xi_n = dx_1 dx_2 \cdots dx_n.$$

Here $dx_1 dx_2 \cdots dx_n$ means strictly the n-fold integral of this quantity over an arbitrary n-fold infinitesimal region G, and $d\xi_1 d\xi_2 \cdots d\xi_n$ means the n-fold integral of the latter quantity over the corresponding region g. Since the theorem will only be applied to the calculation of definite integrals extended over a finite region, and these can always be decomposed into an infinite number of integrals extended over infinitesimal regions, one always obtains the correct result if he writes the equations as follows:

$$\frac{1}{D} d\xi_1 d\xi_2 \cdots d\xi_n = dx_1 dx_2 \cdots dx_n$$

$$F(\xi_1, \xi_2 \cdots \xi_n) = f(x_1, x_2 \cdots x_n),$$

hence

$$F(\xi_1, \xi_2 \cdots \xi_n) \cdot \frac{1}{D} d\xi_1 d\xi_2 \cdots d\xi_n = f(x_1, x_2 \cdots x_n) dx_1 dx_2 \cdots dx_n$$

and hence finally

$$\iint \cdots F(\xi_1, \xi_2 \cdots \xi_n) \cdot \frac{1}{D} d\xi_1 d\xi_2 \cdots d\xi_n$$

$$= \iint \cdots f(x_1, x_2 \cdots x_n) dx_1 dx_2 \cdots dx_n.$$

The first of these equations has the following meaning. Each n-fold definite integral extended over all x can be decomposed into infinitely many integrals over n-fold infinitesimal regions. If one wishes to introduce the ξ as new variables of integration, then in each of the latter, and hence also in the entire region of integration, he has to replace the product $dx_1 dx_2 \cdots dx_n$ by

$$\frac{1}{D} d\xi_1 d\xi_2 \cdots d\xi_n.$$

[4] Kirchhoff, *Vorles. über Theorie d. Wärme* (Teubner, 1894), p. 143.

§28. Application to the formulas of §26.

If one wishes to use these more correct expressions in §26, then, instead of saying "for a certain system the initial values of the coördinates and momenta lie between

$$P_1 \text{ and } P_1 + dP_1 \cdots Q_\mu \text{ and } Q_\mu + dQ_\mu,\text{"}$$

he must say instead, "each initial value lies in the 2μ-fold infinitesimal region

$$G = \int dP_1 dP_2 \cdots dP_\mu dQ_1 dQ_2 \cdots dQ_\mu.\text{"}$$

Instead of saying "then at time t the values lie between p_1 and p_1+dp_1, $\cdots$, q_μ and $q_\mu+dq$," he must use the corresponding expression, "they lie in the corresponding region

$$g = \int dp_1 dp_2 \cdots dp_\mu dq_1 dq_2 \cdots dq_\mu.\text{"}$$

Here the integration over the entire appropriate region is indicated by a single integral sign for the sake of brevity. The region g corresponding to the region G includes all combinations of values that the variables assume after the time t (considered constant), if they initially have a set of values lying inside the region G. All the conclusions of the preceding section will then remain valid, except that in place of the simple product of differentials there will always occur the integral of this product of differentials over a region infinitesimal on all sides. Equation (52) then runs, in this more precise formulation:

$$(55)\qquad \begin{cases} \int dp_1 dp_2 \cdots dp_\mu dq_1 dq_2 \cdots dq_\mu \\ = \int dP_1 dP_2 \cdots dP_\mu dQ_1 dQ_2 \cdots dQ_\mu. \end{cases}$$

One therefore sees that the conclusion is not in the least changed, except that one has an integral sign in front of each differential expression, expressing the integration over a corresponding infinitesimal region.

If an example is needed, one can think of the x as the spatial polar coördinates r, θ, φ and the ξ as the rectangular coördinates x, y, z of a point. The region of values for which x, y, z lie between

$$x \text{ and } x + dx, \quad y \text{ and } y + dy, \quad z \text{ and } z + dz$$

is determined by a parallelepiped. We shall assign different pairs of values to the variables ϑ and φ, corresponding to all points lying within this parallelepiped. The limits r and $r+dr$, between which r must lie for all points in the parallelepiped, will by no means be the same for all these pairs of values. In the equation

$$dxdydz = \begin{vmatrix} \frac{\partial x}{\partial r}, & \frac{\partial y}{\partial r}, & \frac{\partial z}{\partial r} \\ \frac{\partial x}{\partial \vartheta}, & \frac{\partial y}{\partial \vartheta}, & \frac{\partial z}{\partial \vartheta} \\ \frac{\partial x}{\partial \varphi}, & \frac{\partial y}{\partial \varphi}, & \frac{\partial z}{\partial \varphi} \end{vmatrix} drd\vartheta d\varphi = r^2 \sin \vartheta drd\vartheta d\varphi,$$

which is not valid if one understands by dr the greatest difference of the values of r for points lying inside the parallelepiped, which of these possible values of dr is meant? The same problem arises for $d\vartheta$ and $d\varphi$. The equation has the following meaning: the definite integral

$$\iiint dxdydz,$$

extended over an arbitrary triply infinitesimal region, has the same value as the integral

$$\iiint r^2 \sin \vartheta drd\vartheta d\varphi,$$

extended over the corresponding region; in particular, both are equal to the volume filled by all the points of the region, so that it would obviously be completely wrong to set

$$\iiint dxdydz = \iiint drd\vartheta d\varphi$$

if the two integrals are extended over corresponding regions.

The special examples that have been worked out to clarify the meaning of the mechanical equations (52) and (55) will only be briefly discussed here.[1] A material point with mass 1 moves in the abscissa direction under the influence of a constant force, which likewise has the abscissa direction, and imparts to it the acceleration γ. The variables called x at the beginning of this section will now be the initial abscissa X and the initial velocity U of the material point; the ξ will be the abscissa x and velocity u after a fixed time t has elapsed. We have therefore

$$x = X + Ut + \frac{\gamma t^2}{2}, \quad u = U + \gamma t. \tag{56}$$

Since we now have only two variables, we can represent their initial values as well as their values at time t by points in the plane, whose abscissas are equal to the abscissas, and whose ordinates are equal to the velocities, of the material point. All points of a rectangle with sides dX and dU represent a double infinitesimal region of x, i.e., they represent all possible material points for which the coördinates and velocities initially lie between the limits

$$X \text{ and } X + dX \text{ and } U \text{ and } U + dU.$$

The region g of the ξ corresponding to this region G contains the coördinates x and velocities u that all these material points have after a fixed time t has elapsed. According to Equation (56), u is simply larger than U by a constant γt. On the other hand, the difference x–X is greater the larger U is. One sees easily from this that the region g is an oblique-angled parallelogram, whose base is equal to dX, and whose height is equal to dU; it therefore has the same area as the rectangle $G = dXdU$, as it must according to Equation (52).

§29. Second proof of Liouville's theorem.

We shall now give a second proof of Equation (52) or (55), in which, instead of going directly from time zero to time t, we first go from time t to an infinitesimally close time $t+dt$. However, we shall somewhat generalize the theorem at the same time, in that we shall not assume that our independent variable

[1] Boltzmann, Wien. Ber. **74**, 508 (1876); Bryan, Phil. Mag. [5] **39**, 531 (1895).

(denoted by s) is necessarily the time; we shall leave it unspecified, though for purposes of illustration we can still imagine that time is the independent variable. The arbitrary dependent variables $s_1, s_2, \cdots, s_n$ will be determined as functions of the independent variable s for the following differential equations:

$$\frac{\delta s_1}{\delta s} = \frac{\sigma_1}{\sigma}, \quad \frac{\delta s_2}{\delta s} = \frac{\sigma_2}{\sigma} \cdots \frac{\delta s_n}{\delta s} = \frac{\sigma_n}{\sigma}. \tag{57}$$

The σ will be given explicitly as functions of $s, s_1, s_2, \cdots, s_n$. Since we wish to reserve the letter d for another kind of increment, we have denoted the increment of the independent variable by δs, and the corresponding increments of the dependent variables by $\delta s_1, \delta s_2, \cdots, \delta s_n$.

For a system of material points, δs would be identified with the increment of the time, δt. By $\delta s_1, \delta s_2, \cdots, \delta s_n$ we mean the increments $\delta x_1, \delta y_1, \cdots$ of the coördinates as well as the increments $\delta u_1, \delta v_1, \cdots$ of the velocity components during time δt. For example, $\delta x_1 = u_1 \delta t$.

The values of the dependent variables $s_1, s_2, \cdots, s_n$ will be determined by their corresponding initial values at a definite s, e.g., $s=0$:

$$S_1, S_2 \cdots S_n \tag{58}$$

and by the differential equations (57), given once for all times, as single-valued functions of the independent variable s.

We can now keep in mind all the values of the dependent variables that correspond to all possible values of the independent variable s, for given initial values. We wish to call the set of all these values a series of values. This corresponds to the entire course of motion of a mechanical system starting from a specified initial state.

In the particular series of values that starts from the initial values (58), the dependent variables have the values

$$s_1, s_2 \cdots s_n, \tag{59}$$

for a specified value of s, and for an infinitesimally different value $s + \delta s$ they have the values

$$s_1' = s_1 + \delta s_1, \; s_2' = s_2 + \delta s_2 \cdots s_n' = s_n + \delta s_n. \tag{60}$$

We call (59) the values of the dependent variables "corresponding to the initial values (58) after s." The values (60) will similarly be called the values of the dependent variables corresponding to the initial values (58) after $s+\delta s$. According to the differential equations (57), the values (59) and (60) are related by the equations

$$(61)\quad s_1' = s_1 + \frac{\sigma_1}{\sigma}\delta s,\ s_2' = s_2 + \frac{\sigma_2}{\sigma}\delta s \cdots s_n' = s_n + \frac{\sigma_n}{\sigma}\delta s,$$

where the values (59) of the dependent variables and the corresponding values of the independent variable are to be substituted in the given functions $\sigma_1, \sigma_2, \cdots, \sigma_n$.

Now we shall go even further and imagine all possible series of values that can arise from all possible initial states. Of all these series of values, however, we consider only those for which the initial values lie between the limits

$$S_1 \text{ and } S_1 + dS_1,\ S_2 \text{ and } S_2 + dS_2 \cdots S_n \text{ and } S_n + dS_n$$

or lie in some otherwise bounded n-fold infinitesimal region G,[1] in which the values (58) also lie. Those values of the dependent variables to which all the initial values lying in G "correspond after s" form again an n-fold infinitesimal region, which will be the region g.

On the other hand, we shall denote by g' the region that includes all values of the dependent variables that "correspond after $s+\delta s$" to all the initial values lying in g. The integral of the product $ds_1 ds_2 \cdots ds_n$ of the differentials of all dependent variables, extended over the entire region g, will be denoted simply by

$$\int ds_1 ds_2 \cdots ds_n,$$

while the same integral extended over the region g' will be denoted by

$$\int ds_1' ds_2' \cdots ds_n'.$$

These integral signs therefore indicate an integration over all series of values that originate in the infinitesimal region G. Then according to Equation (54):

[1] The meaning of this expression was discussed in §27.

$$\int ds_1' ds_2' \cdots ds_n' = D \int ds_1 ds_2 \cdots ds_n, \tag{62}$$

where D again means the functional determinant

$$\begin{vmatrix} \frac{\partial s_1'}{\partial s_1}, & \frac{\partial s_2'}{\partial s_1} & \cdots \\ \frac{\partial s_1'}{\partial s_2}, & \frac{\partial s_2'}{\partial s_2} & \cdots \\ \cdot & \cdot \cdot \cdot \cdot & \cdot \end{vmatrix}.$$

In forming the partial derivatives in this functional determinant, we should consider s to be constant as well as $s+\delta s$ and δs, since s has the same value for all series of values over which we shall integrate, as does δs. Hence it follows from Equation (61) that

$$\frac{\partial s_1'}{\partial s} = 1 + \frac{\delta s}{\sigma}\left(\frac{\partial \sigma_1}{\partial s_1} - \frac{\sigma_1}{\sigma}\,\frac{\partial \sigma}{\partial s_1}\right), \quad \frac{\partial s_1'}{\partial s_2} = \frac{\delta s}{\sigma}\left(\frac{\partial \sigma_1}{\partial s_2} - \frac{\sigma_1}{\sigma}\,\frac{\partial \sigma}{\partial s_1}\right), \text{ etc.}$$

If one ignores the terms multiplied by higher powers of the infinitesimal quantity δs, then he obtains

$$\begin{aligned} D &= 1 + \frac{\delta s}{\sigma}\left(\frac{\partial \sigma}{\partial s} + \frac{\partial \sigma_1}{\partial s_1} + \cdots \frac{\partial \sigma_n}{\partial s_n}\right) - \\ &\quad - \frac{\delta s}{\sigma}\left(\frac{\partial \sigma}{\partial s} + \frac{\sigma_1}{\sigma}\,\frac{\partial \sigma}{\partial s_1} + \cdots \frac{\sigma_n}{\sigma}\,\frac{\partial \sigma}{\partial s_n}\right) = \\ &= 1 + \frac{\delta \tau}{\tau} - \frac{\delta \sigma}{\sigma} = \frac{\tau'}{\sigma'} \cdot \frac{\sigma}{\tau}, \end{aligned}$$

where

$$\tau = e\left[\int \frac{\delta s}{\sigma}\left(\frac{\partial \sigma}{\partial s} + \frac{\partial \sigma_1}{\partial s_1} + \cdots \frac{\partial \sigma_n}{\partial s_n}\right)\right]. \tag{63}$$

The prime always means that the value corresponds to $s+\delta s$, while the initial values are those given in Equation (58). Hence one can write Equation (62) in the form:

$$\frac{\sigma'}{\tau'}\int ds_1' ds_2' \cdots ds_n' = \frac{\sigma}{\tau}\int ds_1 ds_2 \cdots ds_n. \tag{64}$$

Similarly, just as we go from s to $s+\delta s$, we can go from $s+\delta s$ to $s+2\delta s$, and so forth, and also from $s-\delta s$ to s. We shall denote all values corresponding to $s+2\delta s$, starting from the initial values (58), by two primes.

Let g'' be the region that includes all values of the dependent variables corresponding to initial values lying in G, after $s+2\delta s$, and let

$$\int ds_1'' ds_2'' \cdots ds_n''$$

be the integral of the product of differentials of all dependent variables over the region g''. Then one finds, in the same way that was used to obtain Equation (64), that

$$\frac{\sigma''}{\tau''}\int ds_1'' ds_2'' \cdots ds_n'' = \frac{\sigma'}{\tau'}\int ds_1' ds_2' \cdots ds_n' = \frac{\sigma}{\tau}\int ds_1 ds_2 \cdots ds_n.$$

Since a similar equation holds for all preceding and following increments of s, one has in general:

$$\frac{\sigma}{\tau}\int ds_1 ds_2 \cdots ds_n = \frac{\sigma_0}{\tau_0}\int dS_1 dS_2 \cdots dS_n. \tag{65}$$

Here σ_0 and τ_0 denote the values of σ and τ for $s=0$; $\int dS_1 dS_2 \cdots dS_n$ is the integral of the product of the differentials of all dependent variables extended over the region G.

One sees at once that Equation (55) is that special case of Equation (65) which one obtains when s means the time and $s_1, s_2, \cdots, s_n$ are the generalized coördinates $p_1, p_2, \cdots, p_\mu$ and momenta $q_1, q_2, \cdots, q_\mu$ of an arbitrary mechanical system. If in particular, as in §25, L and V are the kinetic and potential energies of the mechanical systems, and if one sets $L+V=E$, then the Lagrange equations for the mechanical system run as follows:[2]

$$\frac{dp_i}{dt} = \frac{\partial E}{\partial q_i}, \quad \frac{dq_i}{dt} = -\frac{\partial E}{\partial p_i}. \tag{66}$$

The symbol d has the same meaning that δ had previously. We have to specialize our previous formulas to the case $n=2\mu$, $\sigma=1$, $s=t$,

[2] Jacobi, *Vorlesungen über Dynamik*, 9th lecture, p. 71, Eq. (8). Thomson and Tait, *Treatise on Natural Philosophy*, new edition, Vol. I, Part I, p. 307, Art. 319; German edition, p. 284. Rausenberger, *Mechanik* (Leipzig, 1888), Vol. I, p. 200.

$$s_i = p_i, \quad \sigma_i = \frac{\partial E}{\partial q_i} \text{ for } 1 \leqq i \leqq \mu,$$

$$s_i = q_i, \quad \sigma_i = -\frac{\partial E}{\partial p_i} \text{ for } \mu + 1 \leqq i \leqq 2\mu.$$

Hence

$$\frac{\partial \sigma}{\partial s} + \frac{\partial \sigma_1}{\partial s_1} + \cdots \frac{\partial \sigma_n}{\partial s_n} = 0, \quad \tau = \text{const.}$$

and Equation (65) goes over directly into Equation (55).

Arguments completely analogous to those of this section were first made by Liouville,[3] then by Jacobi[4] (by the latter in his lectures on dynamics, for the purpose of deriving the theorem of the last multiplier). They were first applied to statistical considerations on the temporal course of motion of a system, and a family of simultaneously existing systems, by the author of this book and then by Maxwell.[5]

§30. Jacobi's theorem of the last multiplier.

Since we have the required equations directly at hand, we shall derive the theorem of the last multiplier, though to be sure it is not otherwise very closely related to our subject.

We denote by

$$\varphi_i(s, s_1, s_2 \cdots s_n) = \text{const.}, \quad i = 1, 2 \cdots n$$

the n integrals of the differential equations (57). The initial values (58) should correspond to the values $a_1, a_2, \cdots, a_n$ of the constants of integration, so that therefore

$$\varphi_i(0, S_1, S_2 \cdots S_n) = a_i, \quad i = 1, 2 \cdots n. \tag{67}$$

All the initial values of the dependent variables that lie in G (see last section) will correspond to certain values of the constants of integration a, which again form an n-fold infinitesimal region, which we shall call the region A. Let

[3] Liouville, J. de Math. **3,** 348 (1838).

[4] Jacobi, *Vorlesungen über Dynamik*, p. 93.

[5] Boltzmann, Wien. Ber. **63,** 397, 679 (1871), **58,** 517 (1868). Maxwell, "On Boltzmann's theorem," Trans. Camb. Phil. Soc. **12,** 547 (1879), *Scientific Papers* **2,** 713.

$$\int da_1 da_2 \cdots da_n$$

be the integral of the products of the differentials of the constants of integration, extended over the entire region. On the other hand, as in the preceding section we shall denote by $s_1, s_2, \cdots, s_n$ the values of the dependent variables that correspond to the initial values (58) "after the value s of the independent variable." Hence

$$\varphi_i(s, s_1, s_2 \cdots s_n) = a_i, \quad i = 1, 2 \cdots n, \tag{68}$$

where the a have the same values as in Equation (67). Likewise as in the preceding section, we denote by g the region formed from all values of the dependent variables that correspond to all initial values lying in G after the value s of the independent variable, and by

$$\int ds_1 ds_2 \cdots ds_n$$

the integral of the product of differentials of the dependent variables extended over g, while

$$\int dS_1 dS_2 \cdots dS_n$$

is the corresponding integral over G. Since the a are related to the S by Equation (67) and to $s_1, \cdots, s_n$ by Equations (68), and in the latter s is to be considered constant, then one has:

$$\int da_1 da_2 \cdots da_n = \Delta_0 \int dS_1 dS_2 \cdots dS_n = \Delta \int ds_1 ds_2 \cdots ds_n,$$

where

$$\Delta = \begin{vmatrix} \frac{\partial \varphi_1}{\partial s_1}, & \frac{\partial \varphi_2}{\partial s_1} & \cdots \\ \frac{\partial \varphi_1}{\partial s_2}, & \frac{\partial \varphi_2}{\partial s_2} & \cdots \\ \cdot & \cdot \quad \cdot \quad \cdot & \cdot \quad \cdot \quad \cdot \end{vmatrix}$$

and Δ_0 is the value of Δ for $s=0$. Hence, according to Equation (65),

$$\frac{\Delta \tau}{\sigma} = \frac{\Delta_0 \tau_0}{\sigma_0} = C.$$

Since Δ_0, τ_0, and σ_0 are expressions that depend only on the initial values of the dependent variables—or, if one wishes, on the constants of integration a—but not on the values of the s, then likewise C depends only on these quantities.

We now assume that we already know all the integrals up to the one $\varphi_1 = a_1$. The equation

$$\int da_1 da_2 \cdots da_n = \Delta \int ds_1 ds_2 \cdots ds_n \tag{69}$$

holds for each value of s. We imagine that s has an arbitrary constant value, and introduce into both definite integrals on the right- and left-hand sides the variables $s_1, a_2, a_3, \cdots, a_n$—which, since s is considered as a given constant, are unique functions both of $a_1, a_2, \cdots, a_n$ and of $s_1, s_2, \cdots, s_n$. We thus obtain:

$$\int ds_1 ds_2 \cdots ds_n = \frac{1}{\Delta_1} \int ds_1 da_2 \cdots da_n,$$

where in

$$\Delta_1 = \begin{vmatrix} \dfrac{\partial \varphi_2}{\partial s_2}, & \dfrac{\partial \varphi_3}{\partial s_2} & \cdots \\ \dfrac{\partial \varphi_2}{\partial s_3}, & \dfrac{\partial \varphi_3}{\partial s_3} & \cdots \\ \cdot & \cdot \cdot \cdot & \cdot \cdot \cdot \end{vmatrix}$$

s and s_1 are always to be considered constant during the partial differentiation. In the integral on the left-hand side of Equation (69) one must set:

$$da_1 = ds_1 \cdot \frac{\partial \varphi_1(s, s_1, a_2, a_3 \cdots a_n)}{\partial s_1}.$$

Since the region of integration is n-fold infinitesimal, the last factor can come out in front of the integration sign and one obtains, on dividing through by $\int ds_1 da_2 da_3 \cdots da_n$:

$$\frac{\partial \varphi_1(s, s_1, a_2, a_3 \cdots a_n)}{\partial s_1} = \frac{\Delta}{\Delta_1} = C \frac{\sigma}{\Delta_1 \tau}. \tag{70}$$

However, if all the integrals up to φ_1 are known, and are used, in the last of the differential equations to be integrated,

$$\delta s_1 = \delta s \frac{\sigma_1}{\sigma} \tag{71}$$

to express the quantities $s_2, s_3, \cdots, s_n$ in terms of s, s_1 and the constants $a_2, a_3, \cdots, a_n$, then the expression (70) is the integrating factor of this last differential equation. On multiplying by this factor, its left-hand side is transformed to

$$\frac{\partial \varphi_1(s, s_1, a_2, a_3 \cdots a_n)}{\partial s_1} \delta s_1$$

and hence the right-hand side must be transformed to

$$-\frac{\partial \varphi_1(s, s_1, a_2, a_3 \cdots a_n)}{\partial s} \delta s$$

This is Jacobi's theorem of the last multiplier. Since C depends only on the integration constants, likewise $\sigma/\Delta_1\tau$ is an integrating factor of the differential equation (71).

σ is given. Δ_1 can be calculated if all the integrals up to φ_1 are known. τ is of course in general unknown; still, it can often be found by accident, as for example in the case of mechanical problems, where it reduces to a constant.

If the equations of motion (66) of a material system do not explicitly contain the time, then they also have the form (57) after one has eliminated the differential of the time variable; s is now one of the coördinates, say p_1. Then

$$\sigma = \frac{\partial E}{\partial q_1}, \quad \sigma_1 = \frac{\partial E}{\partial q_2} \cdots \sigma_n = -\frac{\partial E}{\partial p_\mu}$$

and the equation

$$\frac{\partial \sigma}{\partial s} + \frac{\partial \sigma_1}{\partial s_1} + \cdots + \frac{\partial \sigma_n}{\partial s_n} = 0,$$

from which it follows that $\tau = \text{const.}$, always holds. Hence one can find directly the integrating factor of the differential equation, which expresses the differential of the last coördinate in terms of the others and the momenta, if the other coördinates and the momenta have already been found as functions of the constants of integration and the two last coördinates. In most applications that Jacobi makes of the principle of the last multiplier, the general equations are applied in this way.

§31. Introduction of the energy differential.

Before we pass to the special applications to gas theory, we shall develop still one more general theorem.

We return to the infinite family of equivalent mechanical systems already considered in §26. The state of each system will again be determined by the variables introduced in §25. As before, let L be the kinetic and V the potential energy, and let $E=L+V$ be the total energy of one of the systems. We assume that the system is a so-called conservative one—i.e., E remains constant during the entire motion for each system. We must therefore exclude dissipative forces like viscosity, internal resistance, etc.; we must require that only internal forces may be present in each system, or, if external forces are present, they must be produced by fixed masses that do not change with time. The force should in general depend only on position, so that V is a function (and indeed a single-valued function) only of the coördinates $p_1, \cdots, p_\mu$.

The values

$$(72) \qquad p_1, p_2 \cdots p_\mu, q_1 \cdots q_\mu,$$

which the coördinates and momenta assume at time t, when initially (at time zero) they have the values

$$(73) \qquad P_1, P_2 \cdots P_\mu, Q_1 \cdots Q_\mu$$

we call the values corresponding to the initial values. The energy E of the system is also determined by the initial values (73), and this will similarly be called the value of the energy corresponding to the initial values. Since the system is conservative, the energy has, for a particular system at each arbitrary later time t, the same value E as at the initial time.

We shall consider next all systems that start from initial values filling some 2μ-fold infinitesimal region enclosing the values (73). The region filled by the values of the coördinates and momenta of all these systems after a definite time t will be called the region g. The integral of the product of differentials of the coördinates and momenta, extended over G, we denote by

$$(74) \qquad \int dP_1 \cdots dQ_\mu,$$

and the same integral extended over the region g is denoted by

$$\int dp_1 \cdots dq_\mu. \tag{75}$$

Then, according to Equation (55),

$$\int dP_1 \cdots dQ_\mu = \int dp_1 \cdots dq_\mu. \tag{76}$$

In each of these integrals we can replace one of the differentials, e.g., the differential of the first momentum q_1, by the differential of the energy E. Here we have to consider that all the coördinates and all the other momenta are constants. We obtain therefore

$$dE = \frac{\partial E}{\partial q_1} dq_1,$$

where, in the partial differentiation, the above-mentioned quantities (and of course the time) are to be regarded as constant. Now we have:

$$\frac{\partial E}{\partial q_1} = \frac{\partial V}{\partial q_1} + \frac{\partial L}{\partial q_1}.$$

Since V is a function only of the coördinates, the first term vanishes immediately; furthermore, it is well known that if we express the time-derivative by a prime, then[1]

$$\frac{\partial L}{\partial q_1} = p_1'.$$

Therefore

$$\frac{\partial E}{\partial q_1} = p_1'$$

and

$$dE = p_1' dq_1.$$

Likewise,

$$dE = P_1' dQ,$$

[1] Jacobi, *Vorlesungen über Dynamik*, 9th lecture, p. 70, Eq. (4).

where P_1' represents the value of the time derivative of p_1 at the initial time. Substitution of this value into Equation (76) yields:

$$(77) \qquad \begin{cases} \dfrac{1}{p_1'}\displaystyle\int dp_1 \cdots dp_\mu dq_2 \cdots dq_\mu dE = \\ = \dfrac{1}{P_1'}\displaystyle\int dP_1 \cdots dP_\mu dQ_2 \cdots dQ_\mu dE. \end{cases}$$

This equation is valid, whatever may be the region over which the integration is extended, provided only that it is 2μ-fold infinitesimal, and that g is the region corresponding to G. We can therefore choose the region G such that for all values of the other variables the energy lies between the same limits E and $E+dE$, while the other variables, namely

$$(78) \qquad \begin{cases} \text{the coördinates } & p_1, \cdots, p_\mu \\ \text{and the momenta } & q_2, \cdots, q_\mu \end{cases}$$

lie in some arbitrary $(2\mu-1)$-fold infinitesimal region G_1, which includes the values

$$(79) \qquad P_1, P_2 \cdots P_\mu, Q_2 \cdots Q_\mu$$

One of the momenta, Q_1, is omitted, since it is already determined by the values of the variables (78) and the energy.

For all systems satisfying these initial conditions, the energy will lie between the same limits after time t; the $(2\mu-1)$-fold infinitesimal region, which will be filled by the values that the variables (78) have for this system, will be called the region g_1. It encloses, of course, the values of the coördinates and momenta corresponding to the initial values (79) after time t. If one chooses the region in this way, then he sees that dE comes out in front of the integral sign on both sides of Equation (77), so that this equation can be divided through by dE, whereby one obtains:

$$(80) \qquad \begin{cases} \dfrac{1}{p_1'}\displaystyle\int dp_1 \cdots dp_\mu \cdots dq_2 \cdots dq_\mu \\ = \dfrac{1}{P_1'}\displaystyle\int dP_1 \cdots dP_\mu \cdots dQ_2 \cdots dQ_\mu. \end{cases}$$

Here the integral on the left is to be extended over g_1, and that on the right over G_1, while E is constant. Equation (80) therefore has the following meaning: we consider many systems, for all of which the energy has the same value, while the values of the variables (78) lie initially in the $(2\mu - 1)$-fold infinitesimal region G_1, and the missing momentum variable q_1 is determined by E. For all these systems, the energy will have the same value E after time t, but the region filled by the values of the variables (78) after time t will be denoted by g_1, and will be the region corresponding to G_1 after time t. Equation (80) always holds then, if one extends the integral on the right over G_1, and that on the left over g_1.

§32. *Ergoden.**

We now imagine again an enormously large number of mechanical systems, all of which have the same properties described earlier. The total energy E will have the same value for all of them. On the other hand, the initial values of the coördinates and momenta will have different values for different systems. Let

* This word is left untranslated since the English equivalent, "microcanonical ensemble," had not yet come into use when Boltzmann wrote this book. (See Gibbs, *Elementary Principles in Statistical Mechanics* [New York, 1902], chap. x.) *Ergoden* should not be confused with "ergodic systems," *i.e.*, hypothetical mechanical systems that have the (impossible) property that their coördinates and momenta eventually take on every value consistent with their fixed total energy. Boltzmann never used the word *Ergoden* for such systems, but "ergodic" came to be applied to them following the discussion published by P. and T. Ehrenfest in the *Encyklopädie der mathematischen Wissenschaften* (1911). Although the Ehrenfests made a valuable contribution by their critical analysis of the foundations of gas theory, they unfortunately misrepresented the opinions and even the terminology of Boltzmann and Maxwell. Boltzmann did discuss ergodic systems without calling them that (Wien. Ber. **63,** 679 [1871], J. r. ang. Math. **100,** 201 [1887]), but he did not make the foundations of gas theory depend on their existence, nor did he even make a clear distinction between going through every point on the energy surface, and going infinitely close to every point. *Ergoden* were first introduced explicitly in 1884 (Boltzmann, Wien. Ber. **90,** 231 [1884]; J. r. ang. Math. **98,** 68 [1885]) although both Maxwell and Boltzmann had previously used the same device in making calculations. Boltzmann used the word "isodic" for the systems that we now call ergodic. Their impossibility was proved independently by Plancherel and Rosenthal, Ann. Physik [4] **42,** 796, 1061 (1913).

$$f(p_1, p_2 \cdots p_\mu, q_2 \cdots q_\mu, t)dp_1 \cdots dp_\mu dq_2 \cdots dq_\mu$$

be the number of systems for which, at time t, the variables (78) lie between the limits

$$p_1 \text{ and } p_1 + dp_1 \cdots p_\mu \text{ and } p_\mu + dp_\mu, q_2 \text{ and } q_2 + dq_2 \cdots$$
$$q_\mu \text{ and } q_\mu + dq_\mu$$

while of course q_1 is determined by the assumed value of the energy. The number of systems for which the values of the variables (78) fill a $(2\mu-1)$-fold infinitesimal region g_1 enclosing the value

(81) $$p_1 \cdots p_p, q_2 \cdots q_\mu$$

is therefore, at time t,

(82) $$f(p_1 \cdots p_\mu, q_2 \cdots q_\mu, t) \cdot \int dp_1 \cdots dp_\mu dq_2 \cdots dq_\mu,$$

where the integration is to be extended over the region g_1.

Instead of saying that the values of the variables (78) lie within the region g_1 for a certain system, we shall often use the expression: this system has the phase pq. We can therefore say also: the expression (82) gives the number of systems that have phase pq at time t.

The region within which, for all systems that have the phase pq at time t, the values of the variables (78) lie at time zero, shall be the region G_1. Since g_1 includes the values (81), G_1 will of course include the initial values of the variables

(83) $$P_1 \cdots P_\mu, Q_2 \cdots Q_\mu$$

corresponding to the values (81). Instead of saying that the values of the variables lie in G_1 for a system, we shall use the expression: the system has the phase PQ. By analogy with the notation used in Equation (82), the integral of the product of differentials of the variables (78) extended over G_1 shall be denoted by

$$\int dP_1 \cdots dP_\mu dQ_2 \cdots dQ_\mu.$$

Since t can have any arbitrary value in Equation (82),

(84) $$f(P_1 \cdots P_n, Q_2 \cdots Q_n, 0) \int dP_1 \cdots dP_n dQ_2 \cdots dQ_n$$

will be the number of systems that initially have phase PQ. Since these are the same systems as the ones that have phase pq at time t, (82) and (84) must be equal to each other; whence, taking account of Equation (80), it follows that:

$$(85)\quad p_1' f(p_1 \cdots p_n, q_2 \cdots q_n, t) = P_1' f(P_1 \cdots P_n, Q_2 \cdots Q_n, 0).$$

The distribution of states among the systems will be called a stationary one if the number of systems having any arbitrary phase pq does not change with time. Since the number of systems that have phase pq at time t is given by the expression (82), one can express the condition that the distribution of states be stationary by saying that for any arbitrary values of the variables and for an arbitrary region g_1, the value of the expression (82) is completely independent of time, as long as g_1 and the values of the variables (78) remain the same. Thus if one equates the value of (82) at time zero to its value at some other time t, he can divide by the integral over g_1, and the condition that the state distribution be stationary takes the form

$$(86)\quad f(p_1 \cdots p_n, q_2 \cdots q_n, t) = f(p_1 \cdots p_n, q_2 \cdots q_n, 0,)$$

in which the variables p, q can have any arbitrary values but must be the same on both sides. Therefore one can also denote them by the corresponding capital letters, so that Equation (86) takes the form:

$$(87)\quad f(P_1 \cdots P_n, Q_2 \cdots Q_n, t) = f(P_1 \cdots P_n, Q_1 \cdots Q_n, 0).$$

Using the last equation, Equation (85) becomes[1]

$$P_1' f(P_1, P_2 \cdots P_n, Q_2 \cdots Q_n, t) = p_1' f(p_1, p_2 \cdots p_n, q_2 \cdots q_n, t).$$

Since the function f no longer contains the time, it is better to omit t from the function sign and to write:

$$(88)\quad \begin{aligned} &P_1' f(P_1, P_2 \cdots P_n, Q_2 \cdots Q_n) = \\ &= p_1' f(p_1, p_2 \cdots p_n, q_2 \cdots q_n). \end{aligned}$$

Here, $P_1, P_2, \cdots, P_\mu, Q_2, \cdots, Q_\mu$ are completely arbitrary

[1] This (or the equivalent Eq. [88]) is a necessary condition that the distribution be stationary; it is also sufficient, since from it and Eq. (85) one can derive Eq. (87) for any P, Q, or Eq. (86) for any p, q; and these latter two equations are the mathematical expression of the fact that the distribution is stationary.

initial values; $p_1, p_2, \cdots, p_\mu, q_2, \cdots, q_\mu$ are the values of the coördinates and momenta that a system would acquire, starting from these initial values, after an arbitrary time t.

Thus if we imagine a system S, starting from some initial values of the coördinates and momenta, then in the course of its motion the coördinates and momenta will take various different values. The coördinates and momenta are therefore functions of the initial values and of the time. But in general there will be certain functions of the coördinates and momenta (known as invariants) that have constant values during the entire motion: for example, for a free system these would include the velocity components of the center of gravity, and the components of the total angular momentum. Now suppose that we substitute into the expression $p_1' f(p_1, p_2, \cdots, p_\mu, q_2, \cdots, q_\mu)$ first the initial values and then the values of the coördinates and momenta at later times. In order that the distribution be stationary, it is necessary and sufficient that the value of $p_1' f$ remain unchanged —or in other words, $p_1' f$ should contain only those functions of the coördinates and momenta that remain constant throughout the entire motion of a system and depend only on the initial values, but not on the elapsed time. Thus $p_1' f$ should be a function only of the invariants.

The simplest case of a stationary state distribution is obtained by setting $p_1' f(p_1, p_2, \cdots, p_\mu, q_2, \cdots, q_\mu)$ equal to a constant; then

$$\frac{C}{p_1'} \int dp_1 dp_2 \cdots dp_\mu dq_2 \cdots dq_\mu \tag{89}$$

is the number of systems for which the variables (78) lie in the region g_1 over which the integration is to be extended. I once allowed myself to call the distribution of states among an infinite number of systems described by this formula an ergodic one.

§33. Concept of the momentoid.

The state distribution mentioned at the conclusion of the preceding section will be considered further in the following, and indeed we shall introduce other variables in place of the momenta.

The kinetic energy L of a system is a homogeneous quadratic function of the momenta; therefore

$$2L = a_{11}q_1^2 + a_{22}q_2^2 \cdots + 2a_{12}q_1q_2 \cdots,$$

where in general the coefficients a are functions of the generalized coördinates p. As is well known, a linear substitution of the form

$$(90) \qquad \begin{cases} q_1 = b_{11}r_1 + b_{12}r_2 \cdots b_{1\mu}r_\mu \\ q_2 = b_{21}r_1 + b_{22}r_2 \cdots b_{2\mu}r_\mu \\ \cdot \quad \cdot \quad \cdot \quad \cdot \quad \cdot \quad \cdot \quad \cdot \quad \cdot \quad \cdot \quad \cdot \quad \cdot \\ q_\mu = b_{\mu 1}r_1 + b_{\mu 2}r_2 \cdots b_{\mu\mu}r_\mu \end{cases}$$

can always be found, for which one obtains

$$2L = \alpha_1 r_1^2 + \alpha_2 r_2^2 + \cdots \alpha_\mu r_\mu^2.$$

The exceptional cases where L cannot be brought into this form can never occur in mechanical systems, and none of the coefficients α (which in general of course are functions of the coördinates) can be zero or negative, since if they were the kinetic energy would be zero or negative for some motions of the system. By multiplying all the r by the same factor (which may also be a function of the coördinates) one can make the determinant of the b equal to 1. In the following we shall denote by r the quantities that have already been multiplied by such a factor. Of course they can conversely be expressed linearly in terms of the q. I have proposed to call them the momentoids corresponding to the coördinates p.

We shall imagine a μ-fold infinitesimal region H marked off in the q, and in the integral over this region

$$\int dq_1 dq_2 \cdots dq_\mu$$

by means of Equations (90) we introduce the r in place of the q as variables of integration. We consider the p constant. Since the determinant of the b is equal to 1, it follows that

$$(91) \qquad \int dq_1 dq_2 \cdots dq_\mu = \int dr_1 dr_2 \cdots dr_\mu,$$

where the latter integral is to be extended over the region of the r corresponding to the region H—i.e., over that region which includes all combinations of values of the r that (by Eqs. [90]) correspond to all combinations of the values of the q contained in the region H.

We shall now introduce, in the integral on the right-hand side of Equation (76), the r instead of the q as integration variables. This integral becomes, according to Equation (91),

$$\int dp_1 \cdots dp_\mu dq_1 \cdots dq_\mu = \int dp_1 \cdots dp_\mu dr_1 \cdots dr_\mu.$$

The latter integral is to be extended over the region corresponding to the region of integration of the former (previously called g).

Now we shall introduce in this equation (just as we did in the derivation of Equations (77) and (80) from (76)) the integration variable E on the left-hand side instead of q_1, and on the right-hand side instead of r_1. Since

$$\frac{\partial E}{\partial r_1} = \frac{\partial L}{\partial r_1} = \frac{1}{\alpha_1 r_1}$$

if follows that

$$\frac{1}{p_1'} \int dp_1 \cdots dp_\mu dq_2 \cdots dq_\mu dE = \frac{1}{\alpha_1 r_1} \int dp_1 \cdots dp_\mu dr_2 \cdots dr_\mu dE.$$

Just as in deriving Equation (80) from (77), we can now choose the region in such a way that, for all possible values of the other variables, E lies between the same limits E and $E+dE$. Then we can divide through by dE, and we find that, for constant E,

$$\frac{1}{p_1'} \int dp_1 \cdots dp_\mu dq_2 \cdots dq_\mu = \frac{1}{\alpha_1 r_1} \int dp_1 \cdots dp_\mu dr_2 \cdots dr_\mu.$$

If we substitute this in Equation (89), we find that for an ergodic distribution of states the number of systems for which the variables

(92) $$p_1 \cdots p_\mu,\ r_2 \cdots r_\mu$$

lie in an arbitrary $(2\mu-1)$-fold infinitesimal region surrounding this value is equal to

$$\frac{C}{\alpha_1 r_1}\int dp_1 \cdots dp_\mu dr_2 \cdots dr_\mu \tag{93}$$

where the integration is to be extended over this region.

The demarcation of this region is arbitrary. In the following, we shall specify it in the simplest way possible so that the variables (92) will lie between the limits

$$p_1 \text{ and } p_1 + dp_1,\ p_2 \text{ and } p_2 + dp_2 \cdots p_\mu \text{ and } p_\mu + dp_\mu \tag{94}$$

$$r_2 \text{ and } r_2 + dr_2,\ r_3 \text{ and } r_3 + dr_3 \cdots r_\mu \text{ and } r_\mu + dr_\mu. \tag{95}$$

The number of systems for which these conditions are satisfied is, according to (93),

$$dN = \frac{C}{\alpha_1 r_1} dp_1 dp_2 \cdots dp_\mu dr_2 \cdots dr_\mu. \tag{96}$$

If we denote the product of the differentials $dp_1 dp_2 \cdots dp_\mu$ by $d\pi$, and the product $dr_{k+1} dr_{k+2} \cdots dr_\mu$ by $d\rho_k$, then

$$\begin{cases} dN_1 = \dfrac{C d\pi dr_\mu}{\alpha_1}\displaystyle\iint \cdots \frac{1}{r_1} dr_2 dr_3 \cdots dr_{\mu-1} \\ \quad\;\; = \dfrac{C d\pi dr_\mu}{\alpha_1}\displaystyle\int \frac{1}{r_1}\frac{d\rho_1}{dr_\mu} \end{cases} \tag{97}$$

is the number of systems for which the coördinates lie between the limits (94) and r_μ is between

$$r_\mu \text{ and } r_\mu + dr_\mu \tag{98}$$

while the other r can have all possible values consistent with the equation of kinetic energy. The number of systems that are subject to the condition that their coördinates lie between the limits (94) while their momenta are subject to no other condition than the constancy of total energy is

$$dN_2 = \frac{C d\pi}{\alpha_1}\int \frac{1}{r_1} d\rho_1. \tag{99}$$

The total number of all systems is

$$N = C\iint \frac{d\pi d\rho_1}{\alpha_1 r_1}; \tag{100}$$

where, everywhere that a product of several differentials is expressed by a single differential sign, the integration over all values of these differentials is likewise expressed by a single integral sign.

§34. Expression for the probability; average values.

The expressions 1. dN/N; 2. dN_1/N; and 3. dN_2/N are the definitions of the following probabilities: 1. that for a system, the coördinates and momentoids lie between the limits (94) and (95); 2. that the coördinates lie between the limits (94) and the momentoid r_μ lie between the limits (98); 3. that the coördinates lie between the limits (94).

For all systems whose coördinates and momenta lie between the limits (94) and (95), the kinetic energy $\frac{1}{2}\alpha_1 r_1^2$ corresponding to the first momentoid is the same. (The number of such systems is given by Eq. [96].) The mean value of this quantity for all systems whose coördinates remain subject to the conditions (94) is therefore

$$\overline{\frac{\alpha_1 r_1^2}{2}} = \frac{1}{dN_2}\int \frac{\alpha_1 r_1^2}{2} dN = \frac{\alpha_1 \int r_1 d\rho_1}{2\int \frac{1}{r_1} d\rho_1}, \tag{101}$$

where the integral sign expresses simply an integration over all possible values of the momentoids. The mean value of $\frac{1}{2}\alpha_1 r_1^2$ for all systems is in general

$$\overline{\overline{\frac{\alpha_1 r_1^2}{2}}} = \frac{\int d\pi \int r_1 d\rho_1}{2\int \frac{d\pi}{\alpha_1}\int \frac{d\rho_1}{r_1}}. \tag{102}$$

The mean value of the potential function V for all systems, however, is

$$\overline{V} = \frac{\int \frac{V}{\alpha_1} d\pi \int \frac{d\rho_1}{r_1}}{\int \frac{d\pi}{\alpha_1}\int \frac{d\rho_1}{r_1}}. \tag{103}$$

The integration over the momentoids can easily be performed as follows. If we let A and α be constants, then one finds by the substitution

$$r = \sqrt{\frac{2A}{\alpha}} \cdot \sqrt{x}$$

the following equation:

$$(104)\quad \begin{cases} \displaystyle\int_{-\sqrt{2A/\alpha}}^{+\sqrt{2A/\alpha}} \sqrt{A - \frac{\alpha r^2}{2}}^{\lambda} dr \\ \displaystyle = \sqrt{\frac{2}{\alpha}}\, A^{\lambda/2+1/2} \int_0^1 x^{-1/2}(1-x)^{\lambda/2} dx \\ \displaystyle = \sqrt{\frac{2}{\alpha}}\, A^{\lambda/2+1/2} B\left(\frac{1}{2}, \frac{\lambda}{2}+1\right) \\ \displaystyle = \sqrt{\frac{2}{\alpha}}\, \frac{\Gamma\left(\frac{1}{2}\right)\cdot\Gamma\left(\frac{\lambda}{2}+1\right)}{\Gamma\left(\frac{\lambda}{2}+\frac{3}{2}\right)} \cdot A^{\lambda/2+1/2}. \end{cases}$$

B and Γ denote the well-known Euler [beta and gamma] functions.

We next use this formula to calculate the integral

$$J_\kappa = \int r_1^\kappa d\rho_1.$$

We denote the quantity

$$E - V - \frac{\alpha_{k+1} r_{k+1}^2}{2} - \frac{\alpha_{k+2} r_{k+2}^2}{2} - \cdots \frac{\alpha_\mu r_\mu^2}{2}$$

by A_k, and the quantity $E-V$ by A_μ. Then

$$r_1 = \sqrt{\frac{2H_1}{\alpha_1}} = \sqrt{\frac{2}{\alpha_1}} \sqrt{A_2 - \frac{\alpha_2 r_2^2}{2}},$$

hence

$$J_\kappa = \sqrt{\frac{2^\kappa}{\alpha_1}} \int d\rho_2 \int \sqrt{A_2 - \frac{\alpha_2 r_2^2}{2}}^{\kappa} dr_2.$$

The momentoid r_2 takes its most extreme possible values when $r_1=0$, in which case $r_2=\pm\sqrt{2A_2/\mu_2}$. The integration over r_2 is therefore to be taken between these limits. If one performs this integration using Equation (104), he finds:

$$J_\kappa = \left(\frac{2}{\alpha_1}\right)^{\kappa/2} \sqrt{\frac{2}{\alpha_2}} \frac{\Gamma\left(\frac{1}{2}\right)\Gamma\left(\frac{\kappa}{2}+1\right)}{\Gamma\left(\frac{\kappa}{2}+\frac{3}{2}\right)} \int A_2^{\kappa/2+1/2} d\rho_2$$

$$= \left(\frac{2}{\alpha_1}\right)^{\kappa/2} \sqrt{\frac{2}{\alpha_1}} \frac{\Gamma\left(\frac{1}{2}\right)\Gamma\left(\frac{\kappa}{2}+1\right)}{\Gamma\left(\frac{\kappa}{2}+\frac{3}{2}\right)}$$

$$\times \int d\rho_3 \int_{-\sqrt{2A_3/\alpha_3}}^{+\sqrt{2A_3/\alpha_3}} \left(A_3 - \frac{\alpha_3 r_3^2}{2}\right)^{\kappa/2+1/2} dr_3.$$

If one also performs the integration over r_3 according to Equation (104), then it follows that:

$$(105)\quad \begin{cases} J_\kappa = \left(\frac{2}{\alpha_1}\right)^{\kappa/2} \sqrt{\frac{2}{\alpha_2}\cdot\frac{2}{\alpha_3}} \frac{\left[\Gamma\left(\frac{1}{2}\right)\right]^2\Gamma\left(\frac{\kappa}{2}+1\right)}{\Gamma\left(\frac{\kappa}{2}+\frac{4}{2}\right)} \times \\ \int d\rho_4 \int_{-\sqrt{2A_4/\alpha_4}}^{+\sqrt{2A_4/\alpha_4}} \left(A_4 - \frac{\alpha_4 r_4^2}{2}\right)^{\kappa/2+1/2} dr_4. \end{cases}$$

From this result we can find the integral of (97) by leaving out the last differential, performing the other integrations exactly, as we have already done in the expression for J_κ, and then setting $\kappa=-1$. Then it follows that

$$\iint \cdots \frac{1}{r_1} dr_2 dr_3 \cdots dr_{\mu-1}$$

$$= \sqrt{\frac{\alpha_1}{2}\cdot\frac{2}{\alpha_2}\frac{2}{\alpha_3}\cdots\frac{2}{\alpha_{\mu-1}}} \frac{\left(\Gamma\frac{1}{2}\right)^{\mu-1}}{\Gamma\left(\frac{\mu-1}{2}\right)} \left(A_\mu - \frac{\alpha_\mu r_\mu^2}{2}\right)^{(\mu-3)/2}$$

If one denotes the last expression by γ, then the mean value of $\frac{1}{2}\alpha_\mu r_\mu^2$ in all systems for which the coördinates lie between the limits (94) is

$$\frac{\overline{\alpha_\mu r_\mu^2}}{2} = \frac{\int_{-\sqrt{2A_\mu/\alpha_\mu}}^{\sqrt{2A_\mu/\alpha_\mu}} \frac{\alpha_\mu r_\mu^2}{2} \gamma dr_\mu}{\int_{-\sqrt{2A_\mu/\alpha_\mu}}^{\sqrt{2A_\mu/\alpha_\mu}} \gamma dr_\mu}$$

Evaluation of the integral yields

$$\text{(105a)} \qquad \frac{\overline{\alpha_\mu r_\mu^2}}{2} = \frac{A_\mu}{\mu} = \frac{E-V}{\mu}.$$

If one allows κ to be arbitrary in Equation (105) and performs all the integrations, then it follows that:

$$J_\kappa = \int r_1^\kappa d\rho_1$$

$$= \sqrt{\frac{2}{\alpha_1}}^{\kappa} \sqrt{\frac{2}{\alpha_2}\frac{2}{\alpha_3}\cdots\frac{2}{\alpha_\mu}} \frac{\left(\Gamma\frac{1}{2}\right)^{\mu-1}\Gamma\left(\frac{\kappa}{2}+1\right)}{\Gamma\left(\frac{\kappa+\mu+1}{2}\right)} A_\mu^{(\kappa+\mu-1)/2}.$$

By means of the last two formulas the integrations over the r in all the previous expressions can be written down immediately, and dN_1, dN_2, and $\overline{\frac{1}{2}\alpha_1 r_1^2}$ will be calculated in closed form. To perform the integrations over the p, a knowledge of the potential function V would of course be required. One obtains, for example, for the probability that, for a system that satisfies the conditions (94), r_μ lies between r_μ and $r_\mu+dr_\mu$, the value

$$\text{(106)} \qquad \frac{dN_1}{dN_2} = \frac{\Gamma\left(\frac{\mu}{2}\right)}{\Gamma\left(\frac{1}{2}\right)\Gamma\left(\frac{\mu-1}{2}\right)} \cdot \sqrt{\frac{\alpha_\mu}{2}} \frac{\left(A_\mu - \frac{\alpha_\mu r_\mu^2}{2}\right)^{(\mu-3)/2}}{A_\mu^{(\mu-2)/2}} dr_\mu.$$

If one sets $\frac{1}{2}\alpha_\mu r_\mu = x$, then

$$dr_\mu = \frac{1}{2\sqrt{x}} \sqrt{\frac{2}{\alpha_\mu}}\, dx_\mu;$$

hence the probability that, for a system satisfying the conditions (94), r_μ is positive and $\frac{1}{2}\alpha_\mu r_\mu^2$ lies between x and $x+dx$, is:

$$\frac{\Gamma\left(\frac{\mu}{2}\right)}{\Gamma\left(\frac{1}{2}\right)\Gamma\left(\frac{\mu-1}{2}\right)} \frac{(A_\mu - x)^{(\mu-2)/2}}{2A_\mu^{(\mu-2)/2}} \frac{dx}{\sqrt{x}}.$$

Since for negative r_μ an equal value of $\frac{1}{2}\alpha_\mu r_\mu^2$ is equally probable, the probability that $\frac{1}{2}\alpha_\mu r_\mu^2$ lies between x and $x+dx$ for either positive or negative r_μ is equal to

(107) $$\frac{\Gamma\left(\frac{\mu}{2}\right)}{\Gamma\left(\frac{1}{2}\right)\Gamma\left(\frac{\mu-1}{2}\right)} \frac{(A_\mu - x)^{(\mu-3)/2}}{A_\mu^{(\mu-2)/2}} \frac{dx}{\sqrt{x}}.$$

Here r_μ may be any arbitrary momentoid. If μ is very large, and one sets $A_\mu = \mu\xi$, then the above expression approaches the limit

(108) $$e^{-x/2\xi} \frac{dx}{\sqrt{2\pi\xi x}}.$$

From the general formulas one finds, furthermore:

(109) $$\frac{\overline{\alpha_1 r_1^2}}{2} = \frac{\alpha_1 J_1}{2J_{-1}} = \frac{A_\mu}{\mu} = \frac{E - V}{\mu}$$

in agreement with Equation (105a). Since the same holds of course for the part of the kinetic energy corresponding to the other momentoids, it follows that:

(110) $$\frac{\overline{\alpha_1 r_1^2}}{2} = \frac{\overline{\alpha_1 r_2^2}}{2} = \cdots \frac{\overline{\alpha_\mu r_\mu^2}}{2}.$$

Thus however the limits (94) may be chosen, the following theorem will always hold in the case of our assumed (ergodic) distribution of states: we pick out of all systems those for which the coördinates lie between the limits (94). We denote by $\frac{1}{2}\alpha_i r_i^2$ the kinetic energy corresponding to some one momentoid, and calculate the average of its values for all specified systems at some time t. This average always comes out to be the same for all times and all values of the index i. It is equal to the μ'th part of the energy $E-V$, which has the form of a kinetic energy in this case.

The integration over the coördinates can of course only be indicated formally, and one finds for the mean value of $\frac{1}{2}\alpha_i r_i^2$ for all systems, for all values of the index i,

$$\overline{\overline{\frac{\alpha_i r_i^2}{2}}} = \frac{\int \overline{\frac{\alpha_i r_i^2}{2}}\, dN_2}{N} = \frac{\int (E-V)\frac{d\pi}{\alpha_1}}{\mu \int \frac{d\pi}{\alpha_1}} \tag{111}$$

Of course this equality of the mean values of the kinetic energy corresponding to each momentoid has only been proved for the assumed (ergodic) distribution of states. This distribution is certainly a stationary one. In general there can and will be other stationary distributions for which this theorem does not hold.

In the special case that V is a homogeneous quadratic function of the coördinates, as L is of the momenta, the integration over coördinates can be performed by the same method as that used for the momenta. Then one obtains from Equation (103)

$$\overline{V} = \overline{L} = \frac{E}{2}, \tag{111a}$$

if one determines the additive constant in the potential in such a way that V vanishes when all the material points are at their rest positions.

Before I proceed to the application of these theorems to the theory of gases with polyatomic molecules, I will first adduce a completely general argument which is irrelevant from a mathematical viewpoint, but calls on experimental evidence; it may perhaps justify the supposition that the significance of these theorems is not restricted to the theory of polyatomic gas molecules.

§35. General relationship to temperature equilibrium.

We now consider an arbitrary warm body as a mechanical system obeying the laws we have deduced up to now—in other words, as a system of atoms, or molecules, or rather some kind of constituents whose positions are determined by generalized coördinates.

Experience shows that whenever a body has the same thermal energy and is subjected to the same external conditions, it eventually comes to the same state, no matter what its initial state may have been. In the sense of the mechanical view of nature, it thus happens that only certain average values—the mean kinetic energy of a molecule in a finite part of the body, the momentum that a molecule transports on the average through a finite surface in a finite time, and so forth—are accessible to observation. However, these average values are the same in by far the greatest number of possible states. We call each state that has this particular average value a probable state.

Thus if the initial state is not a probable state, the body will soon pass over to a probable state if the external conditions remain fixed, and it will persist in this state during further observations, so that although its state changes progressively and occasionally (within a very long time period exceeding all possibility of observation) it will indeed deviate considerably from a probable state, nevertheless it gives the appearance of having attained a stationary final state, since all observable mean values remain fixed.

The mathematically most complete method would be to take account of the initial conditions from which a given warm body happened to evolve to a particular thermal state, which then persists for a long time. However, since the mean values will always be the same no matter what the initial state may have been, we can also obtain the same mean values if we imagine that instead of a single warm body an infinite number are present, which are completely independent of each other and, each having the same heat content and the same external conditions, have started from all possible initial states. We thus obtain the correct average values if we consider, instead of a single mechanical system, an infinite number of equivalent systems, which started from arbitrary different initial conditions. Now these mean values must be the same at all times, as is certainly the case if the average state of the aggregate of all systems remains stationary, and the state that we consider should not be an individual singular state, but rather all possible states must be included.

These conditions are satisfied if we imagine infinitely many mechanical systems, among which there is initially a state distribution such as the one we called ergodic in §32. We saw there that this state distribution is stationary, and that it includes all possible states consistent with the given kinetic energy.

There is therefore a certain probability that the mean values found in §34 are valid not only for the aggregate of systems but also for the stationary final state of each individual warm body, and that in particular in this case the equality of the mean kinetic energy corresponding to each momentoid is the condition of temperature equilibrium between the different parts of the warm body. That the condition of temperature equilibrium of warm bodies has a very simple mechanical meaning independent of their initial state will thereby be made probable, in that compression, expansion, displacement, etc. of the individual parts does not affect this equilibrium.

If we substitute for our general system a system formed from two different gases separated by a solid heat-conducting dividing wall (which is clearly a special case of the general system considered earlier) then we can interpret one of the r as the velocity component of a molecule multiplied by its mass. According to Equation (110) the mean kinetic energy of the center of gravity of a molecule must be equal for both gases, whence Avogadro's law follows.

This mean kinetic energy must be equal to the average kinetic energy corresponding to an arbitrary momentoid which determines the molecular motion of any body in thermal equilibrium with the gas. Hence, if we use a perfect gas as thermometric substance, the increment of kinetic energy corresponding to each such momentoid must be equal to the temperature increment multiplied by a constant that is the same for all momentoids. The heat present in the form of kinetic energy of molecular motion in any such body would therefore be equal to the product of the absolute temperature and the number of momentoids determining the molecular motion, multiplied by a constant that is the same for all bodies and all temperatures.

If we substitute for one of the mechanical systems a pure gas with compound molecules, which is again a special case, then it follows that for each molecule the mean kinetic energy of the center of gravity must be equal to three times the mean kinetic energy corresponding to any one of the momentoids determining the internal motion of the molecules. We shall derive this theorem (insofar as it concerns gases) in another way in the following sections.

We can assign six of the momentoids r of a system subjected only to internal forces to the three components of total momentum and the three components of total angular momentum, referred to three perpendicular axes. For ergodic systems, the mean kinetic energy corresponding to each of these is equal to that for any other momentoid, and hence is vanishingly small when the system consists of many atoms. Our considerations are therefore relevant to the case of nonrotating bodies at rest, subject to internal forces only.

Just as we have restricted ourselves in §32 to systems in which the energy has the same value, we can still further restrict ourselves to systems in which other quantities that are constant during the entire motion of a system have the same values, for example the velocity components of the center of gravity or the components of total angular momentum, as long as the system is subject only to internal forces. One then has to introduce the differentials of these quantities in place of the differentials of momenta, just as we introduced the energy differential in §31. One thus obtains other stationary state distributions which are not ergodic. The corresponding theorems are not necessarily of no mechanical interest; however, we shall not go into them here, since we shall not need them for the sequel.[1]

---

[1] Cf. Boltzmann, Wien Ber. **63,** 704 (1871). Maxwell, Trans. Camb. Phil. Soc. **12,** 561 (1878), *Scientific Papers* **2,** 730.

CHAPTER IV

Gases with compound molecules.

§36. Special treatment of compound molecules.

We shall now return to the general equations of §26, which are based on no hypotheses other than those from which the principles of mechanics are derived. We apply them to the following special case: a gas is found in a container enclosed on all sides by elastic walls. The molecules themselves need not all be of the same kind; thus we do not exclude the case of a mixture of several gases.

Each molecule will be considered as a mechanical system, as defined in §25. In gas theory, it is usually assumed that the centers of any two molecules are, on the average, so far apart that the time during which a molecule interacts with another one is small compared to the time during which it is subject to no such interaction. However, we shall not exclude here the possibility that two or more molecules may interact for a longer time, as would happen in the case of partially dissociated gases, as long as the number of molecules simultaneously interacting at one place is very small compared to the total number of molecules in the container. One always must conceive of only small individual groups of molecules interacting with each other, whose distance from all other molecules is very large compared to their spheres of action. Consequently each molecule will traverse a very long path between successive interactions, so that the frequencies of different kinds of collisions can be calculated by probability theory.

The position of a molecule of a certain kind, which we shall call the first, as well as the relative positions of its constituents, will be determined by μ generalized coördinates.

$$p_1, p_2 \cdots p_\mu.$$

We call these coördinates, and the corresponding momenta $q_1, q_2, \cdots, q_\mu,$

the variables (112).

The corresponding momentoids will be $r_1, r_2, \cdots, r_\mu$.

Three of the coördinates will determine the absolute position of some point in the molecule, for example its center of gravity. These will be the three coördinates p_1, p_2, and p_3. In order to have a definite representation, we assume that they are the rectangular coördinates of the center of gravity of the molecule in question. The rotation of the molecule about its center of gravity, and the relative positions of its constituents, are therefore determined by the other coördinates.

If there is no external force, each place within the container is equivalent. Hence all possible values of the three coördinates p_1, p_2, and p_3 are equally probable.

However, in order to make the problem as general as possible, we shall not exclude the presence of external forces. Then, aside from the forces between the molecules and the wall of the container, we still have three other kinds of forces: 1. the internal forces of a molecule, i.e., the intramolecular forces which act between different parts of the same molecule; 2. external forces, for example gravity, exerted from outside the container on the molecules of the body; 3. the interaction force which acts between two, or eventually more than two, different molecules, if these come unusually near each other.

The force function of the first two kinds of force should depend only on the coördinates of the molecules concerned; the force function of the third kind of force will depend on the coördinates of all the interacting molecules. We assume that the external forces vary very slowly from point to point inside the container, so that we can decompose this interior region into volume elements $dp_1dp_2dp_3$ with the following property: although each such volume element always contains a large number of molecules of each kind, the external force on a molecule does not change appreciably as it moves around inside this volume element. Just as, when there are no external forces at all, all places inside the container are equivalent, so in this case all places inside each such volume element are equivalent.

§37. Application of Kirchhoff's method to gases with compound molecules.

At the initial time (which we again call time zero) let the number of molecules of the first kind whose centers of gravity are found in any parallelepiped $dP_1dP_2dP_3$, for which the values of the variables

$$p_4 \cdots p_\mu,\ q_1 \cdots q_\mu \tag{113}$$

lie between the limits

$$P_4 \text{ and } P_4 + dP_4 \cdots Q_\mu \text{ and } Q_\mu + dQ_\mu$$

and which are not interacting with any other molecules be:

$$A_1e^{-2hE_1}dP_1 \cdots dQ_\mu.$$

Here A_1 is a constant that is different for the different kinds of molecules, while h is a constant that has the same value for all kinds of molecules. Let E_1 be the value of the sum of the kinetic energy of a molecule and the potential energy of the intramolecular and external forces acting on the molecule at the initial time. The negative partial derivatives of the potential energy function with respect to the coördinates give the components of the force, so that E_1 represents the total energy of a molecule, whose value remains constant as long as the molecule does not interact with others.

The number of molecules of the first kind not interacting with others, for which the variables (112) (defined in the previous section) initially lie in a 2μ-fold infinitesimal region G that includes the value

$$P_1, P_2 \cdots P_\mu,\ Q_1, Q_2 \cdots Q_\mu \tag{114}$$

is therefore

$$dN_1 = A_1e^{-2hE_1}\int dP_1 \cdots dQ_\mu, \tag{115}$$

where the integration is to be extended over the region G. The center of gravity should have enough room to move around in G so that, although all variables are confined within very narrow limits, the expression (115) is still a very large number.

When a molecule of the first kind moves under the influence

of internal and external forces, without interacting with other molecules, and the variables (112) start from the initial values (114), then after time t they will have the values

$$(116) \qquad p_1, p_2 \cdots q_\mu.$$

These will be the actual values of these variables, whereas (112) only gives the names of the variables. Let ϵ_1 be the value of the total energy at time t, so that according to the conservation of energy principle

$$(117) \qquad \epsilon_1 = E_1.$$

If, furthermore, all molecules for which the values of the variables (112) initially fill the region G move without interacting with other molecules, then the values of these variables after time t will fill a region which we call the region g. It includes of course the values (116).

If there were no interaction between the molecules at all, then the molecules whose variables lie in G at time zero must be the same ones whose variables lie in g at time t. If we denote the number of the latter by dn_1, then dn_1 would be equal to the expression (115), hence

$$dn_1 = A_1 e^{-2hE_1} \int dP_1 \cdots dQ_\mu.$$

But according to Equation (55),

$$\int dP_1 \cdots dQ_\mu = \int dp_1 \cdots dq_\mu,$$

where the latter integration is to be extended over the region g, corresponding after time t to the region G. Taking account of this fact, and of Equation (117), one finds:

$$(118) \qquad dn_1 = A_1 e^{-2h\epsilon_1} \int dp_1 \cdots dq_\mu.$$

This expression differs from (115) only in that the values of the variables (116) appear in places of the values (114), ϵ_1 in place of E_1, and g in place of G. However, since Equation (115) should be valid for any values of the variables and any regions enclosing them, (118) must also represent the number of molecules of the first kind for which the values of the variables (112) were initially in g. Hence the number of molecules of the first kind for which the values of the variables (112) lie in g has not changed during this time. Since, finally, the region G and hence also the region g are chosen completely arbitrarily, this must hold for any arbitrary region. In other words, the number of molecules for which the variables (112) lie in any arbitrary region does not change during an arbitrary time t. The distribution of states remains stationary, as long as only the intramolecular motion is considered.

§38. On the possibility that the states of a very large number of molecules can actually lie within very narrow limits.

We have assumed up to now that the regions G and g are very narrowly bounded, and yet at the same time we have assumed that the values of the variables for a very large number of molecules lie within these regions. If there are no external forces, this involves no difficulty. For then all points within the entire gas, when chosen as the position of the center of gravity of a molecule, are equivalent. The region

$$\Gamma = \int\int\int dP_1 dP_2 dP_3,$$

within which the center of gravity of a molecule lies does not then need to be infinitesimal, but rather it can be chosen arbitrarily large, since indeed we may set it equal to the entire volume of the container, and this can be chosen arbitrarily large. It is only the region within which the other variables $p_4, \cdots, q_\mu$ are enclosed—which we call symbolically the region G/Γ—which must be $(2\mu-3)$-fold infinitesimal.

We have therefore two quantities, one of which (namely the region Γ) can be chosen arbitrarily large, whereas the other (G/Γ) has to be made very small; and there is no relation between the size of these two regions. Indeed, the differential $dp_4 \cdots dq_\mu$ expresses the fact that we can choose G/Γ to be as small as we wish. However, for any particular such choice, we can choose Γ so large than a large number of molecules will always lie in G.

However, if external forces are present, then there is an upper limit to the size of the region Γ. In particular, this region must be chosen so small that the external force can be considered constant inside it. Then G and g are to be considered 2μ-fold very small; and the condition that the number of molecules for which the values of the variables lie within one of these regions must be very large can be satisfied only if the number of molecules in unit volume is infinite in the mathematical sense. Hence the satisfaction of the above conditions in this case remains merely an ideal; yet we still expect agreement with experience, for the following reasons.

In the molecular theory we assume that the laws of the phenomena found in nature do not essentially deviate from the limits that they would approach in the case of an infinite number of infinitesimally small molecules. This assumption was already made in Part I, for reasons given in §6. It is indispensable for any application of the infinitesimal calculus to molecular theory; indeed, without it, our model which strictly deals always with a large finite number, would not be applicable to apparently continuous quantities. This assumption will seem best justified to those who have carefully considered experiments for the direct proof of the atomic constitution of matter. Even in the smallest neighborhood of the tiniest particles suspended in a gas, the number of molecules is already so large that it seems futile to hope for any observable deviation, even in a very small time, from the limits that the phenomena would approach in the case of an infinite number of molecules.

If we accept this assumption, then we should also obtain agreement with experience by calculating the limit that the laws of the phenomena would approach in the case of an infinitely increasing number and decreasing size of the molecules. In calculating the latter limit, we again have in fact two quantities, which can independently be made arbitrarily small: the size of the volume element, and the dimensions of the molecules. For any given choice of the former, we can always choose the latter so small that each volume element still contains very many molecules, whose properties are closely defined within the given narrow limits.

If, with Kirchhoff, one interprets the expressions (115) and (118) as simply statements of probabilities, then one can allow them to be fractions or even very small quantities; yet one thereby loses their perspicuousness. We shall come back to this point at the end of the book (§92).

§39. Treatment of collisions of two molecules.

Up to now we have not considered the interactions of two molecules, and we still have to seek the conditions under which the initial distribution of states will not be altered by collisions. For this purpose we must seek the probability of the occurrence of groups of several molecules. We shall first restrict ourselves to the case that the simultaneous interaction of more than two molecules occurs so extraordinarily seldom that it is completely negligible. We can then limit ourselves to the consideration of molecule pairs.

The number of molecules of the first kind for which the variables (112) initially lie in the region G enclosing the values (114), when none of the molecules are assumed to interact with each other, will again be given by Equation (115).

Similarly the coördinates and momenta that determine the position and state of a molecule belonging to another type (called the second) will be denoted by

$$p_{\mu+1},\ p_{\mu+2} \cdots p_{\mu+\nu},\ q_1 \cdots q_{\mu+\nu}. \tag{119}$$

We shall ignore other kinds of molecules for the present. Nevertheless, the extension of our results to simultaneous interactions of several molecules does not offer any difficulty, though the expressions would become more complicated.

The number of molecules of the second kind for which the variables (119) initially lie in a region H enclosing the values

$$P_{\mu+1} \cdots Q_{\mu+\nu} \tag{120}$$

and of which none are assumed to interact with other molecules, is equal to

$$dN_2 = A_2 e^{-2hE_2} \int dP_{\mu+1} \cdots dQ_{\mu+\nu} \tag{121}$$

where the integration is to be extended over the region H. A_2 is a constant. E_2 is the total energy of the molecules of the second kind considered.

The centers of gravity of all these molecules of both kinds shall lie within a space in which the external force can be considered constant, and will be completely randomly distributed. Thus in the probability calculation we can treat as completely independent the two events, that for a molecule of the first kind the variables lie in the region G, and that for a molecule of the second kind they lie in the region H. Hence the number of molecule pairs for which one molecule belongs to the first kind and has its variables in G, while the other belongs to the second kind and has its variables in G, is the product of the two expressions (115) and (121):

$$(122) \quad dN_{12} = A_1 A_2 e^{-2h(E_1+E_2)} \int dP_1 \cdots dQ_\mu dP_{\mu+1} \cdots dQ_{\mu+\nu}.$$

We shall express the integration by a single integral sign, and we call the entire region of integration the region J, comprising the aggregate of G and H.

Expressions similar to (122) will of course hold when both molecules belong to the same kind.

The orders of magnitude of the various regions are to be chosen very different. When no external force acts, the region

$$\Gamma = \iiint dP_1 dP_2 dP_3,$$

within which the center of gravity of the first molecule lies will be chosen equal to the entire interior space of the container enclosing the gas—in other words, arbitrarily large. Then $P_{\mu+1}$ means the difference of the x-coördinates of the centers of gravity of the two molecules, and similarly $P_{\mu+2}$ and $P_{\mu+3}$ are the differences of the corresponding y- and z-coördinates. The validity of Equation (121) is not hereby affected, since in that equation $P_{\mu+1}$, $P_{\mu+2}$, $P_{\mu+3}$ were simply the coördinates of the center of gravity of a molecule of the second kind, for which any place in space is equally probable. With this extension of the region Γ, Equation (115), namely

$$dN_1 = A_1 e^{-2hE_1} \int dP_1 \cdots dQ_\mu$$

gives the number of molecules of the first kind in the entire container, for which the variables (113) lie in the $(2\mu-3)$-fold infinitesimal region $\int dP_4 \cdots dQ_\mu$. To each of these molecules corresponds a volume element, similarly situated with respect to its center of gravity,

$$\int dP_{\mu+1}dP_{\mu+2}dP_{\mu+3};$$

the number of these volume elements is therefore equal to the number dN_1 given by Equation (115), and their total volume is equal to $dN_1\iiint dP_{\mu+1}dP_{\mu+2}dP_{\mu+3}$.

The total number of molecules of the second kind lying in all these volume elements for which the other variables lie in the region

$$\int dP_{\mu+4} \cdots dQ_{\mu+\nu}$$

is therefore, according to Equation (121),

$$dN_1 \cdot A_2 e^{-2hE_2} \int dP_{\mu+1} \cdots dQ_{\mu+\nu}.$$

However, this is equal to the number dN_{12} of molecule pairs for which all variables lie in the region J, which agrees with Equation (122). This formula, which we derived earlier from the law of probability of combinations of several events, is therefore obtained again by a simple enumeration.

If external forces act, then the region

$$\Gamma = \iiint dP_1 dP_2 dP_3$$

must be chosen so small that the external force does not change noticeably therein; on the other hand, it must be larger than the entire space that the spheres of action of two interacting molecules occupy, so that it contains an enormous number of molecule pairs, for which the variables lie in the region J.

The region for the center of gravity of the two molecules must however be considered infinitesimal compared to Γ.

If all regions are infinitesimal, and there is only a finite number of molecules in unit volume, then of course it is impossible for a large number of molecules to have the values of their variables in this region, defined by mathematically infinitesimal narrow limits. We therefore set ourselves the problem of finding the limiting laws of the phenomena for an infinite number of molecules in unit volume, in the presence of external forces, and then assume that the actual phenomena will not noticeably deviate from these limits.

By $p_{\mu+1}$, $p_{\mu+2}$, $p_{\mu+3}$ we mean, instead of the coördinates of the center of gravity of the second molecule, the differences of the coördinates of the centers of gravity of the two molecules. As remarked before, this change does not affect the validity of Equation (121). Then, just as earlier in the absence of external forces, we have to calculate the number dN_{12}, for which one again obtains the result (122).

Since we are at present ignoring the case where more than two molecules interact simultaneously, we have to consider only all molecule pairs that are initially interacting. We first consider a pair in which one molecule belongs to the first kind and the other to the second. The number of such pairs initially interacting, for which the positions and velocities lie in a $2(\mu+\nu)$-fold infinitesimal region J, will be given by the expression

$$(123) \qquad dN'_{12} = A_1 A_2 e^{-2h(E_1+E_2+\Psi)} \int dP_1 \cdots dQ_{\mu+\nu}.$$

This region J will include certain given values of the variables (112) and (119), which we call as before $P_1 \cdots Q_\mu$ and $P_{\mu+1} \cdots Q_{\mu+\nu}$, and as before we call these the values (114) and (120), although of course they do not agree numerically with the values thus denoted earlier, since now there is an interaction where there was none before. In Equation (123) the integration is to be extended over the region J. Ψ is the value of the potential function of the interaction force—i.e., the interaction that occurs during this time between the constituents of the two molecules. The additive constant in Ψ is to be chosen such that this function vanishes for all distances of the molecules at which there is no interaction. We denote by $p_{\mu+1}$, $p_{\mu+2}$, and $p_{\mu+3}$ the differences of the centers of gravity of the two molecules.

For any molecule pair considered, the position of the center of gravity of the first molecule will be equally likely to be at any point within a part of the volume of the container, as long as this part is so small that the external forces can be considered constant within it.

§40. Proof that the distribution of states assumed in §37 will not be changed by collisions.

The formula (123) is to be considered as the most general one, which also includes (122), since if the two molecules are not interacting initially, $\Psi=0$ and the region J decomposes into two separate regions G and H, so that (123) reduces to (122).

A formula similar to (123) will likewise hold if the two interacting molecules belong to the same kind.

We now allow an arbitrary time t to elapse, which however should be so short that one may neglect the possibility that a molecule interacts more than once with another one during this time.

If, for a pair of unlike molecules, the position of the first is initially (114) and the position of the second is (120), then after a time t has elapsed, these same variables will have the values

$$(124) \qquad p_1 \cdots q_\mu, \; p_{\mu+1} \cdots q_{\mu+\nu}.$$

The corresponding values of the total energy (excluding the interaction energy) will be denoted by ϵ_1 for the first molecule, and ϵ_2 for the second. The value of the interaction potential function is ψ. The values (124) are of course again numerically different from the values (116) and (119) although they are denoted by the same letters.

The region filled by the values of the variables characterizing the state of the two molecules at time t, when they originally filled the region J, will be called the region i.

We can consider the aggregate of the two molecules as one mechanical system, for which therefore the equation analogous to Equation (55) holds, so that one has

$$(125) \qquad \int dp_1 \cdots dq_{\mu+\nu} = \int dP_1 \cdots dQ_{\mu+\nu},$$

where the second integral is to be extended over the region J, and the first over the corresponding region i. The validity of this equation is independent of whether the molecules are interacting either initially or at time t. It is also valid if the molecules never interact during t, in which case each of the two regions J and i decomposes into two separate regions, the former into G and H, the latter into g and h. Furthermore, in general, according to the principle of conservation of energy,

$$E_1 + E_2 + \Psi = \epsilon_1 + \epsilon_2 + \psi. \tag{126}$$

The latter equation also holds independently of whether an interaction takes place or not, since if there is no interaction the potential function of the interaction force simply vanishes.

The number of molecule pairs for which the values of the variables determining the positions and states initially fill the region J is given in general by the expression (123). This expression reduces, when one takes account of Equations (125) and (126), to

$$dn'_{12} = A_1 A_2 e^{-2h(\epsilon_1+\epsilon_2+\psi)} \int dp_1 \cdots dq_{\mu+\nu}, \tag{127}$$

where the integration is to be extended over the region i corresponding to the region J.

However, the molecule pairs for which the values of the variables fill the region J at time t are identical with those for which these values fill the region i at time t. Equation (127) therefore gives also the number of the latter type of molecule pair. The calculation of the number of molecule pairs for which the values of the variables initially fill the region i can again be performed with the generally valid formula (123). One simply has to replace E_1, E_2, Ψ and J by ϵ_1, ϵ_2, ψ and i. One thereby obtains again just the expression (127), regardless of whether there is any interaction at time zero or at time t or within this interval. However, since the region J, and hence also the region i determined by J, are completely arbitrary, we see that for an arbitrarily chosen region the number of molecule pairs for which the values of the variables lie within the region is the same at the initial time and at time t. The state distribution therefore remains stationary when one takes account of collisions.

One sees at once that he can apply completely analogous considerations to molecule pairs in which both molecules are of the same kind; and that the same considerations can be extended to the case where more than two kinds of gas are present in the container.

Up to now we have chosen the time t so small that we could ignore molecules that interacted with others twice during this time. But, since we saw that exactly the same distribution of states holds at time t as at time zero, the same method of reasoning can be applied again to another time interval of length t, over and over again. One therefore sees that the state distribution must remain stationary. Also, our assumption that in calculating the probability of a particular kind of encounter of two molecules, the two events that the two molecules are found in certain states can be considered independent, must also be true at all later times. For, according to our assumptions, each molecule moves past a very large number of molecules between successive interactions, so that the state of the gas at the place where the molecule experiences one interaction is completely independent of the state of the gas at the place of its previous interaction, and is determined only by the laws of probability. Naturally one has to remember that laws of probability are just that. The possibility of fluctuations hardly comes into consideration; yet when the number of molecules is finite, the probability of a fluctuation, while very small, is not zero; it can actually be calculated numerically in any particular case by the laws of probability, and it vanishes only for the limiting case of an infinite number of molecules.

§41. Generalizations.

We have still imposed a restriction on ourselves by the assumption that the case when more than two molecules interact plays no role. However, one perceives that this restriction was made only in order to simplify the proof, whose validity is completely independent of it. Likewise, just as we have discussed the probability of occurrence of certain molecule pairs, we can also try to calculate the probability of occurrence of groups of three and more molecules, and it will be found that such interactions of three or more molecules do not change the law that the state distribution represented by a formula similar to (123) is a stationary one. Also, the effect of the wall (not previously mentioned) cannot disturb the stationary character of this state distribution, in the case where the molecules rebound directly from it as if it were another identical gas. Any other assumed property of the wall would of course require new calculations. Yet it is manifest that even then, if the container is large enough, its effect would not extend into the interior.

Of course we have not yet proved that the state distribution expressed by Equation (123) is the only possible stationary one under all circumstances. Indeed such a general proof cannot be given, since in fact there are other special state distributions which can likewise be stationary. Such cases would for example, be encountered if all the gas molecules consisted of material points that all originally move in a plane or in a straight line, and the wall is everywhere perpendicular to this plane or line. But these are special distributions in which all variables take only a relatively small number of their possible values, whereas Equation (123) provides a distribution for which all variables take all their possible values.

It appears scarcely conceivable that there could be any other distributions which are stationary and for which all variables can run through all their possible values. In addition, there is a complete analogy between the state distribution represented by Equation (123) and that for a gas of monatomic molecules. This analogy has a definite foundation.

Just as in the game of Lotto, any particular quintuple is not a hair less probable than the quintuple 12345; the latter is distinguished from the others only by having a definite sequential property lacking in the others. Likewise, the most probable state distribution has this property only because it shares the same observable average value with by far the greatest number of equally possible state distributions.[1] Hence that state distribution is the most probable which, without altering this mean value, allows the greatest number of permutations of the individual values among the individual molecules. I have already shown in Part I, §6, how the mathematical condition for this property leads to Maxwell's distribution in the case of monatomic gas molecules. Without going into that further, I will remark that the validity of the considerations established there is in no way limited to the case of monatomic molecules; on the contrary, similar considerations can also be used in the case of compound molecules. In this case the momentoids corresponding to the generalized coördinates play exactly the same role that was played by the velocity components of the center of gravity in the case of monatomic molecules; the potential function of the internal and external forces plays the same role as was played earlier

[1] The criterion for equal possibility is provided by Liouville's theorem.

by the potential of the external forces alone, so that we obtain Equation (123) directly as a generalization of the formulas found in Part I.

That the formula (123) is the only one corresponding to thermal equilibrium, we shall try to make probable on several grounds in Chapter VII; we shall also give a direct proof in one of the simplest special cases. At this point, however, in order not to exhaust ourselves by too long a sojourn with abstract matters, we shall be content with the arguments here advanced in favor of Equation (123), and draw from it the most important consequences.

§42. Mean value of the kinetic energy corresponding to a momentoid.

We shall next consider a mixture of several gases, none of which is considered to be partly dissociated. At any time the number of molecules interacting with each other will be vanishingly small compared to the rest, and it is therefore permitted to take account of only the noninteracting molecules in calculating average values.

If we introduce instead of the momenta $q_1, q_2, \cdots, q_\mu$ the corresponding momentoids $r_1, r_2, \cdots, r_\mu$, then the number of molecules for which the coördinates and momentoids lie in any region K enclosing the values

$$p_1, p_2 \cdots p_\mu, r_1 \cdots r_\mu \tag{128}$$

is given by the expression

$$dn = Ae^{-2h\epsilon}\int dp_1 \cdots dp_\mu dr_1 \cdots dr_\mu \tag{129}$$

which is valid for any kind of gas in the container, since the determinant for the transformation from the variables q to the variables r is equal to 1. The constant h must have the same value for all gases present in the same container. The constant A, however, can have a different value for each kind of gas. ϵ is the sum of the kinetic energy and the potential energy of intramolecular and external forces of a molecule; the potential function will be called V.

For the kinetic energy of a molecule one has, as we saw, the expression

$$L = \tfrac{1}{2}(\alpha_1 r_1^2 + \alpha_2 r_2^2 \cdots \alpha_\mu r_\mu^2) = \tfrac{1}{2}\sum \alpha r^2,$$

where we again denote by the first term the part of the kinetic energy corresponding to the first momentoid.

If we choose the region K in the simplest way—i.e., so that it includes all combinations of values for which the coördinates lie between the limits

(130) $$p_1 \text{ and } p_1 + dp_1 \cdots p_\mu \text{ and } p_\mu + dp_\mu$$

and the momentoids between the limits

(131) $$r_1 \text{ and } r_1 + dr_1 \cdots r_\mu \text{ and } r_\mu + dr_\mu$$

—then

(132) $$dn = Ae^{-h(2V+\Sigma\alpha r^2)}dp_1 dp_2 \cdots dr_\mu.$$

This is the number of molecules of any particular kind for which the values of the variables are enclosed between the limits (130) and (131).

The mean value of the part $\frac{1}{2}\alpha_i r_i^2$ corresponding to the momentoid r_i has for arbitrary i the value

(133) $$\frac{1}{2}\overline{\alpha_i r_i^2} = \frac{\int \alpha_i r_i^2 dn}{2\int dn} = \frac{\int \alpha_i r_i^2 e^{-h(2V+\Sigma\alpha r^2)} dp_1 \cdots dr_\mu}{2\int e^{-h(2V+\Sigma\alpha r^2)} dp_1 \cdots dr_\mu},$$

where the single integral sign indicates an integration over all possible values of the differentials.

If one integrates over r_i in both numerator and denominator, then he can bring outside the r_i integral in both cases a factor not depending on r_i. These factors cancel in numerator and denominator. The integral over r_i in the numerator is

$$\int \alpha_i r_i^2 e^{-h\alpha_i r_i^2} dr_i,$$

while that in the denominator is

$$2\int e^{-h\alpha_i r_i^2} dr_i.$$

In order to find the limits of integration, we recall that for the velocity p' each coördinate may go through all values from $-\infty$ to $+\infty$. The r are linear functions of the p' and can therefore run through all values from $-\infty$ to $+\infty$ likewise. These are therefore the limits of integration for r_i, and one obtains

$$\int \alpha_i r_i^2 e^{-h\alpha_i r_i^2} dr_i = \frac{1}{2h}\int e^{-h\alpha_i r_i^2} dr_i,$$

as can be seen by integrating by parts in the first integral, or by calculating both integrals using Equation (39), §7 of Part I. One can now put the factor $\frac{1}{2}h$ in front of all the integral signs in the numerator, and the factor 2 in the denominator. The expressions multiplying these factors are the same in both numerator and denominator, so one can cancel them and obtain:

$$\frac{1}{2}\overline{\alpha_i r_i^2} = \frac{1}{4h}, \quad \overline{L} = \frac{\mu}{4h}\,. \tag{134}$$

$\overline{L}$ means the average value of the total kinetic energy of a molecule of the kind considered. Hence the kinetic energy corresponding to each momentoid has the same value on the average, and indeed this value is equal for all kinds of gas, since h has the same value for all kinds of gas. Similarly, as in Part I, §19, this theorem can also be extended to gases that are in thermal equilibrium with each other by means of a heat-conducting partition.

Since we have everywhere integrated over each p and r independently, and in general we have always considered the p as independent variables, we have continually assumed that there is no relation between the generalized coördinates $p_1, p_2, \cdots, p_\mu$. μ is therefore the number of independent variables required to determine the absolute position of all constituents of a molecule in space, as well as their positions relative to each other. One calls μ the number of degrees of freedom of the molecule, conceived as a mechanical system.

One can always choose three of the r to be the three velocity components of the center of gravity of a molecule in the three coördinate directions, since the total kinetic energy of a system is always the sum of the kinetic energies of the motion of the center of gravity and of the motion relative to the center of gravity.[1]

[1] Cf. Boltzmann, *Vorlesungen über die Principe der Mechanik*, Part I, §64, p. 208.

The product of half the total mass of a molecule and the mean square of one of these velocity components of its center of gravity is then the mean kinetic energy corresponding to this momentoid; according to Equation (134), it has the value $\frac{1}{4h}$ for each of the coördinate directions. The sum of the three average kinetic energies for the three coördinate directions is however equal to the product of half the total mass of the molecule and the mean square velocity of its center of gravity. This latter product we shall call the mean kinetic energy of the motion of the center of gravity, or of the progressive motion of the molecule, and denote it by $\overline{S}$. Therefore

$$(135) \qquad \overline{S} = \frac{3}{4h}, \quad \overline{S}:\overline{L} = 3:\mu.$$

The mean kinetic energy of the motion of the center of gravity of a molecule is therefore the same for any gases in thermal equilibrium with each other. From this there follows, as we saw in §7 of Part I, the Boyle-Charles-Avogadro law, which therefore appears to have a kinetic foundation for gases with compound molecules as well.[2]

---

[2] We shall think of a particular given solid or liquid as an aggregate of n material points, which therefore have $3n$ degrees of freedom, perhaps just the $3n$ rectangular coördinates. If it is surrounded by a larger gas mass, then it can be considered in certain respects as a single gas molecule, and the law found in the text can be applied to it. The total kinetic energy is therefore $3n/4h$. If the temperature is increased in such a way that $1/4h$ increases by $d(1/4h)$, then the total heat, measured in mechanical units, that must be supplied to raise the mean kinetic energy is $dQ_i = 3nd(1/4h)$. This heat, per unit of mass and temperature increment, is what Clausius calls the true specific heat. It is invariable in all states and forms of aggregation. It would be the total specific heat if the body were a gas dissociated into its atoms. For all bodies it is proportional to the number of atoms in the body. The total specific heat is also proportional to this number (Dulong-Petit law of chemical elements, or Neumann's law for compounds) if the heat increment dQ_i used to perform internal work is in a constant ratio to that used to increase the kinetic energy, dQ_l. This is always the case if the internal forces acting on each atom are proportional to its distance from its rest position, or still more generally if they are linear functions of its coördinate variations. Then the force function V is a homogeneous quadratic function of the coördinates, just as the kinetic energy L is such a function of the momenta. The integrations in the formula for $\overline{V}$ can then be performed in the same way used to obtain

§43. The ratio of specific heats, κ.

We shall now assume for a moment that only one kind of gas is present in the container, and that the external forces can be ignored, so that the intramolecular and intermolecular interaction forces, and the opposing pressure of the wall of the container against the gas, are the only forces that need be considered.

As in Part I, §8, we denote by dQ_2 the amount of heat used to raise the kinetic energy of motion of the centers of gravity of the molecules, and by dQ_3 the heat used to increase the kinetic and potential energy of the intramolecular motion, when the gas experiences a temperature increase dT. The ratio dQ_3/dQ_2 will be denoted by β, as in Part I, §8. The heat will always be measured in mechanical units.

The quantity dQ_3 can be decomposed naturally into two parts: a part dQ_5, which is used to increase the kinetic energy of intramolecular motion, and a part dQ_6, which is used to increase the value of the potential function of the force acting between the constituents of a molecule (intramolecular force).

We have denoted by $\overline{S}$ the mean kinetic energy of the motion of the center of gravity of a molecule. If n is the total number of molecules in the gas, then the total kinetic energy of the progressive motion of the molecules is $n\overline{S}$; hence $dQ_3 = nd\overline{S}$. Since we have denoted by $\overline{L}$ the total mean kinetic energy of a molecule, then $\overline{L}-\overline{S}$ is the mean kinetic energy of intramolecular motion of a molecule. The kinetic energy of the intramolecular motion of all molecules in the gas is therefore $n(\overline{L}-\overline{S})$ or, according to Equation (135), $n\overline{S}(\mu/3-1)$, whence it follows that

$$dQ_5 = n\left(\frac{\mu}{3}-1\right)d\overline{S} = \left(\frac{\mu}{3}-1\right)dQ_3. \tag{136}$$

Equation (134), and one finds $dQ_i = dQ_l$. (Cf. Eq. [111a] at the end of §45.) The total heat capacity is then twice the true monatomic gas value. The assumed law of action of internal molecular forces holds approximately for most solid bodies. For such bodies, whose heat capacity is smaller than half that predicted by the Dulong-Petit law (e.g., diamond) one must assume that the motions related to certain parameters come into equilibrium with the others so slowly that they do not contribute to the specific heats determined by experiment. (Cf. §35. For the case that the molecule makes approximately pendulum-like motions, see Boltzmann, Wien. Ber. **53,** 219 [1866]; **56,** 686 (1867); **63,** 731 (1871); Richarz, Ann. Physik [3] **48,** 708 [1893]; Staigmüller, Ann. Physik [3] **65,** 670 [1898].)

If we denote the average value of the potential for a molecule by $\overline{V}$, then

$$dQ_6 = nd\overline{V}. \tag{137}$$

The latter quantity cannot be calculated unless we make some special assumption about V. We shall therefore simply set

$$dQ_6 = \epsilon dQ_3$$

in order not to restrict the degree of generality, and we then obtain:

$$\beta = \frac{dQ_5 + dQ_6}{dQ_3} = \frac{\mu}{3} - 1 + \epsilon.$$

The ratio of specific heats of the gas κ will therefore be, according to Equation (56) of §8, Part I,

$$\kappa = 1 + \frac{2}{\mu + 3\epsilon} \cdot \tag{138}$$

§44. Value of κ for special cases.

If the molecules are single material points, then they have no other motion besides the motion of the center of gravity; hence $\epsilon = 0$. To determine their position in space, three rectangular coördinates are sufficient; hence $\mu = 3$, $\kappa = 1\frac{2}{3}$.

Now suppose that the molecules are smooth undeformable elastic bodies; then no variation of the potential of the intramolecular force is permitted; hence $\epsilon = 0$.

If each molecule is constructed absolutely symmetrically with respect to its center of gravity—or, still more generally, if it has the form of a sphere whose center of gravity coincides with its midpoint—then indeed each molecule can make arbitrary rotations around an arbitrary axis passing through its midpoints; but the velocity of this rotation cannot be altered in any way by collisions between molecules. If all molecules were initially not rotating, then they would remain so for all time. On the other hand, if they were initially rotating, then each molecule would retain its rotation independently of all the others, although this rotation would exert no observable action.

Of the variables determining the position of a molecule, only the three coördinates of the center of gravity come into consideration for collisions, and one has again[1]

$$\mu = 3, \ \kappa = 1\tfrac{2}{3}.$$

It is otherwise when the molecules are absolutely smooth undeformable elastic bodies that have either the form of solids of revolution differing from the spherical form, or the form of spheres in which the center of gravity does not coincide with the midpoint. In the former case, it will be assumed that either their mass is completely symmetrically arranged around the axis of rotation, or this axis is at least a principal axis of inertia, that the center of gravity lies on it, and that the moment of inertia of the molecule with respect to each line passing through the center of gravity perpendicular to the axis of rotation is the same. If the bodies are spheres with eccentric centers of gravity, the moments of inertia of the molecule with respect to any line through the center of gravity perpendicular to the line connecting the center of gravity and the midpoints must be the same. Then, only the rotation about the axis of rotation will be without effect on the collisions. All other rotations will be changed by collisions, so that their kinetic energy must be in thermal equilibrium with the kinetic energy of progressive motion.

Five variables are now needed to determine the position of a molecule in space: the three coördinates of its center of gravity and two angles determining the position of its axis of rotation in space. Hence $\mu = 5$ and since ϵ is again zero, one has $\kappa = 1.4$.* If the molecules are absolutely flat undeformable bodies, not constructed by either of the above methods, then their rotation about all possible axes will be modified by collisions. Then, to determine the position of a molecule, three angles determining the total rotation around the center of gravity are necessary as well as the three coördinates of its center of gravity, and one gets

$$\mu = 6, \quad \kappa = 1\tfrac{1}{3}.$$

---

[1] In general, one obtains in these two cases all the formulas developed in Part I for monatomic molecules directly from Equation (118), whether external forces are absent (§7) or present (§19). These formulas are therefore only special cases of (118).

* This explanation of the anomalous specific heats of diatomic gases was first proposed by Boltzmann in December 1876 (Wien. Ber. **74**, 553 [1877]), and independently in April 1877 by R. H. M. Bosanquet (Phil. Mag. [5] **3**, 271 [1877]).

§45. Comparison with experiment.

It is remarkable that for mercury vapor—whose molecules have long been considered monatomic on chemical grounds—according to the researches of Kundt and Warburg,* κ is in fact very nearly equal to the value obtained for simple molecules, $1\frac{2}{3}$. Also for helium, neon, argon, metargon [xenon], and krypton, Ramsay found almost the same value of κ.† The limited chemical activity of these gases is likewise consistent with their being monatomic.

For many gases with very simply constructed compound molecules (possibly for all those for which a variation of κ with temperature has not yet been verified) observation gives values of κ that lie very near to the two other ones that we found, 1.4 and $1\frac{1}{3}$.

Of course the problem is still far from being solved. For many gases, κ has yet smaller values; moreover, Wüllner found that often—and indeed just for these latter gases—κ varies strongly with temperature.‡ Our theory predicts that κ should vary with temperature as soon as the potential of the intramolecular forces, V, becomes important; yet it is easy to see that much remains to be done on the theory of specific heats.

If the molecules are spheres filled with mass symmetrically around their midpoints, then of course there is no possibility that they can be set into rotation by collisions, nor that any initial rotation can be lost. Nevertheless it is improbable that such molecules would remain rotationless throughout all eternity, or that they would always preserve the same amount of rotation. It seems more likely that they possess this property only to a very close approximation, so that their rotational state does not noticeably change during the time in which the specific heat is determined, even though over a long period of time rotation will be equilibrated with other molecular motions, so slowly that such energy exchanges escape our observation.

Similarly one can assume that in gases for which $\kappa = 1.4$, the constituents of the molecule are by no means connected together

---

* Kundt and Warburg, Ann. Physik [2] **157**, 353 (1876).

† Ramsay, C. R. Paris **120**, 1049 (1895). Ramsay and Collie, Proc. R. S. London **60**, 206 (1896). Rayleigh and Ramsay, Phil. Trans. **186**, 187 (1896). Ramsay and Travers, Proc. R. S. London **63**, 405 (1898).

‡ A. Wüllner, Ann. Physik [3] **4**, 321 (1878).

as absolutely undeformable bodies, but rather that this connection is so intimate that during the time of observation these constituents do not move noticeably with respect to each other, and later on their thermal equilibrium with the progressive motion is established so slowly that this process is not accessible to observation. In any case, for air at temperatures where noticeable heat begins to be radiated, it is found that in addition to the five variables determining the state of a molecule, another one begins to participate in the thermal equilibrium during the time of observation, so that κ varies with temperature and becomes smaller than 1.4; the same must be true for all other gases.

Naturally, because of the obscurity of the nature of all molecular processes, all hypotheses about them must be expressed with the greatest caution. The hypotheses proposed here would be confirmed experimentally if it were to be shown that, for any gas for which κ varies with temperature, observations extended over a longer period of time give a smaller value of κ than those of shorter duration.

In spite of the appearance of a potential function for intramolecular forces, κ remains independent of temperature in the case that the constituents of the molecule have fixed rest positions relative to each other, and that the force acting on them when they move away from these rest positions is a linear function of the distance. If then λ is the number of variables on which the relative positions of the constituents of a molecule depend, the coördinates can always be chosen such that the potential function takes the form:

$$\tfrac{1}{2}(\beta_1 p_1^2 + \beta_2 p_2^2 + \cdots \beta_\lambda p_\lambda^2).$$

The quantity $\frac{1}{2}\beta_i p_i^2$ should then be the intramolecular potential energy corresponding to the coördinate p_i. Its mean value can be calculated just as we calculated the mean value of the quantity $\frac{1}{2}\alpha_i r_i^2$ above, and one obtains likewise the value $\frac{1}{4h}$ for each i. Hence

$$\overline{V} = \nu/4h, \quad \overline{S}:\overline{V} = 3:\lambda.$$

Since these equations are completely similar to Equations (135), then[1]

[1] Cf. Staigmüller, Ann. Physik [3] 65, 655 (1898) and footnote 2, §42.

$$dQ_6 = \tfrac{1}{3}\lambda dQ_3, \quad \text{hence} \quad \epsilon = \tfrac{1}{3}\lambda,$$

$$\kappa = 1 + \frac{2}{\mu + \lambda} \cdot$$

As an example we consider a molecule consisting of two simple material points, or of two absolutely smooth spheres filled with mass distributed completely symmetrically around their midpoints. These should exert no force on each other at a certain distance, but at greater distances they attract and at smaller distances they repel each other, the force being proportional to the change of distance in both cases. Then this distance is the only coördinate determining the relative positions, hence $\lambda = 1$.

Five other coördinates are needed to determine the absolute position in space. The total number μ of the p, or the number of degrees of freedom, is then six, and therefore $\kappa = 1\frac{2}{7} = 1.2857$.

Treatment of further special cases would not be difficult, but it seems to me to be superfluous as long as more comprehensive experimental data are not available.

§46. Other mean values.

In the previous sections we have calculated the average kinetic energy for a momentoid obtained by averaging over all values for all molecules of a certain kind in the container. This mean value does not change if we impose at the same time restrictions on one or more of the coördinates, for example when we take only those values of all molecules of this kind whose centers of gravity lie within an arbitrarily small region $\iiint dP_1 dP_2 dP_3$: the temperature is the same everywhere in the gas. This theorem, which is self-evident in the absence of external forces, also holds if external forces of any kind are present.

Nor does this mean value change when we include in the averaging only those molecules for which some other coördinates are restricted to lie within some arbitrary limits. The mean value in question will again be given by Equation (133); except that the integration is to be extended not over all values of the coördinates but only over the specified region, which is of course the same in both numerator and denominator. Then just as in §42 the entire factor independent of the integration over dr_i comes out in front of the integral sign, and after performing the integration over dr_i in both numerator and denominator, one can divide through by this factor, so that one obtains again the value $\frac{1}{4h}$ for the integral.

If the α are constant then, as Equation (132) shows, the relative probability that the value of any momentoid lies between any limits whatever and that it lies between any other limits is completely independent of the position of the molecule in space, and of the relative positions of its constituents. We shall call this theorem, which we shall use later, the *S*-theorem. It refers to the case where the r are quantities proportional to the velocity components of material points or the angular velocities of a rigid body around its principal moments of inertia.

For those kinds of gas to which Equation (129) pertains, one finds the number dn' of molecules for which the values of the coördinates lie between the limits (130), when the values of the momenta are not subjected to any restrictive conditions, by integrating Equation (129) over all the r from $-\infty$ to $+\infty$. One obtains in this way

$$(139)\qquad dn' = \frac{A\pi^{\mu/2}}{h^{\mu/2}\sqrt{\alpha_1\alpha_2\cdots\alpha_\mu}}e^{-2hV}dp_1dp_2\cdots dp_\mu.$$

The mean value $\overline{V}$ of the potential function is $\int Vdn'/\int dn'$, where the integrations are to be extended over all possible values of the coördinates.

The number of molecules for which the coördinates lie in any one μ-fold infinitesimal region is then in the same ratio to the number of molecules for which they are in a likewise μ-fold infinitesimal region F' without any restriction on the values of the momenta, as

$$(140)\qquad \frac{e^{-2hV}}{\sqrt{\alpha_1\alpha_2\cdots\alpha_\mu}}\int dp_1\cdots dp_\mu : \frac{e^{-2hV'}}{\sqrt{\alpha_1'\alpha_2'\cdots\alpha_\mu'}}\int dp_1'\cdots dp_\mu',$$

where the letters without primes indicate the values for the region F, and those with primes indicate those for F'; the first integral is over the former region, and the second one over the latter region.

When there are no intramolecular and external forces, and the molecules consist purely of simple material points, then this is the ratio of the product of the volumes of all volume elements available to the material points in the region F, to the analogous product for the region F'. Then the exponentials are equal to 1, so that

$$\frac{\int dp_1 \cdots dp_\mu}{\sqrt{\alpha_1\alpha_2 \cdots \alpha_\mu}} : \frac{\int dp_1' \cdots dp'_\mu}{\sqrt{\alpha_1' \alpha_2' \cdots \alpha_\mu'}}$$

is always equal to the ratio of these products of volume elements.

§47. Treatment of directly interacting molecules.

In the last section we always proceeded from Equation (129)—i.e., we assumed that the interaction of two molecules lasts so short a time that we can ignore molecules that are momentarily interacting with others when calculating average values. Yet the theorems developed in this and the preceding sections are also valid for interacting molecules. For example, we consider all pairs of unlike molecules for which the coördinates of the two molecules lie within any $(\mu+\nu)$-fold infinitesimal region, without any restriction on the momentoids. If no external forces act, then the region available for the center of gravity of one of the molecules can be extended over the entire interior of the container. If we integrate over all the r in Equation (127) (in which we have previously substituted the r in place of the q) we find the following expression for the number of these molecule pairs:

$$dN = AA_1e^{-h(V+V_1+\psi)}\int dp_1 \cdots dp_{\mu+\nu}\int e^{-h\Sigma\alpha r^2}dr_1 \cdots dr_{\mu+\nu}.$$

Here V is the potential of intramolecular and external forces for the first molecule, and V_1 is the potential for the second molecule. ψ is the potential of the force of interaction. The sum pertains to all the momentoids of both molecules. The p integrals are to be extended over all coördinates of both molecules, in the region D, while the r integrals are over all possible values of these variables from $-\infty$ to $+\infty$. The evaluation of the integral over the r gives

$$(141) \qquad dN = \frac{\pi^{(\mu+\nu)/2}AA_1e^{-2h(V+V_1+\psi)}}{h^{(\mu+\nu)/2}\sqrt{\alpha_1 \cdots \alpha_{\mu+\nu}}}\int dp_1 \cdots dp_{\mu+\nu}$$

One can now calculate the mean value of the kinetic energy $\frac{1}{2}\alpha_i r_i^2$ for one of the momentoids, for all molecules of the first kind in those pairs whose total number is denoted by dN. This again comes out to be $\frac{1}{4h}$. I will not write out the explicit formula from which this follows, since it is completely analogous to the corresponding formulas developed in §42 for single molecules. For all these molecule pairs, the mean kinetic energy of the motion of the center of gravity of each molecule pair is again equal to $\frac{3}{4h}$.

I will only derive one more theorem, which we will need in the theory of dissociation of gases. We shall consider, besides the region D, another arbitrary $(\mu+\nu)$-fold infinitesimal region D' for the coördinates of the two molecules, and denote with primes all the values of the variables pertaining to this second region, so that now the p' are not time derivatives but are other values of the p.

The number of pairs of unlike molecules for which the variables lie in the region D, without any restriction on the values of the momentoids, is (according to Eq. [141]):

$$dN' = \frac{AA_1\pi^{(\mu+\nu)/2}}{h^{(\mu+\nu)/2}\sqrt{\alpha_2' \cdots \alpha_{\mu+\nu}'}} e^{-2h(V'+V_1'+\psi')} \int dp_1' \cdots dp_{\mu+\nu}'.$$

Therefore

$$\text{(142)} \quad \frac{dN'}{dN} = \frac{e^{-2h(V'+V_1'+\psi')}\sqrt{\alpha_1 \cdots \alpha_{\mu+\nu}} \int dp_1' \cdots dp_{\mu+\nu}'}{e^{-2h(V+V_1+\psi)}\sqrt{\alpha_1' \cdots \alpha_{\mu+\nu}'} \int dp_1 \cdots dp_{\mu+\nu}}.$$

This formula and Equation (140) are nothing but generalizations —remarkable for their simplicity and symmetry—of Equation (167), Part I, i.e., the completely trivial formula for the barometric measurement of height, according to which the numbers of molecules found in unit volume at different heights z behave like

$$e^{-2hmgz} = e^{-gz/rT}.$$

Since this formula permits one to calculate the pressure of the saturated vapor and the law of dissociation (cf. §60 and §§62–73) it must be regarded as one of the fundamental formulas of gas theory.

If, other things being equal, the molecule pairs exert no forces on each other, either in D or D', then one has the same expression as before for the quotient dN'/dN, except that he has to set $\psi=\psi'=0$. If it is possible to compute the ratio dN'/dN for this latter case, then one can obtain the corresponding value when there is some interaction by multiplying it by $e^{-2h(\psi'-\psi)}$. Likewise, if there is no interaction in D either time, and there is an interaction in D' one time but not the other, then one should multiply by $e^{-2h\psi'}$. One can also call dN'/dN the relative probability of the two events, that for a molecule pair the values of the variables lie in the region D' or the region D.

A similar theorem holds, as can easily be shown, for the interaction of more than two molecules. The relative probability of two configurations is $e^{-2h(\psi'-\psi)}$ times larger when interactions occur than when they are absent, where ψ and ψ' are the values of the potentials of the interaction for the two configurations.

CHAPTER V

Derivation of van der Waal's equation by means of the virial concept.

§48. Specification of the point at which van der Waals' mode of reasoning requires improvement.

In deriving van der Waals' equation in Chapter I, we followed the method that he himself first used, which is characterized by the greatest simplicity and perspicuousness. It was already remarked (footnote 1, §7) that it is perhaps not completely free of objection.

The first assumption subject to doubt is the one that we made in §3 and also later on: that in the entire container, as well as in the cylinder lying near the boundary of the space—which we called there the cylinder γ—each volume element is equally probable as a position for the midpoint of a molecule, regardless of its distance from another molecule.

If there are no forces except the collision forces, the correctness of this assumption follows directly from Equation (140), in which V is then constant, so that it therefore predicts that the average number of molecules will be the same in each equal volume element.

On the other hand, the van der Waals cohesion force will produce a denser arrangement of the molecules in the interior of the fluid than in the immediate neighborhood of the wall. Van der Waals did not take account of this effect, either in deriving the expression for collisions at the wall or in calculating the dependence of the term a/v^2 on the density of the gas. In both cases, however, correct treatment of the volume elements at the boundary is essential; the quantities being calculated vanish more strongly as the number of particles at the surface becomes small compared to those inside. Thus one cannot, as with the formulas of this chapter, improve the accuracy of the results arbitrarily by making the volume large compared to the surface.

Van der Waals calculated the correction to the Boyle-Charles law due to the finite extension of the rigid core of the molecule, as was explained in Chapter I, just as if the cohesion force were not present. He then calculated the additional term in the external pressure due to the cohesion forces as if the molecules were vanishingly small. Since the validity of this procedure might be doubted, we shall give a second deduction of the van der Waals formula from the theory of the virial (to which van der Waals has likewise already related it) against which this objection cannot be made. This second derivation shows that van der Waals' conclusion is completely correct. Yet we cannot of course obtain by the exact method precisely the reciprocal of $v-b$ which occurs in van der Waals' equation, which van der Waals himself has called inexact; on the contrary we obtain an infinite series in powers of b/v.

§49. More general concept of the virial.

The concept of the virial was introduced into gas theory by Clausius.* Let there be given an arbitrary number of material points. Let m_h be the mass of one point, and let x_h, y_h, z_h, c_h, u_h, v_h, w_h be its rectangular coördinates, its velocity, and its velocity components along the coördinate axes, respectively, at some time t. Let ξ_h, η_h, ζ_h be the components of the total force acting on this material point at the same time. The force should have the property that all the material points can move under its influence for an arbitrarily long time, without any of their coördinates or velocity components increasing without bound. The initial conditions should be such that this can actually be true. No matter how long a time of motion may be chosen, the absolute value of any coördinate or velocity component must remain smaller than a fixed finite quantity, which has the value E for the coördinates and the value ϵ for the velocity components. Such motions—of which all molecular motions giving rise to thermal phenomena in a body of finite extent are clearly examples—we shall call stationary.

* R. Clausius, Sitzber. Niederrhein Ges. (Bonn) **114** (1870), reprinted in Ann. Physik [2] **141**, 124 (1870). For further references see S. G. Brush, Amer. J. Phys. **29**, 593 (1961), footnote 14.

Now let G be the value of any quantity at a particular time t; then we shall call the quantity

$$\frac{1}{\tau}\int_0^\tau G dt = \overline{G}$$

the time average of G during the time of motion τ.

By virtue of the equations of motion of mechanics, we have:

$$m_h \frac{du_h}{dt} = \xi_h.$$

Hence

$$\frac{d}{dt}(m_h x_h u_h) = m_h u_h^2 + x_h \xi_h.$$

If one multiplies this equation by dt, integrates over an arbitrary time (from 0 to τ) and finally divides by τ, then it follows that:

$$m_h \overline{u_h^2} + \overline{x_h \xi_h} = \frac{m_h}{\tau}(\overset{\tau}{x}_h \overset{\tau}{u}_h - \overset{0}{x}_h \overset{0}{u}_h),$$

where the values at time τ are characterized by the upper index τ, and the values at time zero by the upper index 0. By virtue of the stationary character of the motion

$$m_h(\overset{\tau}{x}_h \overset{\tau}{u}_h - \overset{0}{x}_h \overset{0}{u}_h)$$

is smaller than $2m_h E\epsilon$. If one allows the time of the entire motion, τ, to increase beyond any limit, then $2m_h E\epsilon$ remains finite; hence the expression $2m_h E\epsilon/\tau$ approaches zero. If one takes the mean value for a sufficiently long time, then:

$$m_h \overline{u_h^2} + \overline{x_h \xi_h} = 0.$$

Similar equations are obtained for all coördinate directions and all material points. If one adds them all up, it follows that:

$$\sum m_h \overline{c_h^2} + \sum \overline{(x_h \xi_h + y_h \eta_h + z_h \zeta_h)} = 0. \tag{143}$$

$\frac{1}{2}\sum m_h c_h^2$ is the kinetic energy L of the system. The expression

$$\sum (x_h \xi_h + y_h \eta_h + z_h \zeta_h)$$

Clausius calls the virial of the forces acting on the system. Therefore the above equation says that twice the time average of the kinetic energy is equal to the negative time average of the virial of the system during a very long time.

We now assume that between any two material points m_h and m_k, whose distance is r_{hk}, a force $f_{hk}(r_{hk})$ acts in the direction of r_{hk}, which we call the internal force. It has a positive sign when it is repulsive, and a negative sign when it is attractive. Moreover, on each material point m_h an external force acts, which results from causes lying outside the system and whose components along the coördinate directions we denote by X_h, Y_h, and Z_h. Then:

$$\xi_1 = X_1 + \frac{x_1 - x_2}{r_{12}} f_{12}(r_{12}) + \frac{x_1 - x_3}{r_{13}} f_{13}(r_{13}) + \cdots.$$

One sees easily[1] that then Equation (143) becomes:

$$(144) \quad 2\overline{L} + \sum \overline{(x_h X_h + y_h Y_h + z_h Z_h)} + \sum \sum \overline{r_{hk} f_{hk}(r_{hk})} = 0.$$

The first addend is twice the time average of the kinetic energy of the entire system. The second is the external virial, and the third the internal virial. The two virials will be denoted by W_a and W_i, so that Equation (144) reduces to

$$(145) \qquad 2\overline{L} + W_a + W_i = 0.$$

§50. Virial of the external pressure acting on a gas.

We consider as a special case a gas in equilibrium, whose molecules, just as explained in Chapter I, behave according to van der Waals' assumptions. Let the gas be enclosed in any container of volume V; it consists of n similar molecules of mass m and diameter σ, and the mean square velocity of a molecule is $\overline{c^2}$. Then:

$$(146) \qquad 2L = \sum m_h \overline{c_h^2} = nm\overline{c^2}.$$

There are no external forces other than the pressure on the

[1] The simplest way is to calculate the virial of the force acting between any two material points, as well as that of the external force, separately, and then to recall that the virial of several forces is the sum of the virials of the forces taken individually, since the ξ_h, η_h, ζ_h are contained linearly in the expression for the virial.

container, whose intensity on unit surface will be p. The container has the form of a parallelepiped of edges α, β, γ, of which three adjacent ones will be chosen as the x, y, and z axes. The two lateral surfaces having the same surface area $\beta\gamma$ will have abscissas zero and α respectively. The pressure forces $p\beta\gamma$ and $-p\beta\gamma$ respectively will act on these surfaces, in the direction of the positive abscissa axis. For these two lateral surfaces together the sum $\sum x_h X_h$ has the value $-p\alpha\beta\gamma = -pV$. Since the equality holds also for the two other coördinate directions, we have for the entire gas:

$$\sum (x_h X_h + y_h Y_h + z_h Z_h) = -3pV.$$

Since the pressure does not change with time, this is also the mean value of the same quantity; hence it is the external virial W_a.

The same equation can easily be derived for a container of any shape. Let $d\omega$ be a surface element of the projection ω of the surface of the container on the yz plane, and K be the cylinder erected on $d\omega$, perpendicular to it and extended to infinity on both sides. This cylinder cuts from the container surface a series of surface elements do_1, do_2, $\cdots$ whose abscissas are x_1, x_2, $\cdots$ and whose normals drawn into the interior of the gas are N_1, N_2, $\cdots$. The x component of the pressure force acting on do_1 is:

$$pdo_1 \cos (N_1 x) = pd\omega.$$

The same x component has, for the surface element do_2, the value

$$pdo \cos (N_2 x) = -pd\omega$$

and so forth. The sum $\sum x_h X_h$ extends over all surface elements lying within the cylinder K, and therefore has the value:

$$-pd\omega(x_2 - x_1 + x_4 - x_3 + \cdots).$$

The factor of the quantity $-p$ is just the volume cut out by the cylinder from the interior of the container. The sum $\sum x_h X_h$, extended over the entire gas, is found by integrating this expression over all surface elements $d\omega$ of the entire projection ω, whereby one obtains the product of the total volume V of the gas and the quantity $-p$. Since the same considerations are applicable also to the y and z axes, it follows that:

$$(147) \qquad \sum (x_h X_h + y_h Y_h + z_h Z_h) = -3pV = W_a.$$

§51. Probability of finding the centers of two molecules at a given distance.

The internal virial will consist of two parts, of which the first, W_i', arises from the forces acting during the collision of two molecules, and the second, W_i'', from the attractive force assumed by van der Waals.

In order to find W_i' we denote (as earlier) by σ the diameter of a molecule, and call a sphere of radius σ circumscribed around a molecule its covering sphere, so that the volume of the covering sphere is eight times as large as the volume of the molecule itself. The midpoint of a second molecule cannot come closer to the midpoint of our molecule than a distance σ, and we shall first calculate the probability that the center of a particular molecule is at a distance between σ and $\sigma+\delta$ (where δ is infinitesimally small compared to σ) from one of the other molecules. In order to have a precise name we call this other molecule the remaining molecule.

In order to have a concept of probability as free from objection as possible, we imagine that the same gas is present infinitely many times (N times) in identical containers at different positions in space. Our specified molecule will in general be in a different place in each of these N gases. Of all the N gases, let the remaining molecule in N_1 gases be in very nearly the same place relative to the container. N_1 is then very small compared to N, but it should still always be a very large number. The influence of the wall on the interior can in any case be made small by making the container large, and in the interior the van der Waals cohesion forces acting on a molecule from all directions will cancel out. Hence, according to Equation (140), all positions in the container will be equally probable for the center of the specified molecule in all these N_1 gases. Let N_2 be the number of gases in which the center of the specified molecule is at a distance between σ and $\sigma+\delta$ from the remaining molecule. Then the ratio of N_1 to N_2 will be the same as the ratio of the total space available to the center of the specified molecule in one of the N_1 gases to the space in which this center must be found in order that its distance from the center of one of the remaining molecules may lie between σ and $\sigma+\delta$. The latter space we shall call the favorable space.

Since in all N_1 gases the center of each of the remaining molecules has a given position, and since the center of the specified molecule cannot come nearer to it than the distance σ, we obtain the available space for the center of the specified molecule in one of these gases by subtracting from the total volume V of the gas the volume of the covering spheres of all remaining molecules, viz., the quantity

$$\Gamma = \frac{4\pi(n-1)\sigma^3}{3}.$$

Since we can neglect unity compared to n, the total space available for the center of the specified molecule in one of the N_1 gases is

$$V - \frac{4\pi n\sigma^3}{3}. \tag{148}$$

The negative term is the total volume of all covering spheres contained in the volume V, or eight times the volume of the molecules in the volume V. It should be small compared to V, and we denote the degree of its smallness compared to V as "smallness of the first order." In order to find the favorable space, we construct around the center of each of the remaining molecules a spherical shell, which is enclosed between two spherical surfaces concentric with the molecule itself, with radii equal to σ and $\sigma+\delta$ respectively. The sum of the volumes of all these spherical shells is the favorable space. One finds for it the contribution

$$\Delta = 4\pi(n-1)\sigma^2\delta,$$

for which we can also write

$$\Delta = 4\pi n\sigma^2\delta.$$

As a first approximation we can ignore the negative term in (148) and hence the ratio of the favorable space to the available space is

$$4\pi n\sigma^2\delta/V.$$

Therefore in

$$4\pi n\sigma^2 N_1\delta/V$$

of our N_1 gases, the center of the specified molecule finds itself at a distance between σ and $\sigma+\delta$ from the remaining molecule. However, since the state of our N_1 gases can still be chosen completely arbitrarily, the same holds also for all N gases. In

$4\pi n\sigma^2 N\delta/V$ of these, the center of the specified molecule will have this property. Since the same holds for all other molecules, in all N gases there will be a total of

$$4\pi n^2\sigma^2 N\delta/V$$

molecules whose centers are at a distance from the center of any other molecule that lies between σ and $\sigma+\delta$. Hence in each gas,

$$4\pi n^2\sigma^2\delta/V$$

molecules will satisfy this condition, and the number of molecule pairs in a gas for which the distance of centers lies between σ and $\sigma+\delta$ is

$$2\pi n^2\sigma^2\delta/V. \tag{149}$$

If one wishes to take account also of terms of next higher order, then he should not only substitute (148) for V, but also introduce a correction in the numerator. Δ was the total volume of all spherical shells of thickness δ, which we had constructed around the centers of all the remaining molecules. Not all of these should be counted as favorable volumes. In particular, the covering spheres of two molecules can partially overlap. Then one part of such a spherical shell lies within the covering sphere of another molecule, so that it is not available as a location for the center of the specified molecule, and must be subtracted from the favorable space Δ. Strictly speaking the circumstance that two covering spheres can interpenetrate should be taken account of in the calculation of the space Γ that we have subtracted from V, in order to find the available space; but one sees at once that we would have thereby obtained a term that is a second-order infinitesimal compared to V. Hence as long as we are only taking account of first-order infinitesimals, this term can be ignored. The quantity δ/σ is supposed to be of higher order than σ^3/V. Hence all terms containing δ^2 can be neglected—i.e., cases where two or more spherical shells of thickness δ interpenetrate—and neither the simultaneous interpenetration of three covering spheres nor the interaction of three molecules need be included.

We shall now calculate the correction to the quantity Δ. The covering spheres of two molecules will overlap when the distance between their centers lies between σ and 2σ. Let r be a length somewhere between these limits. Then, by analogy with Equation

(149), the number of molecule pairs in a gas whose centers have a separation between r and $r+dr$ is equal to:

$$\nu = 2\pi n^2 r^2 dr/V.$$

By the covering sphere we mean a spherical surface of radius σ around the center of the molecule. Since $\sigma < r < 2\sigma$, the covering spheres of all these ν molecule pairs will overlap, and indeed an easy calculation shows that for each such pair, the part of the surface of a covering sphere that lies within the second one is equal to $\pi\sigma(2\sigma - r)$. There is also a spherical shell of thickness δ surrounding the entire covering sphere. The part of this spherical shell that lies inside the covering sphere of the other molecule, and is therefore not available as a location for the center of the specified molecule, therefore has the volume: $\pi\sigma(2\sigma - r)\delta$. An equal part of the spherical shell pertaining to the other molecule lies within the covering sphere of the first molecule, and hence is likewise not available. Therefore one has to subtract a total volume $2\pi\sigma(2\sigma - r)\delta$ from the two spherical shells of the molecule pair because that volume is not available to the center of the specified molecule. For all ν pairs the space

$$\frac{4\pi^2}{V} n^2 \sigma\delta(2\sigma - r) r^2 dr$$

is to be subtracted. Since r can take all values between σ and 2σ, the total volume to be subtracted from all spherical shells is equal to:

$$\frac{4\pi^2}{V} n^2 \sigma\delta \int_\sigma^{2\sigma} (2\sigma - r) r^2 dr = \frac{11\pi^2 n^2 \sigma^5 \delta}{3V}.$$

The total volume of all spherical shells in a gas would be $\Delta = 4\pi n\sigma^2\delta$. Hence the favorable volume remaining is

$$4\pi n\sigma^2\delta\left(1 - \frac{11}{12}\frac{\pi n\sigma^3}{V}\right).$$

The quotient of this quantity divided by the total available volume,

$$V\left(1 - \frac{4\pi n\sigma^3}{3V}\right),$$

—which, since we ignore terms of second order, can be written

$$\frac{4\pi n\sigma^2\delta}{V}\left(1+\frac{5\pi n\sigma^3}{12V}\right)$$

—gives the probability that the center of the specified molecule is at a distance between σ and $\sigma+\delta$ from another molecule. The further conclusions remain as before. The last expression, multiplied by $\frac{1}{2}n$, provides at an arbitrarily chosen instant of time the number of molecule pairs in a gas for which the centers of the two molecules have a distance between σ and $\sigma+\delta$. The number of these molecule pairs is therefore:

$$\frac{2\pi n^2\sigma^2\delta}{V}\left(1+\frac{5\pi n\sigma^3}{12V}\right). \tag{150}$$

The number of molecules from which these pairs are formed is of course twice as large.

§52. Contribution to the virial resulting from the finite extension of the molecules.

We can determine the average virial in different ways. The simplest way is to make use of Equation (142) in §47. We replace the elasticity of the molecule by a repulsive force $f(r)$ which is a function of the distance of centers, r, which vanishes for $r\geq\sigma$ and increases beyond all limits as soon as r becomes smaller than σ. From now on we shall denote by r a distance slightly less than σ. If the repulsion first began at distances a little less than r, then the number of molecule pairs for which the distance of centers lay between r and $r+\delta$ would, according to Equation (150), be equal to:

$$\frac{2\pi n^2r^2\delta}{V}\left(1+\frac{5\pi nr^3}{12V}\right), \tag{151}$$

where we can replace σ by r in Equation (150) since the latter is only infinitesimally different from the former. We still have to calculate how the number of pairs will be decreased by the repulsive force. If in Equation (142) one replaces the p by the rectangular coördinates of the molecule centers, then he finds that the number of systems for which these lie in a certain volume

$do_1, do_2, \cdots$ is proportional to $e^{-2hV_0}do_1do_2 \cdots$, where V_0 is the potential energy function whose negative derivative with respect to a coördinate is the force acting to increase that coördinate. For our molecule pair, V_0 is just a function of r, and indeed it is equal to the negative integral of $f(r)dr$. As soon as the distance of centers of two molecules is equal to or greater than σ, the repulsion stops, and the potential energy then has the same value that it has at infinity, which we denote by $F(\infty)$. The value of the potential at r is denoted by $F(r)$.

The probability that the distance of centers of two molecules lies between r and $r+\delta$, in the absence of all repulsive forces, is to the probability that it lies between the same limits, when repulsive forces are present, as

$$e^{-2hF(\infty)} : e^{-2hF(r)},$$

and for the number of molecule pairs for which the distance of centers is between r and $r+\delta$ one finds instead of (151) the expression

$$\frac{2\pi n^2 r^2 \delta}{V}\left(1+\frac{5\pi n r^3}{12V}\right)e^{2h[F(\infty)-F(r)]}. \tag{152}$$

Since

$$V_0 = F(r) = -\int f(r)dr$$

therefore

$$F(\infty) - F(r) = -\int_r^\infty f(r)dr.$$

Since moreover δ in Equation (152) represents an infinitesimal increase of r, we can denote it by dr following general usage, and Equation (152) then becomes

$$\frac{2\pi n^2 r^2 dr}{V}\left(1+\frac{5\pi n r^3}{12V}\right)e^{-2h\int_r^\infty f(r)dr}. \tag{153}$$

If we multiply this expression by the virial $rf(r)$ of the molecule pair in question, and integrate over all colliding molecules, we obtain the total virial W_i' arising from the forces acting during collisions. Therefore if $\sigma-\epsilon$ is the smallest permitted distance of approach of two molecules when they impinge on each other with an enormous velocity, then one obtains:

$$W_i' = \frac{2\pi n^2}{V}\int_{\sigma-\epsilon}^{\sigma}\left(1+\frac{5\pi n r^3}{12V}\right)r^3 f(r)dr e^{-2h\int_r^\infty f(r)dr}.$$

Since r is always infinitesimally different from σ, it can always be replaced by σ if it does not occur as an argument of the function f, so that it then comes out in front of the integral sign:

$$W_i' = \frac{2\pi n^2\sigma^3}{V}\left(1+\frac{5\pi n\sigma^3}{12V}\right)\int_{\sigma-\epsilon}^{\sigma} f(r)dr e^{-2h\int_r^\infty f(r)dr}.$$

The last integral can easily be computed if one introduces the new variable

$$x = \int_r^\infty f(r)dr$$

which is equal to zero at the upper limit and to infinity at the lower; indeed it is equal to the kinetic energy with which a molecule must approach a molecule at rest, if their centers are to approach to a distance $\sigma-\epsilon$; hence, on introducing this new variable, we have

$$\int_{\sigma-\epsilon}^{\sigma} f(r)dr e^{-2h\int_r^\infty f(r)dr} = \int_0^\infty e^{-2hx}dx = \frac{1}{2h} = \frac{m\overline{c^2}}{3},$$

since, according to §42, $\frac{1}{4h}$ is the mean kinetic energy $\frac{1}{2}m\overline{u^2}$ corresponding to a momentoid, and each of the components of the total velocity c represents a momentoid. (Cf. also Part I, Eq. [44].) If, as in Equation (20), we substitute

$$b = 2\pi\sigma^3/3m,$$

then it follows that

$$W_i' = nm\overline{c^2}\,\frac{b}{v}\left(1+\frac{5b}{8v}\right), \tag{154}$$

where $v = V/nm$ is the specific volume.

§53. Virial of the van der Waals cohesion force.

The virial due to the attractive force is found without difficulty by using the hypothesis about the nature of the interaction made in §2. If ρ is the density of the gas, and do and $d\omega$ are two volume elements separated by a distance r, at which the molecules attract with a force $F(r)$ and hence repel with a force $-F(r)$, then $\rho do/m$ and $\rho d\omega/m$ are the numbers of molecules in those volume elements, and

$$-\frac{\rho^2 do d\omega}{m^2} rF(r)$$

is the virial of the molecules contained in the two volume elements.

$$-\frac{\rho^2}{m^2} do \int d\omega rF(r)$$

is therefore the total virial of the molecules in *do* interacting with all the others. Since there is a noticeable contribution only for molecules at molecular distances,

$$\frac{1}{m^2}\int d\omega rF(r)$$

has the same value for all volume elements *do* in the interior of the gas. Since this value depends only on the nature of the function F, it must be a constant peculiar to the substance, and we denote it by $3a$. The total virial W_i'' is found by integrating over all volume elements *do*, whereby we obtain the total volume V. The contribution of the molecules lying very near to the surface is infinitesimal. Hence

$$W_i'' = 3\rho^2 aV. \tag{155}$$

The substitution $W_i = W_i' + W_i''$, with the values (154) and (155) for W_i' and W_i'', as well as the values (146) and (147) in the virial equation (145), gives therefore:

$$nm\overline{c^2}\left(1+\frac{b}{v}+\frac{5b^2}{8v^2}\right) - 3aV\rho^2 = 3pV.$$

According to Equation (21) we now set $\overline{c^2} = 3rT$. Furthermore, $V/nm = v = 1/\rho$, so that the above equation becomes

$$p+\frac{a}{v^2} = \frac{rT}{v}\left(1+\frac{b}{v}+\frac{5b^2}{8v^2}\right). \tag{156}$$

If one neglects quantities of order b^2/v^2, the right-hand side is identical with the expression $rT/(v-b)$ found by van der Waals.

But there is already a discrepancy in the terms of order b^2/v^2. Van der Waals himself noted that his equation cannot be valid for arbitrary v, since it predicts that the pressure becomes infinite at $v=b$, whereas actually the pressure does not become infinite until v is much smaller than b.

§54. Alternatives to van der Waals' formulas.

Our present considerations teach us that the expression for the pressure given by van der Waals does not agree with the theoretical one for small values of v, as soon as one takes account of terms of order b^2/v^2. Since the theoretical determination of terms of still higher order would be extraordinarily difficult,* one can try to replace van der Waals' equation by one that agrees with theory at least as far as terms of order b^2/v^2. We saw, moreover, that one can also determine theoretically the smallest permitted value of v, for which the pressure becomes infinite. It is $v=\frac{1}{3}b$ (cf. §6), since for approximately this value of v the molecules are as densely packed as possible, and any further decrease of v must make the molecules interpenetrate. One can therefore construct an equation of state for which p becomes infinite at this value of v.

In order to depart as little as possible from the form of van der Waals' equation, we shall write the equation of state in the following form, where x, y, and z are numbers to be chosen suitably:

* For the coefficient of b^3/v^3 (known as the fourth virial coefficient) see Boltzmann, Verslagen Acad. Wet. Amsterdam [4] **7,** 477 (1899); P. Ehrenfest, Wien. Ber. **112,** 1107 (1903); H. Happel, Ann. Physik [4] **21,** 342 (1906); R. Majumdar, Bull. Calcutta Math. Soc. **21,** 107 (1929). The fifth virial coefficient was estimated by the Monte Carlo method by M. N. and A. W. Rosenblueth, J. Chem. Phys. **22,** 881 (1954). A more accurate value will probably be available very soon. Modern methods for calculating such coefficients are reviewed by G. E. Uhlenbeck and G. W. Ford in *Studies in Statistical Mechanics,* Vol. I (Amsterdam: North-Holland, 1962), and J. E. Mayer, *Handbuch der Physik,* XII, 73 (Berlin: Springer, 1958). From the recent work of Groeneveld, Phys. Letters **3,** 50 (1962), it appears that even if one could calculate all the coefficients in the virial series for the pressure, he would not thereby gain much information about the equation of state at high densities, since the radius of convergence of the series is rather small.

$$(157) \qquad \left(p + \frac{a}{v^2}\right)(v - xb) = rT\left(1 + \frac{yb}{v} + \frac{zb^2}{v^2}\right).$$

We then have the advantage that for given p and T we obtain a third degree equation for v. If we set $y = 1 - x$, $z = \frac{5}{8} - x$, then for small values of b/v the terms of order b^2/v^2 agree with those found theoretically. If we set

$$(158) \qquad x = \frac{1}{3}, \quad \text{hence} \quad y = \frac{2}{3}, \quad z = \frac{7}{24},$$

then the condition that p becomes infinite for $v = \frac{1}{3}b$ is also satisfied. Since all our considerations are only approximate, it would perhaps be more rational not just to choose these values for x, y, and z, but rather to assign to them values that yield the best possible agreement with observations.

If one does not wish to add any factors to rT in the numerator, then a quadratic function of b/v in the denominator, such as a law of the form

$$p + \frac{a}{v^2} = \frac{rT}{v}\left(1 + \frac{xb}{v} + \frac{yb^2}{v^2}\right)^{-1}$$

would hardly be very advisable; for then p, if it is to become infinite at all, and if it is also to be correct to first order in b/v, would have to become infinite for a value of v that is greater than or equal to $\frac{1}{2}b$.

The latter occurs if one sets

$$p + \frac{a}{v^2} = \frac{rT}{v}\left(1 - \frac{b}{2v}\right)^{-2}$$

It would be better to set

$$(158a) \qquad p + \frac{a}{v^2} = \frac{rT}{v - \epsilon b}\cdot$$

and to choose for ϵ a transcendental or algebraic function of higher degree, which is nearly equal to 1 for large v, and for $v = \frac{1}{3}b$ is nearly equal to $\frac{1}{3}$, and also agrees as well as possible with experimental data. I owe Equation (158a) to an oral communication of van der Waals (cf. also the work of Kamerlingh-Onnes cited in footnote 1, §60). Moreover, van der Waals has given us such a valuable tool that it would cost us much trouble to obtain by the subtlest deliberations a formula that would really be any more useful than the one that van der Waals found by inspiration, as it were.

A more general way to bring the original van der Waals formula into better agreement with experience would consist in treating the expressions a/v^2 and $v-b$ in it as empirically chosen functions of volume and temperature instead of constants, or, quite generally, to seek in place of a/v^2 and $v-b$ the functions best fitting the observations. Of course these functions must be chosen so that the theorems on critical quantities and on liquefaction do not come out qualitatively different. Clausius and Sarrau have modified the van der Waals formula in this way.* Even though they have been guided by theoretical ideas (Clausius especially seems to have in mind taking account of the combination of molecules into larger complexes), their equations have more the character of empirical approximations, which I will not go into further, although I would not wish to disparage their practical usefulness.

§55. Virial for any arbitrary law of repulsion of the molecules.

By means of Equation (153) we can also calculate by the same method the quantity W_i' for the case where the molecules do not behave like elastic spheres, but rather like material points that exert an arbitrary central repulsive force $f(r)$ on each other during collisions. Since the time during which two colliding molecules act on each other can then no longer be neglected, the method by which we derived Equation (150) from (149) is no longer correct; however, the latter formula is still correct as a first approximation.

The number of pairs of molecules whose distance lies between r and $r+dr$ is therefore $2\pi n^2 r^2 dr/v$, as soon as no force acts at the distance r. The modification of this number by the action of repulsive forces was previously found according to the general

* Clausius, Ann. Physik [3] 9, 337 (1880); Sarrau, C. R. Paris 101, 941 (1885). See Partington, *Treatise on Physical Chemistry* (London, 1949), Vol. I, pp. 660–729.

formula (142). This formula remains applicable here, and hence as soon as there is a repulsive force at the distance r, the number of pairs of molecules at a distance between r and $r+dr$ is (corresponding to Eq. [153]) equal to:

$$\frac{2\pi n^2 r^2 dr}{v} e^{-2h\int_r^\infty f(r)\,dr}.$$

Each such molecule pair provides a contribution $rf(r)$ to the virial W_i'. The sum of all these contributions is therefore

$$W_i' = \frac{2\pi n^2}{v}\int_\zeta^\sigma r^3 f(r)\,dr\,e^{-2h\int_r^\infty f(r)\,dr}, \tag{159}$$

where ζ is the closest allowed distance of approach of two molecules, and σ is the distance at which they cease to interact. However, one can integrate from zero to infinity instead of from ζ to σ, since for $r<\zeta$ the exponential factor is zero, while for $r>\sigma$, $f(r)=0$.

If one sets $f(r)=K/r^5$ as in Chapter III of Part I, then of course the present assumptions are not strictly satisfied, since properly speaking all molecules are continually repelling each other; but this repulsion decreases so rapidly with increasing distance that the deviations thereby produced in our formulas are probably completely insignificant. Hence we obtain:

$$W_i' = \frac{2\pi n^2}{v} K\int_0^\infty \frac{dr}{r^2} e^{-hK/2r^4} = \frac{2\pi n^2}{v}\sqrt[4]{\frac{2}{h}K^3}\int_0^\infty e^{-x^4}dx.$$

We shall introduce the distance of closest approach of two molecules if one were held fixed and the other approached it with a velocity whose square is equal to the mean square velocity of a molecule (cf. Part I, §24). However, we shall denote this distance here by σ instead of s. Then:

$$K/2m\overline{c^2} = \tfrac{1}{3}hK = \sigma^4.$$

Hence

$$W_i' = \frac{4\pi n^2\sigma^3 m\overline{c^2}}{v}\sqrt[4]{\frac{2}{3}}\int_0^\infty e^{-x^4}dx.$$

If one imagines that the van der Waals attractive force is present, and brings it into the calculation as before, then on substituting all values into Equation (145) he obtains the following equation for unit mass of the gas:

$$rT\left(\frac{1}{v}+\frac{b}{v^2}\right)=p+\frac{a}{v^2},$$

where*

$$b=4\pi n\sigma^3\sqrt[4]{\frac{2}{3}}\int_0^\infty e^{-x^4}dx.$$

However, σ is no longer constant now, but instead is inversely proportional to the fourth root of the absolute temperature. Hence the numerical coefficient b has become temperature-dependent. In particular, b is inversely proportional to the $\frac{3}{4}$ power of the absolute temperature, since σ is inversely proportional to the fourth root of the temperature.

§56. The principle of Lorentz's method.

We have previously derived the internal virial from Equation (154) by means of Equation (142) for the case where the gas molecules behave like elastic spheres. One can also do the same thing in another way without using the latter formula, as was first shown by H. A. Lorentz.† By definition,

$$W_i' = \frac{1}{t}\int_0^t \sum rf(r)dt,$$

where the sum is to be extended over all pairs of molecules that collide during a very long time. Since the state is stationary we can choose unit time $t=1$ instead of a very long time t. If we integrate each term of the sum, it follows that:

$$W_i' = \sum\int_0^1 rf(r)dt, \tag{160}$$

where the sum is to be extended over all pairs of molecules that

* $\int_0^\infty e^{-x^4}dx = \frac{1}{4}\Gamma(\frac{1}{4})$.

† H. A. Lorentz, Ann. Physik [3] 12, 127, 660 (1881).

collide during unit time. However, the relative motion of the molecules takes place just as if one of them were at rest and the other had half its mass. If, in this relative motion, the first molecule is imagined to move toward the second one (which is at rest) with relative velocity g, then the component of g perpendicular to the line of centers of the molecules does not change. The component γ in the direction of the line of centers will be exactly reversed by the force $f(r)$, however. Hence, for each collision:

$$\int f(r)dt = \frac{m}{2}\cdot 2\gamma = m\gamma.$$

$$\int rf(r)dt \text{ is however equal to } \sigma\int f(r)dt,$$

since r is always approximately equal to σ during the collision. If one substitutes this into Equation (160), it follows that:

$$W_i' = m\sigma\sum\gamma, \tag{161}$$

where the sum is to be extended over all pairs of molecules that collide in unit time.

In order to calculate this sum, we shall first ask how many molecules in the gas collide in a particular way during a very short time dt. If the molecules are to collide during dt, then their centers must already be at a distance only slightly larger than σ at the beginning of the interval dt. According to Equation (150), at any arbitrary time and therefore also at the beginning of the interval dt there are

$$2\pi n^2\sigma^2\beta\delta/V$$

pairs of molecules whose centers have a distance between σ and $\sigma+\delta$, where

$$\beta = 1 + \frac{5\pi n\sigma^3}{12V} = 1 + \frac{5b}{8v} \tag{162}$$

and v is the specific volume. These molecule pairs consist of

$$\frac{4\pi n^2\sigma^2\beta\delta}{V} \tag{163}$$

molecules. Each of these molecules lies so close to another one that the distance of their centers lies between σ and $\sigma+\delta$.

Here we must invoke, not indeed Equation (142) in particular, but rather a probability law, in that we assume that, in the space where the molecules whose number is given by Equation (163) find themselves, there is the same average distribution of molecules, and among these the same distribution of states, as in the entire gas. This follows directly from the "S-theorem" of §46. Then, out of those molecules whose number is given by Equation (163),

$$4\pi n^2\sigma^2 \frac{\beta\delta}{V} \varphi(c)dc \tag{164}$$

have a velocity between c and $c+dc$, if, as in Equation (8), $\varphi(c)dc$ is the probability that the velocity of a molecule lies between these limits, so that of all n gas molecules, $n\varphi(c)dc$ have a velocity lying between these limits. Then, as we saw in Part I (cf. also Eq. [8], Part II):

$$\varphi(c) = 4\sqrt{\frac{m^3h^3}{\pi}}\, c^2 e^{-hmc^2}. \tag{165}$$

The expression (164) therefore gives the number of molecules that have a velocity between c and $c+dc$ at the beginning of dt, and whose centers have a distance between σ and $\sigma+\delta$ from any other molecule. Now we shall consider, out of all these molecules, only those for which the velocity of the other molecule lies between c' and $c'+dc'$, and the direction of this velocity forms an angle between ϵ and $\epsilon+d\epsilon$ with that of the velocity of the first molecule. Since the other molecules have a Maxwellian velocity distribution, regardless of the state of the molecules near them, and any direction in space is equally probable for their velocities, we must—in order to find the number of those molecules which we have picked out from all the molecules whose number is given by (164)—multiply the latter expression by

$$\tfrac{1}{2}\varphi(c')dc' \sin\epsilon\, d\epsilon.$$

The product

$$d\mu = 2\pi n^2\sigma^2 \frac{\beta\delta}{V} \varphi(c)\varphi(c') \sin\epsilon\, dc dc' d\epsilon, \tag{166}$$

therefore gives the number of pairs of molecules for which the velocity of one molecule (which we call the c-molecule) is between c and $c+dc$, that of the other molecule (the c'-molecule) is between c' and $c'+dc'$, the angle of the directions of their velocities is between ϵ and $\epsilon+d\epsilon$, and the centers of the molecules are at a distance between σ and $\sigma+\delta$ at the beginning of the time interval dt.

If we draw around the center of each c-molecule a spherical shell bounded by two concentric spherical surfaces, the inner surface having radius σ and the outer surface having radius $\sigma+\delta$, then (166) gives the number of molecules whose centers lie in one of these spherical shells at the beginning of the interval dt, and for which the velocity lies between c' and $c'+dc'$ and forms an angle between ϵ and $\epsilon+d\epsilon$ with the velocity of the c-molecule.

§57. Number of collisions.

Each c'-molecule lies very near to a c-molecule. In order to find how many will actually collide during an infinitesimal time interval dt, we imagine that all the c-molecules are at rest, while each c'-molecule moves with velocity

$$g = \sqrt{c^2 + c'^2 - 2cc' \cos \varphi} \tag{167}$$

relative to the c-molecule near it, so that during time dt it moves a distance gdt in the direction of the relative velocity g.

Furthermore, we again imagine that there is a sphere K of radius σ around the center of each c-molecule. From the center of each of these spheres we draw a line G which has the direction of the relative velocity of that c'-molecule which is in the neighborhood of the c-molecule in question. We extend the line G through the opposite side, and draw all radii of each of the K-spheres that form an angle between ϑ and $\vartheta+d\vartheta$ with the extension. The endpoint of all these radii will form a zone on the surface of each of the K-spheres whose surface area is $2\pi\sigma^2 \sin \vartheta d\vartheta$. From each point of each of these zones we draw a line whose length is gdt and whose direction is opposite to the direction of the relative velocity of the c'-molecule. All the lines drawn from the points on a zone fill a ring-like space of volume $2\pi\sigma^2 g \sin \vartheta \cos \vartheta d\vartheta dt$, and one easily sees that all c'-molecules whose centers lie within this ring-like space at the beginning of dt (whose number we call $d\nu$) will collide with the c-molecule in their neighborhood in such a way that the angle between the line drawn from the center of the c'-molecule to the center of the c-molecule and the relative velocity of c' with respect to c lies between ϑ and $\vartheta+d\vartheta$. However, the ratio of $d\nu$ to the number $d\mu$ (given by Eq. [166]) is the same as the ratio of the volume $2\pi\sigma^2 g \sin\vartheta\cos\vartheta d\vartheta dt$ of one of the ring-like spaces in which the centers of the $d\nu$ molecules must be found, to the volume $4\pi\sigma^2\delta$ of one of the spherical shells in which the centers of the $d\mu$ molecules are found; whence one obtains:

$$d\nu = \frac{2\pi\sigma^2}{V} n^2\beta \sin\vartheta \cos\vartheta g\varphi(c)\varphi(c') \frac{\sin\epsilon d\epsilon}{2} dc dc' d\vartheta dt.$$

If we divide by dt, we find the following expression for the number of pairs of molecules that collide in unit time in such a way that before the collision the velocities are between the limits $(c, c+dc)$ and $(c', c'+dc')$ respectively, the angle between the velocities is between ϵ and $\epsilon+d\epsilon$, and the angle between the line of centers and the relative velocity is between ϑ and $\vartheta+d\vartheta$:

$$(168)\qquad d\mathfrak{n}_{cc'\epsilon\vartheta} = \frac{2\pi\sigma^2}{V} n^2\beta \sin\vartheta \cos\vartheta g\varphi(c)\varphi(c') \frac{\sin\epsilon d\epsilon}{2} dc dc' d\vartheta.$$

If we divide the expression (168) by $n\varphi(c)dc$, then we obtain the number of collisions experienced by an individual c-molecule with c'-molecules subject to the above conditions. If we integrate over ϑ from zero to $\frac{1}{2}\pi$, over ϵ from 0 to π, and over c' from 0 to ∞, we obtain the total number $\mathfrak{n}_c$ of collisions experienced per second by a molecule moving with velocity c. We can call the expression

$$(169)\qquad \overline{g_c} = \int_0^\infty dc'\varphi(c') \int_0^\pi \frac{g \sin\epsilon\, d\epsilon}{2}$$

the mean value of all relative velocities of a molecule moving with velocity c, with respect to all other possible velocities. Hence

$$\mathfrak{n}_c = \frac{\pi\sigma^2 n\beta}{V} \overline{g_c}.$$

If one substitutes in (169) the value (165) for the function φ, and the value (167) for g, then he finds, as we already saw in Part I, §9,

$$\overline{g_c} = 4\sqrt{\frac{m^3h^3}{\pi}}\int_0^\infty c'^2dc'e^{-hmc'^2}\int_0^\pi \frac{\sin\epsilon d\epsilon}{2}\sqrt{c^2+c'^2-2cc'\cos\epsilon}$$

$$= \frac{2}{\sqrt{\pi hm}}\left(e^{-hmc^2} + \frac{2hmc^2+1}{c\sqrt{hm}}\int_0^{c\sqrt{hm}} e^{-x^2}dx\right).$$

Since $\varphi(c)dc$ is the probability that the velocity of a molecule is between c and $c+dc$, and therefore is also the fraction of the time during which its velocity lies between these limits in the entire course of its motion during a long period of time, the total number of collisions that an arbitrary molecule experiences on the average during unit time is equal to

$$\mathfrak{n} = \int_0^\infty \mathfrak{n}_c\varphi(c)dc = \frac{\pi\sigma^2 n\beta}{V}\bar{g}, \tag{170}$$

where

$$\bar{g} = \int_0^\infty \bar{g}_c\varphi(c)dc$$

$$= \frac{8mh}{\pi}\int_0^\infty c^2e^{-hmc^2}dc\left(e^{-hmc^2} + \frac{2hmc^2+1}{c\sqrt{hm}}\int_0^{c\sqrt{hm}} e^{-x^2}dx\right)$$

is the mean value of the relative velocities of all possible molecule pairs in the gas. We have performed a similar integration in Part I, §9; if we do this one in the same way, we obtain:

$$\bar{g} = \bar{c}\sqrt{2} = \frac{2\sqrt{2}}{\sqrt{\pi hm}}.$$

The mean relative velocity is therefore just as large as if the two molecules each moved with their average velocity in perpendicular directions. This theorem is also valid if the two molecules are of different kinds. If one substitutes the value found for g into the equation following (169), he obtains:

$$\mathfrak{n} = \frac{2\sigma^2 n\beta}{v}\sqrt{\frac{2\pi}{hm}}.$$

§58. More exact value of the mean free path. Calculation of W_i' according to Lorentz's method.

Since the mean free path is $\lambda = \bar{c}/\mathfrak{n}$, it follows moreover that

$$\lambda = \frac{V}{\sqrt{2}\pi\sigma^2 n\beta},$$

or, on substituting this value (162) for β and expanding in powers of b/v,

$$\lambda = \frac{V\left(1 - \frac{5b}{8v}\right)}{\sqrt{2}\pi\sigma^2 n} = \frac{\sqrt{2}\sigma v}{3b}\left(1 - \frac{5b}{8v}\right). \tag{171}$$

This is therefore the value of the mean free path, one order of magnitude more exact (with respect to b/v) than that given in Part I, §10.

We can now easily find the mean virial of all forces acting in collisions from Equations (161) and (168). For each of the collisions whose number is given by Equation (168), the component γ of the relative velocity g in the direction of the line of centers has the value $\gamma = g \cos \vartheta$; each of these collisions contributes therefore the term $m\sigma g \cos \vartheta$ to the sum (161). If one multiplies this by the expression (168), then he finds the contribution of all these collisions to the sum (161). If one then integrates over all possible values, he obtains finally the total contribution to the sum, and hence, according to Equation (161), the quantity W_i'. Finally, however, one must divide by 2, since otherwise he would have counted each collision twice—once when the velocity of one molecule lies between c and $c+dc$ and then again when the velocity of the other molecule lies between c and $c+dc$. Hence:

$$W_i' = \frac{\pi\sigma^3 n^2 m\beta}{2V}\int_0^{\pi/2} \sin\vartheta \cos^2\vartheta d\vartheta$$
$$\times \int_0^\infty \varphi(c)dc \int_0^\infty \varphi(c')dc' \int_0^\pi g^2 \sin\epsilon\, d\epsilon.$$

If one substitutes for g the value (167) and recalls that:

$$\int_0^\infty \varphi(c)dc = \int_0^\infty \varphi(c')dc' = 1$$

$$\int_0^\infty c^2\varphi(c)dc = \int_0^\infty c'^2\varphi(c')dc' = \overline{c^2},$$

then he obtains the value already found,

$$W_i' = 2\pi\sigma^3 n^2 m\overline{c^2}\beta/3V.$$

The corrected value (171) for the mean free path was first given by Clausius.[1] The additional term of order b/v in the Boyle-Charles law was first calculated by the foregoing method by H. A. Lorentz;[2] the additional term of order b^2/v^2 was calculated by Jäger[3] and van der Waals;[4] the value found by Jäger agrees with the one calculated here, but the one found by van der Waals does not.*

§59. More exact calculation of the space available for the center of a molecule.

Let there be present, in a container of volume V, a total of n similar molecules, which we consider as spheres of diameter σ. Then we can find the space available for the center of another molecule introduced into the container, given the positions of all n molecules, by subtracting from the total volume V the space occupied by the n molecules: $\Gamma = 4\pi n\sigma^3/3 = 2Gb$ (cf. Eq. [148]). As before, m is the mass of a molecule, $mn = G$ is the total mass of the gas, and

$$b = \frac{2\pi\sigma^3}{3m}$$

[1] Clausius, *Kinetische Gastheorie,* Vol. 3 of *Mechanische Wärmetheorie* (Vieweg, 1889–1891), p. 65.

[2] Lorentz, Ann. Physik [3] **12,** 127, 660 (1881).

[3] Jäger, Wien. Ber. **105,** 15 (16 Jan. 1896).

[4] Van der Waals, Verslagen Acad. Wet. Amsterdam [4] **5,** 150 (31 October 1896).

* Van der Waals later accepted the Boltzmann-Jäger value as correct, after his son (J. D. van der Waals, Jr.) showed that the method used in ref. 4 was incorrect. For references to this dispute see Brush, Am. J. Phys. **29,** 593 (1961), esp. footnotes 31–41.

as in Equation (20) is half the sum of the covering spheres of all molecules found in unit mass of the gas. Here the term of order Γ^2/V^2, corresponding to the effect of interpenetration of the covering spheres of two molecules, is omitted. We shall now calculate this term, although we shall still leave out terms of order Γ^3/V^3.

Let Z be the sum of the volumes of all those parts of the covering spheres of molecules that lie within the covering spheres of any other molecule, so that we have to set:

$$D = V - 2Gb + Z. \tag{172}$$

The case that the covering spheres of two molecules overlap occurs each time that their centers have a distance between σ and 2σ. Let x be such a distance. The covering spheres are spheres of radius σ, which are concentric with the molecule in question. If the centers of two molecules are at a distance x, then the total space that belongs simultaneously to the covering spheres of both molecules has the form of two spherical sections of height $\sigma - x/2$. Such a spherical section has a volume

$$K = \pi \int_{x/2}^{\sigma} (\sigma^2 - y^2)dy = \pi\left(\frac{2\sigma^3}{3} - \frac{\sigma^2 x}{2} + \frac{x^3}{24}\right).$$

We shall construct, for each molecule, a concentric spherical shell of inner radius x and outer radius $x+dx$. The sum of the volumes of these spherical shells, $4\pi n x^2 dx$, has the same ratio to the total volume V of the gas that the number dn_x of molecules whose centers are at a distance between x and $x+dx$ from another one has to the total number n of molecules. Therefore:

$$dn_x = \frac{4\pi n^2 x^2 dx}{V}.$$

Here terms of order $\Gamma dn_x/V$ have been omitted, but one can easily convince himself that in the final result these would only give terms of order Γ^3/V^3.

The number of pairs of molecules for which the distance of centers lies between x and $x+dx$ is $\frac{1}{2}dn_x$. Since, for each such molecule pair, two spherical sections of volume K lie within the covering spheres, all these molecule pairs provide a contribution Kdn_x to Z, and we find the quantity Z itself by integrating these contributions from $x=\sigma$ to $x=2\sigma$. One obtains in this way

$$Z = \frac{\pi^2 n^2}{V}\int_0^{2\sigma}\left(\frac{8\sigma^3}{3} - 2\sigma^2 x + \frac{x^3}{6}\right)x^2 dx = \frac{17}{36}\frac{\pi^2 n^2 \sigma^6}{V} = \frac{17}{16}\frac{G^2 b^2}{V},$$

$$D = V - 2Gb + \frac{17}{16}\frac{G^2 b^2}{V}. \tag{173}$$

§60. Calculation of the pressure of the saturated vapor from the laws of probability.[1]

Now let the liquid and vapor phase of a substance be in contact with each other at a particular temperature T. The total mass of the liquid part will be G_f, and its total volume V_f; the mass of the vapor part will be G_g, and it fills the total volume V_g, so that $v_f = V_f/G_f$, $v_g = V_g/G_g$ are the specific volumes or reciprocals of the densities ρ_f and ρ_g of the liquid and the vapor.

If one now introduces a molecule into the space in which the two phases are present, then according to Equation (173) the space available for this molecule within the liquid is

$$V_f - 2G_f b + \frac{17}{16}\frac{G_f^2 b^2}{V_f},$$

while the space in the vapor is

$$V_g - 2G_g b + \frac{17}{16}\frac{G_g^2 b^2}{V_g}.$$

The ratio of these volumes would be—if the van der Waals cohesion force did not exist—the ratio of the probabilities that, for given positions of all other molecules, the last one lay within the liquid or the vapor. This ratio is (on account of the action of the van der Waals cohesion force) to be multiplied by $e^{-2h\psi_f}$: $e^{-2h\psi_g}$, where ψ_f and ψ_g are the values of the potential of the van der Waals cohesion force for a molecule that finds itself in the liquid or the vapor, respectively. If one determines the constant so that $\psi = 0$ for infinite separation, then $-\psi_f$ is the work needed to bring a molecule of mass m, under the action of the

[1] The same problem has been treated by Kamerlingh-Onnes, Arch. Neerl. **30,** §7, p. 128 (1881).

van der Waals cohesion force, out of the interior of the liquid and remove it to a large distance. In §24 we found the expression $2ma\rho_f = 2ma/v_f$ for this work, while $a\rho_f$ was the total work of separation for all molecules in unit mass. Likewise,

$$-\psi_g = 2ma\rho_g = 2ma/v_g.$$

Taking account of the van der Waals force, we find that the ratio of the probability that the last molecule is in the liquid to the probability that it is in the gas is:

$$\left(V_f - 2G_f b + \frac{17}{16}\frac{G^2b^2}{V_f}\right)e^{4hma/v_f} : \left(V_g - 2G_g b + \frac{17}{16}\frac{G^2b^2}{v_g}\right)e^{4hma/v_g}.$$

In the equilibrium state this must also be equal to the ratio of n_f to n_g; or, if one multiplies n_f and n_g by m, equal to G_f/G_g. If one writes out the proportion, he obtains immediately

$$v_g - 2b + \frac{17}{16}\frac{b^2}{v_g} = \left(v_f - 2b + \frac{17}{16}\frac{b^2}{v_f}\right)e^{4hma[(1/v_f)-(1/v_g)]}.$$

Now according to Equations (21) and (135)[2] (cf. also Part I, Eq. [44]), $2h = 1/mrT$, if r is the gas constant of the vapor at high temperature and low density. If one takes the logarithms, expands in powers of b and retains terms of order b^2, then

$$(174)\quad \frac{1}{v_f} - \frac{1}{v_g} = \frac{r}{2}\frac{T}{a}\left[l\frac{v_g}{v_f} - 2b\left(\frac{1}{v_g} - \frac{1}{v_f}\right) - \frac{15}{16}b^2\left(\frac{1}{v_g^2} - \frac{1}{v_f^2}\right)\right].$$

Naturally this formula can hardly be expected to give more than qualitative agreement, since the assumption that b is small compared to v is incorrect for liquids.

If we introduce the Celsius temperature t, consider ρ_f to be constant and large compared to ρ_g, and assume that the vapor obeys the Boyle-Charles law, i.e., $pv_g = rT$, then there follows from Equation (174) an equation of the form

$$(175)\quad p = \frac{1}{A + Bt}e^{t/(c+Dt)},$$

[2] In particular, $\overline{S} = \frac{1}{2}m\overline{c^2}$.

—an equation that has some practical usefulness, though with a somewhat different meaning of the constants A, B, C, and D.

We can also calculate the pressure of the saturated vapor from the condition found in §16, that the two shaded areas of Figure 2 of that section must be equal. In this figure, the abscissa OJ_1 is the specific volume v_f of the liquid, the abscissa OG_1 is the specific volume v_g of the vapor, while the ordinates $J_1J = G_1G$ are equal to the corresponding saturation pressure. The equality of the shaded areas requires that the rectangle $JJ_1G_1GJ = p(v_g - v_f)$ must be equal to the surface area $\int_{v_f}^{v_g} pdv$, which is bounded above by the curve $JCHDG$, below by the abscissa axis, and on the left and right by the ordinates J_1J and G_1G. One therefore obtains:

$$p(v_g - v_f) = \int_{v_f}^{v_g} pdv. \tag{176}$$

If one starts from the van der Waals equation

$$p = \frac{rT}{v - b} - \frac{a}{v^2} \tag{177}$$

as a basis, then it follows on performing the integration that:

$$p(v_g - v_f) = rTl\frac{v_g - b}{v_f - b} + a\left(\frac{1}{v_g} - \frac{1}{v_f}\right). \tag{178}$$

T and the constants a, b, and r are to be considered as given. The three unknowns p, v_f, and v_g follow from Equation (178) and the two conditions: that v_f is the smallest and v_g the largest root of Equation (177). If one again takes for the vapor the Boyle-Charles law $pv_g = rT$, neglecting b and v_f compared to v_g, and ρ_g compared to ρ_f (the latter is considered to be a linear function of the temperature), then he obtains for the saturation pressure an expression of the form (175) again; yet Equation (178) by no means agrees exactly with Equation (174).

This would not be expected, since Equation (177) is only provisory, and is not an exact consequence of the conditions of the problem.

On the contrary, one must obtain precisely Equation (174) if he uses instead of Equation (177) the equation

$$p = rT\left(\frac{1}{v} + \frac{b}{v^2} + \frac{5b^2}{8v^3}\right) - \frac{a}{v^2} \tag{179}$$

which, on neglecting the terms containing powers of b higher than the second, exactly satisfies the conditions of the problem.

In fact, it then follows from (176), on performing the integration, that

$$(180)\quad \begin{cases} p(v_g - v_f) = rT\left[l\dfrac{v_g}{v_f} - b\left(\dfrac{1}{v_g} - \dfrac{1}{v_f}\right)\right. \\ \left. - \dfrac{5b^2}{16}\left(\dfrac{1}{v_g^2} - \dfrac{1}{v_f^2}\right)\right] + a\left(\dfrac{1}{v_g} - \dfrac{1}{v_f}\right). \end{cases}$$

Since now v_g as well as v_f must satisfy Equation (179), one can calculate pv_g by substituting $v = v_g$ in this equation, and pv_f by setting $v = v_f$. Subtraction of the two values yields

$$p(v_g - v_f) = rT\left[b\left(\frac{1}{v_g} - \frac{1}{v_f}\right) + \frac{5b^2}{8}\left(\frac{1}{v_g^2} - \frac{1}{v_f^2}\right)\right] + a\left(\frac{1}{v_g} - \frac{1}{v_f}\right),$$

which, together with Equation (180) yields exactly Equation (174).

§61. Calculation of the entropy of a gas satisfying van der Waals' assumptions, using the calculus of probabilities.

I will now indicate briefly how one can calculate, according to the principles developed in Part I, §§8 and 19, the entropy of a gas for which the space filled by the molecules is not vanishingly small compared to the entire volume of the gas, and in which also the van der Waals cohesion force acts. The van der Waals cohesion force does not change the velocity distribution among the molecules, but only causes them to draw closer together. Like gravity, it has no direct influence on the entropy, so that the dependence of the entropy on the temperature for such a gas will be obtained just as in the sections cited for an ideal gas; in the present case only a correction for the finite size of the molecules is needed.

The expression found in Part I, §8, for the entropy—which

we shall call S in the following—can easily be put in the form[1]

$$S = RMl\mathfrak{W} = RMl(v^n T^{3n/2}).$$

If M is the mass of a hydrogen atom, then R is the gas constant of dissociated hydrogen, hence it is twice the gas constant of ordinary hydrogen. The exponent of T must be

$$\frac{3n}{2}(1+\beta) \text{ instead of } \frac{3n}{2}$$

when internal motions take place in the molecule, for which the variation of the sum of mean kinetic energy and potential energy stands in a constant ratio $\beta:1$ to the mean progressive kinetic energy. If β is a function of temperature, then

$$\frac{3n}{2}\int (1+\beta)\frac{dT}{T}$$

must take the place of $lT^{3n/2}$.

S is the entropy per unit mass, so that n is the number of molecules in unit mass; v is the volume of unit mass.

If one interprets S as a probability expression, then the quantity v^n occurring therein has the following meaning: it represents the ratio of the probability that all n molecules are simultaneously in volume v, to the probability of some standard configuration, for example one in which the first molecule is in a definite space of volume 1, the second in a completely different space of volume 1, and so forth. This quantity is the only one that is changed when we take into account the extension of the molecules, and indeed there occurs in its place the probability W of the simultaneous occurrence of the following events: the first molecule is found in v, and also the second, third, fourth, etc. are also found there (not, as in §60, simply the probability that one additional molecule is found in v).

The entire volume v is available for the center of the first molecule. The ratio of the probability that it is there to the probability that it is in the given space of volume 1 is thus v. In calculating the probability that the center of the second molecule is

[1] In Part I, §8, n is the number of molecules in unit volume; hence Ωn is the total number of molecules of a gas in volume Ω, which we have denoted here by n.

simultaneously in the space v, we have to subtract the volume of the covering sphere of the first molecule, $\frac{4}{3}\pi\sigma^3 = 2mb$, from v. If there are already ν molecules in the space v, then the space available for the center of a $(\nu+1)$th molecule is, according to Equation (173):

$$v - 2\nu mb + \frac{17\nu^2m^2b^2}{16v}. \tag{181}$$

This expression is therefore equal also to the ratio of the probability that the $(\nu+1)$th molecule finds itself in the space v to the probability that it finds itself in another space of volume v completely separate from the other space. Hence the product

$$W = \prod_{\nu=0}^{\nu=n-1} \left(v - 2\nu mb + \frac{17\nu^2m^2b^2}{16v}\right)$$

represents the ratio of the following probabilities: the probability that all n molecules simultaneously are in the space v, and the probability that each of them lies in a separate space of volume unity.[2] This expression has to occur in S instead of v^n, when we take account of the finite extension of the molecules. Therefore the entropy of unit mass is

$$S = rm\left[\frac{3n}{2}\int (1+\beta)\frac{dT}{T} + \sum_{\nu=0}^{\nu=n-1} l\left(v - 2\nu mb + \frac{17}{16}\frac{\nu^2m^2b^2}{v}\right)\right].$$

Here r is the gas constant of our substance in states that are sufficiently similar to the ideal gas state, so that $rm = RM$.

If one expands the logarithm in powers of b and neglects, as usual, powers of b higher than the second, it follows that:

$$l\left(v - 2\nu mb + \frac{17}{16}\frac{\nu^2m^2b^2}{v}\right) = lv - \frac{2\nu mb}{v} - \frac{15\nu^2m^2b^2}{16v^2}.$$

Since moreover we assume that n is large compared to 1, we can set:

$$\sum_{\nu=0}^{\nu=n-1} \nu = \frac{n^2}{2}, \qquad \sum_{\nu=0}^{\nu=n-1} \nu^2 = \frac{n^3}{3},$$

[2] Naturally terms of order b^3 are omitted, and it should also be recalled that ν is large compared to 1 in all terms except those that are vanishingly small.

so that we obtain, since $nm=1$,

$$S = r\left[\frac{3n}{2}\int (1+\beta)\frac{dT}{T} + lv - \frac{b}{v} - \frac{5b^2}{16v^2}\right].$$

Since the partial derivative of TS with respect to v at constant temperature is equal to the pressure due to molecular collisions alone, we find for this pressure, in agreement with the earlier result, the value

$$rT\left(\frac{1}{v} + \frac{b}{v^2} + \frac{5b^2}{8v^3}\right).$$

The calculation of terms with higher powers of b in this expression can be done most easily by the same method, taking account of these terms in the expressions for S and W.[3]

If the molecules are not spherical, but behave like solid bodies, then the probability that the center of the $(\nu+1)$'th molecule lies in the volume v is still given by an expression of the form

$$v - c_1\nu m - c_2\frac{\nu^2 m^2}{v} \cdots - c_k\frac{\nu^k m^k}{v^{k+1}} \cdots.$$

We assume that the series development yields

$$(182)\quad \begin{cases} l\left(v - c_1\nu m \cdots - c_k\dfrac{\nu^k m^k}{v^{k-1}} \cdots\right) = lv - \dfrac{2b_1\nu m}{v} - \\ \qquad \dfrac{3b_2\nu^2 m^2}{v^2} \cdots - \dfrac{(k+1)b_k\nu^k m^k}{kv^k} - \cdots; \end{cases}$$

---

[3] The expression found here for S would not of course be found comparable with that given in §21 for the entropy, since the calculation of the latter assumed the exact validity of the van der Waals equation. However, we find exactly the same expression for the entropy when in Eq. (38) of §21, from which one finds

$$\int\frac{dQ}{T} = \int\left[\frac{3r}{2}(1+\beta)\frac{dT}{T} + \left(p + \frac{a}{v^2}\right)\frac{dv}{T}\right],$$

we substitute instead of Eq. (22) the equation of state

$$p + \frac{a}{v^2} = rT\left(\frac{1}{v} + \frac{b}{v^2} + \frac{5b^2}{8v^3}\right)$$

which forms the basis of our present calculations.

then

$$S = rm\left[\frac{3n}{2}\int (1+\beta)\,\frac{dT}{T} + \sum_{\nu=0}^{\nu=n+1}\left(lv - \frac{2b_1\nu m}{v}\cdots - \frac{k+1}{k}\,\frac{b_k\nu^k m^k}{v^k}\cdots\right)\right] =$$

$$r\left[\frac{3}{2}\int (1+\beta)\,\frac{dT}{T} + lv - \frac{b_1}{v} - \frac{b_2}{2v^2}\cdots - \frac{b_k}{kv^k}\cdots\right].$$

Hence the pressure due simply to molecular collisions will be

$$\frac{\partial(TS)}{\partial v} = rT\left(\frac{1}{v} + \frac{b_1}{v^2}\cdots + \frac{b_k}{v^{k+1}}\cdots\right)$$

and the total external pressure acting on the gas will be:

$$p = rT\left(\frac{1}{v} + \frac{b_1}{v^2} + \frac{b_2}{v^3}\cdots\frac{b_k}{v^{k+1}}\cdots\right) - \frac{a}{v^2}. \tag{183}$$

If we substitute this into the equation

$$p(v_g - v_f) = \int_{v_f}^{v_g} p\,dv,$$

then we obtain, as a condition that vapor and liquid can coexist:

$$p(v_g - v_f) = rT\left[l\frac{v_g}{v_f} - b_1\left(\frac{1}{v_g} - \frac{1}{v_f}\right) - \cdots - \frac{b_k}{k}\left(\frac{1}{v_g^k} + \frac{1}{v_f^k}\right)\cdots\right] + a\left(\frac{1}{v_g} - \frac{1}{v_f}\right),$$

which, on repeated application of Equation (183), can also be written:

$$\begin{cases} 2a\left(\dfrac{1}{v_f} - \dfrac{1}{v_g}\right) = rT\left[l\,\dfrac{v_g}{v_f} - 2b_1\left(\dfrac{1}{v_g} - \dfrac{1}{v_f}\right) - \right. \\ \left. -\,\dfrac{3b_2}{2}\left(\dfrac{1}{v_g^2} - \dfrac{1}{v_f^2}\right)\cdots - \dfrac{k+1}{k}b_k\left(\dfrac{1}{v_g^k} - \dfrac{1}{v_f^k}\right)\cdots\right]. \end{cases} \tag{184}$$

In agreement with this, one now obtains by the same method by which we previously obtained Equation (174) the following condition for equilibrium of liquid and vapor:

$$\frac{2a}{rT}\left(\frac{1}{v_f}-\frac{1}{v_g}\right)=$$

$$= l\left(v_g - c_1 - \frac{c_2}{v_g}\cdots\right) - l\left(v_f - c_1 - \frac{c_2}{v_f}\cdots\right),$$

which, taking account of Equation (182), again yields Equation (184).

The following additions to Chapter I, which van der Waals communicated orally to me after that chapter was printed, may be inserted here.

1. He never explicitly made the assumption, explained in §2, that the attractive force of the molecules decreases so slowly with increasing distance that it is constant at distances large compared to the average separation of two neighboring molecules; and he believes such a force law to be improbable. Nevertheless I cannot obtain an exact foundation for his equation of state without this assumption.

2. If one conceives the bounding curve JKG of the two-phase region (Fig. 3, §17) to be a parabola or circular arc in the immediate neighborhood of the point K, then he sees that JN will be more nearly equal to NK, the closer N is to K, if N remains continually on the line KK_1. Hence if a substance has exactly the critical volume, and is heated at constant volume, then at the moment when the meniscus vanishes, the volume of the liquid part will be exactly equal to the volume of the vapor part. On the other hand, if the volume differs slightly from the critical volume, then the meniscus always moves progressively a great distance from the middle of the tube containing the substance, before it vanishes.

According to the experiments of Kuenen,* gravity plays an important role in causing deviations from the theoretical behavior.

* See J. P. Kuenen, Comm. Phys. Lab., Leiden, No. 17 (1895), and other works cited therein.

CHAPTER VI

Theory of dissociation.

§62. Mechanical picture of the chemical affinity of monovalent similar atoms.

I have once previously treated the problem of the dissociation of gases, on the basis of the most general possible assumptions, which of course I had to specialize at the end.[1] Since here I prefer perspicuousness to generality, I shall make special assumptions that are as simple as possible. The reader must not misunderstand the following and perhaps believe that I hold the opinion that chemical attraction acts precisely according to the laws of the force assumed here. These laws are rather to be considered the simplest, most perspicuous possible picture of forces which have a certain similarity to chemical forces, and hence in the present case can be substituted for them with a certain degree of approximation.

I will first consider the simplest case of dissociation, for which the dissociation of iodine vapor can serve as a prototype. At not too high temperatures, all molecules consist of two iodine atoms; as the temperature increases, however, more and more molecules decompose into single atoms. We explain the existence of molecules composed of two atoms (the double atom) by an attractive force acting between the atoms, which we call chemical attraction. The facts of chemical valence make it probable that the chemical attraction is by no means simply a function of the distance of centers of the atoms; on the contrary it must be associated with a relatively small region on the surface of the atom. Moreover, it is only with the latter assumption, and not the former, that one can obtain a picture of gas dissociation corresponding to reality.

Both for the sake of simplicity of calculation, and because of

[1] Boltzmann, Wien. Ber. **88**, 861 (18 October 1883); **105**, 701 (1896); Ann. Physik [3] **22**, 39 (1884).

the monovalency of iodine, we assume that the chemical attraction exerted on one iodine atom by another is effective only in a space small compared to the size of the atom, which we shall call the sensitive region. This region will lie on the external surface of the atom and will be firmly connected to it. The line drawn from the center of the atom to a particular point of the sensitive region (e.g., its midpoint, or its center of gravity in the purely geometrical sense) we call the axis of this atom. Only when two atoms are situated so that their sensitive regions are in contact, or partly overlap, will there be a chemical attraction between them. We then say that they are chemically bound to each other. They can touch each other at any other place on the surface without chemical attraction occurring. The sensitive region shall be such a

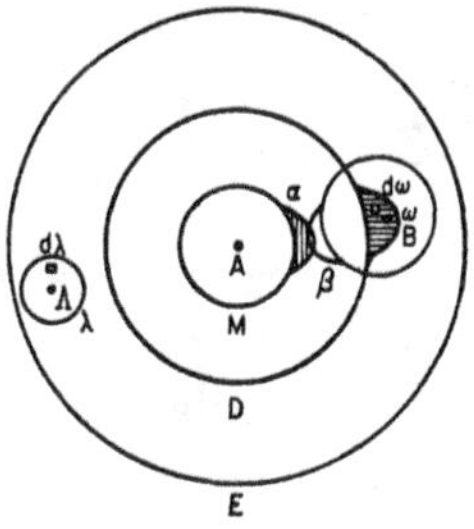

FIG. 4

small part of the entire surface of the atom that the possibility that the sensitive regions of three atoms can be in contact, or partly overlapping, is completely excluded. For the following calculation it is not necessary to assume that the atoms have the form of spheres. Since this is the simplest assumption, however, we shall make it. Let σ be the diameter of the spherical atom.

We consider a particular atom; let it be represented by the circle M of Figure 4. Let A be its center. The shaded region α will be the sensitive region. We need not immediately exclude the case that this lies partly inside the atom, but we draw it as if it were completely outside, as one must naturally assume if one imagines the atom to be completely impenetrable. If a second atom M_1 is chemically bound to the first, then the sensitive region β of the second atom must partially overlap the space α, or at least touch it. We shall again construct the covering sphere of the first atom (a sphere of center A and radius σ), which is indicated in the figure by the circle D. We construct on the surface of the covering sphere D a space (the critical space, the shaded space ω in the figure) with the property that the sensitive regions α and β can never overlap or touch each other unless the center B of the second atom lies in this critical space ω or on its boundary. The converse is not true. If the center B of the second atom lies in the critical space ω, then that atom can still be rotated in such a way that the sensitive regions α and β are widely separated from each other.

In order to define precisely the position that the second atom must have relative to the first if the two are to be chemically bound, we shall next construct in the critical space ω a volume element $d\omega$. At the same time, ω will be the total volume of the critical space.[2] Further, we shall imagine that a concentric sphere of radius 1 is rigidly connected to the first atom; this sphere is indicated by the circle E Figure 4. If the second atom is chemically bound to the first, then the axis of the second atom must not make too great an angle with the line BA, since otherwise the sensitive regions α and β would separate from each other. The line drawn from the point A parallel to and in the same direction as the axis of the second atom should strike the spherical surface E in a point which we shall always call the point Λ. Since the sphere E is rigidly connected to the first atom, the position of the axis of the second atom relative to the first is completely determined by this point Λ, and we can now construct, for each volume element $d\omega$ of the critical space, a surface section λ on the spherical surface E with the following property. If Λ lies within or on the boundary of the surface section λ, then, as soon as the center of the second atom lies within or on the boundary of the volume element $d\omega$, the two sensitive regions α and β will interpenetrate or touch each other. But as soon as Λ lies outside λ, the two sensitive regions α and β will likewise lie outside each other. This surface section λ will of course in general be of a different size according to the position of the volume element $d\omega$ within

[2] In order to exclude the possibility of chemical binding of three atoms, it is therefore necessary and sufficient that, if two atoms are chemically bound, the critical space ω of the first always lies completely within the covering sphere of the second—i.e., that no point of the critical space has a distance from another point of the same space that is equal to or greater than σ.

the critical space ω, and also will have a different position on the sphere E. Now if the center of the second atom lies within or on the boundary of any volume element $d\omega$ of the critical space, and Λ lies within or on the boundary of a surface element $d\lambda$ of that area λ which corresponds to the volume element $d\omega$, then the second atom will be chemically bound to the first—i.e., the two atoms will actively attract each other. The work needed to bring them from this position to a distance at which they no longer interact noticeably with each other will be denoted by χ. This quantity can in general be different according to the position of the volume element $d\omega$ within the critical space, and also according to the position of $d\lambda$ on the corresponding surface λ.

§63. Probability of chemical binding of an atom with a similar one.

Now suppose that a identical atoms are present in a container of volume V at a pressure p and absolute temperature T. Let the mass of an atom be m_1, and that of all atoms $am_1 = G$. We pick out one of the atoms. The others we call again the remaining atoms. For a moment we imagine that the gas is present in infinitely many (N) equivalent but spatially separated containers at the same temperature and pressure. In each of these N gases, let n_1 of the remaining atoms not be bound to the other remaining atoms, while $2n_2$ of the remaining atoms are bound to another one, so that they form n_2 double atoms. We now ask, in how many of the N gases will the specified atom be chemically bound to one of the other atoms, and in how many will this not be the case.

We consider first only one of the N gases. Since we have excluded chemical binding of three atoms, the specified atom, if it is bound at all, can only be united with one of the n_1 not otherwise bound atoms of this gas.

We therefore draw, as in Figure 4, the covering sphere and the concentric sphere E of radius 1 for each of these n_1 atoms; on each of these covering spheres will be found somewhere the critical space ω. In each of the critical spaces corresponding to all n_1 atoms, we draw the volume element $d\omega$, which has exactly the same position relative to the atom in question as the element $d\omega$ of Figure 4 with respect to the atom drawn there, and on each spherical surface E we draw a surface element $d\lambda$, which likewise has the same position relative to the atom in question as has that surface element $d\lambda$ in Figure 4. If the center of the specified atom now finds itself in any one of the volume elements $d\omega$, and moreover the point Λ is in the surface element $d\lambda$ of the surface λ (or its boundary) corresponding thereto, then it is chemically bound to another atom, and indeed is at a completely determined position relative to another atom, so that the quantity denoted by χ has a definite value.

If that attractive force which we call the chemical attraction were not present, then the probability that the center of the specified molecule is in one of the volume elements $d\omega$ would be in the same ratio to the probability w that it is in the arbitrary space Ω in the gas—which is neither a part of the covering sphere of a remaining atom, nor contains one of the spaces ω—as $n_1 d\omega:\Omega$. The space Ω is, according to what has been said, constructed so that the center of the specified atom can be in each point of it without being chemically bound. The probability w_2 that not only does the center of the specified molecule lie in one of the volume elements $d\omega$ but also the point Λ lies within the surface element $d\lambda$ would, in the absence of chemical forces, be in the same proportion to w_1 as $d\lambda:4\pi$, so that therefore

$$w_2 = \frac{d\lambda}{4\pi} \frac{n_1 d\omega}{\Omega} w$$

As a result of chemical attraction, this probability will be increased by a factor $e^{2h\chi}$, according to Equation (142); hence in the presence of chemical attraction, the probability is

$$w_2' = e^{2h\chi} \frac{d\lambda}{4\pi} \frac{n_1 d\omega}{\Omega} w.$$

In order to include all possible positions at which the specified atom is bound to any one of the n_1 remaining individual atoms, we must integrate this expression over all volume elements $d\omega$ of the entire critical space ω, and for each such volume element, over all surface elements $d\lambda$ of the surface λ corresponding to the volume element $d\omega$ concerned; whereby we obtain for the probability that the specified atom is chemically bound at all, the expression

(185)
$$w_3 = \frac{n_1 \omega}{\Omega} \iint \frac{d\omega d\lambda}{4\pi} e^{2h\chi}.$$

If we put

(186) $$k = \iint \frac{d\omega d\lambda}{4\pi} e^{2hx},$$

then

(187) $$w_3 = n_1 wk/\Omega.$$

Here w is the probability that the center of the specified molecule lies in an arbitrary space Ω not containing the covering sphere or critical space of any of the remaining atoms.

We now have to calculate the probability that the specified molecule in the gas considered is not chemically bound. The latter circumstance will always occur if its center is found in the space free of the covering spheres of all remaining atoms and of the critical spaces of the n_1 atoms. The sum of those critical spaces is $n_1\omega$, and the total space filled by covering sphere is, as in §59, equal to Gb. Since the total volume of the gas is V, the space free of covering spheres and critical spaces is $V - Gb - n_1\omega$.[1] The probability, w_4, that the center of the specified atom is found in this space is in the same ratio to the probability that it is in Ω as the volumes of the corresponding spaces, since the latter space is only an arbitrary part cut out of the former. Therefore:

$$w_4 = w(V - Gb - n_1\omega)/\Omega.$$

The specified atom is also not chemically bound when its center is in a volume element of a critical space but the point Λ does not lie on the corresponding surface λ, since then it is rotated so that the sensitive areas do not overlap. For the probability of this latter event one finds, by analogy with Equation (185), the expression

(188) $$w_5 = \frac{wn_1}{\Omega} \iint \frac{dwd\lambda_1}{4\pi},$$

where, however, for each volume element $d\omega$, $d\lambda_1$ means an element of the spherical surface E that does not lie on the surface λ corresponding to $d\omega$. Again one is to integrate over all surface elements that satisfy this condition and over all volume elements $d\omega$. The exponential is now omitted, of course, since in none of the positions now considered does the attractive force act. In each of

[1] The case that two covering spheres or one covering sphere and a critical space overlap is neglected, since it gives only a very small quantity of higher order.

the N gases, therefore, the total probability that the specified atom is not chemically bound is:

$$(189)\quad w_6 = w_4 + w_5 = \left(V - Gb - n_1\omega + n_1 \int\int \frac{d\omega d\lambda_1}{4\pi}\right)\frac{w}{\Omega}\cdot$$

The three quantities Gb, $n_1\omega$, and $n_1\int\int d\omega d\lambda/4\pi$ are completely independent of the chemical attractive forces. The first represents the deviation from the Boyle-Charles law considered by van der Waals, due to the finite extension of the molecules. Since the critical space is very small compared to the covering sphere, the second and third of the above three quantities are small compared to the first. We shall neglect all three quantities compared to V, since we shall calculate the dissociation of a gas that otherwise has the same properties as an ideal gas, so that the deviations from the Boyle-Charles law from causes other than dissociation can be neglected. By the same token we can also forget all terms added to the expression in parentheses in Equation (189) in the search for greater accuracy. Hence Equation (189) reduces to

$$(190)\qquad w_6 = \frac{Vw}{\Omega}\cdot$$

On the other hand, we may not consider the quantity k (given by Eq. [186]) to be small compared to V, since on account of the great intensity of chemical forces the exponential has a very large value. Noticeable dissociation occurs only when the exponential $e^{2h\chi}$ is of the same order of magnitude as $N/n_1\omega$, and hence k and V are of the same order of magnitude. From Equations (187) and (190) it follows that

$$w_6\!:\!w_3 = V\!:\!n_1k.$$

If we now return to the N identical gases, and assume that in N_3 of them the specified atom is chemically bound, and in N_6 it is not bound, then we have also:

$$N_6\!:\!N_3 = w_6\!:\!w_3 = V\!:\!n_1k.$$

Since we could just as well have specified any other atom, this must also be the ratio of the number of unbound atoms, n_1, to the number of bound ones, $2n_2$, in the equilibrium state. Therefore

$$n_1\!:\!2n_2 = V\!:\!n_1k,$$

hence

$$(191)\qquad n_1^2k = 2n_2V.$$

§64. Dependence of the degree of dissociation on pressure.

In determining the two numbers n_1 and n_2, we have of course imagined that one molecule, which we called the specified one, is excluded. However, since these numbers are very large compared to 1, Equation (191) is also valid when n_1 means the total number of all unbound atoms (simple molecules) in the gas, and n_2 means the number of all double atoms (compound molecules). Since a is the total number of all molecules in the gas, one has moreover $n_1+2n_2=a$. Whence it follows that:

$$(192) \qquad n_1 = -\frac{V}{2k} + \sqrt{\frac{V^2}{4k^2} + \frac{Va}{k}}\,.$$

We denote by G the total mass of gas, and by m_1 the mass of one atom, so that $a=G/m_1$. $a/G_1=1/m_1$ is the number of dissociated and chemically bound atoms together in unit mass. We again denote by $v=V/G$ the specific volume, i.e., the volume of unit mass of the partially dissociated gas at the given temperature and pressure, and by $q=n_1/a$ the degree of dissociation, i.e., the ratio of the number of chemically unbound (dissociated) atoms to the total number of atoms. Finally we set

$$(193) \qquad K = \frac{k}{m_1} = \frac{1}{m_1}\int\int e^{2hx}\frac{d\omega d\lambda}{4\pi}\,.$$

Then the above equation reduces to:

$$(194) \qquad q = -\frac{v}{2K} + \sqrt{\frac{v^2}{4K^2} + \frac{v}{K}}\,.$$

For the sake of orientation, we make the following remark: if two single atoms happen to collide in such a way that their sensitive regions interpenetrate, then because of the smallness of these regions and the large relative velocities of the atoms, which are accelerated by the chemical forces, the time during which the sensitive regions overlap will in most cases be small compared to the average time between successive collisions of a molecule. The energy of the double atom is so large that the two atoms can separate from each other again. (We refer to this case as virtual chemical binding.)

The number of these virtually bound atoms is in any case

vanishingly small compared to a, since they always remain together only a very short time. They can therefore make only a vanishingly small contribution to n_2, when n_2 is not very small compared to a. Only when the kinetic energy of the motion of the centers of gravity of the atoms is transformed into internal energy of the atoms (e.g., rotation around their axis, or internal motion) can there be, in the case when the atoms are not solid spheres, a somewhat longer interaction. (First kind of proper chemical binding.) On the other hand, if a third single atom or double atom intervenes while the sensitive regions of the other two atoms are overlapping, the energy can be lowered so much that it no longer suffices to separate the two atoms again, so that they must remain together, at least until another collision occurs. (Second kind of proper chemical binding.) In all cases where our calculation gives the result that the number of double atoms, n_2, is not vanishingly small compared to a, numerous double atoms must remain united with each other for long times. The principal advantage of our general formula is just that it permits us to calculate the number of chemically bound atom pairs without having to take special account of the processes of creation and dissolution of these pairs. In any case, out of the n_2 double atoms found by calculation—in case this number is not very small—all except a vanishingly small number are united with each other for a longer time, so that they are to be considered molecules in the sense of gas theory.

To calculate the pressure we must therefore proceed as if we had a mixture of two gases. The molecule of one of these gases is a single atom, while the molecule of the other is a pair of two atoms. The total pressure of any arbitrary gas mixture is:

$$p = \frac{1}{3V}\,(n_1 m_1 \overline{c_1^2} + n_2 m_2 \overline{c_2^2} \cdots),$$

where $m_1, m_2, \cdots$ are the masses, and $\overline{c_1^2}, \overline{c_2^2}, \cdots$ the mean square velocities of the centers of gravity of the various kinds of molecules. n_1 is the total number of gas molecules of the first kind, n_2 that of the second kind, and so forth (cf. Part I, §2, Eq. [8]). Moreover, if M is the mass of a molecule of the normal gas and $\overline{C^2}$ its mean square velocity at the same temperature T; $\mu_h = m_h/M$ is the atomic weight of one of the other gases, if the molecular weight of the normal gas is set equal to 1; then:

$$m_1\overline{c_1^2} = m_2\overline{c_2^2} = \cdots = M\overline{C^2} = 3MRT = 3/2h,$$

hence

(195) $$p = \frac{MRT}{V}(n_1 + n_2 + \cdots).$$

In our special case,

$$2n_2 = a - n_1, \quad n_3 = n_4 \cdots = 0,$$

hence

$$p = \frac{a + n_1}{2V} MRT = (1 + q)\frac{am_1}{2V}\frac{M}{m_1}RT,$$

therefore, since $v = V/am_1$,

(196) $$p = \frac{1 + q}{2\mu_1 v} RT.$$

If one substitutes the value (194) for q here, then the pressure p is obtained as a function of the specific volume v and the temperature T. K is still a function of temperature, which will be discussed later. In fact, direct observation gives the relation between p, v, and T. The chemist is usually accustomed to state the degree of dissociation q as a function of p and T, however. One can accomplish this by writing Equation (191) in the form:

$$q^2 = \frac{v}{K}(1 - q).$$

Multiplication of this equation by Equation (196) yields

$$q^2 = \frac{RT}{2\mu_1 Kp}(1 - q^2),$$

hence:

(197) $$q = \sqrt{\frac{1}{1 + \frac{2\mu_1 pK}{RT}}}.$$

If one substitutes this value into Equation (196), he obtains v as a function of p, T, and K.

§65. Dependence of the degree of dissociation on temperature.

There remains still the discussion of the quantity K. If we substitute in Equation (193) for h its value $\frac{1}{2}MRT$, it follows that:

$$(198) \qquad K = \frac{1}{m_1}\iint \frac{d\omega d\lambda}{4\pi} e^{\chi/MRT},$$

which in any case is only a function of temperature. At constant temperature, K is therefore a constant, and Equations (194), (196), and (197) give directly the relation between p and v, as well as the dependence of the quantity q on p and v, in which is involved only one new constant K which has to be determined.

Since Equation (198) contains the temperature under the integral sign, the dependence of the quantity K on temperature cannot immediately be given in a simple manner. One must instead make some assumption about the dependence of the function χ on the amount of overlapping of the two sensitive regions. In order not to lose ourselves in vague hypotheses, we shall discuss only the simplest of all assumptions: that χ always has a constant value when the two atoms are chemically bound at all, i.e., as soon as the two sensitive regions overlap at all, no matter how much. This would be the case when, at the instant when the sensitive regions touch, there occurs a strong attraction, equal at all points on the surfaces of these regions. But this attraction at once drops off to zero as soon as the sensitive regions penetrate further into each other. χ would then be the constant work of separation of the two chemically bound atoms, or, conversely, the work done in chemical binding by the chemical attractive forces.

If all the a/G atoms present in unit mass of the gas are initially unbound, and subsequently combine to form $a/2G$ double atoms, then the amount $a\chi/2G$ of work would thereby be done; therefore $\Delta = a\chi/2G$ is the total binding (or also dissociation) heat of unit mass of the gas, measured in mechanical units, and one has:

$$(199) \qquad \chi = 2G\Delta/a, \quad \frac{\chi}{MRT} = \frac{2G\Delta}{aMRT} = \frac{2\Delta\mu_1}{RT}\,.$$

$$(200) \qquad K = e^{2\mu_1\Delta/RT}\frac{1}{m_1}\iint \frac{d\omega d\lambda}{4\pi}$$

In chemistry one calls the mass $2\mu_1$ "one molecule." Therefore $2\mu_1\Delta$ is the heat of dissociation of "one molecule."

One can easily see that χ may not be more than at most logarithmically infinite for any configuration of the atoms, since the probability of that configuration would otherwise be an infinity of order e^{χ} and would thus be so large that the atoms could never separate. Whatever χ may be as a function of the positions of the atoms, one will not obtain a qualitatively different result by replacing χ by its mean value for all positions, whereby he would again arrive at Equation (200). This equation must certainly therefore provide an approximation to reality in the general case.

In the case where χ is constant, no intramolecular work will be done by the relative motion of the chemically bound atoms, as long as they remain bound. The mean kinetic energy however is always the same at a given temperature, whether the atoms are chemically bound or not; hence equal increments of mean kinetic energy correspond to equal increments of temperature, and the specific heat is independent of whether the atoms are chemically bound or not, as long as χ is constant. Of course we mean here the specific heat before the beginning or after the end of the dissociation; in case the degree of dissociation changes, the heat of dissociation is not included in the specific heat.

We shall now introduce for brevity the notation

$$\alpha = \frac{2\mu_1\Delta}{R}, \tag{201}$$

$$\beta = \frac{1}{m_1}\int\int\frac{d\omega d\lambda}{4\pi} = \frac{1}{m_1}\int\frac{\lambda d\omega}{4\pi}, \tag{202}$$

$$\gamma = \frac{2\mu_1}{R}\beta \tag{203}$$

where λ is the surface section on the sphere E, which the point λ may not leave without breaking the chemical bond, when the center B of the second atom lies within $d\omega$ in Figure 4; then it follows from Equations (197), (200), (201), (202), and (203) that

$$q = \sqrt{\frac{1}{1 + \frac{\gamma p}{T}e^{\alpha/T}}}. \tag{204}$$

If q is given experimentally as a function of p and T, then the two constants α and γ can be determined from this formula.

From α there follows at once from Equation (201) the heat of dissociation—or, if one prefers, the heat of combination Δ of unit mass of the gas. From γ, one can determine the quantity β by means of Equation (203). According to Equation (202) this quantity has a significant molecular meaning. For each volume element $d\omega$ of the critical space belonging to an atom, the point Λ must lie on a certain surface section λ of the spherical surface E, in order that chemical binding can occur. We shall now count not all of the volume of each volume element $d\omega$ of a critical space, rather only the fraction of it which is obtained on multiplying the volume element by $\lambda/4\pi$. We call this fraction of the total volume element the reduced volume. $\int\lambda d\omega/4\pi$ is then the sum of the reduced volumes of all volume elements of the critical space belonging to an atom; we call it simply the reduced volume of the critical space.

We shall now use an abbreviated manner of speaking. Instead of saying that the center of the second atom lies in the volume element $d\omega$ and simultaneously the point Λ lies on the corresponding surface λ, we say simply that the second atom lies in the reduced volume element $d\omega$. Instead of saying that it lies in some reduced volume element of the critical space, we say simply that it lies somewhere in the reduced critical space.

Since finally $1/m_1$ is the total number of atoms in unit mass, β is the sum of the reduced volumes of all critical spaces which belong to all the atoms in unit mass. If one wishes to make a definite assumption about the form of the sensitive region, then he can calculate therefrom the form of the surface section λ corresponding to each volume element $d\omega$ of the critical space, and hence not only the reduced volume but also the absolute volume of all the critical spaces belonging to all atoms in unit mass. But we shall not go into this in more detail here.

If one substitutes the expression (204) into Equation (196), then he can express the specific volume v as a function of the pressure p, the temperature T, the gas constant R/μ_1 of the dissociated gas, and the two constants α and γ. If one prefers to express p as a function of v and T, then he can substitute the expression (194) for q into Equation (196), after setting $K=\beta e^{\alpha/T}$ (according to Eq. [220]).

§66. Numerical calculations.

A very short numerical calculation will be inserted here. I have analyzed[1] the experiments of Deville and Troost[2] and of Naumann[3] on the dissociation of hyponitrous acid, and that of Meier and Crafts[4] on the dissociation of iodine vapor, according to Equation (204). The constants there called a and b are related to the ones here called α and γ as follows:

$$\alpha = bl10, \quad \gamma = \frac{1}{a},$$

where l means the natural logarithm. I found that—denoting the quantities pertaining to hyponitrous acid by the index u, and those to iodine vapor by the index j—that:

$$a_u = 1970270 \frac{p_u}{1°\text{C.}}, \quad b_u = 3080.1°\text{C.},$$

$$a_j = 2{\cdot}617 \frac{p_j}{1°\text{C.}}, \quad b_j = 6300.1°\text{C.},$$

whence follows:

$$\alpha_u = 3080.l10.1°\text{C.}, \quad \gamma_u = \frac{1°\text{C.}}{1970270 p_u},$$

$$\alpha_j = 6300.l10.1°\text{C.}, \quad \gamma_j = \frac{1°\text{C.}}{2{\cdot}617 p_j}.$$

p_u is the average pressure used by Deville and Troost in their experiments on the dissociation of hyponitrous acid (about 755.5 mm of mercury); p_j is that used in the experiments of Meier and Crafts on the dissociation of iodine vapor (728 mm). According to Equation (201), the heat of dissociation of a molecule (in the chemical, macroscopic sense) is

$$\Pi = 2\mu_1\Delta = \alpha R.$$

---

[1] Wien. Ber. **88,** 891, 895 (1883).

[2] C. R. Paris **64,** 237 (1867); **86,** 332, 1395 (1878); Jahresber. f. Chem., 177 (1867); Naumann, *Thermochemie,* pp. 115–128.

[3] Ber. d. deutsch. chem. Ges. **11,** 2045 (1878); Jahresber. f. Chem., 120 (1878).

[4] Ber. d. deutsch. chem. Ges. **13,** 851 (1880).

This formula is based on mechanical units for heat. If one uses thermal units, it is necessary to multiply by the appropriate conversion factor J. The heat of dissociation of a chemical molecule in thermal units is therefore

$$P = \alpha RJ. \tag{205}$$

We shall use grams, centimeters, and seconds as units for the mechanical quantities. A weight of 430 kilograms raised one meter produces one kilocalorie. Hence

$$J = \frac{\text{cal}}{430\ \text{gr} \cdot 100\ \text{cm}\ G} \cdot \tag{206}$$

Here $G = 981$ cm/sec^2 is the acceleration of gravity, and "cal" means a gram-calorie. The value of the gas constant r for air is found by substituting the following values in the equation of state $pv = rT$ for air:

$T =$ the absolute temperature 273° corresponding to melting ice,

$p =$ the pressure of one atmosphere $= \dfrac{1033\ \text{gr} \cdot G}{\text{cm}^2}$

$v = \dfrac{1000\ \text{cm}^3}{1.293\ \text{gr}} \cdot$

For a molecule of air, μ_0 is about 28.9 when one sets $H = 1$, $H_2 = 2$. Since $R = r\mu$, we have for monatomic hydrogen:

$$R = \frac{28.9}{273°}\ \frac{1033\ \text{gr}\ G}{\text{cm}^2} \cdot \frac{1000\ \text{cm}^3}{1.293\ \text{gr}} = 84570\ \frac{G \cdot \text{cm}}{1°\text{C}} \cdot \tag{207}$$

Finally it follows that

$$\begin{cases} RJ = \dfrac{28.9}{273°}\ \dfrac{1033\ \text{gr}\ G}{\text{cm}^2} \cdot \dfrac{1000\ \text{cm}^3}{1.293\ \text{gr}} \cdot \dfrac{\text{cal}}{430\ \text{gr} \cdot 100\ \text{cm}\ G} = \\ \quad = 1.967\ \dfrac{\text{cal}}{\text{gr}.1°\text{C}.} \cdot \end{cases} \tag{208}$$

Hence one obtains for hyponitrous acid:

$$P_u = \alpha_u RJ = 13920\ \frac{\text{cal}}{\text{gr}} \cdot \tag{209}$$

On dividing by the molecular weight $2\mu_1 = 92$ of hyponitrous acid

(N_2O_4), one obtains for the heat of dissociation of one gram of it the value

$$D_u = 151.3 \frac{\text{cal}}{\text{gr}}, \tag{210}$$

which is in good agreement with the direct determination of the heat of dissociation of hyponitrous acid by Berthelot and Ogier.[5]

For iodine vapor it follows that:

$$P_j = 28530 \frac{\text{cal}}{\text{gr}}, \qquad D_j = 112.5 \frac{\text{cal}}{\text{gr}}. \tag{211}$$

According to Equations (202) and (203), the sum of the reduced critical spaces of all atoms in unit mass is

$$\beta = \frac{1}{2\mu_1} R\gamma. \tag{212}$$

Now we have

$$\gamma_u = \frac{1°\text{C.}}{1970270 p_u}.$$

p_u corresponds to a column of mercury of 755.5 mm height. Hence[6]

$$p_u = \frac{1033 \text{ gr } G}{\text{cm}^2} \cdot \frac{755.5}{760} = \frac{1027 \text{ gr } G}{\text{cm}^2}. \tag{213}$$

Hence

$$\beta_u = \frac{1}{92} 84570 \frac{G \text{ cm}}{1°\text{C.}} \cdot \frac{1°\text{C.}}{1970270} \cdot \frac{\text{cm}^2}{1027 \text{ gr } G} = \frac{\text{cm}^3}{2200000 \text{ gr}}. \tag{214}$$

This value is three times as small as the one I found previously,[1] due to the fact that there I made the improbable assumption that the four oxygen atoms can be arbitrarily exchanged during the uniting and separation of two NO_2 groups, whereas here I consider such groups to be undecomposable, playing exactly the same role as atoms.

---

[5] C. R. Paris **94,** 916 (1882). Ann. chim. phys. [5] **30,** 382–400 (1883).

[6] If one writes for R the expression $\mu_0 p / {}_0\rho_0 T_0$, where μ_0, p_0, ρ_0 and T_0 refer to an arbitrary gas, e.g., air at 0°C. under normal atmospheric pressure, then it follows that

$$\beta = \frac{\gamma p}{2\rho_0 T_0} \cdot \frac{p_0}{p} \cdot \frac{\mu_0}{\mu_1}$$

and one can use directly the value found for γp, without needing to know that of p.

For iodine vapor I found[1]

$$\gamma_j = \frac{1°\text{C.}}{2.617 p_j} \cdot \tag{215}$$

The experiment was done likewise under atmospheric pressure (728 mm of mercury on the average). Furthermore, the molecular weight $2\mu_1$ of iodine is equal to 253.6. Substitution of these values yields

$$\beta_j = \frac{\text{cm}^3}{8 \text{ gr}} \cdot$$

Neither for hyponitrous acid nor for iodine vapor is the value of van der Waals' constant b known, so that one cannot calculate the space filled by the molecules from the van der Waals equation. This would also give only the order of magnitude since it treats the molecules as nearly undeformable spheres, and makes various other simplifications which can influence the result. There remains finally Loschmidt's method of estimation.* We shall set the density of liquid hyponitrous acid equal to 1.5, and that of solid iodine equal to 5, corresponding to temperatures below that of melting ice. At these temperatures the vapor pressure of both substances is very small; for iodine it is of course much smaller than for hyponitrous acid. But we can leave this out of consideration here, since we only want a rough order-of-magnitude estimation. We assume completely arbitrarily that in these two substances two thirds of the total space is filled up by the molecules. Then the space filled by a gram of molecules of hyponitrous acid would be 0.44 cm^3/gr; the sum of the volumes of the covering spheres of these molecules is eight times as large, viz., 3.55 cm^3/gr. For iodine, the same quantities have the values 0.133 cm^3/gr and 1.07 cm^3/gr. Therefore, for hyponitrous acid the reduced critical space of an NO_2 group, considered as one atom, is only about one 8-millionth part of the covering sphere, whereas for iodine it is an eighth or a ninth part of the covering spheres. The small dissociability of iodine is therefore predominantly due to the relative size of the critical space compared to the covering sphere, whereas the difference in the heats of dissociation per gram of the sub-

* J. Loschmidt, Wien. Ber. **52**, 395 (1865).

stances is relatively small. Iodine vapor would therefore be about as easily dissociated as hyponitrous acid when the former is diluted a millionfold.

§67. Mechanical picture of the affinity of two dissimilar monovalent atoms.

We consider a second simple example. Two kinds of atoms are present in a space V at temperature T and total pressure p. There are a_1 of the first kind and a_2 of the second kind. Let the mass of an atom of the first kind be m_1, and that of an atom of the second kind be m_2. Two atoms of the first kind can be combined into a molecule (double atom of the first kind), likewise two atoms of the second kind (double atom of the second kind). For each of these bonds, exactly the same rules as those established in the preceding section will hold. We shall denote all quantities pertaining to double atoms of the first kind by an index 1, and those to double atoms of the second kind by an index 2. But in addition, chemical binding of an atom of the first kind to one of the second kind, forming what we shall call a mixed molecule, shall be possible. Similar laws will hold for these bonds, and we shall denote the corresponding quantities by two indices, 1 and 2.

In the equilibrium state the following will be present in our gas: first, n_1 single atoms of the first kind, and n_2 single atoms of the second kind; second, n_{11} double atoms of the first kind and n_{22} double atoms of the second kind; third, n_{12} mixed molecules. Chemical combination of more than two atoms will be excluded. The atoms of the first kind shall be impenetrable spheres of diameter σ_1. We shall call a sphere of radius σ_1 around the center of such an atom its covering sphere. To it will be attached the critical space ω_1 for the interaction with a similar atom; $d\omega_1$ denotes the volume element of this critical space. If the center of another atom of the first kind does not lie within ω_1, it is not chemically bound to the first atom. If its center is within $d\omega_1$, then chemical binding takes place only if it is in the reduced volume $d\omega_1$—i.e., when the point Λ_1 lies within a certain surface section λ_1 of the spherical surface E concentric with the first atom. Let $d\lambda_1$ be an element of the surface λ_1. As before, Λ_1 is the point of intersection of a line drawn from the center of the first atom parallel to the axis of the second, with the spherical surface E. Finally, χ_1 is again the work performed when the two atoms come from a great distance to those positions where the center of the second lies within $d\omega_1$ and the point Λ_1 lies within $d\lambda_1$.

If one picks out an atom of the first kind, then he can assume that, of the remaining atoms of the first kind, there will always be n_1 which are bound neither to an atom of the second kind nor to another remaining atom of the first kind. If the specified atom forms a double atom of the first kind, then it can only be bound to one of these n_1 atoms of the first kind, since we exclude triatomic molecules. The probability of this event is to the probability that it remains single as $k_1 n_1 : V$, where

$$k_1 = \iint e^{2h\chi_1} \frac{d\omega_1 d\lambda_1}{4\pi}. \tag{216}$$

This can be found just as in the previous section. The ratio of these two probabilities must however be equal to the ratio $2n_{11}:n_1$, whence

$$k_1 n_1^2 = 2V n_{11}. \tag{217}$$

Likewise one finds for the atoms of the second kind the equation

$$k_2 n_2^2 = 2V n_{22} \tag{218}$$

where all quantities have a similar meaning. Therefore:

$$k_2 = \iint e^{2h\chi_2} \frac{d\omega_2 d\lambda_2}{4\pi}. \tag{219}$$

We still have to discuss the formation of mixed molecules. Since we assumed that the atoms of the first kind are impenetrable spheres of diameter σ_1, and those of the second kind are impenetrable spheres of diameter σ_2, we shall consequently assume that the minimum distance of separation of two unlike atoms is $\frac{1}{2}(\sigma_1+\sigma_2)$. We construct a sphere of radius $\frac{1}{2}(\sigma_1+\sigma_2)$ around the center of an atom of the first kind, and call it the covering sphere of this atom with respect to an atom of the second kind. On the surfaces of both kinds of atoms there will be sensitive regions with the property that attraction takes place only when these regions touch or partly overlap. It seems most probable that the sensitive region for interaction with atoms of a different kind is the same as that for interaction with atoms of the same kind; but this assumption is not strictly necessary. In any case, we can just as before construct a critical space attached to the covering sphere of an atom of the first kind with respect to an atom of the second kind, and call it ω_{12}. For each volume element $d\omega_{12}$ of this critical space we can construct a surface λ_{12} on the unit sphere E concentric with the covering sphere, such that when the center of the second atom lies within $d\omega_{12}$, the two atoms attract only when, as before, a certain point Λ_{12} lies within some surface element $d\lambda_{12}$ of the surface λ_{12}—when the atom is in the reduced volume element $d\omega_{12}$, in other words. In this case χ_{12} is the work of separation.

We again pick out some atom of the second kind. In order that it be a single atom, it should have all but a vanishingly small part of the total volume V available to it. If on the other hand it is in a mixed molecule, then its center must lie within some volume element $d\omega_{12}$ of the critical space belonging to one of the n_1 unbound atoms of the first kind, and the point Λ_{12} should lie in some surface element $d\lambda_{12}$ of the corresponding surface λ_{12}. The probability that its center lies in a certain volume element $d\omega_{12}$ and the point Λ_{12} lies in a certain surface element $d\lambda_{12}$ is, to the probability that for an arbitrary configuration of its axis its center lies within V, as:

$$e^{2h\chi_{12}}\frac{d\omega_{12}d\lambda_{12}}{4\pi}:V.$$

The probability that the specified atom is bound in a mixed molecule at all is, to the probability that it is a single atom, as

$$n_1\iint e^{2h\chi_{12}}\frac{d\omega_{12}d\lambda_{12}}{4\pi}:V.$$

Therefore one obtains, on setting again

$$k_{12}=\iint e^{2h\chi_{12}}\frac{d\omega_{12}d\lambda_{12}}{4\pi} \tag{220}$$

the proportion

$$n_2:n_{12}=V:n_1k_{12},$$

hence

$$Vn_{12}=k_{12}n_1n_2. \tag{221}$$

§68. Dissociation of a molecule into two heterogeneous atoms.

We first consider the special case that k_1 and k_2 are so small compared to k_{12} and V/n_1 respectively that the number of double atoms of the first and second kinds is completely negligible. The gas then consists only of three kinds of molecules: single atoms of the first and second kinds, and mixed molecules.

We specialize further to the case that there is no surplus of single atoms of either kind—i.e., the number of single atoms of the first kind is exactly equal to the number of single atoms of the second kind. Then we set:

$$a_1 = a_2 = a.$$

We have:

$$n_1 = n_2 = a - n_{12}.$$

We again call the quotient $q=(a-n_{12})/a$ the degree of dissociation, so that it follows from Equation (221) that

$$ak_{12}q^2 = V(1-q).$$

Furthermore, one has from Equation (195)

$$p = \frac{MRTa}{V}(1+q).$$

Hence

$$q^2 = \frac{MRT}{k_{12}p}(1-q^2).$$

We again assume that χ_{12} is constant; then

$$\frac{1}{m_1+m_2}\chi_{12} = \Delta_{12}$$

is the heat of dissociation of the original unit mass consisting purely of mixed molecules. Further:

$$k_{12} = e^{(\mu_1+\mu_2)\Delta_{12}/RT}\int \frac{\lambda_{12}d\omega_{12}}{4\pi}.$$

μ_1 is the atomic weight of a gas consisting of single atoms of the first kind, referred to $H_1=1, H_2=2$. μ_2 has a similar meaning for the second kind of gas. The quantity in the exponent, $(\mu_1+\mu_2)\Lambda_{12}=\Pi$, is the heat of dissociation (measured in mechanical units) of a molecule of the undissociated substance in the chemical or macroscopic sense, whose mass (the chemical molecular weight) is $\mu_1+\mu_2$. If we set

$$\frac{(\mu_1 + \mu_2)\Delta_{12}}{R} = \alpha, \qquad \kappa_{12} = \int \frac{\lambda_{12} d\omega_{12}}{4\pi},$$

$$\gamma = \frac{\kappa_{12}}{MR},$$

then we find again

$$q = \sqrt{\frac{1}{1 + \frac{\gamma p}{T} e^{\alpha/T}}}.$$

χ_{12} is the reduced critical space of an atom of the first kind with respect to its interaction with an atom of the second kind. χ_{12} would be the sum of all reduced critical spaces which correspond to all the atoms contained in unit mass of the first kind of gas. χ_{12}/M, on the other hand, is the sum of all critical spaces that correspond to all atoms of the first kind of gas contained in a molecule (in the chemical sense)—i.e., in the mass m_1/M of the first kind of gas, which is chemically equivalent to unit mass of the normal substance.

If one gas is present in excess, then Equation (221) yields

$$(221a) \qquad (a_1 - n_{12})(a_2 - n_{12}) = \frac{V}{k_{12}} n_{12}$$

$$(222) \quad n_{12} = \frac{a_1+a_2}{2} + \frac{V}{2k_{12}} - \sqrt{\frac{(a_1-a_2)^2}{4} + (a_1+a_2)\frac{V}{2k_{12}} + \frac{V^2}{4k_{12}^2}}$$

Since n_{12} can neither be larger than a_1 nor larger than a_2, the root must be taken with the negative sign. If a_1 were very large, then the other factor a_2-n_{12} on the left-hand side of Equation (221a) must be very small, hence n_{12} must be nearly equal to a_2, which also follows from Equation (222) if one considers a_1 to be large compared to a_2. As the number of atoms of the first kind increases, more and more atoms of the second kind will be bound to them, until finally all atoms of the second kind will be chemically bound, which agrees with the law of mass action of Guldberg and Waage.

§69. Dissociation of hydrogen iodide gas.

We now consider the other extreme special case of the general case treated in §67. Again let $a_1 = a_2 = a$, but now let V/a be vanishingly small compared to k_1, k_2, and k_{12}, so that the number of single atoms of the two kinds is vanishingly small. This would be the case, for example, when HI dissociates into I_2 and H_2. We then obtain, on squaring Equation (221), $V^2 n_{12}^2 = k_{12}^2 n_1^2 n_2^2$. If we substitute the values of n_1^2 and n_2^2 from Equations (217) and (218), then it follows that:

$$(223) \qquad n_{12}^2 = \frac{4k_{12}^2}{k_1 k_2} n_{11} n_{22}.$$

If one denotes again by $q = (a - n_{12})/a$ the degree of dissociation, then:

$$n_{11} = n_{22} \frac{aq}{2} \cdot$$

Hence it follows from Equation (223), after extracting the square root, that:

$$1 - q = \frac{k_{12}}{\sqrt{k_1 k_2}} q$$

$$q = \frac{1}{1 + \frac{k_{12}}{\sqrt{k_1 k_2}}} \cdot$$

If one again assumes that χ is constant in the entire reduced critical space, and denotes the reduced critical space for the interaction of two atoms of the first kind with each other, of two atoms of the second kind with each other, and of one atom of the first kind with one of the second, by κ_1, κ_2, and κ_{12}, respectively, then according to Equations (216), (219), and (220):

$$\frac{k_{12}}{\sqrt{k_1 k_2}} = \frac{\kappa_{12}}{\sqrt{\kappa_1 \kappa_2}} e^{h(2\chi_{12} - \chi_1 - \chi_2)}$$

$$= \frac{\kappa_{12}}{\sqrt{\kappa_1 \kappa_2}} e^{(2\chi_{12} - \chi_1 - \chi_2)/2MRT}.$$

On forming two HI molecules from an I_2 and an H_2 molecule, the heat $2\chi_{12}-\chi_1-\chi_2$ will be liberated.

$$\frac{1}{2(m_1+m_2)}(2\chi_{12}-\chi_1-\chi_2)$$

is therefore the heat Δ which is liberated on forming unit mass of HI from ordinary iodine and hydrogen gases. Hence

$$\frac{k_{12}}{\sqrt{k_1 k_2}} = \frac{\kappa_{12}}{\sqrt{\kappa_1 \kappa_2}} e^{(\mu_1+\mu_2)(\Delta/RT)}.$$

$\Pi = 2(\mu_1+\mu_2)\Delta$ is of course the heat of formation of two HI molecules in the chemical sense from a molecule of ordinary iodine vapor and a molecule of ordinary hydrogen gas.

At very high temperatures, q approaches the limit

$$\frac{1}{1+\dfrac{\kappa_{12}}{\sqrt{\kappa_1\kappa_2}}}.$$

From the experiments of Lemoine[1] one calculates for this limit the value $\frac{3}{4}$; however, this may not be completely reliable, on account of a possible false chemical equilibrium. The reduced critical space for the interaction of an iodine atom with a hydrogen atom would then be only about one third of the geometric mean of the reduced critical spaces for the interactions of two iodine and two hydrogen atoms respectively with each other.

§70. Dissociation of water vapor.

We shall now briefly consider a special case, namely the dissociation of two water vapor molecules ($2H_2O$) into two hydrogen molecules ($2H_2$) and one oxygen molecule (O_2). In a volume V at temperature T and pressure p, there will strictly be present all possible molecules that can be formed by combining hydrogen and oxygen atoms. Let: n_{10}, n_{01}, n_{20}, n_{02}, n_{11} and n_{21} molecules of the forms H, O, H_2, O_2, HO, and H_2O be present. We denote by

[1] Ann. Chim. Phys. [5] 12, 145 (1877); cf. also Hautefeuille, C. R. Paris **64**, 608, 704 (1867).

κ_{20}, κ_{02}, κ_{11} and κ_{21} the reduced critical spaces for the combination of two hydrogen atoms, two oxygen atoms, one oxygen atom and one hydrogen atom into the group HO, and such a group with another hydrogen atom into water vapor, respectively. Furthermore, we denote by χ_{20}, χ_{02}, χ_{11}, and χ_{21} the amounts of heat liberated in forming the corresponding compounds, so that:

$$2\chi_{11} + 2\chi_{21} - 2\chi_{20} - \chi_{02}$$

is the heat of formation of two water vapor molecules from two hydrogen molecules and one oxygen molecule. Each of the χ shall be constant in the corresponding critical space.

We next pick out a hydrogen atom. It is paired in a molecule of the form HO when it finds itself in the reduced critical space κ_{11} of one of the n_{01} oxygen atoms. The probability that it is single is, to the probability that it forms a group HO, as

$$V : \kappa_{11} n_{01} e^{2h\chi_{11}}.$$

This however is at the same time equal to the ratio $n_{10} : n_{11}$, whence it follows that:

$$n_{11} V = n_{01} n_{10} \kappa_{11} e^{2\chi_{11}}.$$

If one compares the probability that the specified hydrogen atom remains single with the probability that it is united with an HO molecule into H_2O, it follows likewise that:

$$n_{21} V = n_{10} n_{11} \kappa_{21} e^{2h\chi_{21}}.$$

hence:

$$n_{21} V^2 = n_{10}^2 n_{01} \kappa_{21} \kappa_{11} e^{2h(\chi_{21}+\chi_{11})}. \tag{224}$$

The probability that the specified hydrogen atom remains single is, to the probability that it is united with one of the remaining hydrogen atoms in an H_2 molecule, as

$$V : n_{10} \kappa_{20} e^{2h\chi_{20}}.$$

This however is again the ratio of the number n_{10} of single hydrogen atoms to the number $2n_{20}$ of those bound to another hydrogen atom. Hence:

$$2n_{20} V = n_{10}^2 \kappa_{20} e^{2h\chi_{20}}$$

and likewise:

$$2n_{02} V = n_{01}^2 \kappa_{02} e^{2h\chi_{02}}.$$

Hence, and from Equation (224), it follows that:

$$(225) \qquad n_{21}^2 = n_{20}^2 n_{02} \frac{8}{V} \frac{\kappa_{21}^2 \kappa_{11}^2}{\kappa_{20}^2 \kappa_{02}} e^{2h(2\chi_{21}+2\chi_{11}-2\chi_{20}-\chi_{02})}.$$

We assume that there are originally a water molecules. Of these let n_{21} remain undissociated, so that $a-n_{21}$ are dissociated and indeed are exclusively in molecules of the forms H_2 and O_2. The quotient $q=(a-n_{21})/a$ shall again be called the degree of dissociation.

Since, out of the $a-n_{21}$ water molecules, $a-n_{21}$ hydrogen molecules and $\frac{1}{2}(a-n_{21})$ oxygen molecules would be formed, we have:

$$n_{20} = aq, \qquad n_{02} = \tfrac{1}{2}aq, \qquad n_{21} = a(1-q).$$

$2\chi_{21}+2\chi_{11}-2\chi_{20}-\chi_{02}$ is the heat liberated when two water vapor molecules form two hydrogen molecules and one oxygen molecule. If we denote the heat produced when unit mass of water is formed from the usual explosive gas by Δ, then:

$$\Delta = \frac{2\chi_{21} + 2\chi_{11} - 2\chi_{20} - \chi_{02}}{2(2m_1 + m_2)}.$$

If one sets

$$(226) \qquad \frac{2(2\mu_1 + \mu_2)\Delta}{R} = \alpha, \qquad \frac{8\kappa_{21}^2\kappa_{11}^2}{2(2m_1 + m_2)\kappa_{20}^2\kappa_{02}} = \gamma,$$

then:

$$(227) \qquad (1-q)^2 = q^3 \frac{\gamma}{v} e^{\alpha/T}.$$

Furthermore, according to Equation (195)

$$(228) \quad p = (n_{20} + n_{02} + n_{21}) \frac{MRT}{V} = \left(1 + \frac{q}{2}\right) \frac{RT}{v(2\mu_1 + \mu_2)}.$$

If one eliminates q, then he obtains the relation between p, v, and T. On the other hand, if one eliminates v, then he obtains for the dependence of the degree of dissociation on pressure and temperature the following equation:

$$(1-q)^2\left(1 + \frac{q}{2}\right)\frac{RT}{(2\mu_1 + \mu_2)p} = q^3\gamma e^{\alpha/T}$$

The equation between q, p, and v would be obtained by eliminating T from Equations (227) and (228).

In order to take account of the bivalence of the oxygen atom, one can assume that two equivalent sensitive regions lie on its surface. The critical space for the formation of HO from H and O would then be twice as large as that for the formation of H_2O from HO and H. But then the sensitive regions would not have to lie directly *vis à vis;* or else they must be able to move on the surface of the molecule, in order to make possible double bonding of two oxygen atoms. One would obtain a phenomenon at least partly similar to bivalence by assuming that the critical region of an oxygen atom is not completely covered by the covering sphere of a single oxygen or hydrogen atom chemically bound to it, so that there is still room for the chemical binding of another atom. I certainly am not very hopeful that a more precise formulation of these speculations can be established at the present time; yet perhaps I may be allowed to adduce here the statement of a great scientist, that these general mechanical models will aid more than hinder the knowledge of the facts of chemistry.

§71. General theory of dissociation.

We shall now make some remarks on the most general case of dissociation. Let there be given arbitrarily many atoms of arbitrarily many different kinds. A molecule which contains a_1 atoms of the first, b_1 atoms of the second, c_1 atoms of the third kind, and so forth, will be denoted symbolically by $(a_1b_1c_1 \cdots)$. The aggregate of C_1 molecules of the form $(a_1b_1c_1 \cdots)$, C_2 molecules of form $(a_2b_2c_2 \cdots)$, C_3 molecules of form $(a_3b_3c_3 \cdots)$ and so forth can be transformed into Γ_1 molecules $(\alpha_1\beta_1\gamma_1 \cdots)$, Γ_2 molecules $(\alpha_2\beta_2\gamma_2 \cdots)$, Γ_3 molecules $(\alpha_3\beta_3\gamma_3 \cdots)$ and so forth. Since both aggregates must contain the same atoms, we have the following equations

$$(229) \qquad \begin{cases} C_1a_1 + C_2a_2 + \cdots = \Gamma_1\alpha_1 + \Gamma_2\alpha_2 + \cdots, \\ C_1b_1 + C_2b_2 + \cdots = \Gamma_1\beta_1 + \Gamma_2\beta_2 + \cdots \end{cases}$$

We now assume that all possible combinations of our atoms are present in the gas, even if only in small amounts. We denote by

$n_{100}\ldots$ the number of single atoms of the first kind, by $n_{200}\ldots$ the number of double atoms of the first kind, and so forth. Likewise, let $n_{010}\ldots$ be the number of single atoms, $n_{020}\ldots$ the number of double atoms of the first kind, etc.; let $n_{110}\ldots$ be the number of molecules consisting of one atom of the first kind and one of the second, and so forth. For simplicity we ignore isomers. A double atom of the first kind can be formed only when the center of an atom of the first kind lies within the reduced critical space of another atom of the first kind. Hence, if $\kappa_{200}\ldots$ is this reduced critical space, and $\chi_{200}\ldots$ is the heat of bonding of a double atom of the first kind, then it follows from the principles of our theory that

$$n_{100}\ldots:2n_{200}\ldots = V:n_{100}\ldots\kappa_{200}\ldots e^{2h\chi_{200}\ldots}.$$

Likewise it follows that

$$n_{100}\ldots:3n_{200}\ldots = V:n_{200}\ldots\kappa_{300}\ldots e^{2h\chi_{200}\ldots},$$

where $\chi_{300}\ldots$ is the heat of bonding of a molecule consisting of three atoms of the first kind, formed from a single and a double atom of the first kind. $\chi_{300}\ldots$ is the reduced critical space for this bonding which is available for the single atom in the neighborhood of the double atom. From the two ratios it follows that:

$$n_{300}\ldots = n_{100}^{3}\ldots V^{-2}\kappa'_{300}\ldots e^{2h\psi_{300}\ldots},$$

where $\kappa'_{300}\ldots$ is one sixth of the product of the reduced critical spaces $\kappa_{200}\ldots$ and $\kappa_{300}\ldots$; $\psi_{300}\ldots=\chi_{100}\ldots+\chi_{200}\ldots$ is the heat of bonding of three single atoms of the first kind into a molecule. Continuing in this line of reasoning, one finds easily that

$$n_{a_1 00}\ldots = n_{100}^{a_1}\ldots V^{1-a_1}\kappa'_{a_1 00}\ldots e^{2h\psi_{a_1 00}\ldots},$$

where $\kappa'_{a_1 00}\ldots$ is the product of all the reduced critical spaces, divided by $a!$, and $\psi_{a_1 00}\ldots$ is the heat of bonding of a_1 atoms of the first kind with each other.

Each such molecule will have a definite reduced critical space $\kappa_{a_1 10}\ldots$ for the annexation of an atom of the second kind. Let $\psi_{a_1 10}\ldots$ be the heat of formation of a molecule consisting of a_1 atoms of the first and one atom of the second kind, from its atoms. Then

$$n_{010}\ldots:n_{a_1 10}\ldots = V:n_{a_1 00}\ldots\kappa_{a_1 10}\ldots e^{2h(\psi_{a_1 10}\ldots-\psi_{a_1 00}\ldots)}$$

If one attaches still more atoms of the second and third kinds, and so forth, then it follows finally that

$$n_{a_1b_1c_1\cdots} = n_{100\cdots}^{a_1} n_{010\cdots}^{b_1} n_{001\cdots}^{c_1} V^{1-a_1-b_1-c_1-\cdots} \kappa'_{a_1b_1c_1\cdots} e^{2h\psi_{a_1b_1c_1\cdots}},$$

where $\psi_{a_1b_1c_1\cdots}$ is the heat of formation of the molecule $(a_1b_1c_1 \cdots)$ from its atoms, and $\kappa'_{a_1b_1c_1}\cdots$ is the product of all the critical spaces, divided by $a_1!b_1!c_1!\ldots$

Completely analogous expressions follow of course for $n_{a_2b_2c_2\cdots}$, $n_{a_3b_3c_3\cdots}$, $n_{\alpha_1\beta_1\gamma_1\cdots}$, and so forth. The n's that have a single one, and otherwise all zero indices, can be eliminated by taking account of Equations (229), whence one obtains:

$$(230) \quad \begin{cases} n_{a_1b_1c_1\cdots}^{C_1} n_{a_2b_2c_2\cdots}^{C_2} \cdots = n_{\alpha_1\beta_1\gamma_1\cdots}^{\Gamma_1} n_{\alpha_2\beta_2\gamma_2\cdots}^{\Gamma_2} \cdots \times \\ V^{\Sigma C - \Sigma\Gamma} \kappa e^{2h(C_1\psi_{a_1b_1\cdots} + C_2\psi_{a_2b_2\cdots} + \cdots - \Gamma_1\psi_{\alpha_1\beta_1\cdots} - \Gamma_2\psi_{\alpha_2\beta_2\cdots} - \cdots)} \end{cases}$$

$\kappa = \kappa'^{C_1}_{a_1b_1}\cdots\kappa'^{C_2}_{a_2b_2}\cdots \cdots / \kappa^{\Gamma_1}_{\alpha_1\beta_1}\cdots \cdots$ is the quotient in which occur all the reduced critical spaces of the compounds $(a_1b_1 \cdots)$, $(a_2b_2 \cdots)$, each with its appropriate C as exponent, and all the factorials $(\alpha_1!)^{\Gamma_1}(\beta_1!)^{\Gamma_1} \cdots (\alpha_2!)^{\Gamma_2}(a!)^{\Gamma_2} \cdots$ in the numerator, and the critical spaces of the compounds $(\alpha_1\beta_1 \cdots)$, $(\alpha_2\beta_2 \cdots)$ raised to the powers Γ_1, Γ_2, $\cdots$ as well as the factorials $(a_1!)^{C_1}(b_1!)^{C_1} \cdots (a_2!)^{C_2} \cdots$ in the denominator. $\Gamma_1\psi_{\alpha_1\beta_1\cdots} + \Gamma_2\psi_{\alpha_2\beta_2\cdots} - C_1\psi_{a_1b_1\cdots} - \cdots$ is the heat of reaction which would be liberated if C_1 molecules $(a_1b_1 \cdots)$, C_2 molecules $(a_2b_2 \cdots)$ and so forth were changed into Γ_1 molecules $(\alpha_1\beta_1 \cdots)$, Γ_2 molecules $(\alpha_2\beta_2 \cdots)$ etc. Furthermore,

$$\Sigma C = C_1 + C_2 + \cdots, \quad \Sigma\Gamma = \Gamma_1 + \Gamma_2 + \cdots.$$

We shall now denote by $m_{a_1b_1\cdots}$ the mass of a molecule $(a_1b_1 \cdots)$ in the gas-theoretic sense, and by $\mu_{a_1b_1\cdots}$ the mass of a molecule of this substance in the macroscopic sense, i.e. the quotient $m_{a_1b_1\cdots}/M$. One molecule $(a_1b_1 \cdots)$ in the macroscopic sense therefore contains $1/M$ gas-theoretic molecules. Likewise, in the aggregate of C_1 macroscopic molecules $(a_1b_1 \cdots)$, C_2 macroscopic molecules $(a_2b_2 \cdots)$ etc. there are in all C_1/M gas-theoretic molecules $(a_1b_1 \cdots)$, C_2/M gas-theoretic molecules $(a_2b_2 \cdots)$, etc. Therefore

$$\frac{1}{M}[C_1\psi_{a_1b_1\cdots} + C_2\psi_{a_2b_2\cdots} + \cdots - \Gamma_1\psi_{\alpha_1\beta_1\cdots} - \Gamma_2\psi_{\alpha_2\beta_2\cdots} - \cdots] = \Pi$$

is the heat which would be liberated if this reaction took place with C_1, C_2, $\cdots$ macroscopic molecules and Γ_1, Γ_2, $\cdots$ macroscopic molecules of the specified kinds. Hence one can also write Equation (230) as follows:

$$(231) \quad n_{a_1b_1\ldots}^{C_1} n_{a_2b_2\ldots}^{C_2} \cdots = n_{\alpha_1\beta_1\ldots}^{\Gamma_1} n_{\alpha_2\beta_2\ldots}^{\Gamma_2} \cdots V^{\Sigma C-\Sigma\Gamma} \cdot \kappa e^{\Pi/RT}$$

This equation holds for any possible reaction. We now consider a special case: only one kind of reaction is possible in the gas. Let there be initially a times C_1 (gas-theoretic) molecules of type $(a_1b_1 \cdots)$, likewise a times C_2 molecules of type $(a_2b_2 \cdots)$ and so forth, and no molecules of types $(\alpha_1\beta_1 \cdots)$, $(\alpha_2\beta_2 \cdots)$ etc. At the pressure p and temperature T let there be only $(a-b)\times C_1$ molecules of type $(a_1b_1 \cdots)$, $(a-b)\times C_2$ of type $(a_2b_2 \cdots)$ etc., and $b\times\Gamma_1$ molecules of type $(\alpha_1\beta_1 \cdots)$, $b\times\Gamma_2$ of type $(\alpha_2\beta_2 \cdots)$ etc. present in equilibrium. Then $b/a=q$ is the degree of dissoication. Furthermore,

$$n_{a_1b_1} \cdots = a(1-q)C_1, \quad n_{a_2b_2} \cdots = a(1-q)C_2 \cdots,$$
$$n_{\alpha_1\beta_1} \cdots = aq\Gamma_1, \quad n_{\alpha_2\beta_2} \cdots = aq\Gamma_2 \cdots .$$

Hence Equation (231) takes the form:

$$C_1^{C_1}C_2^{C_2} \cdots (1-q)^{\Sigma C} = \left(\frac{a}{V}\right)^{\Sigma\Gamma-\Sigma C} q^{\Sigma\Gamma}\Gamma_1^{\Gamma_1}\Gamma_2^{\Gamma_2} \cdots \kappa e^{\Pi/RT}.$$

The mass of gas present is $a[C_1m_{a_1b_1\ldots}+C_2m_{a_2b_2\ldots} \cdots]$. If we again denote by v the volume of unit mass then

$$v = \frac{V}{a[C_1m_{a_1b_1\ldots} + C_2m_{a_2b_2\ldots} + \cdots]},$$

and we shall set

$$\gamma = \frac{\kappa \cdot \Gamma_1^{\Gamma_1}\Gamma_2^{\Gamma_2} \cdots}{C_1^{C_1}C_2^{C_2} \cdots [C_1m_{a_1b_1\ldots} + C_2m_{a_2b_1\ldots} + \cdots]^{\Sigma\Gamma-\Sigma C}}.$$

Then the above equation becomes

$$(232) \quad (1-q)^{\Sigma C} = \gamma v^{\Sigma C-\Sigma\Gamma} q^{\Sigma\Gamma} e^{\Pi/RT}.$$

This equation gives the dependence of the degree of dissociation on the temperature and specific volume. γ and (if the heat of reaction is not otherwise known) Π/R are constants to be determined from experiment.

If one wishes to introduce the total pressure p instead of v, then he obtains according to Equation (195)

$$p = (n_{a_1 b_1} \ldots + n_{a_2 b_2} \ldots + n_{\alpha_1 \beta_1} \ldots + n_{\alpha_2 \beta_2} \ldots \cdots) \frac{MRT}{V} =$$

$$= [(1-q)\Sigma C + q\Sigma\Gamma] \frac{aM}{V} RT$$

$$= [(1-q)\Sigma C + q\Sigma\Gamma] \frac{RT}{v(C_1 \mu_{a_1 b_1} \ldots + C_2 \mu_{a_2 b_2} \ldots)} .$$

Since $C_1 \mu_{a_1 b_1} \ldots + C_2 \mu_{a_2 b_2} \ldots + \cdots$ is the molecular weight of the undissociated substance, this agrees with the Boyle-Charles-Avogadro law for $q=0$, and gives, when q is different from zero, the deviations from this law due to dissociation.

If one eliminates from this and from Equation (232) the quantity q, then he obtains again the relation between p, v, and T; if one eliminates v from the same equations, then he obtains the degree of dissociation q as a function of p and T.

More general formulas would follow from the assumption that an atom of the first kind can be bound to one of the second, and then this complex can be bonded to another atom of the first kind, although the atoms of the first kind cannot be bonded to each other alone (isomerism).

All these formulas agree with experience, as far as present observations go.

§72. Relation of this theory to that of Gibbs.

Gibbs[1] has deduced essentially the same formula from the general principles of thermodynamics, without referring to the dynamics of molecules. Yet one should not forget that his deduction is based on the assumption that in a dissociating gas all the constituents are present independently as individual gases, and energy, entropy, pressure, etc. are simply additive. This hypothesis is completely clear from the molecular-theory standpoint,

[1] Conn. Acad. Trans. **3**, 108 (1875); Am. J. Sci. **18**, 277 (1879); German translation, *Thermodynamischen Studien* (Leipzig: Engelmann, 1892); cf. also van der Waals, Verslagen Acad. Wet. Amsterdam **15**, 199 (1880); Planck, Ann. Phys. [3] **30**, 562 (1887), **31**, 189 (1887), **32**, 462 (1887).

since these different molecules are actually present separate from each other; and in many places it seems evident that Gibbs has these molecular-theoretic concepts continuously in mind, even if he does not make use of the equations of molecular mechanics.

On the other hand, if one takes the modern viewpoint, which has been most sharply advocated by Mach[2] and Ostwald,[3] that in chemical binding something completely new is created in place of the constituents, then it has no meaning to assume for example that during the dissociation of water vapor, one has present simultaneously water vapor, hydrogen, and oxygen. On the contrary, one must say that at low temperature there is only water vapor; at intermediate temperature some new substance is present, which finally becomes oxhydrogen gas [Knallgas] at very high temperatures.

The assumption that at these intermediate temperatures the energy and entropy of water vapor and oxhydrogen gas are additive loses any sense; without this assumption, however, the basic equations of dissociation can be derived neither from the first and second laws of thermodynamics nor from any energetic principles whatever. One can only consider them as empirically given.

There is no question that for the calculation of natural processes the mere equations, without their foundation, are sufficient; likewise, empirically confirmed equations have a higher degree of certainty than the hypotheses used in deriving them. But on the other hand, it appears to me that the mechanical basis is necessary to illustrate the abstract equations, in the same way that geometrical constructions illuminate algebraic relations. Just as the latter are not made superfluous by mere algebra, so I believe that one cannot completely dispense with the intuitive representation of the laws valid for the action of macroscopic masses provided by molecular dynamics, even if he doubts the possibility of knowledge of the latter, or indeed the existence of the molecules. A clear understanding is just as important for knowledge as the establishment of results by laws and formulas.

It should still be mentioned that we have discussed here only the simplest relations which would give rise to the so-called theo-

[2] *Populärwissenschaftliche Vorlesungen* (Barth, 1896), Vorl. XI. *Die Ökonomische Natur der phys. Forschung*, p. 219.

[3] Die Ueberwindung des wissenschaftlichtl. Materialismus. Verh. Ges. Naturf. 1, 5, 6 (1895).

retical dissociation equilibrium. A deeper investigation into molecular mechanics can also account for the phenomena which one denotes as false chemical equilibrum.[4] The pertinent facts are as follows. At room temperature, oxhydrogen gas as well as water vapor can exist for an arbitrarily long time, without one being transformed into the other. All the molecules are bound so strongly that in the time of observation no dissociation or reaction of any observable amount of the substance is possible. Of course in a time infinite in the mathematical sense the reaction would take place.*

The phenomena of false chemical equilibria are completely analogous to the phenomena of supercooling and superheating that we have discussed in §15, and have exactly the same basis.

§73. The sensitive region is uniformly distributed around the entire atom.

We shall now, for comparison, consider the simplest case of dissociation on the basis of another mechanical picture, which is actually a special case of the one earlier considered. There are present again equivalent atoms, a in number, with diameter σ. What we called the sensitive region shall now no longer be limited to a small part of the surface of the molecule, but rather shall be uniformly distributed over the entire molecule. The sensitive region therefore has the form of a spherical shell concentric with the molecule, whose inner radius is $\frac{1}{2}\sigma$, and whose outer radius is $\frac{1}{2}(\sigma+\delta)$, where δ is small compared to σ. Whenever the sensitive regions of two molecules touch or overlap, they will be chemically bound. The heat of separation, measured in mechanical units, will be equal to a constant, χ, for all these positions.

The covering sphere is then a sphere of radius σ concentric with the molecule. The critical space, which coincides with the reduced critical space, is a spherical shell which lies between the surface of the covering sphere and a concentric spherical surface

---

[4] Pelabon, Doctordiss. d. Univ. Bordeaux (Paris: Hermann, 1898).

* The fact that hydrogen and oxygen can be mixed without chemical combination at ordinary temperatures was advanced as one of the objections to the equipartition theorem by Crum Brown and others, Nature **32,** 352, 533 (1885).

of radius $\sigma+\delta$. Each time the center of a second atom lies within this critical space, it is chemically bound to the first one, and the heat of separation is a constant equal to χ.

Let there be present n_1 simple and n_2 double atoms, so that

$$n_1:2n_2 = V:4\pi n_1\sigma^2\delta e^{2h\chi},$$

where V is the total volume of the gas. Hence

$$\frac{n_2}{n_1} = \frac{2\pi n_1\sigma^2\delta}{V}\, e^{2h\chi}. \tag{233}$$

The centers of two atoms united in a double atom will be at a distance approximately equal to σ. Of the critical space $4\pi\sigma^2\delta$ of each of the two atoms, the part $3\pi\sigma^2\delta$ lies outside the covering sphere of the second atom, and the part $\pi\sigma^2\delta$ lies inside it. The center of a third atom can lie only inside the former part, which we therefore call "free." The total "free critical space" of the two atoms has therefore the volume $6\pi\sigma^2\delta$. However it is to be noted that a small zone exists where the critical spaces of the two atoms overlap, so that for each double atom there is a narrow ringlike space of volume $2\pi\sigma\delta^2$ which we call "the critical ring" belonging to both critical spaces. In calculating the volume of the total free critical space one should properly subtract twice the volume of the critical ring from $6\pi\sigma^2\delta$. However this effect can be ignored since the volume of the critical ring vanishes compared to the free critical space. The space available for a third atom to be united with the double atom, thus forming a triple atom, consists therefore of two parts: first, the free critical space of the double atom, and second, the critical ring of the double atom. In the former case the work of separation of the third atom from the double atom is χ, while in the latter case it is 2χ. If we denote by n_3 the number of triple atoms, then according to the principles of our theory we obtain the ratio:

$$n_1:3n_3 = V:n_2(6\pi\sigma^2\delta e^{2h\chi} + 2\pi\sigma\delta^2 e^{4h\chi}),$$

hence

$$\frac{n_3}{n_2} = \frac{2\pi\sigma^2 n_1\delta}{V}\, e^{2h\chi} + \frac{2\pi\sigma n_1\delta^2}{3V}\, e^{2h\chi}. \tag{234}$$

Comparison of this formula with Equation (233) shows at once that in any case (n_3/n_2) must be greater than (n_2/n_1). A state of dissociation in which there are many double atoms but very few triple atoms would therefore be impossible in this case, where the critical space is uniformly spread over the entire sphere of action.

Indeed there is still more. We can write the right-hand side of Equation (233) in the form:

$$\frac{2\pi n_1\sigma^3}{V}\cdot\frac{\delta}{\sigma}e^{2h\chi}.$$

Now $4\pi n_1\sigma^3/3$ is the space filled by the covering spheres of n_1 single atoms, while V is the total space occupied by the gas. Hence $2\pi n_1\sigma^3/V$ is in any case a very small quantity. Hence if n_2 is not already very small compared to n_1, so that the gas is almost completely dissociated, then $e^{2h\chi}\cdot\delta/\sigma$ must be very large, hence also in Equation (234) the second term is very large compared to the first. But the first term is equal to n_2/n_1. Therefore n_3/n_2 must be very large compared to n_2/n_1.

Therefore as soon as an appreciable number of single atoms have combined into double atoms, most of the latter must also combine into triple atoms. A pairing of most atoms into double atoms, such as we find in the best known gases, is therefore possible only if the critical space covers a relatively small part of the surface of the covering sphere of each atom.

In the present case, where the critical space is uniformly distributed over the entire surface of the covering sphere, as soon as the atoms begin to combine at all, they will prefer to form aggregates containing a larger number of atoms. Something like liquefaction of a gas would then occur. Unfortunately, except for the case where n_2 is small compared to n_1, further calculations encounter difficulties that can scarcely be overcome, so that it remains undecided whether one can obtain under this assumption laws of liquefaction similar to those provided by the van der Waals equation, which we derived from a hypothesis directly opposed to the present one; for there we proceeded from the assumption that the attraction of the molecules extends to distances large compared to the distances of the centers of two neighboring atoms, but here we assume that the attractive region of an atom is small compared to the space occupied by the atom.

The author of this book once tried to make a mechanical model of the properties of gas molecules in the following way.[1]

[1] Wien. Ber. **89** (2) 714 (1884).

Consider them as material points (single atoms) of mass m and mean square velocity $\overline{c^2}$. At distances greater than or equal to $(\sigma+\epsilon)$ they do not interact with each other, nor do they interact at distances less than or equal to σ. In the intervening region they exert an enormous attraction on each other, so that their kinetic energy increases by χ, as they pass from the distance $\sigma+\epsilon$ to the distance σ. ϵ is assumed to be small compared to σ.

Let $\omega=\frac{4}{3}\pi\sigma^3$ be a sphere of radius σ, let n_1 be the number of single atoms in the volume v, n_2 the number of double atoms—i.e., those for which the distance of centers is smaller than σ. Then one finds, as earlier,

$$\frac{n_2}{n_1}=\frac{n_1\omega}{2v}\,e^{2h\chi}=\frac{n_1\omega}{2v}\,e^{3\chi/m\overline{c^2}}.$$

If for example (as for ordinary air) $n_1\omega/v$ were equal to about 1/1000, and moreover $m\overline{c^2}=\chi$, then n_2 could be rather small compared to n_1; moreover, two atoms would be significantly deflected in encounters, so that on the whole the characteristics of a gas would be preserved. But at an absolute temperature ten times as large, the deflection of the molecules from the rectilinear paths due to encounters would be so slight that the system would scarcely show the properties of a gas any longer. At an absolute temperature ten times smaller, however, n_2 would be much larger than n_1, and as before there would occur a coalescence of larger complexes of atoms in their spheres of attraction, and thus a liquefaction.

Hence although a mechanical system may show the characteristics of a gas at one temperature, it might still be unusable as a mechanical model at all temperatures. The same is probably true of the other model proposed by the author in the same place, based on an attractive force inversely proportional to the fifth power of the distance. If this law held up to zero distance of separation, then all the atoms would coalesce. If the interaction stopped at a certain small distance, then above a certain temperature the deflection by collisions would also be very small. A mechanical model based only on attractive forces, without elastic repulsive forces in collisions, agreeing with all experimental facts for the gaseous and liquid states of aggregation, has not yet been found.

CHAPTER VII

Supplements to the laws of thermal equilibrium in gases with compound molecules.

§74. Definition of the quantity H, which measures the probabilities of states.

We have given (Part I, §3) a proof that the Maxwell velocity distribution for gases of monatomic molecules satisfies the conditions which the stationary state must fulfill; then (Part I, §5) we showed, using the assumption that the molecules move among each other so randomly that the laws of probability can be applied, that it is the only one which satisfies these conditions. Hence, provided that this assumption is correct, it is the only one which can remain stationary in the gas.

Now, in Part II, we have proved that the general distribution of states represented by Equation (118), page 316, satisfies the conditions which one finds for a stationary distribution of states in a gas with compound molecules. The complete proof that it is the only one satisfying these conditions has not yet been given in complete generality. Yet it is possible to carry out this proof in the cases that are the simplest and of the greatest practical importance, to the same extent as for gases with monatomic molecules. Hence in the following I shall carry through in general those steps of the proof for which this is possible, and then add the others at least for some special cases.

Let there be present in a container a gas consisting of identical compound molecules, or a mixture of several different kinds of such cases. They shall have the properties of ideal gases, i.e., the sphere of action of a molecule shall be vanishingly small compared to the mean distance of two neighboring molecules. Let x, y, z be the rectangular coördinates, and

$$u, v, w \tag{235}$$

be the velocity components of the center of gravity of a molecule of the first kind; let $p_1 \cdots p_\nu$ be the generalized coördinates which determine the relative positions of the constituents of such a molecule with respect to three coördinate axes passing through its center of gravity, whose directions in space are fixed; and let $q_1 \cdots q_\nu$ be the corresponding momenta.

The inclusion of external forces would not qualitatively increase the difficulties, but would make the formulas still more complicated. We therefore exclude them, and assume moreover that the mixing ratio and the distribution of states among the molecules is the same in all parts of the container whose volume is so large that they contain many molecules. Let

$$f_1(u, v, w, p_1 \cdots q_\nu, t)du \cdots dq_\nu \tag{236}$$

be the number of molecules of the first kind in unit volume, for which at time t the variables (235), as well as the variables

$$p_1 \cdots p_\nu, \quad q_1 \cdots q_\nu \tag{237}$$

lie between the limits

$$u \text{ and } u + du, \quad v \text{ and } v + dv, \quad w \text{ and } \quad w + dw \tag{238}$$

$$p_1 \text{ and } p_1 + dp_1 \cdots q_\nu \text{ and } q_\nu + dq_\nu. \tag{239}$$

For the sake of brevity we omit the variables from the function symbol, and set

$$H_1 = \int\int \cdots f_1 l f_1 du dv dw dp_1 \cdots dq_\nu, \tag{240}$$

where l means the natural logarithm, and the integration is to be extended over all possible values of the variables.

Since $f_1 du dv dw dp_1 \cdots dq_\nu$ is the number of molecules for which at time t the variables (235) and (237) satisfy the conditions (238) and (239), one obtains the value which the quantity H_1 has at any arbitrary time in the following way: substitute in the function lf_1 the value which the variables (235) and (237) for each molecule of the first kind in the gas have at this time, and add up all the values found for the function lf_1 in this way. We shall therefore also set

$$H_1 = \Sigma l f_1 \tag{241}$$

where the sum is over all molecules of the first kind in the gas at time t. Similarly we define H_2 for the second kind of gas, H_3 for the third kind, and so forth, and set

$$H_1 + H_2 + H_3 + \cdots = H. \tag{242}$$

Between the quantity H and the probability of the corresponding state of the gas there exists a relation completely analogous to the one described in Part I, §6. But we will not go into this here, since we are intentionally putting aside everything which does not lead directly to our goal.

§75. Change of the quantity H through intramolecular motion.

We seek first the change experienced by H as a result of the internal motions of the molecules, leaving out collisions completely. Since each kind of molecule is then completely independent of the other kinds of molecules, it is sufficient to consider only the first kind. The effect of the wall will also be ignored. This is permissible when the container is so large that the thermal equilibrium in the interior is completely independent of the special processes at the wall, or also if each molecule, when it is reflected at the wall, experiences no change in its state of motion aside from the change in the direction of velocity of its center of gravity. Thus for simplicity one imagines the repulsive force at the wall to be such that it acts in the same way on all constituents of an individual molecule, just as, for example, gravity acts in the same way on all parts of a body.

For those molecules for which at time t the variables (237) lie between the limits (239), the variables at some earlier time, the time zero, lie between the limits

(243) $$P_1 \text{ and } P_1 + dP_1 \cdots Q_\nu \text{ and } Q_\nu + dQ_\nu.$$

We shall always use the abbreviations mentioned in §28 for the regions within which the values of the variables lie. The values of u, v, w experience no change with time.

The expression which we obtain when we substitute in the function f_1 the value zero for t and the values

(244) $$P_1 \cdots Q_\nu$$

for $p_1 \cdots q_\nu$, we denote by F_1. Then

(245) $$F_1 du dv dw dP_1 \cdots dQ_\nu$$

is the number of molecules for which at time zero the values of the variables (235) and (237) lie between the limits (238) and (243).

Since these are the same molecules as those for which these variables lie between the limits (238) and (239) at time t, and since the number of the latter molecules is

$$f_1 du dv dw dp_1 \cdots dq_\nu$$

one has:

$$F_1 du dv dw dP_1 \cdots dQ_\nu = f_1 du dv dw dp_1 \cdots dq_\nu. \tag{246}$$

According to Equation (52), however,

$$dp_1 \cdots dq_\nu = dP_1 \cdots dQ_\nu.$$

hence it follows that $F_1 = f_1$ and

$$lF_1 = lf_1. \tag{247}$$

Furthermore, let

$$H_1' = \Sigma F_1 lF_1$$

be the value of the function H_1 at time zero. The molecules for which the variables (235) and (237) lie between the limits (238) and (239) at time t provide a contribution

$$f_1 lf_1 du dv dw dp_1 \cdots dq_\nu. \tag{248}$$

to the sum $H_1 = \sum f_1 lf_1$. For the same molecules, these variables lie between the limits (238) and (243) at time zero. Therefore the same molecules provide the contribution

$$F_1 lF_1 du dv dw dP_1 \cdots dQ_\nu. \tag{249}$$

to H_1'. According to Equations (246) and (247), the expressions (248) and (249) are equal to each other. The same molecules therefore provide exactly the same addends to the sum H_1 as to the sum H_1'. Since this holds in general for all molecules at all times, it is clear that H_1 and hence also H are not changed at all by the intramolecular motion. The latter follows since the same arguments hold for each other kind of molecule.

§76. Characterization of the first special case considered.

Although the change of H by collisions cannot yet be calculated in general for ideal gases with compound molecules, we shall treat here a special case in which this calculation takes an especially simple form.

We consider, as in the two preceding sections, a mixture of any ideal gases with any compound molecules. In each molecule of each kind of gas there is always only one atom that can exert a force on an atom of any other molecule, of the same or a different kind; and the interaction of two such atoms of two different molecules is always the same as that of two completely elastic negligibly deformable spheres. Hence the interaction of two molecules always lasts so short a time that during this time the relative positions of the constituents of the two molecules—and likewise the velocities and directions of velocity of all other atoms except for the ones that collide—change only by an infinitesimal amount.

We shall now calculate the change experienced by H during an infinitesimal time dt as a result of the collision of a molecule of the first kind with one of the second kind. We have characterized the state of a molecule of the first kind by the variables (235) and (237), and its absolute position in space by the coördinates x, y, z of its center of gravity. We shall now retain the variables (237), but instead of the variables (235) we introduce the velocity components

$$u_1, \ v_1, \ w_1 \tag{250}$$

of that atom which collides with an atom of another molecule, and which we shall call the atom A_1. The other atom with which it collides we shall call A_2. The absolute position of the molecule of the first kind in space we determine by the coördinates x_1, y_1, z_1 of the center of the atom A_1. Since the variables (237) give the differences between the velocities of all the atoms and the velocity of the center of gravity, and therefore also the quantities u_1-u, v_1-v, w_1-w, we see that when the variables (237) are kept constant,

$$du_1 = du, \quad dv_1 = dv, \quad dw_1 = dw.$$

Hence

$$f_1 du_1 dv_1 dw_1 dp_1 \cdots dq_\nu \tag{251}$$

is the number of molecules for which the variables (237) and (250) lie between the limits (239) and

$$u_1 \text{ and } u_1 + du_1, \quad v_1 \text{ and } v_1 + dv_1, \quad w_1 \text{ and } w_1 + dw_1. \tag{252}$$

It makes no difference here whether one introduces u_1, v_1, w_1 in f_1 instead of u, v, w, or keeps the old variables.

We denote the coördinates of the center of atom A_2 of the second molecule by x_2, y_2, z_2, and its velocity components by

(253) $$u_2, \quad v_2, \quad w_2$$

and the other generalized coördinates and momenta needed to determine the state of the second molecule by

(254) $$p_{\nu+1} \cdots p_{\nu+\nu'}, \qquad q_{\nu+1} \cdots q_{\nu+\nu'}.$$

Then by analogy with the expression (251), the number of molecules of the second kind for which the variables (253) and (254) lie between the limits

(255) u_2 and $u_2 + du_2$, v_2 and $v_2 + dv_2$, w_2 and $w_2 + dw_2$,

(256) $p_{\nu+1}$ and $p_{\nu+1} + dp_{\nu+1} \cdots q_{\nu+\nu'}$ and $q_{\nu+\nu'} + dq_{\nu+\nu'}$

will be denoted by

(257) $$f_2 du_2 dv_2 dw_2 dp_{\nu+1} \cdots dq_{\nu+\nu'}$$

Following the method of Part I, §3, we can find the number of molecule pairs in which the first molecule belongs to the first kind and the second to the second kind, and which interact in time dt in such a way that atom A_1 of the first molecule collides with atom A_2 of the second, and that at the instant of collision the following conditions are satisfied. The variables (250), (237), (253), and (254) shall lie between the limits (252), (239), (255), and (256), and the line of centers of the two atoms A_1 and A_2 shall be parallel to one of the lines lying within an infinitely narrow cone $d\lambda$. All cases of interaction of two molecules which take place during the time dt in such a way that all these conditions are satisfied, we call the specified collisions.

If σ is the sum of the radii of the two atoms A_1 and A_2, g their relative velocity, and the latter forms, with the line of centers of the colliding atoms at the instant of collision, an angle whose cosine is ϵ, then one finds by the method in the section cited, for the number of specified collisions:

(258) $$dN = \sigma^2 f_1 f_2' g\epsilon du_1 dv_1 dw_1 du_2 dv_2 dw_2 dp_1 \cdots dq_{\nu+\nu'} d\lambda dt.$$

§77. Form of Liouville's theorem in the special case considered.

Since the collisions take place instantaneously, the values of the variables (237) and (254) do not change during collisions.

Also, g, ϵ, and the velocity components ξ, η, ζ of the common center of gravity of A_1 and A_2 have the same values after the collision as before (cf. Part I, §4). Only the values of u_1, v_1, w_1, u_2, v_2, w_2 will be changed. The values of these quantities after the collision will be denoted by the corresponding capital letters; and for given values of g and ϵ, the values of the variables u_1, v_1, w_1, u_2, v_2, w_2, when they lie between the limits (252) and (255) before the collision, will lie afterwards between the limits

$$(259) \quad U_1 \text{ and } U_1 + dU_1, \quad V_1 \text{ and } V_1 + dV_1, \quad W_1 \text{ and } W_1 + dW_1,$$

$$(260) \quad U_2 \text{ and } U_2 + dU_2, \quad V_2 \text{ and } V_2 + dV_2, \quad W_2 \text{ and } W_2 + dW_2,$$

It can then easily be shown from Equation (52)—or even more generally by very simple means, as in Part I, §4—that then

$$(261) \qquad du_1dv_1dw_1du_2dv_2dw_2 = dU_1dV_1dW_1dU_2dV_2dW_2$$

or

$$\sum \pm \frac{\partial U_1}{\partial u_1} \cdot \frac{\partial V_1}{\partial v_1} \cdot \frac{\partial W_1}{\partial w_1} \cdot \frac{\partial U_2}{\partial u_2} \cdot \frac{\partial V_2}{\partial v_2} \cdot \frac{\partial W_2}{\partial w_2} = 1$$

(Previously the letters ξ, η, ζ were used instead of u, v, w, and primed letters were used instead of capital letters.)

The proof given in Part I, §4 contains an error, which was pointed out to me by C. H. Wind[1] and later by M. Segel in Kasan. Therefore I will give this proof again here using a method free of error.

We introduce instead of u_2, v_2, w_2 the components ξ, η, ζ of the velocity of the center of gravity common to A_1 and A_2, if one were to consider these two atoms as one mechanical system. If m_1 and m_2 are their masses, then

$$\xi = \frac{m_1u_1 + m_2u_2}{m_1 + m_2}$$

with two similar equations for the other two coördinate directions. From these equations it follows that, if one leaves the variables u_1, v_1, w_1 unchanged and only introduces ξ, η, ζ for u_2, v_2, w_2, then

$$(262) \quad du_1dv_1dw_1du_2dv_2dw_2 = \left(\frac{m_1 + m_2}{m_2}\right)^3 du_1dv_1dw_1d\xi d\eta d\zeta.$$

[1] Wien. Ber. **106** (2A) 21 (1897).

In the expression on the right-hand side we introduce now instead of u_1, v_1, w_1 the variables U_1, V_1, W_1, while we leave ξ, η, ζ unchanged. It is evident geometrically from Figure 2, page 39 of Part I that if the position of the center of gravity remains fixed, the endpoint of the line which represents in magnitude and direction the velocity of the first atom before the collision describes a volume element, which is congruent to the volume element described by the endpoint of that line which represents the velocity of the same atom after the collision. Hence

$$(263) \qquad du_1dv_1dw_1d\xi d\eta d\zeta = dU_1dV_1dW_1d\xi d\eta d\zeta.$$

Now we introduce for ξ, η, ζ the variables U_2, V_2, W_2, leaving U_1, V_1, W_1 unchanged. Since again we have the equation

$$\xi = \frac{m_1U_1 + m_2U_2}{m_1 + m_2}$$

with two similar equations for the other two coördinate directions, it follows that

$$\left(\frac{m_1 + m_2}{m_2}\right)^3 dU_1dV_1dW_1d\xi d\eta d\zeta = dU_1dV_1dW_1dU_2dV_2dW_2.$$

From this and Equations (262) and (263) there follows at once the equation (261) which was to be proved.

Since the case considered in Part I, §4, is the special case of the one discussed here in which aside from the atoms A_1 and A_2 there are no other atoms in the molecules, we see that our former proof has now been completed.

§78. Change of the quantity H as a consequence of collisions.

In §76 we have called a certain kind of collision the specified kind. It is that kind which takes place between a molecule of the first kind and a molecule of the second kind during time dt, in such a way that at the instant of the beginning of the interaction, the variables (250), (237), (253), and (254) lie between the limits (252), (239), (255), and (256), and that the line of centers of the colliding atoms is, at the instant of collision, parallel to one of the lines within a given infinitesimal cone $d\lambda$. For the same collisions, at the instant of the end of the interaction, the variables (237) and (254) lie between the same limits, but the variables (250) and (253) lie between the limits (259) and (260). Moreover, g, ϵ, and $d\lambda$ will not be changed during the collision.

We now denote as opposite collisions those collisions which take place during time dt in such a way that at the beginning the variables (250) and (253) lie between the limits (259) and (260) and the other variables lie between the same limits as in the case of the specified collisions.

For opposite collisions, in order that they can occur at all, the initial relative positions of the two molecules must be changed so that the second molecule appears displaced relative to the first by a distance which is exactly equal and oppositely directed to the line of centers drawn from atom A_1 to atom A_2.[1] For opposite collisions, conversely, the variables (250) and (253) will lie between the limits (252) and (255) at the end of the interaction.

We now calculate the change experienced by the sum denoted in §74 by H (see Eqs. [241] and [242]) during the time dt, through the combined effects of specified and opposite collisions. Each of the former collisions will decrease by one the number of molecules of the first kind for which the variables (250) and (237) lie between the limits (252) and (239), and hence decrease H_1 by lf_1. Likewise, the number of molecules of the second kind for which the variables (253) and (254) lie between the limits (255) and (256), will decrease by one, and hence H_2 will decrease by lf_2. On the other hand, the same collision will increase by one the number of molecules of the first kind for which the variables (250) and (237) lie between the limits (259) and (239), and will increase by one the number of molecules of the first kind for which the variables (253) and (254) lie between the limits (260) and (256). Hence H_1 increases by lF_1 and H_2 by lF_2, if we write for brevity F_1 and F_2 for $f_1(U_1, V_1, W_1, p_1 \cdots q_\nu, t)$ and $f_2(U_2, V_2, W_2, p_{\nu+1} \cdots q_{\nu+\nu'}, t)$. The number of specified collisions is given by Equation (258); hence all the specified collisions will increase H by

$$(264) \qquad \begin{cases} (lF_1 + lF_2 - lf_1 - lf_2)\sigma^2 g\epsilon f_1 f_2 \\ \times\, du_1 dv_1 dw_1 du_2 dv_2 dw_2 dp_1 \cdots dq_{\nu+\nu'} d\lambda dt. \end{cases}$$

[1] Mun. Ber. 22, 347 (1892); Phil. Mag. [5] 35, 166 (1893).

Conversely, each of the opposite collisions decreases by one the number of molecules of the first kind for which the variables (250) and (237) lie between the limits (259) and (239); while the number of molecules for which the same variables lie between the limits (252) and (239) increases by one. Likewise, the number of molecules of the second kind for which the variables (253) and (254) lie between the limits (260) and (256) will decrease by one, and those for which the same variables lie between the limits (255) and (256) will increase by one. Hence H_1 increases by $lf_1 - lF_1$ as a result of opposite collisions, while H_2 increases by $lf_2 - lF_2$, so that H increases by $lf_1 + lf_2 - lF_1 - lF_2$.

The total number of opposite collisions in time dt, by analogy with Equation (258), is:

$$\sigma^2 g \epsilon F_1 F_2 dU_1 dV_1 dW_1 dU_2 dV_2 dW_2 dp_1 \cdots dq_{\nu+\nu'} d\lambda dt,$$

or, by Equation (261),

$$\sigma^2 g \epsilon F_1 F_2 du_1 dv_1 dw_1 du_2 dv_2 dw_2 dp_1 \cdots dq_{\nu+\nu'} d\lambda dt,$$

so that all the opposite collisions increase H by

$$(lf_1 + lf_2 - lF_1 - lF_2) \cdot \sigma^2 g \epsilon F_1 F_2$$
$$\times\, du_1 dv_1 dw_1 du_2 dv_2 dw_2 dp_1 \cdots dq_{\nu+\nu'} d\lambda dt$$

(Remember that g, ϵ, and $d\lambda$ are unchanged by the collisions.) If one compares this with the expression (264), then he sees that, combining together the specified and the opposite collisions, the quantity H experiences the increment

$$(265) \quad \begin{cases} (lf_1 + lf_2 - lF_1 - lF_2)(F_1 F_2 - f_1 f_2)\sigma^2 g \epsilon \\ \times\, du_1 dv_1 dw_1 du_2 dv_2 dw_2 dp_1 \cdots dq_{\nu+\nu'} d\lambda dt. \end{cases}$$

The value of the latter expression is essentially negative. If one integrates over all possible values of all the differentials except dt, and divides by 2 (since otherwise each collision would be counted twice, once as a specified collision and again as an opposite collision), he obtains the total increment of H during time dt. This is therefore also an essentially negative quantity, provided there is any noticeable change of H at all. Since the same holds for all other kinds of molecules, and similarly for collisions of different molecules of the same kind with each other, we have proved that in this special case the value of H can only decrease as a result of collisions.

For the stationary state, a continual decrease of H is forbidden, so that for such states the expression (265) must in general vanish. Therefore the equation

$$f_1 f_2 - F_1 F_2 = 0 \tag{266}$$

must hold for all kinds of molecules, with similar equations for collisions of molecules of the same kind with each other.

§79. Most general characterization of the collision of two molecules.*

We shall now pass from the special kind of interaction discussed in §76 to the most general case.

We denote by s the distance of the centers of gravity of a molecule of the first and a molecule of the second kind, and we assume that if s is greater than a certain constant b, no perceptible interaction takes place between the two molecules. A sphere of radius b, whose center is the center of gravity of a molecule of the first kind, will be called for short the domain of the molecule in question. Hence we can also say: as soon as the center of gravity of a molecule of the second kind lies outside the domain of a molecule of the first kind, no noticeable interaction takes place between the two molecules. Any process whereby the center of gravity of one of the former molecules penetrates the domain of the latter will be called a collision.

Of course it is possible that even if the center of a molecule of the first kind does penetrate the domain of a molecule of the second kind, it goes out again without any perceptible interaction having occurred, so that the collision does not perceptibly modify the motion of either colliding molecule. But most of the collisions will in fact produce a significant modification of the motion of the two molecules.

Just as in §§75–78, the positions of the constituents relative to the center of gravity, the rotations around the center of gravity, and the velocities of the parts of a molecule of the first kind will be characterized by the variables (250) and (237), and for a

* For more detailed discussion see R. C. Tolman, *The Principles of Statistical Mechanics* (London: Oxford University Press, 1938) Chap. V.

molecule of the second kind by the variables (253) and (254). u_1, v_1, w_1 will now be the velocity components of the center of gravity of a molecule of the first kind, and u_2, v_2, w_2 those of a molecule of the second kind.

We shall call a configuration of the two molecules a critical constellation when the distance of centers is equal to b. We consider critical constellations which satisfy the following conditions: for the first molecule the variables (250) and (237) lie between the limits (252) and (239), for the second molecule the variables (253) and (254) lie between the limits (255) and (256). Finally, the direction of the line of centers shall be parallel to some line lying within an infinitesimal cone of aperture $d\lambda$. The set of these conditions we shall call:

the conditions (267).

When the center of gravity of the second molecule is moving into the domain of the first molecule, the critical constellation represents the beginning of a process of interaction of the two molecules (a collision in the wider sense of the word), and it is then called an initial constellation. If, on the other hand, the second molecule is leaving the domain of the first at this instant, it represents the end of a collision (final constellation). Critical constellations for which the distance of centers of the two molecules attains its minimum at that moment can be ignored, since they represent at the same time the beginning and end of a collision that has no effect on the motion of the molecules.

We denote two constellations as opposite, when the coördinates have the same value in both, while the velocity components have equal magnitude and opposite sign. We denote two critical constellations as corresponding to each other, when the coördinates (237) of the first and (254) of the second molecule and likewise all velocity components have the same magnitude and sign for both collisions, whereas the coördinates of the center of gravity of one molecule with respect to a coördinate axis parallel to a fixed axis drawn through the center of gravity of the other have the same value but opposite signs. The constellation corresponding to any given constellation can therefore be constructed by leaving the first molecule fixed and displacing the second molecule—without changing the configuration and velocities of its constituents—by an amount $2b$ in the direction of the line drawn from its center of gravity to the center of gravity of the first molecule. In other words, one interchanges the positions of the centers of the two molecules without changing their state, and without rotating them.

The following is now immediately clear: if we imagine collected together all the initial constellations and look for all the opposite constellations for each of them, then we obtain all the final constellations, and conversely. Likewise we obtain all the final constellations if we look for all the corresponding initial constellations, and the converse again holds.

§80. Application of Liouville's theorem to collisions of the most general kind.

Now, as before, let the number of molecules of the first kind in the gas, for which at time t the variables (250) and (237) lie between the limits (252) and (239), be given by the expression (251). Likewise, let the number of molecules of the second kind, for which at time t the variables (253) and (254) lie between the limits (255) and (256), be given by the expression (257). If we write the abbreviations

$$d\omega_1 \quad \text{and} \quad d\omega_2$$

for

$$du_1 dv_1 dw_1 dp_1 \cdots dq_\nu \text{ and } du_2 dv_2 dw_2 dp_{\nu+1} \cdots dq_{\nu+\nu'}$$

then

$$dN = f_1 f_2 d\omega_1 d\omega_2 b^2 k d\lambda dt \tag{267a}$$

is the number of collisions that take place during time dt in such a way that their initial constellation is a critical constellation determined by the conditions (267). Here k is the component of the velocity of the center of gravity of the second molecule relative to that of the first, in the direction of the line of centers at the beginning of the collision. For the critical constellations with which all these collisions end, the variables (250) and (237) for the first molecules shall lie between the limits (259) and (243); the variables (253) and (254) for the other molecule shall lie between the limits (260) and

$$P_{\nu+1} \text{ and } P_{\nu+1} + dP_{\nu+1} \cdots Q_{\nu+\nu'} \text{ and } Q_{\nu+\nu'} + dQ_{\nu+\nu'} \tag{268}$$

and the line of centers of the molecules shall be parallel to a line lying within a cone of aperture $d\Lambda$. The set of these conditions we call

the conditions (269).

We again abbreviate the more complicated, though more precise, terminology of §27. We write $d\Omega_1$ and $d\Omega_2$ for

$$dU_1dV_1dW_1dP_1 \cdots dQ_\nu \text{ and } dU_2dV_2dW_2dP_{\nu+1} \cdots dQ_{\nu+\nu'}$$

and let K be the component of the relative velocity of the centers of gravity of the two molecules at the end of the collision in the direction of the line of centers at this instant.

Finally, we denote as before the difference of coordinates of the centers of gravity of the two molecules (drawn from the first) for the initial constellation by ξ, η, ζ, and for the final constellation by Ξ, H, Z. Then Liouville's theorem (Eq. [52]) applied to this case runs as follows:

$$d\xi d\eta d\zeta d\omega_1 d\omega_2 = d\Xi dH dZ d\Omega_1 d\Omega_2. \tag{270}$$

We now replace ξ, η, ζ and Ξ, H, Z by polar coördinates, setting:

$$\xi = s\cos\vartheta, \quad \eta = s\sin\vartheta\cos\varphi, \quad \zeta = s\sin\vartheta\sin\varphi,$$
$$\Xi = S\cos\Theta, \quad H = S\sin\Theta\cos\Phi, \quad Z = S\sin\Theta\sin\Phi.$$

Equation (270) then becomes:

$$s^2\sin\vartheta ds d\vartheta d\varphi d\omega_1 d\omega_2 = S^2\sin\Theta dS d\Theta d\Phi d\Omega_1 d\Omega_2. \tag{271}$$

$\sin\vartheta d\vartheta d\varphi$ and $\sin\Theta d\Theta d\Phi$ are the apertures of the cones within which the lines of centers lie before and after the collision. Since we are denoting the apertures of these cones as before by $d\lambda$ and $d\Lambda$, we have therefore:

$$\sin\vartheta d\vartheta d\varphi = d\lambda \quad \text{and} \quad \sin\Theta d\Theta d\Phi = d\Lambda.$$

We shall also introduce for ds and dS the time differential dt. Let g be the relative velocity of the two centers of gravity, and let s be the line connecting the two centers of gravity before the collision; then the direction cosines of these two lines will be

$$\frac{u_2 - u_1}{g}, \quad \frac{v_2 - v_1}{g}, \quad \frac{w_2 - w_1}{g} \quad \text{and} \quad \frac{\xi}{s}, \quad \frac{\eta}{s}, \quad \frac{\zeta}{s}.$$

The component of relative velocity in the direction of the line s is:

$$k = \frac{1}{s}\left[(u_2 - u_1)\xi + (v_2 - v_1)\eta + (w_2 - w_1)\zeta\right].$$

The corresponding value of this component of relative velocity after the collision will be denoted by K. Therefore we have

$$ds = kdt, \qquad dS = Kdt.$$

On substituting all these values and recalling that $s=b$ at the beginning as well as the end of the collision, Equation (270) takes the form:

$$b^2kd\lambda dtd\omega_1d\omega_2 = b^2Kd\Lambda dtd\Omega_1d\Omega_2,$$

where dt has the same value on the right and left sides of the equation, since t is always considered constant in Liouville's theorem. If we divide the last equation by b^2dt, it follows that:

$$kd\lambda d\omega_1d\omega_2 = Kd\Lambda d\Omega_1d\Omega_2. \tag{272}$$

We shall now keep in mind all the final constellations of those collisions whose number was denoted by dN in Equation (267a). Furthermore, we construct the constellations corresponding to these, and denote by dN' the number of collisions that occur during time dt such that they begin with the corresponding constellations in the manner described. Then

$$dN' = F_1F_2b^2Kd\Omega_1d\Omega_2d\Lambda dt, \tag{273}$$

where F_1 and F_2 are abbreviations for

$$f_1(U_1, V_1, W_1, P_1 \cdots Q_\nu, t) \text{ and } f_2(U_2, V_2, W_2, P_{\nu+1} \cdots Q_{\nu+\nu'}, t)$$

and one has in general $dN=dN'$, if Equation (266) is satisfied for all collisions. Now in each of the dN collisions of a molecule of the first kind, a state in which the variables (250) and (237) lie between the limits (252) and (239) is replaced by one in which these variables lie between the limits (259) and (243). Conversely, in each of the dN' of a molecule of the first kind, the latter state is replaced by the former; similarly for the second kind of molecule and for all other collisions. Hence it follows that the distribution of states is not changed by the collisions when Equation (266) is satisfied, and it is easily proved that this equation is in fact satisfied by the formula (115), so that we have given a second proof that the distribution of states represented by this formula satisfies the conditions which must be satisfied by a stationary distribution of states. In order to prove, insofar as this is at all possible, that it is the only one which satisfies these conditions, we shall again calculate the change of the quantity H.

§81. Method of calculation with finite differences.

In the following we shall need an abstraction which may appear surprising to many, but which must seem natural to anyone who clearly understands that the entire symbolism of the differential and integral calculus is meaningless unless one proceeds first by considering large finite numbers.

We shall assume that the molecule can have only a finite number of states, which we denote by the series 1, 2, 3, and so forth; any arbitrary state can be denoted by 1, any other state by 2, etc. The present representation is related to that of a continuous series of states in such a way that one always considers as being the same all states which fill a region such that they correspond according to Liouville's theorem. Let (a, b) express symbolically a critical constellation of two molecules which have the states a and b; let (b, a) express the corresponding, and $(-a, -b)$ the opposite constellation.

A collision which begins with the constellation (a, b) and ends with the constellation (c, d) shall be denoted by

$$\binom{a, b}{c, d}.$$

Let w_a be the number of molecules in unit volume that have the state a; w_b, etc. have similar meanings. Let

$$C_{c,d}^{a,b} \cdot w_a \cdot w_b$$

be the number of collisions in the gas that begin with the constellation (a, b) and end with the constellation (c, d); then if dw_a means the increment experienced by w_a as a result of collisions during time dt, then

$$\frac{dw_a}{dt} = \Sigma C_{a,z}^{x,y} w_x w_y - \Sigma C_{p,q}^{a,n} w_a w_n,$$

where the sums are to be extended over all possible values of the quantities x, y, z, n, p, q. Now suppose that all the expressions for

$$\frac{dw_1}{dt}, \frac{dw_2}{dt}, \ldots$$

have been written out, and set

$$E = w_1(lw_1 - 1) + w_2(lw_2 - 1) + \cdots,$$

Denote by dE the increment of E during time dt as a result of collisions, and substitute in

$$\frac{dE}{dt} = \frac{dw_1}{dt} lw_1 + \frac{dw_2}{dt} lw_2 + \cdots$$

the above values of

$$\frac{dw_1}{dt}, \frac{dw_2}{dt}, \cdots;$$

l means the natural logarithm. The collision

$$\binom{2, 1}{3, 4},$$

in which 1, 2, 3, 4 can be any states, (2, 1) and (3, 4) any critical constellations, provides in the expression for dw_1, as well as in that for dw_2, the term

$$- C_{3,4}^{2,1} w_1 w_2.$$

However, to the expressions for

$$\frac{dw_3}{dt} \text{ and } \frac{dw_4}{dt}$$

it contributes a positive term. All these terms contribute to dE/dt the sum

$$C_{3,4}^{2,1} w_1 w_2 (lw_3 + lw_4 - lw_1 - lw_2).$$

The corresponding collision

$$\binom{4, 3}{5, 6}$$

—i.e., that one which has as initial constellation the same constellation (4, 3) which corresponds to the final constellation (3, 4) of the previously considered collision—contributes to

$$\frac{dw_3}{dt} \text{ and } \frac{dw_4}{dt}$$

the term

$$C_{5,6}^{4,3} w_3 w_4,$$

and to

$$\frac{dw_5}{dt} \text{ and } \frac{dw_6}{dt}$$

again two equal positive terms.

In the same way one can continue with the collision

$$\begin{pmatrix} 6, 5 \\ 7, 5 \end{pmatrix}$$

which corresponds to the collision

$$\begin{pmatrix} 4, 3 \\ 5, 6 \end{pmatrix},$$

and so forth.

Since we only have a finite number of states, we must eventually arrive at a collision

$$\begin{pmatrix} k, & k-1 \\ x, & y \end{pmatrix}$$

which corresponds to one of the previous ones, and it can be proved that the first collision for which this occurs must correspond to the collision

$$\begin{pmatrix} 2, & 1 \\ 3, & 4 \end{pmatrix}.$$

For if it corresponded, for example, to the collision

$$\begin{pmatrix} 6, & 5 \\ 7, & 8 \end{pmatrix},$$

then (x, y) and (6, 5) must be corresponding collisions, therefore (x, y) and (5, 6) would be identical, and two collisions, one beginning with $(k, k-1)$, and the other with (4, 3) would lead to the same final constellation. But then the initial constellation $(-5, -6)$ would lead to the final constellation $(-4, -3)$ as well as to the final constellation $(-k, -k+1)$. Hence the latter two constellations must be identical, hence

$$\begin{pmatrix} k, & k-1 \\ x, & y \end{pmatrix} \text{ must be identical to } \begin{pmatrix} 4, & 3 \\ 5, & 6 \end{pmatrix}$$

and for the same reason

$$\begin{pmatrix} k-2, & k-3 \\ k-1, & k \end{pmatrix} \text{ must be identical to } \begin{pmatrix} 2, & 1 \\ 3, & 4 \end{pmatrix}.$$

Therefore the cycle must have already been closed previously.

Equation (272) means, in our present notation, that the coefficients

$$C_{c,d}^{a,b} \text{ and } C_{e,f}^{d,c}$$

must be equal to each other, since we have collected together all states for which the variables fill a region that is equal according to Liouville's theorem, and called them a single state. Hence it follows that one can arrange all the terms contained in dE/dt in a cycle of the form:

$$C_{3,4}^{2,1}[w_1w_2(lw_3+lw_4-lw_1-lw_2)+w_3w_4(lw_5+lw_6-lw_3-lw_4)+\cdots$$
$$+w_{k-1}w_k(lw_1+lw_2-lw_{k-1}-lw_k)].$$

If one denotes the expression in square brackets by lX and sets $w_1w_2=\alpha$, $w_3w_4=\beta$, $\cdots$, then:

(274) $$X = \beta^{\alpha-\beta}\gamma^{\beta-\gamma}\delta^{\gamma-\delta}\cdots\alpha^{\sigma-\alpha}.$$

Among the numbers α, β, γ, $\cdots$, there must be at least one, for example γ, which is not larger than its two neighbors β and δ; then

(275) $$X = \left(\frac{\gamma}{\delta}\right)^{\beta-\gamma} Y,$$

where

$$Y = \beta^{\alpha-\beta}\delta^{\beta-\delta}\cdots\alpha^{\sigma-\alpha}$$

has exactly the same form as X but lacks one term.

The factor of Y in Equation (275) is equal to 1 if either $\gamma=\beta$ or $\gamma=\delta$, but otherwise it is always less than 1. If one applies the same considerations to Y again and again, then he can finally reduce X to a product of fractions each of which is less than or equal to 1; they cannot all be equal to 1 unless all the quantities α, β, γ $\cdots$ are equal to each other.

Hence the quantity E—whose time derivative reduces to dH/dt on passing to infinitesimals—can only decrease or remain constant as a result of collisions; and it can remain constant only if for all collisions

$$\begin{pmatrix} a, & b \\ c, & d \end{pmatrix}$$

the equation

$$w_a w_b = w_c w_d$$

is satisfied. Since for the stationary state E cannot decrease any further, the equation

$$w_a w_b = w_c w_d$$

must be satisfied for all possible collisions in the stationary state, and on passage to infinitesimals this is identical with Equation (266).

§82. Integral expression for the most general change of H by collisions.

If one wishes to avoid the transition from a finite number of states to an infinite number, but at the same time make use of differentials, he may use the method outlined here. As in Part I, §18, and Part II, §75–78, one finds

$$(276)\quad \begin{cases} \dfrac{d}{dt}\displaystyle\int f_1 l f_1 d\omega_1 = \\ = \displaystyle\iiint f_1 f_2 (lF_1 + lF_2 - lf_1 - lf_2) d\omega_1 d\omega_2 b^2 g d\lambda, \end{cases}$$

where the single integral denotes an integration over the differentials contained in $d\omega_1$, and the triple integral denotes integration over all the differentials contained in $d\omega_1 d\omega_2 d\lambda$. $d\int f_1 l f_1 d\omega_1$ means the change experienced by this integral merely as a consequence of the collisions of molecules of the first and second kinds. The change due to intramolecular motion is zero. The other quantities have the same meaning as in the preceding sections. We imagine that for each collision the corresponding one has been constructed, whose initial constellation therefore corresponds to the final constellation of the first one. We denote by f_1'' and f_2'' the values which the functions f_1 and f_2 take when one substitutes therein the variables characterizing the state of both molecules at the end of this second collision; furthermore, we shall once again construct the collision corresponding to this second collision, and denote by f_1''' and f_2''' the function values that arise on substituting the values of the variables characterizing the final states of the two molecules for this latter collision, and so forth.

Then the quantity $(d/dt)\int f_1 lf_1 f\omega_1$c an be brought into the form:

$$(277)\quad \begin{cases} b^2 g d\omega_1 d\omega_2 d\lambda \left[f_1 f_2 (lF_1 + lF_2 - lf_1 - lf_2) + \right. \\ + F_1 F_2 (lf_1'' + lf_2'' - lF_1 - lF_2) + \\ \left. + f_1'' f_2'' (lf_1''' + lf_2''' - lf_1'' - lf_2'') + \cdots \right]. \end{cases}$$

If one sets again

$$f_1 f_2 = \alpha, \quad F_1 F_2 = \beta, \quad f_1'' f_2'' = \gamma \text{ etc.},$$

then the expression in square brackets in (277) will be the natural logarithm of

$$(278)\qquad \beta^{\alpha-\beta}\gamma^{\beta-\gamma}\delta^{\gamma-\delta}\cdots$$

This quantity has exactly the same form as the expression (274), except that now the cycle of quantities $\alpha, \beta, \gamma \cdots$ is in general not finite. Nevertheless, if one proceeds far enough along this series, he will eventually come to a term whose base is again very nearly equal to α, so that the difference between (278) and an expression terminated at this point can be made arbitrarily small. As soon as the motion of two molecules is not changed by a collision, then it can of course happen that one of the quantities $\alpha, \beta, \gamma \cdots$ is equal to its neighbor. However, as long as we do not choose b to be so large that this is the case for most of the collisions, then most of these quantities will be completely different from their neighbors, so that most of the fractions being multiplied in the expression (278) will be smaller than 1; the same is true of the factors of Y in (275). Hence dH/dt will be negative, and can be zero only when the condition (266) is satisfied for all collisions.

§83. Detailed specification of the case now to be considered.

We have shown in the preceding sections that, for thermal equilibrium in ideal gases with any kind of compound molecules, equation (266) must be satisfied for all collisions of like or unlike molecules. In carrying out the proofs we have excluded external forces, yet the proofs can still be performed when external forces are permitted. One sees immediately, moreover, that Equation (266) will be satisfied as soon as the distribution of states is determined by the formula (118).

However, the proof that this distribution is the only possible one apparently cannot be carried out in complete generality, so that one still has to provide a proof in each special case. Naturally we shall have to leave all these different special cases to monographs; we can only treat here a rather small number of examples.

The simplest of these is the following special case. Let there be a mixture of any ideal gases on which no external forces act. The atoms of the different molecules shall be held together by arbitrary conservative forces, for which the Lagrange equations hold. The interaction of two different molecules proceeds as if one atom from each molecule collided with the other in the same way as elastic, negligibly deformable spheres.

On account of the negligible deformability, during such a collision neither the position of the colliding atom nor the positions and velocities of the other atoms will change. But since every direction in space is equally probable for the velocity of an individual atom before the collision, one can calculate the probabilities of various kinds of collisions just as in Part I, §3.

For the sake of generality we consider a collision in which the two interacting molecules are of different kinds, and call these the first and second kinds of gas. The same statements will still be valid when both molecules are actually of the same kind.

§84. Solution of the equation valid for each collision.

A particular atom of mass m_1 belonging to the first molecule collides with a particular atom of mass m_2 belonging to the second molecule. We call atoms equivalent to the first atom the m_1-atoms, and all the atoms equivalent to the second atom the m_2-atoms. Let c_1 and c_2 be the velocities of the two colliding atoms just before the collision, and γ_1 and γ_2 the velocities just after the collision. The values of c_1 and c_2 are completely arbitrary. γ_1 can take any value lying between the limits zero and

$$\sqrt{c_1^2 + \frac{m_2 c_2^2}{m_1}} \tag{279}$$

but γ_2 must, according to the conservation of energy, be equal to

$$\sqrt{c_2^2 + \frac{m_1}{m_2}(c_1^2 - \gamma_1^2)}$$

since because of the shortness of the duration of the collision, the energy of the molecules does not change noticeably.

The number of those m_1-atoms in the entire gas for which the three components of velocity of their center in the three coördinate directions lie between the limits

$$u_1 \text{ and } u_1 + du_1, \quad v_1 \text{ and } v_1 + dv_1, \quad w_1 \text{ and } w_1 + dw_1 \tag{280}$$

while all other vȧriables determining the state of motion of the molecules can have any possible values, we denote by

$$f(c_1)du_1dv_1dw_1$$

Since there is no preferred direction in space for the velocity of this atom, the coefficient of the product of differentials is clearly a function only of c_1, and hence will be called $f_1(c_1)$.

Similarly the number of m_2-atoms whose velocity components lie between the limits

(281 $\quad u_2 \text{ and } u_2 + du_2, \quad v_2 \text{ and } v_2 + dv_2, \quad w_2 \text{ and } w_2 + dw_2$

will be denoted by

$$f_2(c_2)du_2dv_2dw_2.$$

For this collision, Equation (266) reduces to

$$f_1(c_1)f_2(c_2) = f_1(\gamma_1)f_2\left(\sqrt{c_2^2 + \frac{m_1}{m_2}(c_1^2 - \gamma_1^2)}\right). \tag{282}$$

Since this equation must be satisfied for all possible values of the variables which fulfill the condition of conservation of energy, it follows that, as a simple calculation shows (cf. also Part I, §7),

$$f_1(c_1) = A_1 e^{-hm_1c_1^2}, \quad f_2(c_2) = A_2 e^{-hm_2c_2^2}.$$

These formulas, together with the condition that all directions of space are equally probable for the velocities, completely determine the probability of the various velocity components. If all the atoms of all the molecules can collide with each other, then

h must have the same value for all. Hence the mean kinetic energy is the same for all atoms, and, as can easily be proved from the equivalence of all directions of velocity, the mean kinetic energy of progressive motion of the center of gravity is the same for all molecules, and is equal to the mean kinetic energy of an atom. The coefficients A_1 and A_2 are constants; however, they would depend on the other variables determining the state of the molecules and the limits assumed for these variables, if all values were not allowed for these variables but only values lying between given limits.

A special case of the one considered is that of diatomic molecules whose atoms are solid spheres connected with rods so that they form a solid system like the so-called gymnastic dumbbells.[1] If the connecting rods were considered to be elastic, then the atoms could perform radial vibrations back and forth. However, we can go to the limiting case where the deformability of the rod becomes zero, hence the amplitude of these vibrations is so small that, just like the rotation around the line connecting the centers of the atoms, they do not come into thermal equilibrum with the other motions in the time of observation.

The result then agrees completely with that obtained earlier, where we found the value 1.4 for the ratio of specific heats.

Another special case is a molecule consisting of three or more spheres rigidly bound together. We then have the case for which we previously obtained the value $1\frac{1}{3}$ for the ratio of specific heats. One could treat this case extensively without much difficulty, finding the probability of various combinations of values of the coördinates as before. We shall not discuss these cases further, but instead give an example of the method of treating more difficult cases.

§85. Only the atoms of a single type collide with each other.

Let there be given an ideal gas, all of whose molecules are the same. Each molecule consists of two different atoms of masses m_1 and m_2 (to be called the atoms of the first and second

[1] Cf. Ramsay, *Les gaz de l'atmosphère* (Paris: Carré, 1898), p. 172.

kinds). The two atoms of a molecule shall behave, with respect to their intramolecular motion, like material points concentrated at the centers of the atoms, exerting on each other a force in the direction of their connecting line, which is a function of their distance. The intramolecular motion will therefore be ordinary central motion.[1] The interaction of two different molecules is as follows: the two atoms of the first kind collide like negligibly deformable elastic spheres, but there is no interaction between atoms of the second kind, or between an atom of the first kind and an atom of the second kind.*

On applying the same arguments as before, we obtain for the atoms of the first kind an equation analogous to Equation (282), from which it follows that:

$$f_1(c_1) = Ae^{-hm_1c_1^2}. \tag{283}$$

As before, u_1, v_1, w_1 are the velocity components for an atom of the first kind; $f_1(c_1)du_1dv_1dw_1$ is the number of atoms of the first kind for which u_1, v_1, and w_1 lie between the limits (280); A can still depend on the limits imposed on the states of the molecule.

However, the same method of reasoning cannot be applied to the atoms of the second kind, since they never collide with atoms of other molecules. We must therefore introduce the probability of orbits and phases of motion of the central motion.

---

[1] Boltzmann, *Vorlesungen über die Principe der Mechanik*, §§20–24.

* This model is somewhat similar to one proposed by Lord Kelvin (Proc. R. S. London **50**, 79 [1891]) as a test case for which the equipartition theorem was alleged to be violated. Kelvin's model was a system of "molecules" each of which consisted of a globule enclosed in a hollow spherical shell; thus the globule can interact with its own shell, but not with any other globule or shell. Kelvin had proposed similar models for vibratory molecules imbedded in the lumeniferous ether, and he doubted whether the globules could equilibrate their energy when they did not interact directly. P. G. Tait had also asserted that a necessary condition for equipartition is the possibility of collisions between all parts of the system (Trans. R. S. Edinburgh, **33**, 65 [1886]). Boltzmann's considerations in §§85–86 are also relevant to subsequent studies based on the Bohr model of the atom, in which one might like to know whether nuclei could equilibrate their energy if they could not interact directly but only through their coupling to the orbital electrons. See also the discussion at the British Association meeting, reported in Nature **32**, 533–535 (1885).

§86. Determination of the probability of a particular kind of central motion.

We have already denoted by c_1 and c_2 the absolute velocities of the first and second atoms at any time. Let ρ be the distance of centers of the two atoms at the same time; let α_1 and α_2 be the angles formed by the directions of c_1 and c_2 with the line drawn from the first to the second atom; finally, let β be the angle between the two planes passing through the line ρ, one with the same direction as c_1, the other with the same direction as c_2.

The total energy of the molecule is

$$(284)\qquad L = \frac{m_1 c_1^2}{2} + \frac{m_2 c_2^2}{2} + \varphi(\rho),$$

where φ is the potential function of the central force. Twice the angular velocity of m_2 relative to m_1 in the orbital plane is:

$$(285)\quad K = \rho\sqrt{c_1^2 \sin^2\alpha_1 + c_2^2 \sin^2\alpha_2 - 2c_1c_2 \sin\alpha_1 \sin\alpha_2 \cos\beta},$$

the velocity of the center of gravity of the molecule, multiplied by m_1+m_2, is

$$(286)\quad G = \sqrt{m_1^2c_1^2 + m_2^2c_2^2 + 2m_1m_2c_1c_2(\cos\alpha_1\cos\alpha_2 + \sin\alpha_1\sin\alpha_2\cos\beta)}$$

and its component perpendicular to the orbital plane is

$$(287)\quad H = \frac{c_1c_2\sin\alpha_1\sin\alpha_2\sin\beta}{\sqrt{c_1^2\sin^2\alpha_1 + c_2^2\sin^2\alpha_2 - 2c_1c_2\sin\alpha_1\sin\alpha_2\cos\beta}}.$$

The number of molecules in unit volume for which K, L, G, H lie between the limits

$$K \text{ and } K + dK,\quad L \text{ and } L + dL,\quad G \text{ and } G + dG,\quad H \text{ and } H + dH$$

will be denoted by

$$\Phi(K, L, G, H)dKdLdGdH.$$

The number of molecules for which ρ lies between ρ and $\rho+d\rho$ is

$$\Phi\cdot dKdLdGdH\cdot\frac{d\rho}{\sigma} : \int_{\rho_0}^{\rho_1}\frac{d\rho}{\sigma} = \Psi dKdLdG\cdot dH\,\frac{d\rho}{\sigma}.$$

Here

$$\sigma = \frac{d\rho}{dt} : \int_{\rho_0}^{\rho_1} \frac{d\rho}{\sigma}$$

is the time elapsed from the perigee to the apogee, and is therefore a given function of K, L, G, and H;

$$\Psi = \Phi : \int_{\rho_0}^{\rho_1} \frac{d\rho}{\sigma}$$

is therefore a given function of these four quantities. We limit ourselves to those molecules for which, first, the line of apses of the path forms with a line in the orbital plane parallel to a fixed plane an angle between ϵ and $\epsilon+d\epsilon$; second, the two planes through the velocity of the center of gravity normal to the orbital plane and parallel to a fixed line Γ form an angle between ω and $\omega+d\omega$; and third, the velocity direction of the center of gravity lies within a cone of specified direction and infinitesimal aperture $d\lambda$. Then we have to multiply by $d\epsilon d\omega d\lambda : 16\pi^3$. The number of molecules in the gas satisfying all these conditions is therefore

$$\Psi \cdot \frac{1}{16\pi^3\sigma} dKdLdGdHd\rho d\epsilon d\omega d\lambda. \tag{288}$$

If we denote by g and $g+dg$, h and $h+dh$, k and $k+dk$ the limits between which the velocity components of the center of gravity of these molecules relative to fixed rectangular coördinate axes lie, then

$$G^2 dG d\lambda = dg dh dk.$$

Now keep g, h, and k fixed, and construct through the center of the first atom a rectangular coördinate system whose z-axis has the direction of G. Denote the coördinates and velocity components of the second atom relative to this system by x_3, y_3, z_3, u_3, v_3, w_3, and transform these six variables into K, L, H, ρ, ϵ, ω. For this purpose we construct through the center of the second atom a second coördinate system, with respect to which the coördinates and velocity components of the second atom shall be called x_4, y_4, z_4, u_4, v_4, w_4. The z-axis of the second system shall be perpendicular to the orbital plane, and the x-axis shall lie in its intersection line with the old xy plane. Then

$$H = G \sin \vartheta,$$

where $90° - \vartheta$ is the angle between the two z-axes; hence, since G is constant,

$$dH = G \cos \vartheta d\vartheta.$$

Finally, we denote the angle between the two x-axes by ω, since it differs from the angle previously so denoted only by an amount which we always now consider constant. We find:

$$z_4 = x_3 \cos \vartheta \sin \omega + y_3 \cos \vartheta \cos \omega + z_3 \sin \vartheta$$
$$w_4 = u_3 \cos \vartheta \sin \omega + v_3 \cos \vartheta \cos \omega + w_3 \sin \vartheta,$$

both of which expressions must vanish, since the x_4y_4 plane is the orbital plane. By means of these two equations, keeping x_3, y_3, u_3 and v_3 constant, one can introduce ϑ, ω in place of z_3, w_3, and find

$$dz_3dw_3 = (y_3u_3 - x_3v_3)\frac{\cos \vartheta}{\sin^3 \vartheta} d\vartheta d\omega.$$

Now furthermore

$$x_4 = x_3 \cos \omega - y_3 \sin \omega$$
$$y_4 \sin \vartheta = x_3 \sin \omega + y_3 \cos \omega$$

and similar equations follow for u_4, v_4. Hence it follows that

$$y_3u_3 - x_3v_3 = \sin \vartheta(y_4u_4 - x_4v_4) = K \sin \vartheta$$

and for constant ϑ, ω,

$$dx_4dy_4 \sin \vartheta = dx_3dy_3; \qquad du_4dv_4 \sin \vartheta = du_3dv_3,$$

hence

$$dx_3dy_3dz_3du_3dv_3dw_3 = K \cos \vartheta dx_4dy_4du_4dv_4d\vartheta d\omega.$$

Now we denote by σ and τ the velocity components of the motion of the second atom relative to the first in the direction of ρ and perpendicular thereto, as before; then, for constant x_4 and y_4,

$$d\sigma d\tau = du_4dv_4$$

$$K = \rho\tau, \quad L = L_g + \frac{m_1m_2}{2(m_1 + m_2)}(\sigma^2 + \tau^2) + \varphi(\rho)$$

$$dKdL = \frac{m_1m_2}{m_1 + m_2}\sigma\rho d\sigma d\tau,$$

where L_g is the energy of motion of the center of gravity, now considered constant. Finally, if ψ is the angle between ρ and the last line of apses,

$$x_4 = \rho \cos (\epsilon + \psi), \quad y_4 = \rho \sin (\epsilon + \psi):$$

where ψ is a function of ρ, K, and L. But the last two are now constant, hence

$$\rho d\rho d\epsilon = dx_4 dy_4.$$

Collecting all these together, we see that:

$$dx_3dy_3dz_3du_3dv_3dw_3 = \frac{m_1 + m_2}{m_1m_2}\frac{K}{\sigma}dKdLdHd\rho d\omega d\epsilon$$

and one sees at once that, if x, y, z are the coördinates of the second atom with respect to a coördinate system going through the center of the first atom, whose axes are parallel to the originally chosen completely arbitrary coördinate axes, then likewise

$$dxdydzdu_2dv_2dw_2 = \frac{m_1 + m_2}{m_1m_2}\frac{K}{\sigma}dKdLdHd\rho d\omega d\epsilon.$$

If we introduce this into the expression (288) and recall that for constant u_2, v_2, w_2 we have

$$dgdhdk = \frac{m_1^3}{(m_1 + m_2)^3}du_1dv_1dw_1$$

then we find

$$(289)\qquad \frac{1}{16\pi^3}\frac{m_1^4m_2}{(m_1 + m_2)^4}\frac{\Psi}{KG^2}dxdydzdu_1dv_1dw_1du_2dv_2dw_2$$

as the number of molecules in unit volume for which the variables $x \cdots w_2$ lie between x and $x+dx \cdots w_2$ and w_2+dw_2. We shall set this number equal to

$$(290)\qquad F = Be^{-h(m_1c_1^2+m_2c_2^2+2\vartheta(\rho))}.$$

According to Equation (283), on the one hand B can be only a function of c_2, ρ, and α_2, while on the other hand according to Equation (289) it can be only a function of K, L, G, and H. B must therefore be a function of these latter variables which is completely independent of the values of c_1, α_1 and β, and is merely a function of c_2, ρ, and α_2. If we therefore set $B=f(K, L, G, H)$, then this function must be completely independent of c_1, α_1, and β when one substitutes for K, L, G, H the values (284) to (287). Since this must hold for all values of c_2, ρ, and α_2, we shall first set $c_2=0$; then

$$K = \rho c \sin\alpha,\quad L = \frac{mc^2}{2} + \varphi(\rho),\quad G = mc,\quad H = 0,$$

therefore

$$B = f\left(\rho c_1 \sin \alpha_1, \frac{mc_1^2}{2} + \varphi(\rho), \; mc_1, \; 0\right).$$

Since this must be independent of c_1 and α_1, K cannot occur at all in f, and L and G can occur only in the combination $2mL-G^2$. The latter can be seen at once by substituting in f, instead of the two variables L and G, $2mL-G^2$ and G. We obtain therefore

$$B = f(2mL - G^2, H)$$

and inserting the general values (284) to (287)

$$B = f\Bigg(m_1(m_1 - m_2)c_1^2 + 2m_1\varphi(\rho)$$

$$- 2m_1m_2c_1c_2(\cos \alpha_1 \cos \alpha_2 + \sin \alpha_1 \sin \alpha_2 \cos \beta),$$

$$\frac{c_1^2c_2^2 \sin^2 \alpha_1 \sin^2 \alpha_2 \sin^2 \beta}{c_1^2 \sin^2 \alpha_1 + c_2^2 \sin^2 \alpha_2 - 2c_1c_2 \sin \alpha_1 \sin \alpha_2 \cos \beta}\Bigg).$$

This must be completely independent of c_1, α_1, and β. One sees at once that then both quantities under the function sign must be completely independent of each other, and hence B must be a constant. But then Equation (290) becomes in fact just a special case of (118).

Another special case, whose treatment involves no particular difficulty, would be molecules which are arbitrary solid bodies, either having the form of solids of revolution or not. However, I fear that I have already spent too much time on complicated calculations for special cases, and I will therefore leave the rest of these special problems for doctoral dissertations.

§87. Characterization of our assumption about the initial state.

When a gas is enclosed in a rigid container, and initially one part of it has a visible motion with respect to the rest, then it soon comes to rest as a consequence of viscosity. When two kinds of gas are initially unmixed, but in contact with each other, then they mix, even if the lighter one was originally on top. In general, when a gas or a system of several kinds of gas has initially some improbable state, then it passes to the most probable state under the given external conditions, and remains there during all observable later times. In order to prove that this is a necessary consequence of the kinetic theory of gases, we used the quantity H defined and discussed in this chapter. We proved that it continually decreases as a result of the motion of the gas molecules among each other. The one-sidedness of this process is clearly not based on the equations of motion of the molecules. For these do not change when the time changes its sign. This one-sidedness rather lies uniquely and solely in the initial conditions.

This is not to be understood in the sense that for each experiment one must specially assume just certain initial conditions and not the opposite ones which are likewise possible; rather it is sufficient to have a uniform basic assumption about the initial properties of the mechanical picture of the world, from which it then follows with logical necessity that, when bodies are always interacting, they must always be found in the correct initial conditions. In particular, our theory does not require that each time when bodies are interacting, the initial state of the system they form must be distinguished by a special property (ordered or improbable) which relatively few states of the same mechanical system would have under the external mechanical conditions in question. Hereby the fact is clarified that this system takes in the course of time states which do not have these properties, and which one calls disordered. Since by far most of the states of the system are disordered, one calls the latter the probable states.

The ordered initial states are not related to the disordered ones in the way that a definite state is to the opposite state (arising from the mere reversal of the directions of all velocities), but rather the state opposite to each ordered state is again an ordered state.

The self-regulating most probable state—which we call the Maxwell velocity distribution since Maxwell first found its mathematical expression in a special case—is not some kind of special singular state which is contrasted to infinitely many more non-Maxwellian distributions. Rather it is, on the contrary, characterized by the fact that by far the largest number of possible states have the characteristic properties of the Maxwell distribution, and compared to this number, the number of possible velocity distributions which significantly deviate from the Maxwellian is vanishingly small. The criterion of equal possibility or equal probability is provided by Liouville's theorem.

In order to explain the fact that the calculations based on this assumption correspond to actually observable processes, one must assume that an enormously complicated mechanical system represents a good picture of the world, and that all or at least most of the parts of it surrounding us are initially in a very ordered—therefore very improbable—state. When this is the case, then whenever two or more small parts of it come into interaction with each other, the system formed by these parts is also initially in an ordered state, and when left to itself it rapidly proceeds to the disordered most probable state.

§88. On the return of a system to a former state.

We make the following remarks: 1. It is by no means the sign of the time which constitutes the characteristic difference between an ordered and a disordered state. If, in the "initial states" of the mechanical picture of the world, one reverses the directions of all velocities, without changing their magnitudes or the positions of the parts of the system; if, as it were, one follows the states of the system backwards in time, then he would likewise first have an improbable state, and then reach ever more probable states. Only in those periods of time during which the system passes from a very improbable initial state to a more probable later state do the states change in the positive time direction differently than in the negative.

2. The transition from an ordered to a disordered state is only extremely improbable. Also, the reverse transition has a definite calculable (though inconceivably small) probability, which approaches zero only in the limiting case when the number of molecules is infinite. The fact that a closed system of a finite number of molecules, when it is initially in an ordered state and then goes over to a disordered state, finally after an inconceivably long time must again return to the ordered state,* is therefore not a refutation but rather indeed a confirmation of our theory.

* H. Poincaré, Acta Math. 13, 67 (1890); E. Zermelo, Ann. Phys. [3] 57, 485 (1896).

One should not however imagine that two gases in a $\frac{1}{10}$ liter container, initially unmixed, will mix, then again after a few days separate, then mix again, and so forth. On the contrary, one finds by the same principles which I used[1] for a similar calculation that not until after a time enormously long compared to $10^{10^{10}}$ years will there be any noticeable unmixing of the gases. One may recognize that this is practically equivalent to *never*, if one recalls that in this length of time, according to the laws of probability, there will have been many years in which every inhabitant of a large country committed suicide, purely by accident, on the same day, or every building burned down at the same time—yet the insurance companies get along quite well by ignoring the possibility of such events. If a much smaller probability than this is not practically equivalent to impossibility, then no one can be sure that today will be followed by a night and then a day.

We have looked mainly at processes in gases and have calculated the function H for this case. Yet the laws of probability that govern atomic motion in the solid and liquid states are clearly not qualitatively different in this respect from those for gases, so that the calculation of the function H corresponding to the entropy would not be more difficult in principle, although to be sure it would involve greater mathematical difficulties.

§89. Relation to the second law of thermodynamics.

If therefore we conceive of the world as an enormously large mechanical system composed of an enormously large number of atoms, which starts from a completely ordered initial state, and even at present is still in a substantially ordered state, then we obtain consequences which actually agree with the observed facts; although this conception involves, from a purely theoretical—I might say philosophical—standpoint, certain new aspects which contradict general themodynamics based on a purely phenomenological viewpoint. General thermodynamics proceeds from the fact that, as far as we can tell from our experience up to now, all natural processes are irreversible. Hence according to the principles of phenomenology, the general thermodynamics of the

[1] Ann. Phys. [3] **57**, 783 (1896).

second law is formulated in such a way that the unconditional irreversibility of all natural processes is asserted as a so-called axiom, just as general physics based on a purely phenomenological standpoint asserts the unconditional divisibility of matter without limit as an axiom.

Just as the differential equations of elasticity theory and hydrodynamics based on this latter axiom will always remain the basis of the phenomenological description of a large group of natural phenomena, since they provide the simplest approximate expression of the facts, so likewise will the formulas of general thermodynamics. No one who has fallen in love with the molecular theory will approve of its being given up completely. But the opposite extreme, the dogma of a self-sufficient phenomenology, is also to be avoided.

Just as the differential equations represent simply a mathematical method for calculation, whose clear meaning can only be understood by the use of models which employ a large finite number of elements,[1] so likewise general thermodynamics (without prejudice to its unshakable importance) also requires the cultivation of mechanical models representing it, in order to deepen our knowledge of nature—not in spite of, but rather precisely because these models do not always cover the same ground as general thermodynamics, but instead offer a glimpse of a new viewpoint. Thus general thermodynamics holds fast to the invariable irreversibility of all natural processes. It assumes a function (the entropy) whose value can only change in one direction—for example, can only increase—through any occurrence in nature. Thus it distinguishes any later state of the world from any earlier state by its larger value of the entropy. The difference of the entropy from its maximum value—which is the goal [Treibende] of all natural processes—will always decrease. In spite of the invariance of the total energy, its transformability will therefore become ever smaller, natural events will become ever more dull and uninteresting, and any return to a previous value of the entropy is excluded.*

---

[1] Boltzmann, Die Unentbehrlichkeit der Atomistik i.d. Naturwissenschaft. Wien. Ber. **105** (2) 907 (1896); Ann. Phys. [3] **60**, 231 (1897). Ueber die Frage nach der Existenz der Vorgänge in der unbelebten Natur, Wien. Ber. **106** (2) 83 (1897).

* W. Thomson (Lord Kelvin), On a universal tendency in nature to

One cannot assert that this consequence contradicts our experience, for indeed it seems to be a plausible extrapolation of our present knowledge of the world. Yet, with all due recognition to the caution which must be observed in going beyond the direct consequences of experience, it must be granted that these consequences are hardly satisfactory, and the discovery of a satisfactory way of avoiding them would be very desirable, whether one may imagine time as infinite or as a closed cycle. In any case, we would rather consider the unique directionality of time given to us by experience as a mere illusion arising from our specially restricted viewpoint.

§90. Application to the universe.

Is the apparent irreversibility of all known natural processes consistent with the idea that all natural events are possible without restriction? Is the apparent unidirectionality of time consistent with the infinite extent or cyclic nature of time? He who tries to answer these questions in the affirmative sense must use as a model of the world a system whose temporal variation is determined by equations in which the positive and negative directions of time are equivalent, and by means of which the appearance of irreversibility over long periods of time is explicable by some special assumption. But this is precisely what happens in the atomic view of the world.

One can think of the world as a mechanical system of an enormously large number of constituents, and of an immensely long period of time, so that the dimensions of that part containing our own "fixed stars" are minute compared to the extension of the universe; and times that we call eons are likewise minute compared to such a period. Then in the universe, which is in thermal equilibrium throughout and therefore dead, there will occur here and there relatively small regions of the same size as

---

the dissipation of mechanical energy, Proc. R. S. Edinburgh **3**, 139 (1852), and later papers by many authors, too numerous to cite. For the generalized "dissipation of energy" version of the second law, see F. O. Koenig, On the history of science and of the second law of thermodynamics, p. 57 in *Men and Moments in the History of Science*, ed. H. M. Evans (Seattle: University of Washington Press, 1959), and other articles cited therein.

our galaxy (we call them single worlds) which, during the relative short time of eons, fluctuate noticeably from thermal equilibrium, and indeed the state probability in such cases will be equally likely to increase or decrease. For the universe, the two directions of time are indistinguishable, just as in space there is no up or down. However, just as at a particular place on the earth's surface we call "down" the direction toward the center of the earth, so will a living being in a particular time interval of such a single world distinguish the direction of time toward the less probable state from the opposite direction (the former toward the past, the latter toward the future). By virtue of this terminology, such small isolated regions of the universe will always find themselves "initially" in an improbable state. This method seems to me to be the only way in which one can understand the second law—the heat death of each single world—without a unidirectional change of the entire universe from a definite initial state to a final state.

Obviously no one would consider such speculations as important discoveries or even—as did the ancient philosophers—as the highest purpose of science. However it is doubtful that one should despise them as completely idle. Who knows whether they may not broaden the horizon of our circle of ideas, and by stimulating thought, advance the understanding of the facts of experience?

That in nature the transition from a probable to an improbable state does not take place as often as the converse, can be explained by assuming a very improbable initial state of the entire universe surrounding us, in consequence of which an arbitrary system of interacting bodies will in general find itself initially in an improbable state. However, one may object that here and there a transition from a probable to an improbable state must occur and occasionally be observed. To this the cosmological considerations just presented give an answer. From the numerical data on the inconceivably great rareness of transition from a probable to a less probable state in observable dimensions during an observable time, we see that such a process within what we have called an individual world—in particular, our individual world—is so unlikely that its observability is excluded.

In the entire universe, the aggregate of all individual worlds, there will however in fact occur processes going in the opposite direction. But the beings who observe such processes will simply reckon time as proceeding from the less probable to the more probable states, and it will never be discovered whether they reckon time differently from us, since they are separated from us by eons of time and spatial distances $10^{10^{10}}$ times the distance of Sirius—and moreover their language has no relation to ours.*

Very well, you may smile at this; but you must admit that the model of the world developed here is at least a possible one, free of inner contradiction, and also a useful one, since it provides us with many new viewpoints. It also gives an incentive, not only to speculation, but also to experiments (for example on the limit of divisibility, the size of the sphere of action, and the resulting deviations from the equations of hydrodynamics, diffusion, and heat conduction) which are not stimulated by any other theory.

§91. Application of the probability calculus in molecular physics.

Doubts have been expressed as to the permissibility of the applications made of the probability calculus to this subject. Since however the probability calculus has been verified in so many special cases, I see no reason why it should not also be applied to natural processes of a more general kind. The applicability of the probability calculus to the molecular motion in gases cannot of course be rigorously deduced from the differential equations for the motion of the molecules. It follows rather from the great number of the gas molecules and the length of their paths, by virtue of which the properties of the position in the gas where a molecule undergoes a collision are completely independent of the place where it collided the previous time. This independence is of course attained only for a finite number of gas molecules during an arbitrarily long time. For a finite number of molecules in a rigid container with completely smooth walls it is never completely exact, so that the Maxwell velocity distribution cannot hold throughout all time.[1]

---

* For further discussion see H. Reichenbach, *The Direction of Time* (Berkeley and Los Angeles: University of California Press, 1956).

[1] Of the relevant literature, I cite only: Loschmidt, Ueber den Zu-

In practice, however, the walls are continually undergoing perturbations, which will destroy the periodicity resulting from the finite number of molecules. In any case, the applicability of probability theory to gas theory is not refuted but rather is confirmed by the periodicity of motion of a finite closed system in the course of eons of time, and since it leads to a model of the world which not only agrees with experience but also stimulates speculations and experiments, it should be retained in gas theory.

Moreover we see that the probability calculus plays yet another role in physics. The calculation of errors by Gauss's famous method is confirmed in purely physical processes, like the calculation of insurance premiums for statistical ones. We have to thank the laws of probability for the fact that in an orchestra the sounds regularly reinforce each other in unison rather than cancelling out by interference; and the same laws also clarify the nature of unpolarized light. Since today it is popular to look forward to the time when our view of nature will have been completely changed, I will mention the possibility that the fundamental equations for the motion of individual molecules will turn out to be only approximate formulas which give average values, resulting according to the probability calculus from the interactions of many independent moving entities forming the surrounding medium—as for example in meteorology the laws are valid only for average values obtained by long series of observations using the probability calculus. These entities must of course be so numerous and must act so rapidly that the correct average values are attained in millionths of a second.

---

stand des Warmegleichgewichts eines Systems von Körpern mit Rucksicht auf die Schwere, Wien. Ber. **73,** 139 (1876). Boltzmann, Wien. Ber. **75** (2) 67 (1877); **76,** 373 (1878); **78,** 740 (1878). Wien. Alm. 1886; Nature **51,** 413 (1895). *Vorlesungen über Gastheorie,* Part I, §6. Ann. Phys. [3] **57,** 773 (1896); **60,** 392 (1897). Math. Ann. **50,** 325 (1898). Bürbury, Nature **51,** 78 (1894). Bryan, Am. J. Math. **19,** 283 (1897). Zermelo, Ann. Phys. [3] **57,** 485 (1896); **59,** 793 (1896). [See also: P. and T. Ehrenfest, *The Conceptual Foundations of the Statistical Approach in Mechanics* (Ithaca: Cornell University Press, 1959) (translated from Enc. Math. Wiss. **4,** 2, II, [Leipzig: B. G. Teubner, 1911]). D. ter Haar, *Elements of Statistical Mechanics* (New York: Rinehart, 1954), Appendix I; R. Dugas, *La Théorie Physique au sens de Boltzmann* (Neuchatel-Suisse: Griffon, 1959) —Tr.]

§92. Derivation of thermal equilibrium by reversal of the time direction.

These considerations are connected with the method of derivation of Equation (266) which was first indicated by Maxwell[1] and further developed by Planck.[2] We imagine a mixture of arbitrarily many ideal gases having any properties desired, enclosed by fixed solid walls, which we will call our mechanical system. The positions of all the parts of a molecule of one kind of gas, which we call the first, will be specified by μ coördinates p_1, $p_2 \cdots p_\mu$, that of a molecule of another kind (the second) by ν coördinates $p_{\mu+1}, p_{\mu+2} \cdots p_{\mu+\nu}$. The corresponding momenta will be $q_1, q_2 \cdots q_{\mu+\nu}$.

We assume that, with the exception of a few singular states, all initial states pass gradually to probable states, in which the system then remains for a time long compared to the time during which it had an improbable state. For all probable states, the mean values of the various quantities in each small region shall be equal, even though the individual molecules are distributed over many different states in different ways.

We mean by

(290a) $$f(p_1, p_2 \cdots q_\mu)dp_1dp_2 \cdots dq_\mu$$

the probability that for a molecule of the first kind the variables

(291) $$p_1, p_2 \cdots q_\mu$$

lie between the limits

(292) $$p_1 \text{ and } p_1 + dp_1,\ p_2 \text{ and } p_2 + dp_2 \cdots q_\mu \text{ and } q_\mu + dq_\mu$$

and we define it as follows: we consider our system during a long time T, during which it is continually in probable states. The ratio of the sum of all time intervals during which the variables (291) for some molecules of the first kind lie within the limits (292), to the total time T multiplied by the number of molecules of the first kind, is then the definition of the probability that the variables (291) for a molecule of the first kind lie between the limits (292).

---

[1] Maxwell, Phil. Mag. [4] **35,** 187 (1868); *Scientific Papers*, Vol. **2,** p. 45.

[2] Planck, Mun. Ber. **24,** 391 (1894); Ann. Phys. [3] **55,** 220 (1895).

The time T may include times during which the system has an improbable state, since these are very rare anyway. Only singular states that continually deviate from probable states must be excluded. If, during a short time, the variables (291) for 2 or 3 molecules of the first kind simultaneously lie between the limits (292), then these time intervals are to be counted doubly or triply in the sum.

Similarly let

(293) $$f_2(p_{\mu+1}, p_{\mu+2} \cdots q_{\mu+\nu})dp_{\mu+1}dp_{\mu+1} \cdots dq_{\mu+\nu}$$

be the probability that for a molecule of the second kind the variables

(294) $$p_{\mu+1}, p_{\mu+2} \cdots q_{\mu+\nu}$$

lie between the limits

(295) $$\begin{cases} p_{\mu+1} \text{ and } p_{\mu+1} \cdots + dp_{\mu+2}, p_{\mu+2} \text{ and} \\ p_{\mu+2} + dp_{\mu+2} \cdots q_{\mu+\nu} \text{ and } q_{\mu+\nu} + dq_{\mu+\nu} \end{cases}$$

Now let the limits (292) and (293) be chosen so that the two molecules are not interacting but will soon be. We shall call the type of interaction which comes about in this way a collision with property A. Then (cf. Eq. [123])

(296) $$f_1(p_1 \cdots q_\mu)f_2(p_{\mu+1} \cdots q_{\mu+\nu})dp_1 \cdots dq_{\mu+\nu}$$

is the probability that for a molecule pair[3] the variables (291) and (294) lie between the limits (292) and (295), which we call the probability of a collision with property A, for brevity.

From the instant when the variables (291) and (294) of a molecule pair (consisting of molecule B of the first kind and molecule C of the second kind) first fall within the limits (292) and (295), let there elapse a certain time t, which is longer than the time during which two molecules interact in any collision. The values of the variables (291) and (294) for molecules B and C will lie, at the end of time t, between the limits

(297) $$P_1 \text{ and } P_1 + dP_1 \cdots Q_{\mu+\nu} \text{ and } Q_{\mu+\nu} + dQ_{\mu+\nu}.$$

Now at the end of the time denoted above by T we reverse the directions of the velocities of all the constituents of all the

[3] When we speak of a molecule pair, we shall always in the following mean a pair consisting of a molecule of the first kind and a molecule of the second kind.

molecules, without changing the magnitude of these velocities or the positions of the constituents. The system will then pass through the same sequence of states in the opposite order, which we shall call the inverse process, in contrast to the originally considered variation of states during time T, which we call the direct process.

During the inverse process, it will just as often happen that the variables (291) and (294) lie between the limits

$$(298)\quad \begin{cases} P_1 \text{ and } P_1 + dP_1 \cdots P_{\mu+\nu} \text{ and } P_{\mu+\nu} + dP_{\mu+\nu}, \\ -Q_1 \text{ and } -Q_1 - dQ_1 \cdots -Q_{\mu+\nu} \text{ and } Q_{\mu+\nu} - dQ_{\mu+\nu} \end{cases}$$

as during the direct process it happened that they lay between the limits (292) and (295).

We next assume that in our system two states in which all the coördinates and all magnitudes of the velocities are the same, but the directions of the latter are opposite, are equally probable. We shall call this assumption A. It is obviously true when the molecules are simple material points or solid bodies of arbitrary shape, and in many other cases. However, in certain cases it needs to be proved.

Therefore in the inverse process the variables will lie between the limits (297) as many times as they lie between the limits (292) and (295) in the direct process. But the inverse process likewise consists of a very long sequence of states in which the variables can assume many different values. These states therefore cannot consist exclusively or predominantly of singular states, but rather for the most part they must be probable states. Hence the various mean values must be the same for the inverse process as for the direct process, and the probability that for a molecule pair the values of the variables lie between the limits (297) must be given by the expression similar to (296):

$$f_1(P_1 \cdots Q_\mu) f_2(P_{\mu+1} \cdots Q_{\mu+\nu}) dP_1 \cdots dQ_{\mu+\nu}$$

which according to the aforesaid must be equal to the expression (296). But according to Liouville's theorem,

$$dp_1 \cdots dq_{\mu+\nu} = dP_1 \cdots dQ_{\mu+\nu},$$

hence one obtains finally the equation

$$(299)\quad f_1(r_1 \cdots q_\mu) f_2(p_{\mu+1} \cdots q_{\mu+\nu}) = f_1(P_1 \cdots Q_\mu) f_2(P_{\mu+\nu} \cdots Q_{\mu+\nu})$$

whereby Equation (266) is proved for all kinds of possible collisions.

§93. Proof for a cyclic series of a finite number of states.

If one does not wish to make assumption A, which in fact is by no means obvious in all cases, then the proof, as in §81, must be carried out by means of recurrent cycles. For the sake of simplicity we assume that all molecules are of the same kind, and we consider the series of collisions

$$\binom{2,\ 1}{3,\ 4},\ \binom{4,\ 3}{5,\ 6},\ \binom{6,\ 5}{7,\ 8}\cdots\binom{a,\ a-1}{1,\ 2}. \tag{300}$$

The notation is the same as that of §81. The probability of the first of these collisions is

$$C_{3,4}^{2,1}w_1w_2,$$

that of the next is

$$C_{5,6}^{4,3}w_3w_4, \text{ etc.}$$

Now by virtue of Liouville's theorem,

$$C_{3,4}^{2,1} = C_{5,6}^{4,3} = C_{7,8}^{6,5} \cdots.$$

If we denote the common value of all these coefficients by C, then the probabilities of the different collisions in (300) are:

$$Cw_1w_2, \quad Cw_3w_4, \quad Cw_5w_6 \cdots, \quad Cw_{a-1}w_a.$$

Now suppose we reverse the entire process of variation of states of our system in the way described above. We must again obtain a stationary distribution of states. Hence the probability of any particular collision must be the same in the reversed sequence of states as in the original one. Now in the reversed sequence, the probability of the last collision in the series (300) is Cw_1w_2, that of the next to last is $Cw_{a-1}w_a$, and so forth. Therefore we must have

$$w_1w_2 = w_{a-1}w_a = w_{a-3}w_{a-2} \cdots = w_3w_4$$

Since these equations must hold for all collisions, Equation (266) is again proved.

I believe that I have here developed *in extenso* the idea indicated by Maxwell[1] in the passage beginning with the words, "This is therefore a possible form of the final distribution of velocities; it is also the only form."

---

[1] Maxwell, Phil. Mag. [4] **35**, 187 (1868); *Scientific Papers*, Vol. **2**, p. 45.

BIBLIOGRAPHY

Details of the works cited in the text and in the Translator's Introduction and notes are given here, together with other papers by Boltzmann on the kinetic theory, and a few biographical articles about him. In most cases the style of abbreviation of journal titles is close enough to the standard style to be easily recognizable; see for example the Royal Society *Catalogue of Scientific Literature.* "Wien. Ber." means *Sitzungsberichte der kaiserlichen Akademie der Wissenschaften in Wien, Klasse IIa.* The journal *Annalen der Physik* was edited by Poggendorff from 1824 to 1876, by Wiedeman from 1877 to 1899, and by Drude from 1900 to 1928, hence the common abbreviations "Pogg. Ann.," "Wied. Ann.," and "Drude's Ann."; we have used instead the abbreviations Ann. Physik [2], [3], [4] respectively, for the sake of consistency with the modern style. "C. R. Paris" means *Comptes Rendus hebdomadaires des Séances de l'Académie des Sciences, Paris.* The places where a work was cited in the text are indicated at the end of each reference by (I, §6) meaning Part I, Section 6, etc.; TI means Translator's Introduction and F means Foreword.

Alder, B. J., and Wainwright, T. E.

Phase transition for a hard sphere system. J. Chem. Phys. **27,** 1208 (1957). (II, §6)

Studies in molecular dynamics. I. General method. II. Behavior of a small number of elastic spheres. J. Chem. Phys. **31,** 459 (1959); **33,** 1439 (1960). (II, §6)

Phase transition in elastic disks. Phys. Rev. **127,** 359 (1962). (II, §6)

Aliotta, A.

La reazione idealistica contro la scienza. Palermo, Casa Editrice "Optima," 1912. English translation by Agnes McCaskill: *The Idealistic Reaction against Science.* London, Macmillan, 1914. (TI; II, F)

Benndorff, H.

Weiterführung der Annäherungsrechnung in der Maxwell'schen Gastheorie. Wien. Ber. **105,** 646 (1896). (I, §17)

Bernoulli, D.

Hydrodynamica. Argentorati, 1738.

A translation of the derivation of the gas law (Sectio Decima) is given by Magie, *A Source Book in Physics*, New York, 1935, and also by Newman, *The World of Mathematics*, New York, 1956, Vol. 2, p. 774. German translation by P. du Bois-Reymond in Ann. Physik [2] **107,** 490 (1859). (TI)

Berthelot, M., and Ogier.

Sur la chaleur specifique du gaz hypoazotique. C. R. Paris **94,** 916 (1882); Bull. Soc. Chim. Paris **37,** 434 (1882); Ann. Chim. Phys. [5] **30,** 382 (1883). (II, §66)

Bogolyubov, N. N., and Sanochkin, Yu. V.

Ludwig Boltzmann. (Address delivered 5 September 1956 at the session of the Department of Physico-mathematical Sciences of the USSR Academy of Sciences dedicated to the 50th anniversary of the death of Ludwig Boltzmann.) Uspekhi Fiz. Nauk **61,** 7 (1957); translation AEC-tr-3971, p. 7 (U. S. National Science Foundation). (TI)

Boltzmann, L.

Wissenschaftliche Abhandlungen (edited by F. Hasenohrl). Leipzig, J. A. Barth, 1909, 3 vols. (Hereafter cited as "Abh.")

Populäre Schriften. Leipzig, J. A. Barth, 1905. (Hereafter cited as "Schr.")

Über die mechanische Bedeutung des zweiten Hauptsatzes der Wärmetheorie. Wien. Ber. **53,** 195 (1866); Wien. Anz. **3,** 36 (1866); Abh. **1,** 9. (II, §42)

Über die Anzahl der Atome in den Gasmolekülen und die innere Arbeit in Gasen. Wien. Ber. **56,** 682 (1867); Wien. Anz. **4,** 235 (1867); Abh. **1,** 34. (II, §42)

Studien über das Gleichgewicht der lebendigen Kraft zwischen bewegten materiellen Punkten. Wien. Ber. **58,** 517 (1868); Wien. Anz. **5,** 196 (1868); Abh. **1,** 49. (II, §29)

Lehre von den Gasen und Dämpfen. Fortschritte der Physik **26,** 470 (1870).

Boiling-points of organic bodies. Phil. Mag. [4] **42,** 393 (1871); Abh. **1,** 199.

Zur Prioritat der Auffindung der Beziehung zwischen dem zweiten Hauptsatze der mechanischen Wärmetheorie und dem Prinzip der kleinsten Wirkung. Ann. Physik [2] **143,** 211 (1871); Abh. **1,** 228.

Über das Wärmegleichgewicht zwischen mehratomigen Gasmolekülen. Wien. Ber. **63,** 397 (1871); Wien. Anz. **8,** 46 (1871); Abh. **1,** 237. (II, §29)

Einige allgemeine Sätze über Wärmegleichgewicht. Wien. Ber. **63,**

679 (1871); Wien. Anz. **8,** 55 (1871); Abh. **1,** 259.
(TI; II, §29, 32, 35)

Analytischer Beweis des zweiten Hauptsatzes der mechanischen Wärmetheorie aus den Sätzen über das Gleichgewicht der lebendigen Kraft. Wien. Ber. **63,** 712 (1871); Wien. Anz. **8,** 92 (1871); Abh. **1,** 288. (II, §42)

Über das Wirkungsgesetz der Molekularkräfte. Wien. Ber. **66,** 213 (1872); Wien. Anz. **9,** 134 (1872); Abh. **1,** 309. (I, §12)

Weitere Studien über das Wärmegleichgewicht unter Gasmolekulen. Wien. Ber. **66,** 275 (1872); Wien. Anz. **9,** 23 (1872); Abh. **1,** 316. (TI; I, §5, 23)

Über das Wärmegleichgewicht von Gasen, auf welche äussere Kräfte wirken. Wien. Ber. **72,** 427 (1875); Wien. Anz. **12,** 174 (1875); Phil. Mag. [4] **50,** 495 (1875); Abh. **2,** 1.

Bemerkungen über Wärmeleitung der Gase. Wien. Ber. **72,** 458 (1876); Wien. Anz. **12,** 174 (1875); Ann. Physik [2] **157,** 457 (1876); Phil. Mag. [4] **50,** 495 (1875); Abh. **2,** 31.

Über die Aufstellung und Integration der Gleichungen, welche die Molecularbewegung in Gasen bestimmen. Wien. Ber. **74,** 503 (1876); Wien. Anz. **13,** 204 (1876); Abh. **2,** 55. (II, §28)

Über die Natur der Gasmolecüle. Wien. Ber. **74,** 553 (1877); Ann. Physik [2] **160,** 175 (1877); Wien. Anz. **13,** 204 (1876); Phil. Mag. [5] **3,** 320 (1877). (I, §19, 44)

Bemerkungen über einige Probleme der mechanischen Wärmetheorie. Wien. Ber. **75,** 62 (1877); Wien. Anz. **14,** 9 (1877); Abh. **2,** 112. (II, §91)

Über die Beziehung zwischen dem zweiten Hauptsatze der mechanischen Wärmetheorie und der Wahrscheinlichkeitsrechnung, respective den Sätzen über das Wärmegleichgewicht. Wien. Ber. **76,** 373 (1877); Wien. Anz. **14,** 196 (1877); Ann. Physik Beibl. **3,** 166 (1879); Phil. Mag. [5] **6,** 236 (1878); Abh. **2,** 164.
(I, §6; II, §91)

Über die Beziehung der Diffusionsphänomene zum zweiten Hauptsatze der mechanischen Wärmetheorie. Wien. Ber. **78,** 733 (1878); Wien. Anz. **15,** 115, 177 (1878); Phil. Mag. [5] **6,** 236 (1878); Abh. **2,** 289. (II, §91)

Weitere Bemerkungen über einige Probleme der mechanischen Wärmetheorie. Wien. Ber. **78,** 7 (1879); Wien. Anz. **15,** 115 (1878); Abh. **2,** 250. (I, §3)

Erwiderung auf die Notiz des Herrn. O. E. Meyer: "Ueber eine veränderte Form usw." Ann. Physik [3] **11,** 529 (1880); Abh. **2,** 358.

Zur Theorie der Gasreibung. Wien. Ber. **81,** 117 (1880); **84,** 40,

1230 (1882); Wien. Anz. **17,** 11, 213 (1880); **18,** 148 (1881); Abh. **2,** 388, 431, 523. (I, §11)

Über einige das Wärmegleichgewicht betreffende Sätze. Wien. Ber. **84,** 136 (1881); Wien. Anz. **18,** 148 (1881); Abh. **2,** 572.

Referat über die Abhandlung von J. C. Maxwell "Über Boltzmann's Theorem betreffend die mittlere Verteilung der lebendigen Kraft in einem System materiellen Punkte." Ann. Physik Beibl. **5,** 403 (1881); Phil. Mag. [5] **14,** 299 (1882); Abh. **2,** 582.

Zu K. Strecker's Abhandlungen: Die specifische Wärme der gasförmigen zweiatomigen Verbindungen von Chlor, Brom, Jod u.s.w. Ann. Physik [3] **18,** 309 (1883); Abh. **3,** 64.

Über die Arbeitsquantum, welches bei chemischen Verbindungen gewonnen werden kann. Wien. Ber. **88,** 861 (1883); Ann. Physik [3] **22,** 39 (1884); Wien Anz. **20,** 204 (1883); Abh. **3,** 66. (II, §62, 66)

Über die Möglichkeit der Begründung einer kinetischen Gastheorie auf anziehende Kräfte allein. Wien. Ber. **89,** 714 (1884); Ann. Physik [3] **24,** 37 (1885); Exner's Repertorium **21,** 1 (1885); Wien. Anz. **21,** 100 (1884); Abh. **3,** 101. (I, §12, 21; II, §73)

Über die Eingenschaften monocyclischer und anderer damit verwandter Systeme. J. r. ang. Math. **98,** 68 (1885); Wien. Ber. **90,** 231 (1884); Wien. Anz. **21,** 153, 171 (1884); Abh. **3,** 122. (TI; II, §32)

Der zweite Hauptsatze der mechanische Wärmetheorie. Almanach K. Akad. Wiss. Wien **36,** 225 (1886); Schr. 25. (II, §91)

Neuer Beweis eines von Helmholtz aufgestellten Theorems betreffend die Eigenschaften monocyclischer Systeme. Göttingen Nachr. 209 (1886); Abh. **3,** 176.

Über die zum theoretischen Beweise des Avogadroschen Gesetzes erforderlichen Voraussetzungen. Wien. Ber. **94,** 613 (1887); Phil. Mag. [5] **23,** 305 (1887); Wien. Anz. **23,** 174 (1886); Abh. **3,** 225. (There is an appendix that appears only in the English version in Phil. Mag., reprinted in Abh. **3,** 255.) (I, §4)

Über die mechanischen Analogien des zweiten Hauptsatzes der Thermodynamik. J. r. ang. Math. **100,** 201 (1887); Abh. **3,** 258. (TI)

Über einige Fragen der kinetischen Gastheorie. Wien. Ber. **96,** 891 (1888); Phil. Mag. [5] **25,** 81 (1888); Wien. Anz. **24,** 228 (1887); Abh. **3,** 293. (I, §10)

Neuer Beweis zweier Sätze über das Wärmegleichgewicht unter mehratomigen Gasmolekülen. Wien. Ber. **95,** 153 (1887); Wien. Anz. **24,** 25 (1887); Abh. **4,** 272.

Gustav Robert Kirchhoff: Festrede zur Feier des 301. Gründungstages der Karl-Franzens-Universität zu Graz. Leipzig, J. A. Barth, 1888; Schr. 51.

Über das Gleichgewicht der lebenden Kraft zwischen progressiver und Rotations-Bewegung bei Gasmolekülen. Berlin Ber. 1395 (1888); Abh. **3,** 366.

III Teil der Studien über Gleichgewicht der lebendigen Kraft. Mun. Ber. **22,** 329 (1892); Phil. Mag. [5] **35,** 153 (1893). (TI; II, §78)

Über die Bestimmung der absoluten Temperatur. Mun. Ber. **23,** 321 (1894). Ann. Physik [3] **53,** 948 (1894); Abh. **3,** 490. (II, §10)

Über die Methoden der theoretischen Physik. (From the Catalogue of the Mathematical Exhibition held by the Association of German Mathematicians at Munich, 1893.) Phil. Mag. [5] **36,** 37 (1893); Proc. Phys. Soc. London **12,** 336 (1894); Schr. 1.

On the application of the determinantal relation to the kinetic theory of polyatomic gases. Brit. Assoc. Rept. **64,** 102 (1894); Abh. **3,** 520.

On Maxwell's method of deriving the equations of hydrodynamics from the kinetic theory of gases. Brit. Assoc. Rept. **64,** 579 (1894); Abh. **3,** 526.

Über den Beweis des Maxwellschen Geschwindigkeitsverteilungsgesetzes unter Gasmolekülen. Mun. Ber. **24,** 207 (1894); Ann. Physik [3] **53,** 955 (1894); Proc. Phys. Soc. Abstr. **1,** 96 (1895); Abh. **3,** 528.

On certain questions of the theory of gases. Nature **51,** 413, 581 (1895). (TI; I, F, §6; II, §91)

Nochmals das Maxwell'sche Verteilungsgesetz der Geschwindigkeiten. Mun. Ber. **25,** 25 (1896); Ann. Physik [3] **55,** 223 (1895); Abh. **3,** 532.

On the minimum theorem in the theory of gases. Nature **52,** 221 (1895); Abh. **3,** 546.

Über die Berechnung der Abweichungen der Gase vom Boyle-Charles'schen Gesetz und der Dissoziation derselben. Wien. Ber. **105,** 695 (1896); Abh. **3,** 547. (II, §58)

Ein Wort der Mathematik an der Energetik. Ann. Physik [3] **57,** 39 (1896); Proc. Phys. Soc. Abstr. **2,** 72 (1896); Schr. 104. (I, §1)

Sur la théorie des gaz. C. R. Paris **122,** 1173, 1314 (1896); Proc. Phys. Soc. Abstr. **2,** 398 (1896); Abh. **3,** 564, 566.

Zur Energetik. Ann. Physik [3] **58,** 595 (1896); Proc. Phys. Soc. Abstr. **3,** 122 (1897); Schr. 137. (I, §1)

(Letter to P. G. Tait) Proc. R. S. Edinburgh **21,** 123 (1896);

Tait's *Scientific Papers*, Cambridge University Press, Vol. **2**, p. 431 (1900).

Entgegnung auf die Wärmetheoretischen Betrachtungen des Hrn. E. Zermelo. Ann. Physik [3] **57,** 773 (1896); Proc. Phys. Soc. Abstr. **2,** 245 (1896); Abh. **3,** 567. (II, §88, 91)

Vorlesungen über Gastheorie. Leipzig, J. A. Barth; Part I, 1896; Part II, 1898. Reprinted 1910–1912. French translation by A. Gallotti and H. Bénard, with an introduction and notes by M. Brillouin: *Leçons sur la theorie des gaz,* Paris, Gauthier-Villars, 1902 and 1905. Russian translation, edited with introduction and notes by B. I. Davydov: Лекции по теории газов. Москва, государственное издательство технико-теоретической литературы, 1956.

Über einen mechanischen Satz Poincaré's. Wien. Ber. **106,** 12 (1897); Wien. Anz. **34,** 3 (1897); Proc. Phys. Soc. Abstr. **3,** 350 (1897); Abh. **3,** 587.

Vorlesungen über die Principe der Mechanik. Leipzig, J. A. Barth; Vol. I, 1897; Vol. II, 1904: Vol. III, 1920. (II, §42, 85)

Über die Unentbehrlichkeit der Atomistik in der Naturwissenschaft. Wien. Ber. **105,** 907 (1896); Proc. Phys. Soc. Abstr. **3,** 90 (1897); Schr. 141. (I, §1; II, §89)

Zu Hrn. Zermelo's Abhandlung "Ueber die mechanische Erklärung irreversibler Vorgänge." Ann. Physik [3] **60,** 392, 776 (1897); Proc. Phys. Soc. Abstr. **3,** 211 (1897); Abh. **3,** 579. (II, §91)

Ueber die Frage nach der objectiven Existenz der Vorgänge in der unbelebten Natur. Wien. Ber. **106,** 83 (1897); Schr. 162. (I, §1; II, §89)

Ueber einige meiner bekannten Abhandlungen über Gastheorie und deren Verhältnis zu derselben. Jahresb. deutsch. Mathematiker-Vereinigung **6,** 130 (1899); published as a separate by F. C. W. Vogel, Leipzig, 1897. Abh **3,** 598.

Nochmals über die Atomistik. Ann. Physik [3] **61,** 790 (1897); Schr. 158. (I, §1)

Ueber die kinetische Ableitung der Formeln für den Druck des gesättigen Dampfes, für den Dissociationsgrad von Gasen und für die Entropie eines das van der Waals'sche Gesetz befolgenden Gases. Verhandl. Gesellschaft deutscher Naturforscher und Aerzte, Th. 2, Hälfte 1, p. 74 (1898); Abh. **3,** 642. (II, §61)

Sur le rapport des deux chaleurs specifiques des gaz. C. R. Paris **127,** 1009 (1898); Abh. **3,** 645.

Zur Energetik. Verhandl. Gesellschaft deutscher Naturforscher und Aerzte, Th. 2, Hälfte 1, p. 65 (1898). (I, §1)

Über die sogennante H-curve. Math. Ann. **50,** 325 (1898); Abh. **3,** 629. (II, §91)

(with H. Mache) Ueber eine Modification der van der Waals'schen Zustandsgleichung. Ann. Physik [3] **68,** 350 (1899); Wien. Anz. **36,** 87 (1899); Abh. **3,** 651.

(with H. Mache) Über die Bedeutung der Constante *b* des van der Waals'schen Gesetzes. Trans. Cambridge Phil. Soc. **18,** 91 (1900); Abh. **3,** 654.

Ueber die Zustandsgleichung v. d. Waals'. Verslagen Acad. Wet. Amsterdam [4] **7,** 477 (1899); Proc. Sect. Sci. Amsterdam **1,** 398 (1899); Abh. **3,** 658. (II, §49)

Über die Entwicklung der Methoden der theoretischen Physik in neuer Zeit. Verhandl. Gesellschaft deutscher Naturforscher und Aerzte, Th. 1, p. 99 (1899); Jahresb. deutsch. Mathematiker-Vereinigung **8,** 71 (1900); The Monist **11,** 226 (1901); Schr. 198. (I, §1)

Festrede, gehalten am 5. November 1899 anlässlich der Enthüllung des Denkmals des Universitätsprofessors Dr. Joseph Loschmidt; Gedenkrede auf Joseph Loschmidt gehalten am 29. Oktober 1895 in der Chemisch-Physikalischen Gesellschaft in Wien. Physik. Zeits. **1,** 169, 180, 254, 264 (1900); Schr. 228.

Über die Grundprinzipien und Grundgleichungen der Mechanik, I (Lecture at Clark University, 1899), Schr. 253; English translation "Theories as representations" in *Philosophy of Science,* New York, Meridian Books, 1960.

Notiz über die Formel für den Drucke der Gase. Recueil de Travaux offerts par les Auteurs à H. A. Lorentz, Arch. Neerl. [2] **5,** 76 (1900); Abh. **3,** 671.

Über die Prinzipien der Mechanik. I. Antritts-Vorlesung, gehalten im November 1900; II. Antritts-Vorlesung, gehalten in Wien im Oktober 1902; Schr. 308.

Über statistische Mechanik. Vortrag gehalten beim wissenschaftlichen Kongress in St. Louis, 1904. Schr. 345.

(with J. Nabl) Kinetische Theorie der Materie. *Encyklopädie der mathematischen Wissenschaften,* **5**: 1, Art. V 8. Leipzig, B. G. Teubner, 1905.

Boltzmann, L.: see also Bryan, G. H.

Bosanquet, R. H. M.

Notes on the theory of sound. Phil. Mag. [5] **3,** 271 (1877). (II, §44)

Boyle, R.

The Works of the Honourable Robert Boyle. Revised edition by Thomas Birch, London, 1772, 6 vols. (Cited as "Works".)

New Experiments physico-mechanical, touching the spring of the air, and its effects; made, for the most part, in a new pneumatical engine. Oxford, 1660. Works **1,** 1. (TI)

A Defence of the doctrine touching the spring and weight of the air, proposed by Mr. R. Boyle in his new physico-mechanical experiments; against the objections of Franciscus Linus. Wherewith the objector's funicular hypothesis is also examined. Oxford, 1662. Works **1,** 118. (TI)

Broda, E.

Ludwig Boltzmann: Mensch, Physiker, Philosoph. Berlin, Deutscher Verlag der Wissenschaften, 1955. (TI)

Brown, A. Crum

Difficulties connected with the dynamical theory of gases. Nature **32,** 352 (1885). (II, §72, 85)

Brush, S. G.

The development of the kinetic theory of gases. I. Herapath. II. Waterston. III. Clausius. IV. Maxwell. Annals of Science **13,** 188, 273 (1957); **14,** 185, 243 (1958). (TI)

John James Waterston and the kinetic theory of gases. American Scientist **49,** 202 (1961). (TI)

Development of the kinetic theory of gases. V. The equation of state. Amer. J. Phys. **29,** 593 (1961). (TI; II, §6, 49, 58)

Development of the kinetic theory of gases. VI. Viscosity. Amer. J. Phys. **30,** 269 (1962). (TI; I, §9)

Theories of liquid viscosity. Chem. Revs. **62,** 513 (1962). (I, §23)

Bryan, G. H.

Researches related to the connection of the second law with dynamical principles. Brit. Assoc. Rept. **61,** 85 (1891). (I, §6)

The laws of distribution of energy and their limitations. Brit. Assoc. Rept. **64,** 64 (1894).

(with L. Boltzmann) Über die mechanische Analogie des Wärmegleichgewichtes zweier sich berührender Körper. Wien. Ber. **103,** 1125 (1894); Proc. Phys. Soc. **13,** 485 (1895); Nature **51,** 454 (1895); Boltzmann's *Wissenschaftliche Abhandlungen* **3,** 510. (I, §19)

Prof. Boltzmann and the kinetic theory of gases. Nature **51,** 31 (1894). (I, F)

The kinetic theory of gases. Nature **51,** 152, 176, 319, **52,** 244 (1894–5). (I, F)

On Maxwell's law of partition of energy. Nature **51,** 262 (1895). (I, F)

Note on a simple graphic illustration of the determinantal relation of dynamics. Phil. Mag. [5] **39,** 531 (1895); Proc. Phys. Soc. London **13,** 481 (1895). (II, §28)

The assumptions in Boltzmann's minimum theorem. Nature **52,** 29 (1895). (I, §6)
Obituary of Boltzmann. Nature **74,** 569 (1906).
Obituary notice of Boltzmann. Proc. Roy. Soc. **A80,** xi (1908).

Burbury, S. H.
Boltzmann's minimum function. Nature **51,** 78, 320, **52,** 104 (1894–1895). (I, §3; II, §91)
The kinetic theory of gases. Nature **51,** 175, **52,** 316 (1894–1895). (I, F)

Carus, P.
Ludwig Boltzmann; an obituary. The Open Court **20,** 759 (1906).

Chapman, S.
On the law of distribution of molecular velocities, and on the theory of viscosity and thermal conduction, in a non-uniform simple monatomic gas. Phil. Trans. **A216,** 279 (1916); Phys. Abstr. **19,** 290 (1916). (TI; I, §14)
On the kinetic theory of a gas, Part II, a composite monatomic gas, diffusion, viscosity, and thermal conduction. Phil. Trans. **A217,** 115 (1907); Phys. Abstr. **20,** 408 (1917). (TI; I, §14)
(with F. W. Dootson) Note on thermal diffusion. Phil. Mag. [6] **33,** 248 (1917). (I, §14)
On certain integrals occurring in the kinetic theory of gases. Mem. Proc. Manchester Lit. Phil. Soc. **66,** No. 1 (1922). (I, §21, 24)
(with T. G. Cowling) *The Mathematical Theory of Non-Uniform Gases.* London, Cambridge University Press, 1939; second ed. 1952. (I, §14)

Clausius, R.
Abhandlungen über die mechanische Wärmetheorie. Braunschweig, F. Vieweg, 1864 and 1867; second edition, Vols. I and II, 1879; Vol. III, *Die Kinetische Theorie der Gase,* 1889–91. Third edition, Vol. I, 1887. (Hereafter cited as "Abh.") English translations (not including Vol. III) by T. A. Hirst, with an introduction by J. Tyndall: *The Mechanical Theory of Heat,* London, J. Van Voorst, 1867; and by W. R. Browne, same title, London, Macmillan, 1879. (I, §1, 14; II, §16, 58)
Ueber die Art der Bewegung, welche wir Wärme nennen. Ann. Physik [2] **100,** 353 (1857); Ann. Chim. Phys. [3] **50,** 497 (1857); Phil. Mag. [4] **14,** 108 (1857); Z. Math. Phys. **2,** 170 (1857); Arch. Sci. Phys. (Geneve) **36,** 293 (1857); Nuovo Cimento **6,** 435 (1857); Abh. **3,** 1. (TI)
Ueber die mittlere Länge der Wege, welche bei der Molecularbewegung gasförmiger Körper von den einzelnen Molecülen

zurückgelegt werden, nebst einigen anderen Bemerkungen über die mechanischen Wärmetheorie. Ann. Physik [2] **105,** 239 (1858); Phil. Mag. [4] **17,** 81 (1859); Arch. Sci. Phys. (Geneve) **4,** 341 (1859); Abh. **3,** 46. (I, §9)

On the dynamical theory of gases. Phil. Mag. [4] **19,** 434 (1860). (I, §9)

Ueber die Wärmeleitung gasformiger Körper. Ann. Physik [2] **115,** 1 (1862); Phil. Mag. [4] **23,** 417, 512 (1862); Presse Sci. deux Mondes 24 (1862) (2); Arch. Sci. Phys. (Geneve) **16,** 134 (1862); Z. gesammten Naturwiss. (Berlin) **20,** 216 (1862); Abh. **3,** 105. (I, §14)

Ueber einen auf die Wärme anwendbaren mechanischen Satz. Sitzber. Niederrhein Ges. (Bonn) 114 (1870); Ann. Physik [2] **141,** 124 (1870); Carl's Repertorium **6,** 197 (1870); Phil. Mag. [4] **40,** 122 (1870); Z. Math. Phys. **17,** 82 (1872). (TI; II, §49)

Ueber das Verhalten der Kohlensäure in Bezug auf Druck, Volumen, und Temperatur. Ann. Physik [3] **9,** 337 (1880); Phil. Mag. [5] **9,** 393 (1880); Ann. Chim. [4] **30,** 358 (1883); Abh. **3,** 184. (II, §54)

Ueber einige neue Untersuchungen über die mittlere Weglänge der Gasmolecüle. Ann. Physik [3] **10,** 92 (1880); Abh. **3,** 204. (I, §14)

Crookes, W.

On the action of heat on gravitating masses. Proc. R. S. London **22,** 37 (1874); Phil. Trans. **164,** 501 (1874); Chem. News **29,** 1 (1874), **31,** 1, 11, 23, 33, 43, 53 (1875). (I, §1)

Culverwell, E. P.

Note on Boltzmann's kinetic theory of gases, and on Sir W. Thomson's address to Section A, British Association, 1884. Phil. Mag. [5] **30,** 95 (1890). (I, §6)

Dr. Watson's proof of Boltzmann's theorem on permanence of distributions. Nature **50,** 617 (1894). (I, F)

The kinetic theory of gases. Nature **51,** 78 (1894). (I, F)

Boltzmann's minimum theorem. Nature **51,** 105, 246, **52,** 149 (1894–1895). (I, F)

Prof. Boltzmann's letter on the kinetic theory of gases. Nature **51,** 581 (1895). (I, F, §6)

Davydov, B. I.

A great physicist (on the 50th anniversary of Ludwig Boltzmann's death). Uspekhi Fiz. Nauk **61,** 17 (1957); translation AEC-tr-3971, p. 20 (U. S. National Science Foundation).

Deville: see Saint-Claire Deville.

Dugas, R.
La Théorie Physique au sens de Boltzmann et ses prolongements modernes. Neuchatel-Suisse, Editions du Griffon, 1959. (TI; I, §1; II, §91)

Duhem, P. M. M.
La Théorie Physique, son objet et sa structure. Paris, Chevalier et Riviera, 1906; second edition, 1914; third edition, 1933; English translation: *The Aim and Structure of Physical Theory*, Princeton University Press, 1954. (II, F)

Traité d'energetique; ou, De Thermodynamique Générale. Paris, Gauthier-Villars, 1911. (I, §1)

Ehrenfest, P.
Collected Scientific Papers. Amsterdam, North-Holland, 1959 (hereafter cited as "Papers").

Zur Berechnung der Volumkorrektion in der Zustandsgleichung von van der Waals. Wien. Ber. **112,** 1107 (1903); Papers, 77. (II, §54)

Ludwig Boltzmann. Math.-Naturw. Blatter **3** (1906); Papers, 131.

(with T. Ehrenfest) Begriffliche Grundlagen der statistischen Auffassung in der Mechanik. *Encyklopädie der mathematischen Wissenschaften*, Vol. 4, Part 32; Leipzig, B. G. Teubner, 1911. Papers, 213. French translation by E. Borel, *Encyclopédie des Sciences Mathematiques*, **4**:1:1, Paris, Gauthier-Villars, 1915 (with a supplement by Borel). English translation by M. J. Moravcsik, with a new preface by T. Ehrenfest: *The Conceptual Foundations of the Statistical Approach in Mechanics*, Cornell University Press, 1959. (TI; II, §32, 91)

Enskog, D.
Kinetische Theorie der Vorgänge in mässig verdünnten Gasen. (Inaugural Dissertation.) Uppsala, Almqvist and Wiksell, 1917. (TI; I, §14)

Euler, L.
Tentamen explicationis phaenomenorum aeris. Comm. Acad. Sci. Imp. Petropol. **2,** 347 (1727). [See R. Hooykaas, Arch. Int. d'Hist. Sci. **2,** 180 (1948).] (TI)

Fitzgerald, G. F.
Scientific Writings of the late George Francis Fitzgerald. Dublin, Hodges, Figgis, 1902 (cited as "Writings").

On the mechanical theory of Crookes's force. Sci. Trans. Royal Dublin Soc. **1,** 57 (1878); Phil. Mag. [5] **7,** 15 (1879); Writings, 18. (I, §14)

The kinetic theory of gases. Nature **51,** 221 (1895); Writings, 321. (I, F)

Flamm, L.

Die Persönlichkeit Boltzmanns. Wiener Chem.-Zeitung **47,** 28 (1944).

In memory of Ludwig Boltzmann. Uspekhi Fiz. Nauk **61,** 3 (1957); translation AEC-tr-3971, p. 3 (U. S. National Science Foundation).

Füchtbauer, C., and Hoffmann, W.

Über Maximalintensität, Dämpfung und wahre Intensitätsverteilung von Serienlinien in Absorption. Ann. Physik [4] **43,** 96 (1914). (I, §9)

Gay-Lussac, L. J.

Premier essai pour déterminer les variations de temperature qu'éprouvent les gaz en changeant de densité, et considérations sur leur capacité pour le calorique. Mem. Phys. Chim. Soc. d'Arcueil (Paris) **1,** 180 (1807); J. f. Chem. und Physik (Gehlen) **6,** 392 (1808); Ann. Physik **30,** 249 (1808). (II, §1)

Gibbs, J. W.

The Scientific Papers of J. Willard Gibbs. London and New York, Longmans, Green, 1906. Second edition: *The Collected Works of J. Willard Gibbs,* 1928 (cited as "Works"). Reprinted by Yale University Press, 1948, and by Dover, New York, 1960.

On the equilibrium of heterogeneous substances. Trans. Conn. Acad. **3,** 108, 343 (1875); Amer. J. Sci. **16,** 441 (1878); Works **1,** 55. German translation by W. Ostwald, *Thermodynamische Studien,* Leipzig, Engelmann, 1892. French translation by G. Matisse, *L'Equilibre des substances hétérogénes,* Paris, Gauthier-Villars, 1919. (II, F, §72)

Elementary principles in statistical mechanics, developed with especial reference to the rational foundation of thermodynamics. New York, Scribner, 1902; Works **2,** 1. German translation by E. Zermelo: *Elementare Grundlagen der statistische Mechanik, entwickelt besonders im Hinblick auf eine rationelle Begrundung der Thermodynamik,* Leipzig, J. A. Barth, 1905. French translation by F. Cosserot and J. Rossignol, with an introduction by M. Brillouin: *Principes élémentaires de mécanique statistique, developpés plus particulièrement en vue d'obtenir une base rationelle de la thermodynamique,* Paris, J. Hermann, 1926. Russian translation by K. V. Nikol'skogo: Основные принципы статистической Механики, излагаемые со специапным применением к рациональному обоснованию термодинамики, Москва, огиз, гос. изд.-во технико—теоретической литературы, 1946. (TI; II, §32)

Groeneveld, J.
Two theorems on classical many-particle systems. Physics Letters **3,** 50 (1962). (II, §54)

ter Haar, D.
Elements of Statistical Mechanics. New York, Rinehart, 1954. (TI; II, §91)
Foundations of statistical mechanics. Rev. Mod. Phys. **27,** 289 (1955). (TI)

Happel, H.
Zur Theorie und Prufung der Zustandsgleichung. Ann. Physik [4] **21,** 342 (1906). (II, §54)

Hautefeuille, P.
Action de la chaleur sur l'acide iodhydrique. C. R. Paris **64,** 608 (1867); Bull. Soc. Chim. Paris **7,** 203 (1867). (II, §69)
Sur quelques réactions inverses. C. R. Paris **64,** 704 (1867); Bull. Soc. Chim. Paris **7,** 200 (1867). (II, §69)

Heine, E.
Handbuch der Kugelfunctionen, Theorie und Anwendungen. Berlin, G. Reimer, 1861; second edition, 1878–1881. (I, §22)

Helm, G.
Die Lehre von der Energie historisch-kritisch entwickelt. Leipzig, A. Felix, 1887. (I, §1)
Grundzüge der mathematischen Chemie; Energetik der chemischen Erscheinungen. Leipzig, W. Engelmann, 1894. English translation by J. Livingston and R. Morgan, *The Principles of Mathematical Chemistry: The Energetics of Chemical Phenomena*, New York, Wiley, 1897. (I, §1)
Zur Energetik. Ann. Physik [3] **57,** 646 (1896). (I, §1)
Die Energetik nach ihrer geschichtlichen Entwickelung. Leipzig, Veit, 1898. (I, §1)

Herapath, J.
A mathematical inquiry into the causes, laws and principal phenomena of heat, gases, gravitation, etc. Ann. Phil. [2] **1,** 273, 340, 401 (1821). (TI)
Tables of temperature, and a mathematical development of the causes and laws of phaenomena which have been adduced in support of hypotheses of 'Calorific capacity,' 'Latent heat,' etc. Ann. Phil. [2] **2,** 50, 89, 201, 257, 363, 435, **19,** 16 (1821). (TI)
Mathematical Physics. London, Whittaker, 1847. (TI)

Hertz, H.
Die Prinzipien der Mechanik. Leipzig, J. A. Barth, 1894. English translation by D. E. Jones and J. T. Whalley, *The Principles of Mechanics*, London, Macmillan, 1899; reprinted by Dover,

New York, 1956, with a new introduction by R. S. Cohen. (I, §1)

Hirn, G.-A.

Recherches expérimentales sur la relation qui existe entre la resistance de l'air et sa temperature. Consequences physiques et philosophiques qui découlent de ces experiences. Mem. Acad. Roy. Sci. Belgique **43** (1881); Ann. Chim. Phys. [6] **7**, 289 (1886); also published as a separate by Gauthier-Villars, Paris. (II, F)

La cinetique modern et le dynamisme de l'avenir. Réponse à diverse critiques faites par M. Clausius aux conclusions [des] travaux précédents. Recherches expérimentales et analytiques sur les lois de l'écoulement et du choc des gaz en fonction de la temperature. Mem. Acad. Roy. Sci. Belgique **46** (1886). (II, F)

Réflexions relatives à la note precedente de M. Ladislas Natanson. C. R. Paris **107**, 166 (1888). (II, F)

Hirschfelder, J. O., Curtiss, C. F., and Bird, R. B.

Molecular Theory of Gases and Liquids. New York, Wiley, 1954. (I, §12)

Holtsmark, J.

Ueber die Verbreiterung von Spektrallinien. Ann. Physik [4] **58**, 577 (1919); Phys. Zeits. **20**, 162 (1919). (I, §9)

Jacobi, K. G. J.

Vorlesungen über Dynamik. (Gehalten an der Universität zu Konigsberg im Wintersemester 1842–1843 und nach einem von C. W. Borchardt ausgearbeiteten Hefte hrsg. von A. Clebsch.) Berlin, G. Reimer, 1866. Second edition, 1884 (as *Supplementband, Gesammelte Werke*). (II, §25, 29, 31)

Jäger, G.

Über die kinetische Theorie der inneren Reibung der Flüssigkeiten. Wien. Ber. **102**, 253 (1893). (I, §14)

Die Gasdruckformel mit Berücksichtigung des Molecularvolumens. Wien. Ber. **105**, 15 (1896). (II, §58)

Über den Einfluss des Molecularvolumens auf die mittlere Weglänge der Gasmolekeln. Wien. Ber. **105**, 97 (1896). (I, §1)

Über den Einfluss des Molecularvolumens auf innere Reibung der Gase. Wien. Ber. **108**, 447 (1899); **109**, 74 (1900). (I, §14)

Zur kinetischen Theorie der inneren Reibung der Gase. Wien. Ber. **127**, 849 (1918); Phys. Abstr. **23**, 40 (1920). (I, §14)

Jaffé, G.

Recollections of three great laboratories. J. Chem. Educ. **29**, 230 (1952). (TI)

Jeans, J. H.

The persistence of molecular velocities in the kinetic theory of gases. Phil. Mag. [6] **8,** 700 (1904). (I, §14)

The Dynamical Theory of Gases. Cambridge University Press, 1904; second edition, 1916; third edition, 1921; fourth edition, 1925, reprinted by Dover, New York, 1954. (I, §14)

Joule, J. P.

The Scientific Papers of James Prescott Joule. London, The Physical Society, 1884. (Cited as "Papers.")

On the changes of temperature produced by the rarefaction and condensation of air. Phil. Mag. [3] **26,** 369 (1845); Proc. R. S. London **5,** 517 (1844); Ann. Chim. **35,** 118 (1852); Papers, 171. German translation by J. W. Spengel: *Das mechanische Warmeäquivalent. Gesammelte Abhandlungen,* Braunschweig, F. Vieweg, 1872. (II, §1)

Some remarks on heat, and the constitution of elastic fluids. Mem. Manchester Lit. Phil. Soc. **9,** 107 (1851) (read 1848); Phil. Mag. [4] **14,** 211 (1857) (reprint with corrections); Ann. Chim. [3] **50,** 381 (1857); Arch. Sci. Phys. (Geneve) **36,** 349 (1857); Papers, 290; Amer. J. Phys. **17,** 63 (1949). (TI)

(with W. Thomson) On the thermal effects of fluids in motion. Phil. Trans. **144,** 321 (1854), **152,** 579 (1862); Kelvin's *Mathematical and Physical Papers,* Cambridge University Press, Vol. I, p. 333 (1882). *Joint Scientific Papers of James Prescott Joule,* London, The Physical Society, 1887, p. 247, 342. (II, §1)

Kamerlingh Onnes, H.

Théorie Générale de l'état fluide. Arch. Neerl. **30,** 101 (1897). (II, §60)

Kelvin, W. Thomson (Lord)

Mathematical and Physical Papers. Cambridge University Press, 1882–1911. Second edition. (Cited as "Papers.")

On a universal tendency in nature to the dissipation of mechanical energy. Phil. Mag. [4] **4,** 304 (1852); Proc. R. S. Edinburgh **3,** 139 (1857); Papers **1,** 511. (II, §89)

(with P. G. Tait) *Treatise on Natural Philosophy,* Vol. I. Clarendon Press, Oxford, 1867. Second edition (1 vol. in 2 parts), Cambridge University Press, 1879, 1883. Reprinted by Dover, New York, 1962, as *Principles of Mechanics and Dynamics.* German translation by H. Helmholtz and G. Wertheim: *Handbuch der theoretischen Physik,* Braunschweig, F. Vieweg, 1871, 1874. (II, §29)

The size of atoms. Nature **1,** 551 (1870); Amer. J. Sci. **50,** 38 (1870); Proc. Manchester Lit. Phil. Soc. **9,** 136 (1870); Les Mondes **22,** 701 (1870); **27,** 616 (1872); Ann. Chem. Pharm. **157,** (1871); Rev. Sci. [2] **2,** 896 (1872); *Treatise on Natural*

Philosophy, second edition, Part II, p. 495; Papers, second edition, **5**, 289 (1911). (I, §12)
On some test cases for the Maxwell-Boltzmann doctrine regarding distribution of energy. Proc. R. S. London **50**, 79 (1891); Nature **44**, 355 (1891); Papers **4**, 484.

Kennard, E. H.
Kinetic theory of gases with an introduction to statistical mechanics. New York, McGraw-Hill, 1938.

Kirchhoff, G. R.
Vorlesungen über mathematische Physik. Leipzig, B. G. Teubner.
I. Mechanik. (1874). (TI; I, §1)
IV. Vorlesungen über die Theorie der Wärme (1894). (I, F, §3, 14, 17, 20, 23, 24)

Knudsen, M.
Kinetic Theory of Gases. London, Methuen, 1934.

Koenig, F. O.
On the history of science and of the second law of thermodynamics. *Men and Moments in the History of Science* (ed. H. M. Evans), Seattle, University of Washington Press, 1959, p. 57. (II, §89)

Krönig, A.
Grundzüge einer Theorie der Gase. Ann. Physik [2] **99**, 315 (1856); L'Institut **24**, 408 (1856); Chem. Cent. 725 (1856); Arch. Sci. Phys. **33**, 137 (1856); Ann. Chim. [3] **50**, 491 (1857); Nuovo Cimento **6**, 435 (1857).

Kuenen, J. P.
On the influence of gravitation on the critical phenomena of simple substances and of mixtures. Comm. Phys. Lab., Leiden, No. 17 (1895). (II, §61)

Kundt, A., and Warburg, E.
Ueber Reibung and Wärmeleitung verdünnten Gase. Ann. Physik [2] **155**, 337, 525, **156**, 177 (1875); Phil. Mag. [4] **50**, 53 (1875). Monatsber. Akad. Berlin 160 (1875). (I, §1, 12, 13)
Ueber die specifische Wärme des Quecksilbergases. Ber. Chem. Ges. Berlin **8**, 945 (1875); Ann. Physik [2] **157**, 353 (1876). (I, §8, II, §45)

Kutta, M. W.
Zur Theorie des Stefan'schen Calorimeters. Ann. Physik [3] **54**, 104 (1895). (I, §13)

Laplace, P. S. (Marquis de)
Traité de mécanique célèste, Paris, J. B. M. Duprat, 1798–1825, 5 vols. Supplément au Xe Livre: Sur l'action capillaire. Livre XII: De l'attraction et de la répulsion des sphères et des lois

l'équilibre et du mouvement des fluids élastiques. Reprinted in *Oeuvres Complètes*, Paris, Gauthier-Villars, 1878–1912, Vols. 1–5. (TI; II, §23)

Larmor, J.

The kinetic theory of gases. Nature **51**, 152 (1894). (I, §6)

Lemoine, G.

Équilibres chimiques entre l'hydrogène et l'iode gazeux. Ann. Chim. Phys. [5] **12**, 145 (1877). (II, §69)

Lenin, V. I.

Материализм и эмпириокритицизм; критические заметки об одной реакционной философии. Moscow, Izdanie "Zveno," 1909; second edition, Gos. Izd.-vo, 1920. English translation by A. Fineberg: (based on second edition). *Materialism and Empirio-Criticism: Critical Comments on a Reactionary Philosophy*. Moscow, Cooperative Pub. Soc. of Foreign Workers in the USSR, 1937; reprinted by Foreign Languages Pub. House, Moscow, 1947. (Also many other editions.) (TI; II, F)

Liouville, J.

Note sur la théorie de la variation des constantes arbitraires. J. de Math. **3**, 342 (1838). (II, §29)

Loeb, L.

Kinetic Theory of Gases. New York, McGraw-Hill, 1922; second edition, 1934. (I, §1, 14)

Lorentz, H. A.

Collected Papers. The Hague, M. Nijhoff, 1934–1939. (Cited as "Papers")

Abhandlungen über theoretische Physik. Leipzig, B. G. Teubner, 1907. (Cited as "Abh.")

Über die Anwendung des Satzes vom Virial in der kinetischen Theorie der Gase. Ann. Physik [3] **12**, 127, 660 (1881); Papers **6**, 40; Abh. 114. (II, §56, 58)

Über das Gleichgewicht der lebendigen Kraft unter Gasmolekülen. Wien. Ber. **95**, 115 (1887); Papers **6**, 74; Abh. 124. (I, §5)

Ludwig Boltzmann (Gedächtnisrede, gehalten in der Sitzung der Deutschen Physikalischen Gesellschaft, 17 May 1907). Verhandl. Deutsch. Phys. Ges. **9**, 206 (1907).

Loschmidt, J.

Zur Grösse der Luftmolecule. Wien. Ber. **52**, 395 (1865); Wien. Anz. **2**, 162 (1865); Z. Math. Phys. **10**, 511 (1865). (I, §12; II, §66)

Experimental-Untersuchungen über die Diffusion von Gasen ohne pörose Scheidewände. Wien. Ber. **61**, 367, **62**, 468 (1870). (I, §13, 24)

Über den Zustand des Wärmegleichgewichtes eines Systems von Körpern mit Rucksicht auf die Schwerkraft. Wien. Ber. **73,** 128, 366 (1876); **75,** 287, **76,** 209 (1877). (TI; I, §6; II, §91)

Mach, E.

Die Geschichte und die Wurzel des Satzes von der Erhaltung der Arbeit. (Vortrag gehalten in der k. böhm. Ges. der Wiss. am 15 Nov. 1871) Prague, Calve, 1872. English translation, by P. E. B. Jourdain, *History and Root of the Principle of Conservation of Energy,* Chicago, Open Court, 1911. (TI; II, F)

On the principle of the conservation of energy. ("In part a re-elaboration" of the above treatise.) The Monist **5,** 167 (1894); reprinted in *Popular Scientific Lectures,* Chicago, Open Court, 1894; fifth edition, 1943; *Popular-wissenschaftliche Vorlesungen,* Leipzig, J. A. Barth, 1897; fourth edition 1910. (II, F)

Die ökonomische Natur der physikalischen Forschung (Vortrag). Wien. Almanach 293 (1882); also published as a separate by Gerold's Sohn, Wien. The Open Court **8,** 4263, 4271 (1894); *Popular Scientific Lectures,* fifth edition, p. 186. (II, F, §72)

Die Prinzipien der Wärmelehre, historisch-kritisch entwickelt. Leipzig, J. A. Barth, 1896; second edition, 1900. (II, F, §1)

Majumdar, R.

Equation of state. Bull. Calcutta Math. Soc. **21,** 107 (1929). (II, §54)

Maxwell, J. C.

The Scientific Papers of James Clerk Maxwell. Cambridge University Press, 1890; reprinted by Hermann, Paris, 1927, and by Dover, New York, 1952. (Cited as "Papers.")

Illustrations of the dynamical theory of gases. I. On the motions and collisions of perfectly elastic spheres. II. On the process of diffusion of two or more kinds of moving particles among one another. III. On the collision of perfectly elastic bodies of any form. Phil. Mag. [4] **19,** 19, **20,** 21, 33 (1860); Brit. Assoc. Rept. **29** (2) 9 (1859); Athenaeum, p. 468 (Oct. 8, 1859); L'Institut 364 (1859); Papers **1,** 377. (TI; I, §9, 14)

Viscosity or internal friction of air and other gases. Phil. Trans. **156,** 249 (1866); Proc. R. S. London **15,** 14 (1867); Papers **2,** 1. (I, §12)

On the dynamical theory of gases. Phil. Trans. **157,** 49 (1867); Proc. R. S. London **15,** 146 (1867) (read May 1, 1866); Phil. Mag. [4] **32,** 390 (1866); **35,** 129, 185 (1868); Papers **2,** 26. (TI; I, §12, 17, 20, 21, 22, 23, 24; II, §92, 93)

A discourse on molecules. Nature **8,** 437 (1873); Phil. Mag. [4] **46,** 453 (1873); Les Mondes **32,** 311, 409 (1873); Pharmaceut. J. **4,** 404, 492, 511 (1874); Papers **2,** 361. (I, §12)

On the dynamical evidence of the molecular constitution of bodies (lecture). Nature **11,** 357, 374 (1875); J. Chem. Soc. **13,** 493 (1875); Gazz. Chim. Ital. **5,** 190 (1875); Papers **2,** 418. (II, §16)

On stresses in rarified gases arising from inequalities of temperature. Phil. Trans. **170,** 231 (1880); Proc. R. S. London **27,** 304 (1878); Papers **2,** 681. (I, §1, 14, 22, 23)

On Boltzmann's theorem on the average distribution of energy in a system of material points. Trans. Cambridge Phil. Soc. **12,** 547 (1879); Phil. Mag. [5] **14,** 299 (1882); Ann. Physik Beibl. **5,** 403 (1881); Papers **2,** 713.

Mayer, J. R.

Bemerkungen über die Kräfte der unbelebten Natur. Ann. Chem. Pharm. **42,** 233 (1842); Phil. Mag. [4] **24,** 371 (1862); Isis **13,** 27, 35 (1929); *Die Mechanik der Wärme in gesammelten Schriften,* Stuttgart, J. G. Cotta, 1867; third edition, 1893. (TI; II, F)

Meier, F., and Crafts, J. M.

Sur la densité de vapeur de l'iode. Bull. Soc. Chim. Paris **33,** 501, 550; **34,** 2 (1880); Arch. Sci. Phys. (Geneve) **4,** 132 (1880); Amer. Chem. J. **2,** 108 (1880); C. R. Paris **92,** 39 (1881); Ber. deutsch. chem. Ges. **13,** 851 (1880). (II, §66)

Mendoza, E.

A sketch for a history of the kinetic theory of gases. Physics Today, **14,** 36 (March 1961). (TI)

The surprising history of the kinetic theory of gases. Memoirs Manchester Lit. Phil. Soc. (in press). (TI)

Meyer, L.

Ueber die Molecularvolumina chemische Verbindungen. Ann. Chem. Pharm. **5** (Suppl.) 129 (1867). (I, §12)

Meyer, O. E.

Ueber die innere Reibung der Gase. Ann. Physik [2] **148,** 203 (1873). (I, §12)

Die kinetische Theorie der Gase, in elementarer Darstellung mit mathematischen Zusatzen. Breslau, Maruschke und Berendt, 1877; second edition, 1899. English translation by R. E. Baynes: *The Kinetic Theory of Gases,* London, Longmans, Green, 1899. (I, F, §12, 13, 14)

Nagaoka, H.

On two constants A_1 and A_2 in the kinetic theory of gases. Nature **69,** 79 (1903). (I, §21, 24)

Natanson, L.

O znaczeniu kinetycznem funkcyi dysypacyjnej [On the kinetic interpretation of the dissipation function]. Rozprawy Spraw. Pos. Wyd. Mat.-Prz. Akad. Um. Kraków **9,** 171 (1895); C. R. Paris **117,** 539 (1893); Bull. Acad. Sci. Cracovie 348 (1893);

Phil. Mag. [5] **39,** 455 (1895); Z. phys. Chem. **13,** 437 (1894); Proc. Phys. Soc. Abstr. **1,** 6 (1895). (I, §24)

O energii kinetycznej ruchu ciepła i o funckcyi dysypacyjnej odpowiedniej [On the kinetic energy of the motion of heat and the corresponding dissipation function]. Rozprawy Spraw. Pos. Wyd. Mat.-Prz. Akad. Um. Kraków **7,** 273 (1895); Bull. Acad. Sci. Cracovie **2,** 295 (1894); Phil. Mag. [5] **39,** 501 (1895); Z. Phys. Chem. **16,** 289 (1895); Proc. Phys. Soc. Abstr. **1,** 267 (1895). (I, §24)

Nature (anonymous correspondent)

Discussion on the kinetic theory of gases, British Association meeting. Nature **32,** 533 (1885). [Remarks of Crum Brown, Liveing, W. Thomson, J. J. Thomson, Hicks, Reynolds, Dixon.]

Naumann, A. (II, §72)

Grundriss der Thermochemie, oder Der Lehre von den Beziehung zwischen Wärme und chemischen Erscheinungen vom Standpunkt der mechanischen Wärmetheorie dargestellt. Braunschweig, F. Vieweg, 1869. (II, §66)

Ueber Dichte und Zersetzung des Dampfs der Untersalpetersäure unterhalb ihres Siedepunkts bei verschiedenen Druck. Ber. deutsch. chem. Ges. **11,** 2045 (1878); Jahresb. Fortschr. Chem. 120 (1878). (II, §66)

Newton, I.

Philosophiae Naturalis Principia Mathematica, London, 1687. Prop. XIX, Theorem XIV, Section V, Book II. For an English translation, see for example Florian Cajori's revision of Andrew Motte's translation of 1729: Sir Isaac Newton's *Mathematical Principles of Natural Philosophy and his System of the World,* Berkeley and Los Angeles: University of California Press, 1946, p. 290–292. (TI)

Ostwald, W.

Abhandlungen und Vorträge allgemeinen Inhaltes (1887–1903). Leipzig, Veit, 1904 (cited as "Abh.").

Fortschritte der physikalischen Chemie in den letzten Jahren (Vortrag, gehalten in der vereinigten Abteilung für Physik und Chemie der 64. Versammlung Deutscher Naturforscher und Aerzte zu Halle am 24 Sept. 1891) Abh. 34. (II, F)

Die Ueberwindung des wissenschaftlichen Materialismus. Verhandl. Ges. Deutscher Naturforscher und Aerzte (Leipzig), Th. 1, p. 155 (1895); Abh. 220. Science Progress **4,** 419 (1896). (I, §1; II, §72)

Studien zur Energetik. Berichte Verh. k. Sächsischen Ges. Wiss. (Leipzig) Math.-Phys. Classe, **43,** 271 (1891), **44,** 211 (1892). (I, §1)

Zur Energetik. Ann. Physik [3] **58,** 154 (1896). (I, §1)

Individuality and Immortality. (Ingersoll Lecture, Harvard) Boston, Houghton, Mifflin, 1906. (I, §1)

Energetische Grundlagen der Kulturwissenschaft. Leipzig, Klinkhardt, 1909. French translation by E. Philippi: *Les Fondements énergétiques de la science de la civilisation,* Paris, V. Giard et E. Brière, 1910. (I, §1)

Grundriss der allgemeinen Chemie. Leipzig, Engelmann, fourth edition, 1909, Vorbericht. English translation by W. W. Taylor: *Outlines of General Chemistry,* London, Macmillan, 1912. (TI)

Die Forderung des Tages. Leipzig, Akademische Verlagsgesellschaft m. b. H., 1910. (I, §1)

Monistische Sonntagspredigten. Leipzig, Akademische Verlagsgesellschaft, m. b. H., 1911. (I, §1)

Die energetische Imperativ. Leipzig, Akademische Verlagsgesellschaft m. b. H., 1912. (I, §1)

Die Energie. Leipzig, J. A. Barth, 1912. (I, §1)

Die Philosophie der Werte. Leipzig, A. Kröner, 1913. (I, §1)

Partington, J. R.

An Advanced Treatise on Physical Chemistry. Vol. I. Fundamental Principles. The Properties of Gases. London, Longmans, Green, 1949. (II, §54)

Pearson, K.

The Grammar of Science. London, Block, 1892; second edition, 1900; third edition, 1911. (II, F)

Pelabon, H.

Sur la dissociation de l'acide selenhydride (Thèses presentées à la Faculté des Sciences de Bordeaux, pour obtenir le grade de docteur ès sciences physiques). Paris, Hermann, 1898. Mem. Soc. Sci. Bordeaux **3,** 141 (1899). (II, §72)

Plancherel, M.

Beweis der Unmoglichkeit ergodischer mechanischer Systeme. Ann. Physik [4] **42,** 1061 (1913). Translation UCRL Trans. 874 (L) (Lawrence Radiation Laboratory, Livermore, California). (II, §32)

Planck, M.

Physikalische Abhandlungen und Vorträge. Braunschweig, F. Vieweg, 1958 (cited as "Abh.").

Ueber das Princip der Vermehrung der Entropie. Ann. Physik [3] **30,** 562, **31,** 189, **32,** 462 (1887); **44,** 385 (1891); Abh. **1,** 196, 217, 232, 382. (II, §72)

Allgemeines zur neueren Entwickelung der Wärmetheorie. (Vortrag, gehalten in der vereinigten Sitzung der Abt. für Physik und Chemie der 64. Versammlung der Gesellschaft deutscher

Naturforscher und Aerzte in Halle am 24 Sept. 1891). Z. phys. Chem. **8**, 647 (1891); Abh. **1**, 372. (I, §1)

Ueber den Beweis des Maxwell'schen Geschwindigkeitsvertheilungsgesetzes unter Gasmolekülen. Mun. Ber. **24**, 391 (1895); Ann. Physik [3] **55**, 220 (1895); Abh. **1**, 442. (I, §6)

Gegen die neuere Energetik. Ann. Physik [3] **57**, 72 (1896); Proc. Phys. Soc. Abstr. **2**, 68 (1896); Abh. **1**, 459. (I, §1)

Zur Theorie des Gesetzes der Energieverteilung im Normalspektrum. Verh. Deutsch. Phys. Ges. **2**, 237 (1900); Abh. **1**, 698. (TI)

Über das Gesetz der Energieverteilung im Normalspektrum. Ann. Physik [4] **4**, 553 (1901); Abh. **1**, 717. (TI)

Über die Elementarquanta der Materie und der Elektrizität. Ann. Physik [4] **4**, 564 (1901); Abh. **1**, 728. (TI)

Poincaré, H.

Oeuvres. Paris, Gauthier-Villars, 1951–54.

Sur le problème des trois corps et les équations de la dynamique. Acta Math. **13**, 1 (1890); Oeuvres **7**, 262. (TI; II, §88)

Thermodynamique. Paris, Gauthier-Villars, 1892. German translation by J. Blondin: *Thermodynamik,* Berlin, J. Springer, 1893. (I, F; II, F)

Poincaré's "Thermodynamics." Nature **45**, 414, 485 (1892). (I, F; II, F)

Sur une objection à la théorie cinétique des gaz. C. R. Paris **116**, 1017 (1893); Oeuvres **10**, 240. (I, §20, 23)

Poisson, S. D.

Sur la chaleur des gaz et des vapeurs. Ann. Chim. **23**, 337 (1823): Ann. Physik **76**, 269 (1824); English translation with notes by J. Herapath, Phil. Mag. **62**, 328 (1823). (TI; I, §20)

Mémoire sur la théorie du son. Nouv. Bull. Sci. Soc. Philomat. Paris **1**, 19 (1807); J. École Polyt. (Paris) **7**, 14ᵉ Cah., 319 (1808). (TI)

Nouvelle Théorie de l'action capillaire. Paris, Bachelier, 1831. (TI; II, §23)

Théorie mathematique de la chaleur. Paris, Bachelier, 1835. (TI)

Ramsay, W.

Sur l'argon et l'helium. C. R. Paris **120**, 1049 (1895). (II, §45)

Helium, a gaseous constituent of certain minerals. Proc. R. S. London **58**, 81 (1895). (I, §8)

(with J. N. Collie) The homogeneity of helium and argon. Proc. R. S. London **60**, 206 (1896). (II, §45)

The Gases of the atmosphere; the History of their discovery. London, Macmillan and Co., Ltd., 1896; second edition, 1902; third edi-

tion, 1905; fourth edition, 1915. French translation: *Les gaz de l'atmosphère*, Paris, Carré, 1898. (II, F, §84)
(with M. W. Travers) On a new constituent of atmospheric air. Proc. R. S. London **63,** 405 (1898). (II, §45)
(with M. W. Travers) On the companions of argon. Proc. R S London **63,** 437 (1898). (II, F)
(with M. W. Travers) The homogeneity of helium. Proc. R. S. London **62,** 316 (1898). (II, F)
The newly discovered elements; and their relation to the kinetic theory of gases. Mem. Proc. Manchester Lit. Phil. Soc. **43,** No. 4 (1900) (read March 28, 1899). (I, §8; II, F)

Ramsay, W.: see also Rayleigh

Rausenberger, O.
Lehrbuch der analytischen Mechanik. Leipzig, B. G. Teubner, 1888; second edition, 1893. (II, §29)

Rayleigh, J. W. Strutt (Lord)
Scientific Papers. Cambridge University Press, 1899–1920. (Cited as "Papers.")
Some general theorems relating to vibrations. Proc. London Math. Soc. **4,** 357 (1873); Papers **1,** 170. (I, §24)
On the flow of viscous liquids, especially in two dimensions. Phil. Mag. [5] **36,** 354 (1893). (I, §24)
(with W. Ramsay) Argon, a new constituent of the atmosphere. Phil. Trans. **186,** 187 (1896); *Smithsonian Contributions to Knowledge,* Washington, 1896, Vol. 29, Art. 4; Papers **4,** 130. (I, §8; II, §45)
On the viscosity of argon as affected by temperature. Proc. R. S. London **66,** 68 (1900); Papers **4,** 452. (I, §14)
Address to the Royal Society [including obituary of Boltzmann]. Proc. R. S. London **79A,** 1 (1907).

Reichenbach, H.
The Direction of Time. Berkeley and Los Angeles, University of California Press, 1956. (TI; II, §90)

Rey, A.
La Théorie de la Physique chez les Physiciens contemporains. Paris, F. Alcan, 197; second edition, 1927. (I, §1)

Reynolds, O.
On the forces caused by evaporation from, and condensation at, a surface. Proc. R. S. London **22,** 401 (1874); Chem. News **30,** 11 (1874). (I, §1)
On certain dimensional properties of matter in the gaseous state. I. Experimental researches on thermal transpiration of gases through porous plates, and on the laws of transpiration and impulsion, including an experimental proof that a gas is not a

continuous plenum. II. On an extension of the dynamical theory of gas which includes the stresses, tangential and normal, caused by a varying condition of the gas, and affords an explanation of the phenomena of transpiration and impulsion. Proc. R. S. London **28,** 304 (1879); Phil. Trans. **170,** 727 (1880); Nature **19,** 435 (1879). (I, §14)

Richarz, F.
Ueber das Gesetz von Dulong und Petit. Ann. Physik [3] **48,** 708 (1893). (II, §42)

Rosenblueth, M. N., and Rosenblueth, A. W.
Further results on Monte Carlo equations of state. J. Chem. Phys. **22,** 881 (1954). (II, §54)

Rosenthal, A.
Beweis der Unmöglichkeit ergodischer Gassysteme. Ann. Physik [4] **42,** 796 (1913). Translation UCRL Trans. 873(L) (Lawrence Radiation Laboratory, Livermore, California). (II, §32)

Royal Society of London
Catalogue of Scientific Papers. (1) 1800–1863; (2) 1864–1873; (3) 1874–1883; (4) 1884–1900. London, Eyre and Spottiswoode, Clay and Sons, Cambridge University Press, 1867–1925. (II, §1)

Sageret, J.
La vague mystique. Henri Poincaré—Energetisme (W. Ostwald)—Néo-thomisme (P. Duhem)—Bergsonisme—Pragmatisme—Émile Boutroux. Paris, E. Flammarion, 1920. (I, §1)

Saint-Claire Deville, E. H., and Troost, L.
Sur le coefficient de dilatation et la densité de vapeur de l'acide hypoazotique. C. R. Paris **64,** 237 (1867); Jahresb. Fortschr. Chem. 177 (1867). (II, §66)

Sarrau, E.
Sur la compressibilité des fluides. C. R. Paris **101,** 941 (1885). (II, §54)

Schlömilch, O. X.
Compendium der höheren Analysis. Braunschweig, F. Vieweg, 1874. (I, §6)

Schuster, A.
On the nature of the force producing the motion of a body exposed to rays of heat and light. Proc. R. S. London **24,** 391 (1876); Phil. Trans. **166,** 715 (1877); Phil. Mag. [5] **2,** 313 (1876); Nature **13,** 458 (1876). (I, §1)

Smoluchowski, M. v.
Pisma Marjana Smoluchowskiego. Drukarnia Uniwersytetu Jagiellonskiego, Krakowie, 1924 (cited as "Pisma")
Ueber den Temperatursprung bei Wärmeleitung in Gasen. Wien.

Ber. **107,** 304 (1898), **108,** 5 (1899); Pisma **1,** 113, 199. (II, F)

O sredniej swobodnej drodze cząsteczek gazu i o jej związku teorją dyfuzji. [On the mean path traversed by the molecules in a gas, and its relation to the theory of diffusion.] Rozprawy Spraw. Pos. Wyd. Mat.-Prz. Akad. Um. Kraków **A46,** 129 (1906); Bull. Acad. Sci. Cracovie 202 (1906); Pisma **1,** 468, 479. (I, §14)

Sommerfeld, A.

Das Werk Boltzmanns. Wiener Chem.-Zeitung **47,** 25 (1944). (TI)

Staigmüller, H.

Beiträge zur kinetischen Theorie mehratomiger Gase. Ann. Physik [3] **65,** 655 (1898); Phys. Abstr. **1,** 696 (1898). (II, §45)

Versuch einer theoretischen Ableitung der Constanten des Gesetzes von Dulong und Petit. Ann. Physik [3] **65,** 670 (1898). (II, §42)

Stallo, J. B.

The Concepts and Theories of Modern Physics. New York, Appleton, 1882. German translation by H. Kleinpeter, with a preface by E. Mach: *Die Begriffen und Theorien der modernen Physik,* Leipzig, J. A. Barth, 1901. (For further information see S. Drake, "J. B. Stallo and the critique of classical physics," in *Men and Moments in the History of Science,* (ed. H. M. Evans) Seattle, University of Washington Press, 1959, p. 22). (II, F)

Stankevitsch, B. W.

Zur dynamischen Gastheorie. Ann. Physik [3] **29,** 153 (1886). (I, §4)

Stefan, J.

Untersuchungen über die Wärmeleitung in Gasen. Wien. Ber. **65,** 45 (1872); **72,** 69 (1876); Wien. Anz. **9,** 42 (1872), **12,** 131 (1875); Carl's Repertorium **8,** 138 (1872); J. Chem. Soc. **25,** 591 (1872), **30** (2) 37 (1876); Chem. Cent. **3,** (177 (1872), **6,** 529 (1875). (I, §13)

Über die dynamische Theorie der Diffusion der Gase. Wien. Ber. **65,** 323 (1872). (I, §12, 14)

Ueber die Beziehung zwischen den Theorien der Capillarität und der Verdampfung. Wien. Ber. **94,** 4 (1886); Ann. Physik [3] **29,** 655 (1886). (II, §24)

Stoney, G. J.

The internal motions of gases compared with the motions of waves of light. Phil. Mag. [4] **36,** 132 (1868). (I, §12)

On Crookes's radiometer. Phil. Mag. [5] **1,** 177, 305 (1876); Nature **13,** 420 (1876). (I, §14)

On the mechanical theory of Crookes's (or polarization) stress in gases. Sci. Trans. Roy. Sci. Dublin **1,** 39 (1877); Phil. Mag. [5] **6,** 401 (1878). (I, §14)

Curious consequences of a well-known dynamical theorem. Sci. Proc. Roy. Dublin Soc. **5,** 448 (1887); Phil. Mag. [5] **23,** 544 (1887). (I, §6)

Sutherland, W.

Thermal transpiration and radiometer motion. Phil. Mag. [5] **42,** 373, 476 (1896), **44,** 52 (1897). (I, §14)

Tait, P. G.

Scientific Papers. Cambridge University Press, 1898–1900. (Cited as "Papers").

On the foundations of the kinetic theory of gases. Trans. R. S. Edinburgh **33,** 65 (1886); Phil. Mag. [5] **21,** 343 (1886); Papers **2,** 124. (I, §10, 14; II, §85)

On the foundations of the kinetic theory of gases. II. Trans. R. S. Edinburgh **33,** 251 (1887); Phil. Mag. [5] **23,** 141, 433 (1887); Papers **2,** 153. (I, §11, 14)

On some questions in the kinetic theory of gases. Reply to Prof. Boltzmann. Phil. Mag. [5] **25,** 172 (1888). (I, §14)

On the mean free path, and the average number of collisions per particle per second in a group of equal spheres. Proc. R. S. Edinburgh **15,** 225 (1889). (I, §14)

On the foundations of the kinetic theory of gases. III. Trans. R. S. Edinburgh **35,** 1029 (1890) (read 1888); Papers **2,** 179. (I, §14)

Thomson, W.: see Kelvin

Tolman, R. C.

The principles of statistical mechanics. Oxford University Press, 1938. (I, §4; II, §79)

Troost, L. J.

Sur les densités de vapeur. C. R. Paris **86,** 331, 1395 (1878); J. de Pharm. **28,** 293 (1878). (II, §66)

Uhlenbeck, G. E., and Ford, G. W.

The theory of linear graphs with applications to the theory of the virial development of the properties of gases. *Studies in Statistical Mechanics,* Vol. I, Part B (ed. J. de Boer and G. E. Uhlenbeck), Amsterdam, North-Holland, 1962. (II, §54)

van der Waals, J. D.

Over de continuiteit van den gas- en vloeistoftoestand. Dissertation, Leiden, 1873. German translation: *Ueber den Uebergangs-Zustand zwischen Gas und Flussigkeit,* Leiden, A. W. Sijthoff, 1873. Second edition, with additions, translated by F. Roth: *Die Continuität des Gasformigen und Flussigen Zustandes,* Leipzig, J. A. Barth, 1881. English translation by R. Threlfall and J. F. Adair, in *Physical Memoirs*, Vol. I, Part 3, London, Taylor and Francis (for the Physical Society), 1890. (TI; I, §12; II, §1)

Over het betrekkelijk aantal botsingen, dat een molekuul ondergaat, wanneer het zich beweegt door bewegende molekulen of door molekulen, die men onderstelt stil te staan; alsmede over den invloed van de admetingen der molekulen volgens de richting der relative beweging op het aantal dier botsingen. [On the relative number of collisions experienced by a molecule as it moves through a medium of molecules in motion or a medium of molecules assumed at rest; and on the influence which the dimensions of the molecules in the direction of relative motion exert on the number of collisions.] Verslagen Akad. Wet. Amsterdam [2] **10,** 321 (1876); Arch. Neerl. **12,** 201, 217 (1877). (II, §1)

De betrekking tusschen spanning, volumen en temperatur bij dissociatie. [On the pressure-volume-temperature relation in the case of dissociation.] Verslagen Akad. Wet. Amsterdam [2] **15,** 199 (1880). (II, §72)

Eene bijdrage tot de kennis der toestandsvergelijking. [A contribution to the knowledge of the equation of state.] Verslagen Akad. Wet. Amsterdam [4] **5,** 150 (1896).

De verandering van de grootheid *b* der toestandsvergelijking als quasi-verkleining van het molekul. [The variation of the quantity *b* in the equation of state, considered as an apparent diminution of the molecule.] *Festschrift Ludwig Boltzmann gewidmet zum sechzigsten Geburtstage 20. Februar 1904*, Leipzig, J. A. Barth, 1904, p. 305. (II, §9)

Eenige merkwaardige betrekkingen, hetzij exacte of approximatieve, bij verschillende stoffen [Some remarkable relations, either exact or approximate, for different substances.] Verslagen Akad. Wet. Amsterdam [4] **21,** 800 (1912); Proc. Sect. Sci. Amsterdam **15,** 903 (1913). (II, §9)

Wald, F.

Die Energie und ihre Entwerthung. Studien über den zweiten Hauptsatz der mechanischen Wärmetheorie. Leipzig, Engelmann, 1889. (II, F)

Ward, J.

Naturalism and Agnosticism. New York, Macmillan, 1899. (II, F)

Waterston, J. J.

On the physics of media that are composed of free and perfectly elastic molecules in a state of motion. Phil. Trans. **183A,** 5 (1893) (read March 5, 1846); Proc. R. S. London **5,** 604 (1846). *The Collected Scientific Papers of John James Waterson* (ed.

with biography by J. S. Haldane), Edinburgh, Oliver and Boyd, 1928, p. 207. (TI)

Watson, H. W.

Boltzmann's minimum theorem. Nature **51,** 105 (1894). (I, §6)

Wind, C. H.

Über den dem Liouville'schen Satze entsprechenden Satz der Gastheorie. Wien. Ber. **106,** 21 (1897).

Woodruff, A. E.

Action at a distance in nineteenth-century electrodynamics. Isis **53,** 439 (1962). (I, §1)

Wüllner, A.

Ueber die Abhängigkeit der specifischen Wärme der Gase bei constantem Volumen von der Temperatur und die Wärmeleitungsfähigkeit der Gase. Ann. Physik [3] **4,** 321 (1878). (II, §45)

Wroblewski, Z. F.

Sur la densité de l'air atmosphérique et de ses composants, et sur le volume atomique de l'oxygène et de l'azote. C. R. Paris **102,** 1010 (1886). (I, §12)

Zermelo, E.

Ueber einen Satze der Dynamik und die mechanische Wärmetheorie. Ann. Physik [3] **57,** 485 (1896). (TI; II, §88, 91)

Ueber mechanische Erklarungen irreversibler Vorgange. Eine Antwort auf Hrn. Boltzmann's "Entgegnung." Ann. Physik [3] **59,** 793 (1896). (TI; II, §91)

INDEX

(Bibliographic references are not included except where they have not been cited in text, Translator's Introduction, or notes.)

A CATALOG OF SELECTED

DOVER BOOKS

IN SCIENCE AND MATHEMATICS

Astronomy

BURNHAM'S CELESTIAL HANDBOOK, Robert Burnham, Jr. Thorough guide to the stars beyond our solar system. Exhaustive treatment. Alphabetical by constellation: Andromeda to Cetus in Vol. 1; Chamaeleon to Orion in Vol. 2; and Pavo to Vulpecula in Vol. 3. Hundreds of illustrations. Index in Vol. 3. 2,000pp. 6⅛ x 9¼.
Vol. I: 0-486-23567-X
Vol. II: 0-486-23568-8
Vol. III: 0-486-23673-0

EXPLORING THE MOON THROUGH BINOCULARS AND SMALL TELESCOPES, Ernest H. Cherrington, Jr. Informative, profusely illustrated guide to locating and identifying craters, rills, seas, mountains, other lunar features. Newly revised and updated with special section of new photos. Over 100 photos and diagrams. 240pp. 8¼ x 11. 0-486-24491-1

THE EXTRATERRESTRIAL LIFE DEBATE, 1750–1900, Michael J. Crowe. First detailed, scholarly study in English of the many ideas that developed from 1750 to 1900 regarding the existence of intelligent extraterrestrial life. Examines ideas of Kant, Herschel, Voltaire, Percival Lowell, many other scientists and thinkers. 16 illustrations. 704pp. 5⅜ x 8½. 0-486-40675-X

THEORIES OF THE WORLD FROM ANTIQUITY TO THE COPERNICAN REVOLUTION, Michael J. Crowe. Newly revised edition of an accessible, enlightening book re-creates the change from an earth-centered to a sun-centered conception of the solar system. 242pp. 5⅜ x 8½. 0-486-41444-2

ARISTARCHUS OF SAMOS: The Ancient Copernicus, Sir Thomas Heath. Heath's history of astronomy ranges from Homer and Hesiod to Aristarchus and includes quotes from numerous thinkers, compilers, and scholasticists from Thales and Anaximander through Pythagoras, Plato, Aristotle, and Heraclides. 34 figures. 448pp. 5⅜ x 8½.
0-486-43886-4

A COMPLETE MANUAL OF AMATEUR ASTRONOMY: TOOLS AND TECHNIQUES FOR ASTRONOMICAL OBSERVATIONS, P. Clay Sherrod with Thomas L. Koed. Concise, highly readable book discusses: selecting, setting up and maintaining a telescope; amateur studies of the sun; lunar topography and occultations; observations of Mars, Jupiter, Saturn, the minor planets and the stars; an introduction to photoelectric photometry; more. 1981 ed. 124 figures. 25 halftones. 37 tables. 335pp. 6½ x 9¼. 0-486-42820-8

AMATEUR ASTRONOMER'S HANDBOOK, J. B. Sidgwick. Timeless, comprehensive coverage of telescopes, mirrors, lenses, mountings, telescope drives, micrometers, spectroscopes, more. 189 illustrations. 576pp. 5⅝ x 8¼. (Available in U.S. only.)
0-486-24034-7

STAR LORE: Myths, Legends, and Facts, William Tyler Olcott. Captivating retellings of the origins and histories of ancient star groups include Pegasus, Ursa Major, Pleiades, signs of the zodiac, and other constellations. "Classic."—Sky & Telescope. 58 illustrations. 544pp. 5⅜ x 8½. 0-486-43581-4

Chemistry

THE SCEPTICAL CHYMIST: THE CLASSIC 1661 TEXT, Robert Boyle. Boyle defines the term "element," asserting that all natural phenomena can be explained by the motion and organization of primary particles. 1911 ed. viii+232pp. $5^3/_8$ x $8^1/_2$.
0-486-42825-7

RADIOACTIVE SUBSTANCES, Marie Curie. Here is the celebrated scientist's doctoral thesis, the prelude to her receipt of the 1903 Nobel Prize. Curie discusses establishing atomic character of radioactivity found in compounds of uranium and thorium; extraction from pitchblende of polonium and radium; isolation of pure radium chloride; determination of atomic weight of radium; plus electric, photographic, luminous, heat, color effects of radioactivity. ii+94pp. $5^3/_8$ x $8^1/_2$. 0-486-42550-9

CHEMICAL MAGIC, Leonard A. Ford. Second Edition, Revised by E. Winston Grundmeier. Over 100 unusual stunts demonstrating cold fire, dust explosions, much more. Text explains scientific principles and stresses safety precautions. 128pp. $5^3/_8$ x $8^1/_2$. 0-486-67628-5

MOLECULAR THEORY OF CAPILLARITY, J. S. Rowlinson and B. Widom. History of surface phenomena offers critical and detailed examination and assessment of modern theories, focusing on statistical mechanics and application of results in mean-field approximation to model systems. 1989 edition. 352pp. $5^3/_8$ x $8^1/_2$. 0-486-42544-4

CHEMICAL AND CATALYTIC REACTION ENGINEERING, James J. Carberry. Designed to offer background for managing chemical reactions, this text examines behavior of chemical reactions and reactors; fluid-fluid and fluid-solid reaction systems; heterogeneous catalysis and catalytic kinetics; more. 1976 edition. 672pp. $6^1/_8$ x $9^1/_4$. 0-486-41736-0 $31.95

ELEMENTS OF CHEMISTRY, Antoine Lavoisier. Monumental classic by founder of modern chemistry in remarkable reprint of rare 1790 Kerr translation. A must for every student of chemistry or the history of science. 539pp. $5^3/_8$ x $8^1/_2$. 0-486-64624-6

MOLECULES AND RADIATION: An Introduction to Modern Molecular Spectroscopy. Second Edition, Jeffrey I. Steinfeld. This unified treatment introduces upper-level undergraduates and graduate students to the concepts and the methods of molecular spectroscopy and applications to quantum electronics, lasers, and related optical phenomena. 1985 edition. 512pp. $5^3/_8$ x $8^1/_2$. 0-486-44152-0

A SHORT HISTORY OF CHEMISTRY, J. R. Partington. Classic exposition explores origins of chemistry, alchemy, early medical chemistry, nature of atmosphere, theory of valency, laws and structure of atomic theory, much more. 428pp. $5^3/_8$ x $8^1/_2$. (Available in U.S. only.) 0-486-65977-1

GENERAL CHEMISTRY, Linus Pauling. Revised 3rd edition of classic first-year text by Nobel laureate. Atomic and molecular structure, quantum mechanics, statistical mechanics, thermodynamics correlated with descriptive chemistry. Problems. 992pp. $5^3/_8$ x $8^1/_2$.
0-486-65622-5

ELECTRON CORRELATION IN MOLECULES, S. Wilson. This text addresses one of theoretical chemistry's central problems. Topics include molecular electronic structure, independent electron models, electron correlation, the linked diagram theorem, and related topics. 1984 edition. 304pp. $5^3/_8$ x $8^1/_2$. 0-486-45879-2

Engineering

DE RE METALLICA, Georgius Agricola. The famous Hoover translation of greatest treatise on technological chemistry, engineering, geology, mining of early modern times (1556). All 289 original woodcuts. 638pp. $6^3/_4$ x 11. 0-486-60006-8

FUNDAMENTALS OF ASTRODYNAMICS, Roger Bate et al. Modern approach developed by U.S. Air Force Academy. Designed as a first course. Problems, exercises. Numerous illustrations. 455pp. $5^3/_8$ x $8^1/_2$. 0-486-60061-0

DYNAMICS OF FLUIDS IN POROUS MEDIA, Jacob Bear. For advanced students of ground water hydrology, soil mechanics and physics, drainage and irrigation engineering and more. 335 illustrations. Exercises, with answers. 784pp. $6^1/_8$ x $9^1/_4$. 0-486-65675-6

THEORY OF VISCOELASTICITY (SECOND EDITION), Richard M. Christensen. Complete consistent description of the linear theory of the viscoelastic behavior of materials. Problem-solving techniques discussed. 1982 edition. 29 figures. xiv+364pp. $6^1/_8$ x $9^1/_4$. 0-486-42880-X

MECHANICS, J. P. Den Hartog. A classic introductory text or refresher. Hundreds of applications and design problems illuminate fundamentals of trusses, loaded beams and cables, etc. 334 answered problems. 462pp. $5^3/_8$ x $8^1/_2$. 0-486-60754-2

MECHANICAL VIBRATIONS, J. P. Den Hartog. Classic textbook offers lucid explanations and illustrative models, applying theories of vibrations to a variety of practical industrial engineering problems. Numerous figures. 233 problems, solutions. Appendix. Index. Preface. 436pp. $5^3/_8$ x $8^1/_2$. 0-486-64785-4

STRENGTH OF MATERIALS, J. P. Den Hartog. Full, clear treatment of basic material (tension, torsion, bending, etc.) plus advanced material on engineering methods, applications. 350 answered problems. 323pp. $5^3/_8$ x $8^1/_2$. 0-486-60755-0

A HISTORY OF MECHANICS, René Dugas. Monumental study of mechanical principles from antiquity to quantum mechanics. Contributions of ancient Greeks, Galileo, Leonardo, Kepler, Lagrange, many others. 671pp. $5^3/_8$ x $8^1/_2$. 0-486-65632-2

STABILITY THEORY AND ITS APPLICATIONS TO STRUCTURAL MECHANICS, Clive L. Dym. Self-contained text focuses on Koiter postbuckling analyses, with mathematical notions of stability of motion. Basing minimum energy principles for static stability upon dynamic concepts of stability of motion, it develops asymptotic buckling and postbuckling analyses from potential energy considerations, with applications to columns, plates, and arches. 1974 ed. 208pp. $5^3/_8$ x $8^1/_2$. 0-486-42541-X

BASIC ELECTRICITY, U.S. Bureau of Naval Personnel. Originally a training course; best nontechnical coverage. Topics include batteries, circuits, conductors, AC and DC, inductance and capacitance, generators, motors, transformers, amplifiers, etc. Many questions with answers. 349 illustrations. 1969 edition. 448pp. $6^1/_2$ x $9^1/_4$. 0-486-20973-3

CALCULUS OF VARIATIONS WITH APPLICATIONS, George M. Ewing. Applications-oriented introduction to variational theory develops insight and promotes understanding of specialized books, research papers. Suitable for advanced undergraduate/graduate students as primary, supplementary text. 352pp. $5^3/_8$ x $8^1/_2$. 0-486-64856-7

MATHEMATICIAN'S DELIGHT, W. W. Sawyer. "Recommended with confidence" by *The Times Literary Supplement,* this lively survey was written by a renowned teacher. It starts with arithmetic and algebra, gradually proceeding to trigonometry and calculus. 1943 edition. 240pp. $5^3/_8$ x $8^1/_2$. 0-486-46240-4

ADVANCED EUCLIDEAN GEOMETRY, Roger A. Johnson. This classic text explores the geometry of the triangle and the circle, concentrating on extensions of Euclidean theory, and examining in detail many relatively recent theorems. 1929 edition. 336pp. $5^3/_8$ x $8^1/_2$. 0-486-46237-4

COUNTEREXAMPLES IN ANALYSIS, Bernard R. Gelbaum and John M. H. Olmsted. These counterexamples deal mostly with the part of analysis known as "real variables." The first half covers the real number system, and the second half encompasses higher dimensions. 1962 edition. xxiv+198pp. $5^3/_8$ x $8^1/_2$. 0-486-42875-3

CATASTROPHE THEORY FOR SCIENTISTS AND ENGINEERS, Robert Gilmore. Advanced-level treatment describes mathematics of theory grounded in the work of Poincaré, R. Thom, other mathematicians. Also important applications to problems in mathematics, physics, chemistry and engineering. 1981 edition. References. 28 tables. 397 black-and-white illustrations. xvii + 666pp. $6^1/_8$ x $9^1/_4$. 0-486-67539-4

COMPLEX VARIABLES: Second Edition, Robert B. Ash and W. P. Novinger. Suitable for advanced undergraduates and graduate students, this newly revised treatment covers Cauchy theorem and its applications, analytic functions, and the prime number theorem. Numerous problems and solutions. 2004 edition. 224pp. $6^1/_2$ x $9^1/_4$. 0-486-46250-1

NUMERICAL METHODS FOR SCIENTISTS AND ENGINEERS, Richard Hamming. Classic text stresses frequency approach in coverage of algorithms, polynomial approximation, Fourier approximation, exponential approximation, other topics. Revised and enlarged 2nd edition. 721pp. $5^3/_8$ x $8^1/_2$. 0-486-65241-6

INTRODUCTION TO NUMERICAL ANALYSIS (2nd Edition), F. B. Hildebrand. Classic, fundamental treatment covers computation, approximation, interpolation, numerical differentiation and integration, other topics. 150 new problems. 669pp. $5^3/_8$ x $8^1/_2$. 0-486-65363-3

MARKOV PROCESSES AND POTENTIAL THEORY, Robert M. Blumental and Ronald K. Getoor. This graduate-level text explores the relationship between Markov processes and potential theory in terms of excessive functions, multiplicative functionals and subprocesses, additive functionals and their potentials, and dual processes. 1968 edition. 320pp. $5^3/_8$ x $8^1/_2$. 0-486-46263-3

ABSTRACT SETS AND FINITE ORDINALS: An Introduction to the Study of Set Theory, G. B. Keene. This text unites logical and philosophical aspects of set theory in a manner intelligible to mathematicians without training in formal logic and to logicians without a mathematical background. 1961 edition. 112pp. $5^3/_8$ x $8^1/_2$. 0-486-46249-8

INTRODUCTORY REAL ANALYSIS, A.N. Kolmogorov, S. V. Fomin. Translated by Richard A. Silverman. Self-contained, evenly paced introduction to real and functional analysis. Some 350 problems. 403pp. $5\frac{3}{8} \times 8\frac{1}{2}$. 0-486-61226-0

APPLIED ANALYSIS, Cornelius Lanczos. Classic work on analysis and design of finite processes for approximating solution of analytical problems. Algebraic equations, matrices, harmonic analysis, quadrature methods, much more. 559pp. $5\frac{3}{8} \times 8\frac{1}{2}$. 0-486-65656-X

AN INTRODUCTION TO ALGEBRAIC STRUCTURES, Joseph Landin. Superb self-contained text covers "abstract algebra": sets and numbers, theory of groups, theory of rings, much more. Numerous well-chosen examples, exercises. 247pp. $5\frac{3}{8} \times 8\frac{1}{2}$. 0-486-65940-2

QUALITATIVE THEORY OF DIFFERENTIAL EQUATIONS, V. V. Nemytskii and V.V. Stepanov. Classic graduate-level text by two prominent Soviet mathematicians covers classical differential equations as well as topological dynamics and ergodic theory. Bibliographies. 523pp. $5\frac{3}{8} \times 8\frac{1}{2}$. 0-486-65954-2

THEORY OF MATRICES, Sam Perlis. Outstanding text covering rank, nonsingularity and inverses in connection with the development of canonical matrices under the relation of equivalence, and without the intervention of determinants. Includes exercises. 237pp. $5\frac{3}{8} \times 8\frac{1}{2}$. 0-486-66810-X

INTRODUCTION TO ANALYSIS, Maxwell Rosenlicht. Unusually clear, accessible coverage of set theory, real number system, metric spaces, continuous functions, Riemann integration, multiple integrals, more. Wide range of problems. Undergraduate level. Bibliography. 254pp. $5\frac{3}{8} \times 8\frac{1}{2}$. 0-486-65038-3

MODERN NONLINEAR EQUATIONS, Thomas L. Saaty. Emphasizes practical solution of problems; covers seven types of equations. ". . . a welcome contribution to the existing literature. . . ."—*Math Reviews*. 490pp. $5\frac{3}{8} \times 8\frac{1}{2}$. 0-486-64232-1

MATRICES AND LINEAR ALGEBRA, Hans Schneider and George Phillip Barker. Basic textbook covers theory of matrices and its applications to systems of linear equations and related topics such as determinants, eigenvalues and differential equations. Numerous exercises. 432pp. $5\frac{3}{8} \times 8\frac{1}{2}$. 0-486-66014-1

LINEAR ALGEBRA, Georgi E. Shilov. Determinants, linear spaces, matrix algebras, similar topics. For advanced undergraduates, graduates. Silverman translation. 387pp. $5\frac{3}{8} \times 8\frac{1}{2}$. 0-486-63518-X

MATHEMATICAL METHODS OF GAME AND ECONOMIC THEORY: Revised Edition, Jean-Pierre Aubin. This text begins with optimization theory and convex analysis, followed by topics in game theory and mathematical economics, and concluding with an introduction to nonlinear analysis and control theory. 1982 edition. 656pp. $6\frac{1}{8} \times 9\frac{1}{4}$. 0-486-46265-X

SET THEORY AND LOGIC, Robert R. Stoll. Lucid introduction to unified theory of mathematical concepts. Set theory and logic seen as tools for conceptual understanding of real number system. 496pp. $5\frac{3}{8} \times 8\frac{1}{4}$. 0-486-63829-4